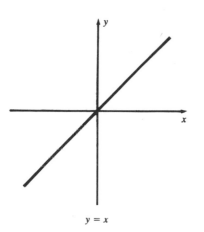

$y = x$

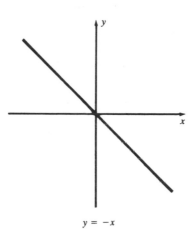

$y = -x$

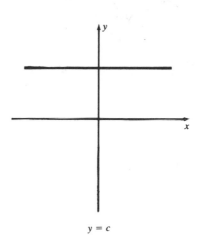

$y = c$

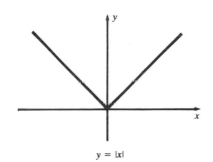

$y = |x|$

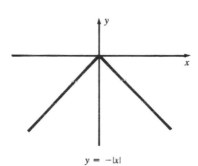

$y = -|x|$

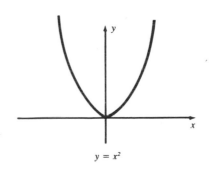

$y = x^2$

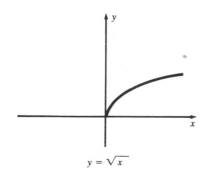

$y = \sqrt{x}$

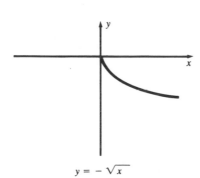

$y = -\sqrt{x}$

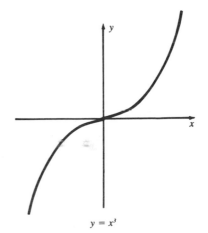

$y = x^3$

PRECALCULUS
FUNCTIONS AND GRAPHS
SECOND EDITION

PRECALCULUS
FUNCTIONS AND GRAPHS
SECOND EDITION

BERNARD KOLMAN
Drexel University

ARNOLD SHAPIRO

Harcourt Brace Jovanovich, Publishers
and its subsidiary, Academic Press
San Diego   New York   Chicago   Austin   Washington, D.C.
London   Sydney   Tokyo   Toronto

To the Memory of Lillie
B. K.

PREFACE

The authors are indebted to the faculty at those institutions that have adopted and have steadfastly continued to use the first edition of this text. Your letters and comments are genuinely appreciated and will always receive a prompt response.

NEW IN THIS EDITION

Significant additions and improvements have been made in this edition. They include:

- Division of the text into three units, with **Unit Cumulative Progress Tests.**
- A **Guide to the Use of Calculators** in the student preface. Calculators and calculator key sequences are emphasized and the use of logarithms as a calculational device has been dropped.
- A new section, **Maxima and Minima,** with an introduction to applied max/min problems similar to those tackled in a calculus course.
- A new section, **Approximating the Roots of a Polynomial Equation,** providing an opportunity to use either a calculator or computer.
- Added stress on **forms and techniques needed for calculus courses:** the difference quotient, the notation $(a, f(a))$ to designate a point in the plane, factoring with fractional exponents, and rationalizing of numerators.
- **Vignettes** discussing recent news items that pertain to mathematics.
- Tabular format for **Key Ideas for Review.**
- Tabular **summaries of various rules.**

OBJECTIVES OF THIS TEXT

The rapid technological progress of the past several decades has led to the conclusion that sophisticated mathematical concepts are required for advanced study in many fields. As a result, many colleges have included a course in calculus as a prerequisite for those pursuing a degree in a variety of disciplines. Our goal is to provide a complete, self-contained presentation of the basic mathematical techniques and ideas required for the successful completion of a calculus course.

Our classroom experience has shown that understanding the concept of a function, using function notation, and being able to sketch graphs of functions with ease constitute the most important skills that students must master in preparation for the study of calculus. Chapter 2, "Functions and Graphs," forms the cornerstone of this book. In it, attention is given not only to the concept of a function and to function notation but also to graphing techniques. We have insisted that students become familiar with the so-called standard graphs, and we have included techniques for quickly sketching a wide variety of functions (see "Assists in Graphing").

PEDAGOGIC DEVICES

We have employed many of the pedagogic devices that instructors found useful in the first edition and have added a few new stimuli.

UNIT ORGANIZATION

The text is divided into three units:

Unit 1: The Algebra of Functions
Unit 2: Trigonometry: The Circular Functions
Unit 3: Topics in Algebra

PRESENTATION

The style is informal, supportive, and "user-friendly." Many algebraic procedures are described with the aid of a "split screen" that displays simultaneously both the steps of an algorithm and a worked-out example.

 PROGRESS CHECKS

At carefully selected places, problems similar to those worked in the text have been inserted (with upside-down answers) to enable the student to test his or her understanding of the material just described.

CUMULATIVE PROGRESS TESTS

Each unit is concluded with a **Unit Cumulative Progress Test.**

 WARNINGS

To help eliminate misconceptions and prevent bad mathematical habits, we have inserted numerous **Warnings** (indicated by the symbol shown in the margin) that point out the incorrect practices most commonly found in homework and exam papers.

VIGNETTES

In each chapter we have inserted one or two vignettes, elements that are independent of the text yet are often related to the mathematical concepts. The vignettes are intended to catch the attention of the student and heighten interest in the material. (We hope they will provide interesting reading for the instructor as well.)

END-OF-CHAPTER MATERIAL

Every chapter contains a summary, including

Terms and Symbols   with appropriate page references;
Key Ideas for Review   to stress the concepts;
Review Exercises   to provide additional practice;
Progress Tests   to provide self-evaluation and reinforcement.

ANSWERS

The answers to all **Review Exercises** and **Progress Tests** appear in the back of the book.

SOLUTIONS

Worked-out solutions to selected **Review Exercises** appear in a separate section at the back of the book. The solved problems provide one more level of reassurance to the student using the **Review Exercises** in preparation for the **Progress Tests.** In addition, a *Student's Solution Manual* containing fully worked-out solutions to selected exercises will be made available to the bookstore.

EXERCISES

Abundant, carefully graded exercises provide practice in the mechanical and conceptual aspects of algebra and trigonometry. Exercises requiring a calculator are indicated by the symbol shown in the margin. Answers to odd-numbered exercises appear at the back of the book.

SUPPLEMENTARY MATERIAL

Instructor's Solutions Manual by Gail Edinger.

Student Solutions Manual by Gail Edinger.

Computerized Testbank (Micro-Pac Genie) by Microsystems Software, Ltd., is the most complete test-generating and authoring system with graphics on the market. This system is available for IBM PC, XT or compatible systems and the MACINTOSH.

Testbank by Michael Levitan.

Apple II and *IBM Software* by The Math Lab.

ACKNOWLEDGMENTS

We thank the following for their review of the manuscript and their helpful comments: Professor Terry Herdman at Virginia Polytechnic Institute and State University, Professor Monty Strauss at Texas Tech University, Professor Patricia Lock at St. Lawrence University, Professor William Blair at Northern Illinois University, Professor Jack Goldberg at the University of Michigan, Professor David Price at Tarrant County Junior College, Professor Istvan Kovacs at the University of South Alabama, Professor Bill Echoles at the University of Houston, Professor Joan Dykes at Edison Community College, and Professor Linda Holden at Indiana University.

The staff at Harcourt Brace Jovanovich has provided us with extensive and unflagging support. Our acquisitions editor, Michael Johnson, has provided a stream of new ideas and constant stimulation. Pamela Whiting, associate editor, has responded promptly to our sundry requests. Finally, our thanks to the editorial and production staff for their consistent attention to the myriad details involved in the birth of a book: Diane Pella, designer; Candy Young, art editor; Mandy Van Dusen, production manager; Julia Ann Ross, manuscript editor; and Sheila Spahn, production editor.

TO THE STUDENT

If you are like the great majority of students using this book, you are doing so to prepare for a calculus course. Calculus was developed in the seventeenth century by Sir Isaac Newton in England and by Gottfried von Leibniz in Germany. One of the outstanding developments in the history of science, calculus is the mathematics of change. Since everything around us is constantly changing, calculus is used in the solution of many commonplace problems in almost every discipline.

Contrary to what you may have heard, calculus is not a difficult subject, *provided you have prepared yourself* by learning the needed algebra and trigonometry. This book provides you with all the background material necessary for the successful study of calculus. This book was written for you. It gives you every possible chance to succeed—if you use it properly.

We would like to have you think of mathematics as a challenging game—but not as a spectator sport. This wish leads to our primary rule: *Read this textbook with pencil and paper handy*. Every new idea or technique is illustrated by fully worked-out examples. As you read the text, carefully follow the examples and then do the **Progress Checks.** The key to success in a math course is working problems, and the **Progress Checks** are there to provide immediate practice with the material you have just learned.

Your instructor will assign homework from the extensive selection of exercises that follows each section in the book. *Do the assignments regularly, thoroughly, and independently*. By doing lots of problems you will develop the necessary skills in algebra, and your confidence will grow. Since algebraic techniques and concepts build on previous results, you can't afford to skip any of the work.

To help prevent or eliminate improper habits and to help you avoid those errors that we see each semester as we grade papers, we have interspersed **Warnings** throughout the book. The **Warnings** point out common errors and emphasize the proper method.

There is important review material at the end of each chapter. The **Terms and Symbols** should all be familiar by the time you reach them. If your understanding of a term or symbol is hazy, use the page reference to find the place in the text where it is introduced. Go back and read the definition.

It is possible to become so involved with the details of techniques that you lose track of the broader concepts. The list of **Keys Ideas for Review** at the end of each chapter will help you focus on the principal ideas.

The **Review Exercises** can be used as part of your preparation for examinations. The section covering each exercise is indicated so that, if needed, you can go back to restudy the material. Answers to all review exercises are in the **Answers** section in the back of the book; solutions to the review exercises whose numbers are in color are in the **Solutions** section, also in the back of the book.

You are then ready to try **Progress Test A.** By checking your answers against the ones in the **Answers** section, you will soon pinpoint your weak spots. Go back for

further review and more exercise in those areas. Only then should you proceed to
Progress Test B.

We believe that the eventual "payoff" in studying calculus is an improved ability to
tackle practical problems in your field of interest. Since the mathematics presented in
this book is widely used in almost all fields and is the basic stepping-stone to success in
calculus, the study of this material is well worth your effort.

GUIDE TO THE USE OF CALCULATORS

If you don't already own a *scientific calculator* we strongly suggest that you purchase
one. When used sensibly, these tools have a place in a course in precalculus by easing
the burden associated with lengthy calculations. Calculators also provide a viable
alternative to the tedium of using tables such as those provided in the Tables Appendix
to this book.

First, a word of caution. *Don't reach for your calculator and begin pushing the keys
until you have first thought about the problem and the necessity for using a calculator.*
The answers to many problems can be determined immediately simply by thinking
about them. For example, if you are asked to evaluate $64^{2/3}$, you should be able to
apply the definition of a rational exponent and produce the answer of 16. Or, if you are
asked to evaluate

$$\arctan(\tan 0.64)$$

you should (after studying this topic!) recognize that the answer is 0.64.

Because it is helpful to have a means of indicating a sequence of keystrokes and the
result that is displayed, we have placed a box around each keystroke, except for
numeric input. For example, the sequence

Display

25 $\boxed{1/x}$ 0.04

means "calculate the reciprocal of 25." The answer displayed is 0.04.

All scientific calculators share a common problem: there are more functions to be
performed than there are keys. The solution has been to invest some keys with both a
primary and a secondary function. A special key is provided in the upper left of the
keypad which we will designate by

$$\boxed{\text{INV}}$$

for *inverse.* (Many calculators use $\boxed{\text{2ND}}$ to designate this key.) When you depress this
key, you are initiating the secondary function. For example, there is a square root key
marked

$$\boxed{\sqrt{}}$$

If you enter

9 $\boxed{\sqrt{}}$

the answer displayed is 3, which is indeed the square root of 9. Now, if you enter

9 $\boxed{\text{INV}}$ $\boxed{\sqrt{}}$

the answer displayed is 81, which is the square of 9. The key combination $\boxed{\text{INV}}$ $\boxed{\sqrt{}}$ selects the alternate function which in this case squares the input value. You will find that the INV key is also related to the concept of inverse functions that we will explore in detail in Chapters 4 and 5.

Many excellent scientific calculators are available at reasonable prices. These calculators differ somewhat in their mode of operation but the similarities outweigh the differences. A scientific calculator should provide you with the facilities listed in the following table:

KEY	FUNCTION
$+$ $-$ $\times$ $\div$	The standard arithmetic operations.
( )	Used to group operations as in algebra. For example, to calculate $$\frac{3.12}{4.57 - 7.98}$$ you would enter $$3.12 \; \boxed{\div} \; \boxed{(} \; 4.57 \; \boxed{-} \; 7.98 \; \boxed{)}$$
$=$	Completes pending operations.
$+/-$	Negates the value in the display.
$1/x$	Reciprocal of the value in the display.
$\sqrt{}$	Square root of the value in the display.
y^x	Exponential key; for example, $2^{1.6}$
log ln	Logarithmic functions
10^x e^x	Exponential functions.
sin cos tan	Trigonometric functions.
$\sin^{-1}$ $\cos^{-1}$ $\tan^{-1}$	Inverse trigonometric functions.
π	Displays 3.1415927 as an approximation to π. Saves many keystrokes.
STO	Stores the display in memory.
RCL	Recalls the value in memory for use as an operand.

An ever-growing number of students has had some exposure to computers. In this edition we have provided several vignettes dealing with suggestions for computational tasks that exceed the capacity of a calculator but are suitable for beginning programmers. We have also introduced numerical methods for finding the roots of a polynomial equation with specific suggestions for computer implementation.

TABLE OF CONTENTS

PRECALCULUS

FUNCTIONS AND GRAPHS

SECOND EDITION

© Frank Siteman/Stock, Boston

THE ALGEBRA OF FUNCTIONS

A mathematician, like a painter or a poet, is a maker of patterns. If his patterns are more permanent than theirs, it is because they are made with ideas.

G. H. Hardy

The concept of function is of the greatest importance, not only in pure mathematics but also in practical applications.

Richard Courant and Herbert Robbins

Scientific discoveries of each generation are made possible by those of the prior generations. Understanding that which came before is crucial to participating in that which is yet to come.

For the individual who wishes to participate in the adventure of mathematics, the fundamental building-block is *algebra*. Wherever you turn in higher mathematics and especially in calculus, algebraic skills are needed. Chapter 1 of this Unit reviews the fundamentals of algebra.

The word *function* is not new to mathematics. (It was first used by Leibniz in 1694.) The concept of function has achieved a central position in the study of modern mathematics. Accordingly, Chapter 2 of this Unit is devoted to a serious introduction to functions with special emphasis on (a) the use of graphs to provide understanding of their behavior and (b) a constant hint of the role of functions in the study of calculus.

The remainder of Unit 1 deals with two important classes of functions: polynomial functions (Chapter 3) and exponential and logarithmic functions (Chapter 4).

THE FOUNDATIONS OF ALGEBRA

Algebra provides us with the basic tools used in all branches of mathematics. A firm grasp of algebraic concepts and techniques is essential for the student who intends to begin a study of calculus, the gateway to advanced mathematics.

It is easy to show that $1 + 2 = 2 + 1$ by using specific items, such as matchsticks or marbles. Algebra enables us to abstract this result and to say that

$$x + y = y + x$$

for all positive integers. The key to this general rule is the introduction of symbols such as x and y to represent a variety of numbers. Symbols are the language of algebra. They allow us to write relationships in a symbolic manner, which facilitates the process of finding solutions.

In general, the numbers denoted by a symbol are restricted to a particular number system, and this is where our study begins. We will also review the fundamentals of polynomials, exponents, and radicals. Much of the work of this chapter is devoted to a study of inequalities, since this topic is not often treated adequately in secondary school mathematics.

1.1
THE REAL NUMBER
SYSTEM

SETS

We will need to use the notation and terminology of sets from time to time. A **set** is simply a collection of objects or numbers, which are called the **elements** or **members** of the set. The elements of a set are written within braces so that

$$A = \{4, 5, 6\}$$

tells us that the set A consists of the numbers 4, 5, and 6. If every element of set A is also a member of set B, then A is a **subset** of B. For example, the set of all robins is a subset of the set of all birds.

THE REAL NUMBER SYSTEM

Much of our work in algebra deals with the real number system and we must review the composition of this number system.

The numbers $1, 2, 3, \ldots$, used for counting, form the set of **natural numbers.** The set of **integers**

$$\ldots, -2, -1, 0, 1, 2, \ldots$$

consists of the positive integers (the natural numbers), the negative integers, and zero.

The set of **rational numbers** consists of those numbers that can be written as a ratio of two integers p/q where q is not equal to zero. Examples of rational numbers are

$$0 \qquad \frac{2}{3} \qquad -4 \qquad \frac{7}{5} \qquad \frac{-3}{4}$$

We make use of rational numbers when we divide two apples equally among four people, since each person then gets half, or $\frac{1}{2}$, an apple. By writing an integer n in the form $n/1$, we see that every integer is a rational number. The decimal number 1.3 is also a rational number, since $1.3 = \frac{13}{10}$.

We have now seen three fundamental number systems: the natural number system, the system of integers, and the rational number system. Each later system includes the previous system or systems, and each is more complicated than the one before. However, the rational number system is still inadequate for sophisticated uses of mathematics, since there exist numbers that are not rational, that is, numbers that cannot be written as the ratio of two integers. These are called **irrational numbers.** It can be shown that the number a that satisfies $a \cdot a = 2$ is such a number. The number π, which is the ratio of the circumference of a circle to its diameter, is also such a number.

The decimal form of a rational number will either terminate, as

$$\frac{3}{4} = 0.75 \qquad -\frac{4}{5} = -0.8$$

or will form a repeating pattern, as

$$\frac{2}{3} = 0.666\ldots \qquad \frac{1}{11} = 0.090909\ldots \qquad \frac{1}{7} = 0.1428571\ldots$$

Remarkably, the decimal form of an irrational number *never* forms a repeating pattern.

The rational and irrational numbers together form the **real number system.**

THE REAL NUMBER LINE

There is a simple and very useful geometric interpretation of the real number system. Draw a horizontal straight line; pick a point on this line, label it with the number 0, call it the origin, and denote it by O. Designate the side to the right of the origin as the **positive direction** and the side to the left as the **negative direction.**

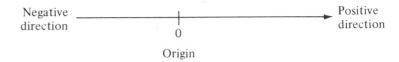

Next, select a unit of length for measuring distance. With each positive real number r we associate the point that is r units to the right of the origin, and with each negative number $-r$ we associate the point that is r units to the left of the origin. Thus, the set of real numbers is identified with all possible points on a straight line. For every point on the line there is a real number and for every real number there is a point on the line. The line is called the **real number line,** and the number associated with a point is called its **coordinate.** We can now show some points on this line.

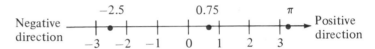

The numbers to the right of zero are called **positive,** the numbers to the left of zero are called **negative.** The positive numbers and zero together are called the **nonnegative** numbers.

We will frequently use the real number line to help picture the results of algebraic computations. For this purpose, we are only concerned with relative locations on the line. For example, it is adequate to show π slightly to the right of 3 since π is approximately 3.14.

INEQUALITIES

If a and b are real numbers, we can compare their positions on the real number line by using the relations of **less than, greater than, less than or equal to,** and **greater than or equal to,** denoted by the **inequality symbols** $<$, $>$, $\leq$, and $\geq$, respectively. Table 1 describes both algebraic and geometric interpretations of the inequality symbols.

TABLE 1   The Inequality Symbols

Algebraic Statement	Equivalent Statement	Geometric Statement
$a > 0$	a is positive	a lies to the right of the origin
$a < 0$	a is negative	a lies to the left of the origin
$a > b$	$a - b$ is positive	a lies to the right of b
$a < b$	$a - b$ is negative	a lies to the left of b
$a \geq b$	$a - b$ is zero or positive	a coincides with b or lies to the right of b
$a \leq b$	$a - b$ is zero or negative	a coincides with b or lies to the left of b

Expressions involving inequality symbols, such as $a < b$ and $a \geq b$, are called **inequalities.** We often combine these expressions so that $a \leq b < c$ means both $a \leq b$ and $b < c$. For example, $-5 \leq x < 2$ is equivalent to $-5 \leq x$ and $x < 2$.

FINDING ALL THE PRIMES UP TO *N*: THE SIEVE OF ERATOSTHENES

Prime Integers Less Than or Equal to 100

2 3 ~~4~~ 5 ~~6~~ 7 ~~8~~
~~9~~ ~~10~~ 11 ~~12~~ 13 ~~14~~ ~~15~~
~~16~~ 17 ~~18~~ 19 ~~20~~ ~~21~~ ~~22~~
23 ~~24~~ ~~25~~ ~~26~~ ~~27~~ ~~28~~ 29
~~30~~ 31 ~~32~~ ~~33~~ ~~34~~ ~~35~~ ~~36~~
37 ~~38~~ ~~39~~ ~~40~~ 41 ~~42~~ 43
~~44~~ ~~45~~ ~~46~~ 47 ~~48~~ ~~49~~ ~~50~~
~~51~~ ~~52~~ 53 ~~54~~ ~~55~~ ~~56~~ 57
~~58~~ 59 ~~60~~ 61 ~~62~~ ~~63~~ ~~64~~
~~65~~ ~~66~~ 67 ~~68~~ ~~69~~ ~~70~~ 71
~~72~~ 73 ~~74~~ ~~75~~ ~~76~~ ~~77~~ ~~78~~
79 ~~80~~ ~~81~~ ~~82~~ 83 ~~84~~ ~~85~~
~~86~~ 87 ~~88~~ 89 ~~90~~ ~~91~~ ~~92~~
~~93~~ ~~94~~ ~~95~~ ~~96~~ 97 ~~98~~ ~~99~~
~~100~~

An integer $p > 1$ is called a **prime** if the only positive integers that divide p are p and 1. For example, 3, 5, 11, and 2 are primes. The number 2 is the only even prime, since every even integer greater than 2 is divisible by 2. A positive integer that is not a prime is said to be a **composite.** For example, 4, 10, and 15 are composite integers.

A method for listing all the primes up to a given integer N was developed by the Greek scientist and mathematician Eratosthenes (275–194 B.C.), who was a friend of Archimedes. We will describe this method, called the **Sieve of Eratosthenes,** and apply it to the accompanying table, which lists the positive integers less than or equal to 100.

Step 1. Make a list of all integers from 2 to N.

Step 2. Since 2 is the first prime, cross out all multiples of 2. The next integer in the list that has not been crossed out is 3, which is a prime. Now cross out all multiples of 3. The next integer in the list that has not been crossed out is 5, which is a prime. Next, cross out all multiples of 5. Repeat the process until the list is exhausted.

Step 3. The numbers that have not been crossed out are the primes less than N.

Clearly, the process can be terminated when you reach $N/2$ since there will be no multiples less than N. In fact, it can be shown that you can stop when you reach a number K such that $K^2 > N$.

The number of computations required for executing the "sieve" rises dramatically as N increases. For this reason, the "sieve" has become a favorite benchmark program for comparing computer hardware and software.

PROGRESS CHECK

Verify that the following inequalities are true by using either the "Equivalent Statement" or "Geometric Statement" of Table 1.

(a) $-1 < 3$ (b) $2 \le 2$ (c) $-2.7 < -1.2$

(d) $-4 < -2 < 0$ (e) $-\dfrac{7}{2} < \dfrac{7}{2} < 7$

The real numbers satisfy the following useful properties of inequalities.

Properties of Inequalities

Let a, b, and c be real numbers.

(i) One and only one of the following relations holds:

$$a < b, \ a > b, \ a = b \quad \textbf{Trichotomy property}$$

(ii) If $a < b$ and $b < c$, then $a < c$. **Transitive property**

(iii) If $a < b$, then $a + c < b + c$ and $a - c < b - c$.

(iv) If $a < b$ and $c > 0$, then $ac < bc$. When an inequality is multiplied by a positive number, the sense of the inequality is preserved.

(v) If $a < b$ and $c < 0$, then $ac > bc$. When an inequality is multiplied by a negative number, the sense of the inequality is reversed.

EXAMPLE 1   Properties of Inequalities

Replace the "?" in each of the following by the appropriate inequality symbol and state the property of inequalities that is exhibited.

(a) Since $-2 < 4$ and $4 < 5$, then -2 ? 5.

(b) Since $-2 < 5$, $-2 + 3$? $5 + 3$, or 1 ? 8.

(c) Since $2 < 5$, $2(3)$? $5(3)$, or 6 ? 15.

(d) Since $-3 < 2$, $(-3)(-2)$? $2(-2)$, or 6 ? -4.

SOLUTION

(a) $-2 < 5$ (transitive property)

(b) $-2 + 3 < 5 + 3$, or $1 < 8$ (if $a < b$, then $a + c < b + c$)

(c) $2(3) < 5(3)$, or $6 < 15$ (if $a < b$ and $c > 0$, then $ac < bc$)

(d) $(-3)(-2) > 2(-2)$, or $6 > -4$ (if $a < b$ and $c < 0$, then $ac > bc$)

■

ABSOLUTE VALUE

When we are interested in the magnitude of a number a, and don't care about the direction or sign, we use the concept of **absolute value,** which we write as $|a|$. The formal definition of absolute value is stated this way.

$$|a| = \begin{cases} a & \text{when} & a \geq 0 \\ -a & \text{when} & a < 0 \end{cases}$$

This definition tells us that $|a|$ is always nonnegative. For, if a is itself nonnegative, then $|a| = a$; if a is negative, then $|a| = -a$ must be positive.

Since distance is independent of direction and is always nonnegative, we can provide an alternative definition of absolute value.

If $a \neq 0$ is a real number, then $|a|$ is the distance on the real number line from the origin to either the point whose coordinate is a or the point whose coordinate is $-a$.

EXAMPLE 2   Absolute Value

(a) $|4| = 4$ $|-4| = 4$ $|0| = 0$

(b) The distance on the real number line between the point labeled 3.4 and the origin is $|3.4| = 3.4$. Similarly, the distance between point -2.3 and the origin is $|-2.3| = 2.3$.

■

In working with the notation of absolute value, it is important to perform the operations within the bars first. Here are some examples.

EXAMPLE 3   Absolute Value

(a) $|2 - 5| = |-3| = 3$ (b) $|3 - 5| - |8 - 6| = |-2| - |2| = 2 - 2 = 0$

The following properties of absolute value follow from the definition.

Properties of Absolute Value	For all real numbers a and b, (i) $\|a\| \geq 0$ (ii) $\|a\| = \|-a\|$ (iii) $\|a - b\| = \|b - a\|$

We can use absolute value to denote the distance between *any* two points a and b on the real number line. In Figure 1, the distance between the points labeled 2 and 5 is 3

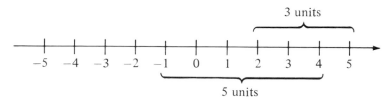

FIGURE 1   **Distance on the Real Number Line**

units and can be obtained by evaluating either $|5 - 2|$ or $|2 - 5|$. Similarly, the distance between the points labeled -1 and 4 is given by either

$$|4 - (-1)| = 5 \qquad \text{or} \qquad |-1 - 4| = 5$$

Using the notation $\overline{AB}$ to denote the distance between the points A and B, we provide the following definition.

Distance on the Real Number Line	The **distance** $\overline{AB}$ between points A and B on the real number line, whose coordinates are a and b, respectively, is given by $$\overline{AB} = \|b - a\|$$

Property (iii) then tells us that $\overline{AB} = |b - a| = |a - b|$. Viewed another way, Property (iii) states that the distance between any two points on the real number line is independent of the direction. (A trip from New York to Miami is the same mileage as a trip from Miami to New York.)

EXAMPLE 4   Distance on the Real Number Line

Let points A, B, and C have coordinates -4, -1, and 3, respectively, on the real number line. Find the following distances. (The symbol O denotes the origin.)
(a) $\overline{AB}$ (b) $\overline{CB}$ (c) $\overline{OB}$

SOLUTION
Using the definition, we have
(a) $\overline{AB} = |-1 - (-4)| = |-1 + 4| = |3| = 3$
(b) $\overline{CB} = |-1 - 3| = |-4| = 4$
(c) $\overline{OB} = |-1 - 0| = |-1| = 1$ ∎

☑ **PROGRESS CHECK**

The points P, Q, and R on the real number line have coordinates -6, 4, and 6, respectively. Find the following distances.

(a) $\overline{PR}$ (b) $\overline{QP}$ (c) $\overline{PQ}$

ANSWERS

(a) 12   (b) 10   (c) 10

EXERCISE SET 1.1

In Exercises 1–14 determine whether the given statement is true (T) or false (F).

1. -14 is a natural number.

2. $-\frac{4}{5}$ is a rational number.

3. $\pi/3$ is a rational number.

4. $1.75/18.6$ is an irrational number.

5. -1207 is an integer.

6. 0.75 is an irrational number.

7. $\frac{4}{5}$ is a real number.

8. 3 is a rational number.

9. 2π is a real number.

10. The sum of two rational numbers is always a rational number.

11. The sum of two irrational numbers is always an irrational number.

12. The product of two rational numbers is always a rational number.

13. The product of two irrational numbers is always an irrational number.

14. The difference of two irrational numbers is always an irrational number.

Indicate the following sets of numbers on a real number line.

15. The natural numbers less than 8.

16. The natural numbers greater than 4 and less than 10.

17. The integers that are greater than 2 and less than 7.

18. The integers that are greater than -5 and less than or equal to 1.

In Exercises 19 and 20 list the real numbers in ascending order. Locate each on a real number line.

19. (a) $\dfrac{474}{217}$ (b) $\dfrac{3405}{2719}$ (c) $\sqrt{2}$

     (d) $-\dfrac{243}{88}$ (e) $\dfrac{\sqrt{272}}{4}$

20. (a) $\dfrac{\sqrt{54}}{2}$ (b) $-\dfrac{292}{117}$ (c) $-\dfrac{61}{135}$

     (d) $\dfrac{\pi}{2}$ (e) $-\sqrt{7}$

In Exercises 21–40 express each statement as an inequality.

21. a is nonnegative.

22. -6 is less than -2.

23. b is greater than or equal to 5.

24. a is between 3 and 7.

25. b is less than or equal to -4.

26. a is between $\frac{1}{2}$ and $\frac{1}{4}$.

In Exercises 27–32 state a property of inequalities that justifies each of the following statements.

27. Since $-3 < 1$, then $-1 < 3$.

28. Since $-5 < -1$ and $-1 < 4$, then $-5 < 4$.

29. Since $14 > 9$, then $-14 < -9$.

30. Since $5 > 3$, then $5 \neq 3$.

31. Since $-1 < 6$, then $-3 < 18$.

32. Since $6 > -1$, then 7 is a positive number.

In Exercises 33–40 find the value of each of the following.

33. $|2 - 3|$

34. $|2 - 2|$

35. $|2 - (-2)|$

36. $|2| + |-3|$

37. $\dfrac{|14 - 8|}{|-3|}$

38. $\dfrac{|2 - 12|}{|1 - 6|}$

39. $\dfrac{|3| - |2|}{|3| + |2|}$

40. $\dfrac{|3 - 2|}{|3 + 2|}$

In Exercises 41–46 the coordinates of points A and B are given. Find $\overline{AB}$.

41. 2, 5   42. $-3, 6$ 43. $-3, -1$ 44. $-4, \frac{11}{2}$

45. $-\frac{4}{5}, \frac{4}{5}$ 46. 2, 2

In Exercises 47–52 express each statement using absolute value notation.

47. The distance between x and 2 is greater than 5.

48. The distance between x and 6 is no more than 2.

49. The distance between x and -4 is at least 1.

50. The distance between x and $-1/2$ is less than 4.

51. a is less than 3 units from b.

52. b is at least 6 units from a.

1.2 POLYNOMIALS AND FACTORING

In the introduction to this chapter we said that "symbols are the language of algebra." Unfortunately, we have to introduce many definitions so that statements can distinguish precisely between different types of symbols and their relationships. Fortunately, you have seen these definitions before and this section will serve to refresh your memory.

A **variable** is a symbol to which we can assign values. For example, in Section 1.1 we defined a rational number as one that can be written as p/q, where p and q are integers (and q is not zero). The symbols p and q are variables, since we can assign values to them. A variable can be restricted to a particular number system (for example, p and q must be integers) or to a subset of a number system (note that q cannot be zero).

If we invest P dollars at an annual interest rate of 6%, then we will earn $0.06P$ dollars interest per year, and we will have $P + 0.06P$ dollars at the end of the year. We call $P + 0.06P$ an **algebraic expression.** Note that an algebraic expression involves **variables** (in this case P), **constants** (such as 0.06), and **algebraic operations** (such as $+$, $-$, $\times$, $\div$). Virtually everything we do in algebra involves algebraic expressions, sometimes as simple as our example and sometimes very involved.

An algebraic expression takes on a **value** when we assign a specific number to each variable in the expression. Thus, the expression

$$\frac{3m + 4n}{m + n}$$

is **evaluated** when $m = 3$ and $n = 2$ by substituting these values for m and n:

$$\frac{3(3) + 4(2)}{3 + 2} = \frac{9 + 8}{5} = \frac{17}{5}$$

We often need to write algebraic expressions in which a variable multiplies itself repeatedly. We use the notation of exponents to indicate such repeated multiplication. Thus,

$$a^1 = a \qquad a^2 = a \cdot a \qquad a^n = \underbrace{a \cdot a \cdots \cdot a}_{n \text{ factors}}$$

where n is a natural number and a is a real number. We call a the **base** and n the **exponent** and say that a^n is the nth **power of** a. When $n = 1$, we simply write a rather than a^1.

It is convenient to define a^0 for all real numbers $a \neq 0$ by having $a^0 = 1$. We will provide motivation for this seemingly arbitrary definition in Section 1.3.

WARNING

Note the difference between

$$(-3)^2 = (-3)(-3) = 9$$

and

$$-3^2 = -(3 \cdot 3) = -9$$

Later in this chapter we will need an important rule of exponents. Observe that if m and n are natural numbers and a is any real number, then

$$a^m \cdot a^n = \underbrace{a \cdot a \cdots a}_{m \text{ factors}} \cdot \underbrace{a \cdot a \cdots a}_{n \text{ factors}}$$

Since there are a total of $m + n$ factors on the right side, we conclude that

$$a^m a^n = a^{m+n}$$

EXAMPLE 1   Multiplication with Natural Number Exponents
Multiply.
(a) $x^2 \cdot x^3$ (b) $(3x)(4x^4)$

SOLUTION
(a) $x^2 \cdot x^3 = x^{2+3} = x^5$
(b) $(3x)(4x^4) = 3 \cdot 4 \cdot x \cdot x^4 = 12x^{1+4} = 12x^5$

∎

POLYNOMIALS

A polynomial is an algebraic expression of a certain form. Polynomials play an important role in the study of algebra, since many word problems translate into equations or inequalities that involve polynomials. We first study the manipulative and mechanical aspects of polynomials; this will serve as background for dealing with their applications in later chapters.

Let x denote a variable and let n be a nonnegative integer. The expression ax^n, where a is a constant real number, is called a **monomial in x**. A **polynomial in x** is an expression that is a sum of monomials and has the general form

$$P = a_n x^n + a_{n-1} x^{n-1} + \cdots + a_1 x + a_0, \quad a_n \neq 0 \qquad (1)$$

Each of the monomials in Equation (1) is called a **term** of P, and $a_0, a_1, \ldots, a_n$ are constant real numbers that are called the **coefficients** of the terms of P. Note that a polynomial may consist of just one term; that is, a monomial is also considered to be a polynomial.

EXAMPLE 2 Polynomial Expressions

(a) The following expressions are polynomials in x:

$$3x^4 + 2x + 5 \qquad 2x^3 + 5x^2 - 2x + 1 \qquad \frac{3}{2}x^3$$

Notice that we write $2x^3 + 5x^2 + (-2)x + 1$ as $2x^3 + 5x^2 - 2x + 1$.

(b) The following expressions are not polynomials in x:

$$2x^{1/2} + 5 \qquad 3 - \frac{4}{x} \qquad \frac{2x-1}{x-2}$$

Remember that each term of a polynomial in x must be of the form ax^n where a is a real number and n is a nonnegative integer.

∎

The **degree of a monomial in x** is the exponent of x. Thus, the degree of $5x^3$ is 3. A monomial in which the exponent of x is 0 is called a **constant term** and is said to be of **degree zero.** The nonzero coefficient a_n of the term in P with highest degree is called the **leading coefficient** of P and we say that P is a **polynomial of degree n.** A special case is the polynomial all of whose coefficients are zero. Such a polynomial is called the **zero polynomial,** is denoted by 0, and is said to have no degree.

EXAMPLE 3 Vocabulary of Polynomials

Given the polynomial

$$P = 2x^4 - 3x^2 + \frac{4}{3}x - 1$$

The terms of P are
$$2x^4, \quad 0x^3, \quad -3x^2, \quad \frac{4}{3}x, \quad -1.$$

The coefficients of the terms are
$$2, \quad 0, \quad -3, \quad \frac{4}{3}, \quad -1.$$

The degree of P is 4 and the leading coefficient is 2.

∎

OPERATIONS WITH POLYNOMIALS

If P and Q are polynomials in x, then the terms cx^r in P and dx^r in Q are said to be **like terms;** that is, like terms have the same exponent in x. For example, given

$$P = 4x^2 + 4x - 1$$

and

$$Q = 3x^3 - 2x^2 + 4$$

then the like terms are $0x^3$ and $3x^3$; $4x^2$ and $-2x^2$; $4x$ and $0x$; -1 and 4.

We define equality of polynomials in this way.

Two polynomials are equal if all like terms are identical.

EXAMPLE 4 Equality of Polynomials
Find A, B, C, and D if

$$Ax^3 + (A + B)x^2 + Cx + (C - D) = -2x^3 + x + 3$$

SOLUTION
Equating the coefficients of like terms, we have

$$A = -2 \qquad A + B = 0 \qquad C = 1 \qquad C - D = 3$$

and, by substitution,

$$B = 2 \qquad\qquad\qquad D = -2$$

∎

If P and Q are polynomials in x, the **sum** $P + Q$ is obtained by forming the sums of all pairs of like terms. The sum of cx^r in P and dx^r in Q is $(c + d)x^r$. Similarly, the **difference** $P - Q$ is obtained by forming the differences, $(c - d)x^r$, of like terms.

EXAMPLE 5 Polynomial Addition and Subtraction
(a) Add $2x^3 + 2x^2 - 3$ and $x^3 - x^2 + x + 2$.
(b) Subtract $2x^3 + x^2 - x + 1$ from $3x^3 - 2x^2 + 2x$.

SOLUTION
(a) Adding the coefficients of like terms,

$$(2x^3 + 2x^2 - 3) + (x^3 - x^2 + x + 2) = 3x^3 + x^2 + x - 1$$

(b) Subtracting the coefficients of like terms,

$$(3x^3 - 2x^2 + 2x) - (2x^3 + x^2 - x + 1) = x^3 - 3x^2 + 3x - 1$$

∎

Multiplication of polynomials is based on the rule for exponents developed earlier in this section,

$$a^m a^n = a^{m+n}$$

and on the **distributive laws**

$$a(b + c) = ab + ac$$
$$(a + b)c = ac + bc$$

EXAMPLE 6 Polynomial Multiplication
Multiply $(x + 2)(3x^2 - x + 5)$.

SOLUTION
$(x + 2)(3x^2 - x + 5)$
$\quad = x(3x^2 - x + 5) + 2(3x^2 - x + 5)$ Distributive law
$\quad = 3x^3 - x^2 + 5x + 6x^2 - 2x + 10$ Distributive law and $a^m a^n = a^{m+n}$
$\quad = 3x^3 + 5x^2 + 3x + 10$ Adding like terms

∎

PROGRESS CHECK

Multiply.

(a) $(x^2 + 2)(x^2 - 3x + 1)$ (b) $(x^2 - 2xy + y)(2x + y)$

ANSWERS

(a) $x^4 - 3x^3 + 3x^2 - 6x + 2$ (b) $2x^3 - 3x^2y + 2xy - 2xy^2 + y^2$

The multiplication in Example 6 can be carried out in long or vertical form as follows.

$$3x^2 - x + 5$$
$$x + 2$$
$$\overline{}$$
$$3x^3 - x^2 + 5x \qquad = x(3x^2 - x + 5)$$
$$6x^2 - 2x + 10 \qquad = 2(3x^2 - x + 5)$$
$$\overline{3x^3 + 5x^2 + 3x + 10} \qquad = \text{sum of above lines}$$

In Example 6, the product of polynomials of degrees one and two is seen to be a polynomial of degree three. From the multiplication process it is easy to derive the following useful rule.

> The degree of the product of two nonzero polynomials is the sum of the degrees of the polynomials.

Products of the form $(2x + 3)(5x - 2)$ or $(2x + y)(3x - 2y)$ occur often, and we can handle them mentally by the FOIL (First Outer Inner Last) method:

$$
\begin{array}{c}
10x^2 \qquad -6 \\
(2x + 3)\,(5x - 2) \qquad = 10x^2 + 11x - 6 \\
15x \\
-4x \\
\hline
\text{Sum} = 11x
\end{array}
$$

A number of special products occur frequently and it is worthwhile knowing them.

Special Products

$$(a + b)^2 = (a + b)(a + b) = a^2 + 2ab + b^2$$
$$(a - b)^2 = (a - b)(a - b) = a^2 - 2ab + b^2$$
$$(a + b)(a - b) = a^2 - b^2$$

FACTORING

Now that we can find the product of two polynomials, let's consider the reverse problem: Given a polynomial, can we find factors whose product will yield the given polynomial? This process, known as **factoring,** is one of the basic tools of algebra. In this chapter a polynomial with *integer* coefficients is to be factored as a product of polynomials of lower degree with *integer* coefficients; a polynomial with *rational* coefficients is to be factored as a product of polynomials of lower degree with *rational* coefficients.

The simplest type of factoring involves the removal of a common factor. For example, given the polynomial

$$x^2 + x$$

we see that the factor x is common to both terms and write

$$x^2 + x = x(x + 1)$$

Note that x and $x + 1$ are both polynomials of degree one. We have thus factored a polynomial of second degree with integer coefficients into the product of polynomials of lower degree with integer coefficients.

Before proceeding to other methods of factoring, we offer a useful rule.

> Always remove common factors before attempting any other factoring techniques.

To factor a second-degree polynomial

$$ax^2 + bx + c$$

where a, b, and c are integers and $a \neq 0$, we must have

$$ax^2 + bx + c = (rx + u)(sx + v) = (rs)x^2 + (rv + su)x + uv$$

where r, s, u, and v are integers. Equating the coefficients of like terms,

$$rs = a \quad rv + su = b \quad uv = c$$

These three equations give candidates for r, s, u, and v. The final choices from among the candidates are determined by trial and error, which is made easier by using mental multiplication.

EXAMPLE 7 Factoring Second-Degree Polynomials

Factor $2x^2 - x - 6$.

SOLUTION

The term $2x^2$ can result only from the factors $2x$ and x, so the factors must be of the form

$$2x^2 - x - 6 = (2x \quad)(x \quad)$$

The constant term, -6, must be the product of factors of opposite signs, so we may write

$$2x^2 - x - 6 = \begin{cases} (2x + \quad)(x - \quad) \\ \text{or} \\ (2x - \quad)(x + \quad) \end{cases}$$

The integer factors of 6 are

$$1 \cdot 6 \quad 6 \cdot 1 \quad 2 \cdot 3 \quad 3 \cdot 2$$

By trying these we find that

$$2x^2 - x - 6 = (2x + 3)(x - 2)$$

∎

PROGRESS CHECK

Factor.

(a) $3x^2 - 16x + 21$ **(b)** $2x^2 + 3x - 9$

ANSWERS

(b) $(2x - 3)(x + 3)$ (a) $(3x - 7)(x - 3)$

WARNING

The polynomial $x^2 - 6x$ can be written as

$$x^2 - 6x = x(x - 6)$$

and is then a product of two polynomials of positive degree. Students often fail to consider x to be a "true" factor.

EXAMPLE 8   Common Factors and Grouping

Factor.

(a) $2x^3 - 12x^2 + 16x$ **(b)** $2ab + b + 2ac + c$

SOLUTION

(a) We first remove the common factor $2x$.

$$2x^3 - 12x^2 + 16x = 2x(x^2 - 6x + 8)$$
$$= 2x(x - 2)(x - 4)$$

(b) It is sometimes possible to discover common factors by first grouping terms. Grouping those terms containing b and those terms containing c,

$$
\begin{aligned}
2ab + b + 2ac + c &= (2ab + b) + (2ac + c) &&\text{Grouping} \\
&= b(2a + 1) + c(2a + 1) &&\text{Common factors } b, c \\
&= (2a + 1)(b + c) &&\text{Common factor } 2a + 1
\end{aligned}
$$

∎

PROGRESS CHECK

Factor.

(a) $2x^3 - 2x^2y - 4xy^2$ **(b)** $2m^3n + m^2 + 2mn^2 + n$

ANSWERS

(b) $(2mn + 1)(m^2 + n)$ (a) $2x(x + y)(x - 2y)$

SPECIAL FACTORS

There are three expressions that occur frequently and deserve special attention. The first is the difference of two squares, the second is the sum of two cubes, and the third is the difference of two cubes.

Special Factors	
	$a^2 - b^2 = (a + b)(a - b)$
	$a^3 + b^3 = (a + b)(a^2 - ab + b^2)$
	$a^3 - b^3 = (a - b)(a^2 + ab + b^2)$

When using these formulas, be careful with the placement of plus and minus signs.

EXAMPLE 9   Special Factors

Factor.

(a) $4x^2 - 25$ **(b)** $\dfrac{1}{27}u^3 + 8v^3$

SOLUTION

(a) Since

$$4x^2 - 25 = (2x)^2 - (5)^2$$

we may use the formula for the difference of two squares with $a = 2x$ and $b = 5$. Thus,

$$4x^2 - 25 = (2x + 5)(2x - 5)$$

(b) Note that

$$\frac{1}{27}u^3 + 8v^3 = \left(\frac{1}{3}u\right)^3 + (2v)^3$$

and then use the formula for the sum of two cubes.

$$\frac{1}{27}u^3 + 8v^3 = \left(\frac{u}{3} + 2v\right)\left(\frac{u^2}{9} - \frac{2}{3}uv + 4v^2\right)$$

∎

PROGRESS CHECK

Factor.

(a) $16x^2 - 9$ **(b)** $8s^3 - 27t^3$ **(c)** $125r^3 + \dfrac{1}{125}s^3$

ANSWERS

(a) $(4x + 3)(4x - 3)$ **(b)** $(2s - 3t)(4s^2 + 6t + 9t^2)$ **(c)** $\left(5r + \dfrac{s}{5}\right)\left(25r^2 - rs + \dfrac{s^2}{25}\right)$

Are there polynomials with integer coefficients that cannot be written as products of polynomials of lower degree with integer coefficients? The answer is yes. Examples are the polynomials $x^2 + 1$ and $x^2 + x + 1$. A polynomial is said to be **prime** or **irreducible** if it cannot be written as a product of two polynomials of positive degree. Thus, $x^2 + 1$ is irreducible over the integers.

"NO FUSS" FACTORING FOR SECOND-DEGREE POLYNOMIALS

Factoring involves a certain amount of trial and error, which can become frustrating, especially when the leading coefficient is not 1. You might want to try a rather neat scheme that will greatly reduce the number of candidates.

We'll demonstrate the method for the polynomial

$$4x^2 + 11x + 6 \qquad (1)$$

Using the leading coefficient of 4, write the pair of incomplete factors

$$(4x \quad)(4x \quad) \qquad (2)$$

Next, multiply the coefficient of x^2 and the constant term in (1) to produce $4 \cdot 6 = 24$. Now find two integers whose product is 24 and whose sum is 11, the coefficient of the middle term of (1). It's clear that 8 and 3 will do nicely, so we write

$$(4x + 8)(4x + 3) \qquad (3)$$

Finally, within each parenthesis in (3) discard any common divisor. Thus $(4x + 8)$ reduces to $(x + 2)$ and we write

$$(x + 2)(4x + 3) \qquad (4)$$

which is the factorization of $4x^2 + 11x + 6$.

Will the method always work? Yes—if you first remove all common factors in the original polynomial. That is, you must first write

$$6x^2 + 15x + 6 = 3(2x^2 + 5x + 2)$$

and apply the method to the polynomial $2x^2 + 5x + 2$.

(For a proof that the method works, see M. A. Autrie and J. D. Austin, "A Novel Way to Factor Quadratic Polynomials," *The Mathematics Teacher 72*, no. 2 [1979].)

We'll use the polynomial $2x^2 - x - 6$ of Example 7 to demonstrate the method when some of the coefficients are negative.

Try the method on these second-degree polynomials.

$3x^2 + 10x - 8$

$6x^2 - 13x + 6$

$4x^2 - 15x - 4$

$10x^2 + 11x - 6$

Factoring $ax^2 + bx + c$	Example: $6x^2 + 7x - 3$
Step 1. Use the lead coefficient a to write the incomplete factors $\quad (ax \quad)(ax \quad)$	*Step 1.* The lead coefficient is 6, so we write $$(6x \quad)(6x \quad)$$
Step 2. Multiply a and c, the coefficients of x^2 and the constant term.	*Step 2.* $a \cdot c = (6)(-3) = -18$
Step 3. Find integers whose product is $a \cdot c$ and whose sum equals b. Write these integers in the incomplete factors of Step 1.	*Step 3.* Two integers whose product is -18 and whose sum is 7 are 9 and -2. We then write $$(6x + 9)(6x - 2)$$
Step 4. Discard any common factor *within each parenthesis* in Step 3. The result is the desired factorization.	*Step 4.* Reducing $(6x + 9)$ to $(2x + 3)$ and $(6x - 2)$ to $(3x - 1)$, we have $$6x^2 + 7x - 3 = (2x + 3)(3x - 1)$$

EXERCISE SET 1.2

1. Evaluate $\frac{2}{3}r + 5$ when $r = 12$.

2. Evaluate $\frac{9}{5}C + 32$ when $C = 37$.

3. If P dollars are invested at a simple interest rate of r percent per year for t years, the amount on hand at the end of t years is $P + Prt$. Suppose \$2000 is invested at 8% per year ($r = 0.08$). How much money is on hand after
   (a) one year?   **(b)** half a year?
   (c) 8 months?

In Exercises 7 and 8 evaluate the given expressions.

7. $\dfrac{|a - 2b|}{2a}$ when $a = 1$, $b = 2$

In Exercises 9–14 carry out the indicated operations.

9. $b^5 \cdot b^2$

10. $x^3 \cdot x^5$

13. $\left(\dfrac{3}{2}x^3\right)(-2x)$

14. $\left(-\dfrac{5}{3}x^6\right)\left(-\dfrac{3}{10}x^3\right)$

15. Which of the following expressions are not polynomials?
    (a) $-3x^2 + 2x + 5$ **(b)** $-3x^2y$
    (c) $-3x^{2/3} + 2xy + 5$ **(d)** $-2x^{-4} + 2xy^3 + 5$

In Exercises 17–20 indicate the leading coefficient and the degree of each polynomial.

17. $2x^3 + 3x^2 - 5$

18. $-4x^5 - 8x^2 + x + 3$

19. $\dfrac{3}{5}x^4 + 2x^2 - x - 1$

20. $-1.5 + 7x^3 + 0.75x^7$

21. An investor buys x shares of G.E. stock at \$55 per share, y shares of Exxon stock at \$45 per share, and z shares of A.T.&T. stock at \$60 per share. What does the polynomial $55x + 45y + 60z$ represent?

22. A field consists of a rectangle and a square arranged as shown in Figure 2.
    What does each of the following polynomials represent?
    (a) $x^2 + xy$ **(b)** $2x + 2y$ **(c)** $4x$
    (d) $4x + 2y$

In Exercises 23–32 perform the indicated operations.

23. $(2x^2 + 3x + 8) - (5 - 2x + 2x^2)$

24. $(4x^2 + 3x + 2) + (3x^2 - 2x - 5)$

25. $3xy^2z - 4x^2yz + xy + 3 - (2xy^2z + x^2yz - yz + x - 2)$

26. $(x^2 + 3)(2x^2 - x + 2)$

27. $(2y^2 + y)(-2y^3 + y - 3)$

28. $(x^2 + 2x - 1)(2x^2 - 3x + 2)$

29. $(a^2 - 4a + 3)(4a^3 + 2a + 5)$

30. $(2a^2 + ab + b^2)(3a - b^2 + 1)$

4. The perimeter of a rectangle is given by the formula $P = 2(L + W)$, where L is the length and W is the width of the rectangle. Find the perimeter if
   (a) $L = 2$ feet, $W = 3$ feet
   (b) $L = \frac{1}{2}$ meter, $W = \frac{1}{4}$ meter

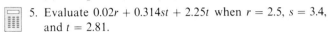

5. Evaluate $0.02r + 0.314st + 2.25t$ when $r = 2.5$, $s = 3.4$, and $t = 2.81$.

6. Evaluate $10.421x + 0.821y + 2.34xyz$ when $x = 3.21$, $y = 2.42$, and $z = 1.23$.

8. $\dfrac{|x| + |y|}{|x| - |y|}$ when $x = -3$, $y = 4$

11. $(4y^3)(-5y^6)$

12. $(-6x^4)(-4x^7)$

16. Which of the following expressions are not polynomials?
    (a) $4x^5 - x^{1/2} + 6$ **(b)** $\dfrac{2}{5}x^3 + \dfrac{4}{3}x - 2$
    (c) $4x^5y$ **(d)** $x^{4/3}y + 2x - 3$

FIGURE 2 Exercise 22

31. $(-3a + ab + b^2)(3b^2 + 2b + 2)$

32. $5(2x - 3)^2$

33. An investor buys x shares of IBM stock at \$260 per share at Thursday's opening of the stock market. Later in the day, he sells y shares of G&W stock at \$13 per share and z shares of Holiday Inn stock at \$17 per share. Write a polynomial that expresses the money transactions for the day.

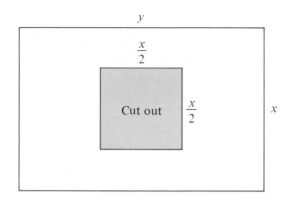

FIGURE 3 Exercise 34

34. An artist takes a rectangular piece of cardboard whose sides are x and y and cuts out a square of side $x/2$ (Figure 3) to obtain a mat for a painting. Write a polynomial giving the area of the mat.

35. A zoo has available 200 feet of stockade fencing with which to enclose a rectangular region that abuts an existing building as shown in Figure 4. Write a polynomial in terms of x for the area A of the region.

36. A racetrack has the shape of a rectangle with semicircular ends as shown in Figure 5. Write polynomials in terms of x for the area A and perimeter p if
    (a) y is half the length of x

    (b) y is two units less than x

37. The can shown in Figure 6 has a radius of r inches and a height of h inches. Write a polynomial in terms of r

for the volume V of the can if

(a) $h = 2r$ (b) $h = \dfrac{5}{2}r$

(*Hint:* For a cylinder of radius r and height h, $V = \pi r^2 h$.)

38. The top of the can shown in Figure 6 has been removed. Write a polynomial in terms of the radius r for the surface area A of the can if

(a) $h = \dfrac{3}{2}r$ (b) $h = 3r$

(*Hint:* The surface area A of cylinder whose radius is r and whose height is h is $A = 2\pi rh$. Don't forget to include the bottom of the can.)

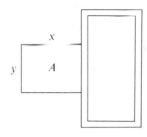

FIGURE 4 Exercise 35

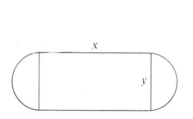

FIGURE 5 Exercise 36

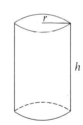

FIGURE 6 Exercise 37

In Exercises 39–76 factor completely.

39. $5bc + 25b$
40. $2x^4 + x^2$
41. $-3y^2 - 4y^5$
42. $3abc + 12bc$
43. $3x^2 + 6x^2y - 9x^2z$
44. $9a^3b^3 + 12a^2b - 15ab^2$
45. $x^2 + 4x + 3$
46. $x^2 + 2x - 8$
47. $x^2 - 5x - 14$
48. $x^2 - 8x - 20$
49. $2x^2 - 3x - 2$
50. $2x^2 + 7x + 6$
51. $10x^2 - 13x - 3$
52. $9x^2 + 24x + 16$
53. $6x^2 + 13x + 6$
54. $4x^2 + 20x + 25$
55. $4(x + 1)(y + 2) - 8(y + 2)$
56. $2(x + 1)(x - 1) + 5(x - 1)$
57. $3(x + 2)^2(x - 1) - 4(x + 2)^2(2x + 7)$
58. $4(2x - 1)^2(x + 2)^3(x + 1) - 3(2x - 1)^5(x + 2)^2(x + 3)$
59. $y^2 - \dfrac{1}{9}$
60. $4a^2 - b^2$
61. $16 - 9x^2y^2$
62. $x^{12} - 1$
63. $x^4 - y^4$
64. $a^4 - 16$
65. $x^3 + 27y^3$
66. $8x^3 + 125y^3$
67. $27x^3 - y^3$
68. $64x^3 - 27y^3$
69. $a^3 + 8$
70. $8r^3 - 27$
71. $\dfrac{1}{8}m^3 - 8n^3$
72. $8a^3 - \dfrac{1}{64}b^3$
73. $(x + y)^3 - 8$
74. $27 + (x + y)^3$
75. $8x^6 - 125y^6$
76. $a^6 + 27b^6$

77. The expression

$$\frac{x^2 - 9}{x - 3}$$

is not defined when $x = 3$ but is defined for all $x \neq 3$, including values of x as close to 3 as you like.

(a) Use a calculator to complete the tables:

x	2.9	2.99	2.999	2.9999
$\dfrac{x^2 - 9}{x - 3}$				

x	3.1	3.01	3.001	3.0001
$\dfrac{x^2 - 9}{x - 3}$				

(b) Examine the tables you completed in part (a). The expression should appear to get closer and closer to a finite value that is called the **limit** of the expression as x approaches 3. What is this value?

(c) Factor the numerator $x^2 - 9$. When $x \neq 3$, you can simplify the expression. What is the value of the simplified expression when $x = 3$? How does this relate to the limit?

1.3
EXPONENTS, RADICALS, AND COMPLEX NUMBERS

In Section 1.2 we defined a^n for a real number a and a positive integer n as

$$a^n = \underbrace{a \cdot a \cdot \cdots \cdot a}_{n \text{ factors}}$$

POSITIVE INTEGER EXPONENTS

and we showed that if m and n are positive integers then $a^m a^n = a^{m+n}$. The method we used to establish this rule was to write out the factors a^m and a^n and count the total number of occurrences of a. The same method can be used to establish the rest of the following rules when m and n are positive integers.

Rules for Exponents

$$a^m a^n = a^{m+n} \qquad (a^m)^n = a^{mn}$$

$$(ab)^m = a^m b^m \qquad \left(\frac{a}{b}\right)^m = \frac{a^m}{b^m}, \ b \neq 0$$

If $a \neq 0$,

$$\frac{a^m}{a^n} = \begin{cases} a^{m-n} & \text{when} & m > n \\ \dfrac{1}{a^{n-m}} & \text{when} & n > m \\ 1 & \text{when} & m = n \end{cases}$$

A rigorous proof of the rules listed above requires the method of mathematical induction (see Chapter 11).

EXAMPLE 1   Operations with Positive Integer Exponents
Simplify the following.

(a) $(4a^2b^3)(2a^3b)$ **(b)** $(2x^2y)^4$ **(c)** $\dfrac{x^{2n+1}}{x^n}$

SOLUTION

(a) $(4a^2b^3)(2a^3b) = 4 \cdot 2 \cdot a^2a^3b^3b = 8a^5b^4$

(b) $(2x^2y)^4 = 2^4(x^2)^4y^4 = 16x^8y^4$

(c) $\dfrac{x^{2n+1}}{x^n} = x^{2n+1-n} = x^{n+1}$

■

ZERO AND NEGATIVE EXPONENTS

We now repeat the definition of a^0 and define a^{-m}.

$$a^0 = 1 \qquad a^{-m} = \frac{1}{a^m}, a \neq 0 \text{ and } m \text{ a positive integer}$$

We can now restate the definition of a^m/a^n in this way:

$$\frac{a^m}{a^n} = a^{m-n}$$

For example,

$$3^0 = 1 \qquad 2^{-3} = \frac{1}{8} \qquad \left(\frac{1}{4}\right)^{-1} = 4 \qquad \frac{x^3}{x^4} = x^{-1} = \frac{1}{x}$$

These definitions for zero and negative integer exponents were chosen because they are compatible with the rules for positive integer exponents. For instance, il $a \neq 0$, we would expect the rule $a^ma^n = a^{m+n}$ to produce

$$a^ma^0 = a^{m+0} = a^m$$

Using the definition $a^0 = 1$, we have

$$a^ma^0 = a^m \cdot 1 = a^m$$

which demonstrates that this rule holds with $a^0 = 1$. In a similar way, you can verify that *all* of the rules for exponents hold for our definition of zero and negative exponents.

EXAMPLE 2   Operations with Zero and Negative Exponents
Simplify and write the answers using only positive exponents.

(a) $4(xy)^0$ (b) $(x^2y^{-3})^{-5}$ (c) $\dfrac{9a^2b^{-3}}{3a^{-2}b^5}$

SOLUTION

(a) $4(xy)^0 = 4(1) = 4$

(b) $(x^2y^{-3})^{-5} = (x^2)^{-5}(y^{-3})^{-5} = x^{-10}y^{15} = \dfrac{y^{15}}{x^{10}}$

(c) $\dfrac{9a^2b^{-3}}{3a^{-2}b^5} = \dfrac{9}{3}\dfrac{a^2}{a^{-2}}\dfrac{b^{-3}}{b^5} = 3a^4b^{-8} = \dfrac{3a^4}{b^8}$

■

WARNING

Don't confuse negative numbers and negative exponents.

(a) $2^{-4} = \dfrac{1}{2^4}$

Note that $2^{-4} \neq -2^4$.

(b) $(-2)^{-3} = \dfrac{1}{(-2)^3} = \dfrac{1}{-8} = -\dfrac{1}{8}$

Note that $(-2)^{-3} \neq \dfrac{1}{2^3} = \dfrac{1}{8}$.

RATIONAL EXPONENTS AND RADICALS

If a and b are real numbers and n is a natural number, we define a root by saying

$$a \text{ is an } \textbf{\textit{n}th root} \text{ of } b \text{ if } a^n = b$$

When $n = 2$, we say that a is a **square root** of b, and when $n = 3$, we say that a is a **cube root** of b. Thus, 5 and -5 are square roots of 25 since $(5)^2 = (-5)^2 = 25$; -2 is a cube root of -8 since $(-2)^3 = -8$. More generally, if $b > 0$ and a is a square root of b, then $-a$ is also a square root of b. If $b < 0$, there is no real number a such that $a^2 = b$, since the square of a real number is always nonnegative. The cases are summarized in Table 2.

TABLE 2   *n*th Roots of a Real Number

b	n	Number of nth roots of b such that $b = a^n$	Form of nth roots	b	Examples
>0	Even	2	$a, -a$	4	Square roots are 2, -2
<0	Even	None	None	-1	No square roots
>0	Odd	1	$a > 0$	8	Cube root is 2
<0	Odd	1	$a < 0$	-8	Cube root is -2
0	All	1	0	0	Square root is 0

We would like to define rational exponents in a manner that will be consistent with the rules for integer exponents. If the rule $(a^m)^n = a^{mn}$ is to hold, then we must have

$$(b^{1/n})^n = b^{n/n} = b$$

But a is an nth root of b if $a^n = b$. Then for every natural number n, we say that

$$b^{1/n} \text{ is an } n\text{th root of } b$$

If n is even and b is positive, Table 2 indicates that there are two numbers, a and $-a$, that are nth roots of b. To avoid ambiguity we always *choose* the positive number to be the nth root and call it the **principal *n*th root** of b. Thus, $b^{1/n}$ denotes the principal nth root of b.

WHEN IS A PROOF NOT A PROOF?

Books of mathematical puzzles love to include "proofs" that lead to false or contradictory results. Of course, there is always an incorrect step hidden somewhere in the proof. The error may be subtle, but a good grounding in the fundamentals of mathematics will enable you to catch it.

Examine the following "proof."

$$1 = 1^{1/2} \tag{1}$$
$$- [(-1)^2]^{1/2} \tag{2}$$
$$= (-1)^{2/2} \tag{3}$$
$$= (-1)^1 \tag{4}$$
$$= -1 \tag{5}$$

The result is obviously contradictory: we can't have $1 = -1$. Yet each step seems to be legitimate. Did you spot the flaw? The rule

$$(b^m)^{1/n} = b^{m/n}$$

used in going from (2) to (3) doesn't apply when n is even and b is negative. Any time the rules of algebra are abused the results are unpredictable!

EXAMPLE 3 nth Roots of a Real Number
Evaluate.

(a) $144^{1/2}$ **(b)** $(-8)^{1/3}$ **(c)** $(-25)^{1/2}$ **(d)** $-\left(\dfrac{1}{16}\right)^{1/4}$

SOLUTION
(a) $144^{1/2} = 12$ **(b)** $(-8)^{1/3} = -2$

(c) $(-25)^{1/2}$ is not a real number   **(d)** $-\left(\dfrac{1}{16}\right)^{1/4} = -\dfrac{1}{2}$

We are now prepared to define $b^{m/n}$ for restricted values of b, m, and n. ∎

If m is an integer, n a natural number, and b a real number, we define $b^{m/n}$ by

$$b^{m/n} = (b^{1/n})^m = (b^m)^{1/n}$$

where b must be positive when n is even.

With these restrictions, all the rules of exponents continue to hold when the exponents are rational numbers.

EXAMPLE 4   Operations with Rational Exponents

Simplify.

(a) $(-8)^{4/3}$ (b) $x^{1/2} \cdot x^{3/4}$ (c) $(x^{3/4})^2$ (d) $(3x^{2/3}y^{-5/3})^3$

SOLUTION

(a) $(-8)^{4/3} = [(-8)^{1/3}]^4 = (-2)^4 = 16$

(b) $x^{1/2} \cdot x^{3/4} = x^{1/2 + 3/4} = x^{5/4}$

(c) $(x^{3/4})^2 = x^{(3/4)(2)} = x^{3/2}$

(d) $(3x^{2/3}y^{-5/3})^3 = 3^3 \cdot x^{(2/3)(3)}y^{(-5/3)(3)} = 27x^2 y^{-5} = \dfrac{27x^2}{y^5}$

■

PROGRESS CHECK

Simplify.

(a) $27^{4/3}$ (b) $(a^{1/2}b^{-2})^{-2}$ (c) $\left(\dfrac{x^{1/3}y^{2/3}}{z^{5/6}}\right)^{12}$

ANSWERS

(a) 16   (b) $\dfrac{b^4}{a}$ (c) $\dfrac{x^4 y^8}{z^{10}}$

RADICALS

The symbol $\sqrt{b}$ is an alternative way of writing $b^{1/2}$; that is, $\sqrt{b}$ denotes the nonnegative square root of b. The symbol $\sqrt{}$ is called a **radical sign** and $\sqrt{b}$ is called the **principal square root** of b. Thus,

$$\sqrt{25} = 5 \qquad \sqrt{0} = 0 \qquad \sqrt{-25} \text{ is undefined}$$

In general, the symbol $\sqrt[n]{b}$ is an alternative way of writing $b^{1/n}$, the principal nth root of b. Of course, we must apply the same restrictions to $\sqrt[n]{b}$ that we established for $b^{1/n}$. In summary,

$$\sqrt[n]{b} = b^{1/n} = a \qquad \text{where} \qquad a^n = b$$

with these restrictions:
- if n is even and $b < 0$, $\sqrt[n]{b}$ is not a real number;
- if n is even and $b \geq 0$, $\sqrt[n]{b}$ is the *nonnegative* number a satisfying $a^n = b$.

WARNING

Many students are accustomed to writing $\sqrt{4} = \pm 2$. This is incorrect, since the symbol $\sqrt{}$ indicates the *principal* square root, which is nonnegative. Get in the habit of writing $\sqrt{4} = 2$. If you want to indicate *all* square roots of 4, write $\pm\sqrt{4} = \pm 2$.

In short, $\sqrt[n]{b}$ is the **radical form** of $b^{1/n}$. We can switch back and forth from one form to the other. For instance,

$$\sqrt[3]{7} = 7^{1/3} \qquad (11)^{1/5} = \sqrt[5]{11}$$

Finally, we treat the radical form of $b^{m/n}$ as follows.

$$b^{m/n} = (b^m)^{1/n} = \sqrt[n]{b^m}$$

and

$$b^{m/n} = (b^{1/n})^m = (\sqrt[n]{b})^m$$

where m is an integer and n is a natural number.

EXAMPLE 5   Radicals and Rational Exponents
Change from radical form to rational exponent form or vice versa.

(a) $(2x)^{-3/2}$ **(b)** $\dfrac{1}{\sqrt[7]{y^4}}$ **(c)** $(-3a)^{3/7}$ **(d)** $\sqrt{x^2 + y^2}$

SOLUTION

(a) $(2x)^{-3/2} = \dfrac{1}{(2x)^{3/2}} = \dfrac{1}{\sqrt{8x^3}}$ **(b)** $\dfrac{1}{\sqrt[7]{y^4}} = \dfrac{1}{y^{4/7}} = y^{-4/7}$

(c) $(-3a)^{3/7} = \sqrt[7]{-27a^3}$ **(d)** $\sqrt{x^2 + y^2} = (x^2 + y^2)^{1/2}$

∎

PROGRESS CHECK
Change from radical form to rational exponent form or vice versa.

(a) $\sqrt[4]{2rs^3}$ **(b)** $(x + y)^{5/2}$ **(c)** $y^{-5/4}$ **(d)** $\dfrac{1}{\sqrt[4]{m^5}}$

ANSWERS

(a) $(2r)^{1/4}s^{3/4}$ **(b)** $\sqrt{(x + y)^5}$ **(c)** $\dfrac{1}{\sqrt[4]{y^5}}$ **(d)** $m^{-5/4}$

WARNING
Note that

$$\sqrt{16} + \sqrt{9} \neq \sqrt{25}$$

and, in general,

$$\sqrt{a} + \sqrt{b} \neq \sqrt{a + b}$$

Since radicals are just another way of writing exponents, the properties of radicals can be derived from the properties of exponents.

Properties of Radicals	If n is a natural number, a and b are real numbers, and all radicals denote real numbers, then

(1) $\sqrt[n]{b^m} = (b^m)^{1/n} = (b^{1/n})^m = (\sqrt[n]{b})^m$

(2) $\sqrt[n]{a} \cdot \sqrt[n]{b} = a^{1/n} \cdot b^{1/n} = (ab)^{1/n} = \sqrt[n]{ab}$

(3) $\dfrac{\sqrt[n]{a}}{\sqrt[n]{b}} = \dfrac{a^{1/n}}{b^{1/n}} = \left(\dfrac{a}{b}\right)^{1/n} = \sqrt[n]{\dfrac{a}{b}}, \; b \neq 0$

(4) $\sqrt[n]{a^n} = \begin{cases} a & \text{if } n \text{ is odd} \\ |a| & \text{if } n \text{ is even} \end{cases}$

Here are some examples using these properties.

EXAMPLE 6   Operations with Radicals
Simplify.
(a) $\sqrt{18}$ **(b)** $\sqrt[3]{-54}$ **(c)** $2\sqrt[3]{8x^3y}$ **(d)** $\sqrt{x^6}$ **(e)** $(2\sqrt{3} - 2)(\sqrt{3} + 4)$

SOLUTION
(a) $\sqrt{18} = \sqrt{9 \cdot 2} = \sqrt{9}\sqrt{2} = 3\sqrt{2}$

(b) $\sqrt[3]{-54} = \sqrt[3]{(-27)(2)} = \sqrt[3]{-27}\sqrt[3]{2} = -3\sqrt[3]{2}$

(c) $\sqrt[3]{8x^3y} = \sqrt[3]{8}\sqrt[3]{x^3}\sqrt[3]{y} = 2x\sqrt[3]{y}$

(d) $\sqrt{x^6} = \sqrt{x^2} \cdot \sqrt{x^2} \cdot \sqrt{x^2} = |x| \cdot |x| \cdot |x| = |x|^3$

(e) $(2\sqrt{3} - 1)(\sqrt{3} + 4) = 2(3) - \sqrt{3} + 8\sqrt{3} - 4 = 2 + 7\sqrt{3}$

■

PROGRESS CHECK
Simplify. All variables represent positive real numbers.
(a) $\sqrt{4xy^5}$ **(b)** $\sqrt[3]{16x^4y^6}$ **(c)** $\sqrt[4]{16x^8y^5}$

ANSWERS
(a) $2y^2\sqrt{xy}$ **(b)** $2xy^2\sqrt[3]{2x}$ **(c)** $2x^2y\sqrt[4]{y}$

WARNING
The properties of radicals state that

$$\sqrt{x^2} = |x|$$

It is a common error to write $\sqrt{x^2} = x$, but this leads to the conclusion that $\sqrt{(-6)^2} = -6$. Since the symbol $\sqrt{}$ represents the principal or nonnegative square root of a number, the result cannot be negative. It is therefore essential to write $\sqrt{x^2} = |x|$ (and, in fact, $\sqrt[n]{x^n} = |x|$ whenever n is even) unless we know that $x \geq 0$, in which case we can write $\sqrt{x^2} = x$.

**RATIONALIZING
RADICALS**

The process of rewriting an expression so that it is free of radicals in either the numerator or denominator is known as **rationalizing.** In algebra, it has long been traditional to write answers that are free of radicals in the denominator. In calculus, it is sometimes helpful to rationalize the numerator.

The rationalization process involves multiplying the given expression by some cleverly chosen form of unity. We frequently make use of the fact that the product of $\sqrt{m} + \sqrt{n}$ by its **conjugate** $\sqrt{m} - \sqrt{n}$ is free of radicals. That is,

$$(\sqrt{m} + \sqrt{n})(\sqrt{m} - \sqrt{n}) = m - n$$

has no radical.

EXAMPLE 7   Rationalizing Denominators

Rationalize the denominator. Assume all radicals denote real numbers.

(a) $\sqrt{\dfrac{x}{y}}$ (b) $\dfrac{4}{\sqrt{5} - \sqrt{2}}$ (c) $\dfrac{5}{\sqrt{x} + 2}$ (d) $\dfrac{5}{\sqrt{x + 2}}$

SOLUTION

(a) Since we are told that all radicals denote real numbers, we know that x/y must be positive. Therefore, x and y are either both positive or both negative. With this caution in mind, we can write

$$\sqrt{\frac{x}{y}} = \frac{\sqrt{|x|}}{\sqrt{|y|}} = \frac{\sqrt{|x|}}{\sqrt{|y|}} \cdot \frac{\sqrt{|y|}}{\sqrt{|y|}} = \frac{\sqrt{xy}}{\sqrt{y^2}} = \frac{\sqrt{xy}}{|y|}$$

(b)

$$\frac{4}{\sqrt{5} - \sqrt{2}} = \frac{4}{\sqrt{5} - \sqrt{2}} \cdot \frac{\sqrt{5} + \sqrt{2}}{\sqrt{5} + \sqrt{2}} = \frac{4(\sqrt{5} + \sqrt{2})}{5 - 2} = \frac{4}{3}(\sqrt{5} + \sqrt{2})$$

(c) Since all radicals denote real numbers, x must be nonnegative and we need not be concerned with the use of absolute value.

$$\frac{5}{\sqrt{x} + 2} = \frac{5}{\sqrt{x} + 2} \cdot \frac{\sqrt{x} - 2}{\sqrt{x} - 2} = \frac{5(\sqrt{x} - 2)}{x - 4}$$

(d) Again, we see that $x + 2$ must be nonnegative and we need not worry about the use of absolute value.

$$\frac{5}{\sqrt{x + 2}} = \frac{5}{\sqrt{x + 2}} \cdot \frac{\sqrt{x + 2}}{\sqrt{x + 2}} = \frac{5\sqrt{x + 2}}{x + 2}$$

∎

☑ **PROGRESS CHECK**

Rationalize the denominator. Assume all radicals denote real numbers.

(a) $\dfrac{-9xy^3}{\sqrt{3xy}}$ (b) $\dfrac{-6}{\sqrt{2} + \sqrt{6}}$ (c) $\dfrac{4}{\sqrt{x} - \sqrt{y}}$

ANSWERS

(a) $-3y^2\sqrt{3xy}$ (b) $\dfrac{3}{2}(\sqrt{2} - \sqrt{6})$ (c) $\dfrac{4(\sqrt{x} + \sqrt{y})}{x - y}$

EXAMPLE 8 **Rationalizing Numerators**

Rationalize the numerator of the expression $\dfrac{\sqrt{x}-2}{x-4}$. Assume all radicals denote real numbers.

SOLUTION

$$\frac{\sqrt{x}-2}{x-4} = \frac{\sqrt{x}-2}{x-4} \cdot \frac{\sqrt{x}+2}{\sqrt{x}+2} \qquad [\text{the conjugate of } \sqrt{x}-2 \text{ is } \sqrt{x}+2]$$

$$= \frac{x-4}{(x-4)(\sqrt{x}+2)} = \frac{1}{\sqrt{x}+2}, \qquad x \neq 4$$

FACTORING WITH FRACTIONAL EXPONENTS

You may work through a problem in calculus and obtain the answer

$$(x-1)^{1/2} + (x-1)^{3/2}$$

When you look in the back of your text, you find the answer given as

$$x\sqrt{x-1}$$

Naturally, you assume that you've made a serious blunder. However, it is possible that the answer in the text is an alternative form of your answer. The alternative answer is obtained by factoring and simplifying using the following rule:

> To factor common expressions with fractional exponents, factor out the expression with the smallest exponent.

We'll outline the steps for dealing with positive fractional exponents in the next example. Note that the process makes use of the laws of exponents and of radical notation.

EXAMPLE 9 **Factoring with Positive Fractional Exponents**

Factor and simplify: $(x-1)^{1/2} + (x-1)^{3/2}$

SOLUTION

Factoring with Fractional Exponents	
Step 1. Find the common expression with the *smallest* exponent.	*Step 1.* The expression with the smallest exponent is $$(x-1)^{1/2}$$
Step 2. Factor out the expression chosen in Step 1, applying the rule $$a^m a^n = a^{m+n}$$ to the fractional exponents.	*Step 2.* $$(x-1)^{1/2} + (x-1)^{3/2}$$ $$= (x-1)^{1/2}[1 + (x-1)]$$

Step 3. Simplify the result of Step 2. You may choose to use radicals in place of fractional exponents.	*Step 3.* $\quad= (x - 1)^{1/2}(x)$ $\qquad\quad= x\sqrt{x - 1}$

■

EXAMPLE 10 Factoring with Negative Fractional Exponents
Factor and simplify: $(x + 1)^{-1/2} + (x + 1)^{-3/2}$

SOLUTION
Careful: the *smallest* exponent of the common expressions is $-3/2$. Then,

$$(x + 1)^{-1/2} + (x + 1)^{-3/2} = (x + 1)^{-3/2}[(x + 1) + 1]$$
$$= (x + 1)^{-3/2}(x + 2)$$
$$= \frac{x + 2}{\sqrt{(x + 1)^3}}$$

■

PROGRESS CHECK
Factor and simplify.
(a) $(x^2 + 1)^{1/3} - (x^2 + 1)^{4/3}$ (b) $(2x - 1)^{-1/3} + (2x - 1)^{-4/3}$

(a) $-x^2\sqrt[3]{x^2 + 1}$ (b) $\dfrac{2x}{\sqrt[3]{(2x - 1)^4}}$

RADICALS, EXPONENTS AND CALCULATORS

There are two keys on a scientific calculator that are used to evaluate expressions involving radicals and exponents. The *square root key* $\boxed{\sqrt{}}$ can be used to find the square root of an argument. For all other radicals, it is best to convert to exponential form and use the *exponential key* $\boxed{y^x}$ as shown in the following example.

EXAMPLE 7 Using a Calculator to Evaluate Radicals and Exponents
Use a calculator to evaluate:
(a) $3^{2.72}$ (b) $5^{-3/7}$ (c) $\sqrt[3]{-15}$

SOLUTION
(a) To use the exponential key, you must first enter the base (3), then press the exponential key $\boxed{y^x}$ followed by the exponent (2.72), and terminate the sequence with the $\boxed{=}$ key.

Display

$3 \boxed{y^x} 2.72 \boxed{=} \qquad 19.850425$

(b) A good way to handle this problem requires the use of the *parentheses keys* and the *negation key* $\boxed{+/-}$.

$$5\boxed{y^x}\boxed{(}\,3\boxed{\div}7\boxed{)}\boxed{+/-} \qquad \begin{array}{l} Display \\ -0.4285714 \end{array}$$
$$\boxed{=} \qquad\qquad\qquad\qquad\quad 0.5016969$$

An alternative method is to rewrite the problem using positive exponents

$$5^{-3/7} = \frac{1}{5^{3/7}}$$

and then make use of the *reciprocal key* $\boxed{1/x}$:

$$5\boxed{y^x}\boxed{(}\,3\boxed{\div}7\boxed{)}\boxed{=} \qquad \begin{array}{l} Display \\ 1.9932353 \end{array}$$
$$\boxed{1/x} \qquad\qquad\qquad\qquad\quad 0.5016969$$

(c) In exponential form we need to evaluate $(-15)^{1/3}$.

$$15\boxed{+/-}\boxed{y^x}\boxed{(}\,1\boxed{\div}3\boxed{)}\boxed{=} \qquad \begin{array}{l} Display \\ -2.4662121 \end{array}$$

(How would you handle the problem on a calculator that will not permit the base y to be negative?)

∎

PROGRESS CHECK

Use a calculator to evaluate

(a) $\sqrt{17.44}$ (b) $\sqrt[3]{-100}$ (c) $12^{-3/2}$

ANSWERS

(a) 4.1761226   (b) −4.6415888   (c) 0.0240562

COMPLEX NUMBERS

One of the central problems in algebra is that of finding solutions to a given polynomial equation. This problem will be discussed in later chapters of this book. However, observe at this point that there is no real number that satisfies a simple polynomial equation such as

$$x^2 = -4$$

since the square of a real number is always nonnegative.

To resolve this problem, mathematicians created a new number system built upon an "imaginary unit" i defined by $i = \sqrt{-1}$. This number i has the property that when we square both sides of the equation we have $i^2 = -1$, a result that cannot be obtained with real numbers. By definition,

$$i = \sqrt{-1}$$
$$i^2 = -1$$

We also assume that i behaves according to all the algebraic laws we have already developed (with the exception of the rules for inequalities for real numbers). This allows us to simplify higher powers of i. Thus,

$$i^3 = i^2 \cdot i = (-1)i = -i$$
$$i^4 = i^2 \cdot i^2 = (-1)(-1) = 1$$

Now it's easy to simplify i^n when n is any natural number. Since $i^4 = 1$, we simply seek the highest multiple of 4 that is less than or equal to n. For example,

$$i^5 = i^4 \cdot i = (1) \cdot i = i$$
$$i^{27} = i^{24} \cdot i^3 = (i^4)^6 \cdot i^3 = (1)^6 \cdot i^3 = i^3 = -i$$

EXAMPLE 11   The Imaginary Unit i

Simplify.

(a) i^{51} **(b)** $-i^{74}$

SOLUTION

(a) $i^{51} = i^{48} \cdot i^3 = (i^4)^{12} \cdot i^3 = (1)^{12} \cdot i^3 = i^3 = -i$

(b) $-i^{74} = -i^{72} \cdot i^2 = -(i^4)^{18} \cdot i^2 = -(1)^{18} \cdot i^2 = -(1)(-1) = 1$

∎

It is easy also to write square roots of negative numbers in terms of i. For example,

$$\sqrt{-25} = i\sqrt{25} = 5i$$

and, in general, we define

$$\boxed{\quad \sqrt{-a} = i\sqrt{a} \qquad \text{for} \qquad a > 0 \quad}$$

Any number of the form bi, where b is a real number, is called an **imaginary number.**

WARNING

$$\sqrt{-4}\sqrt{-9} \neq \sqrt{36}$$

The rule $\sqrt{a} \cdot \sqrt{b} = \sqrt{ab}$ holds only when $a \geq 0$ and $b \geq 0$. Always convert the square root of a negative value to imaginary form before proceeding with any algebraic operations. Then

$$\sqrt{-4}\sqrt{-9} = 2i \cdot 3i = 6i^2 = -6$$

Having created imaginary numbers, we next combine real and imaginary numbers. We say that $a + bi$, where a and b are real numbers, is a **complex number.** The number a is called the **real part** of $a + bi$ and b is called the **imaginary part.** The following are examples of complex numbers.

$$3 + 2i \qquad 2 - i \qquad -2i \qquad \frac{4}{5} + \frac{1}{5}i$$

Note that every real number a can be written as a complex number by choosing $b = 0$. Thus,

$$a = a + 0i$$

We see therefore that the real number system is a subset of the complex number system. Once again we have established a number system that incorporates all of the previous number systems and is itself more complicated than the earlier systems.

EXAMPLE 12   Complex Numbers $a + bi$
Write as a complex number.

(a) $-\dfrac{1}{2}$ (b) $\sqrt{-9}$ (c) $-1 - \sqrt{-4}$

SOLUTION

(a) $-\dfrac{1}{2} = -\dfrac{1}{2} + 0i$

(b) $\sqrt{-9} = i\sqrt{9} = 3i = 0 + 3i$

(c) $-1 - \sqrt{-4} = -1 - i\sqrt{4} = -1 - 2i$

∎

We next seek to define operations with complex numbers in such a way that the rules for real numbers and the imaginary unit i continue to hold. We begin with equality and say that two complex numbers are **equal** if their real parts are equal and their imaginary parts are equal; that is,

$$a + bi = c + di \quad \text{if} \quad a = c \quad \text{and} \quad b = d$$

EXAMPLE 13   Equality of Complex Numbers
Solve the equation $x + 3i = 6 - yi$ for x and for y.

SOLUTION
Equating the real parts, we have $x = 6$; equating the imaginary parts, $3 = -y$ or $y = -3$.

∎

Complex numbers are added and subtracted by adding or subtracting the real parts and by adding or subtracting the imaginary parts. That is,

$$(a + bi) + (c + di) = (a + c) + (b + d)i$$
$$(a + bi) - (c + di) = (a - c) + (b - d)i$$

Note that the sum or difference of two complex numbers is again a complex number.

EARLY MATHEMATICIANS' VIEWS OF COMPLEX NUMBERS

When mathematicians in the middle of the sixteenth century tried to solve certain quadratic equations by completing the square, they found themselves, much to their distress, having to deal with the square root of a negative quantity. For example, in 1545 Girolamo Cardano solved the problem of dividing the number 10 into two parts whose product is 40. If one of the parts is x, then the other part is $10 - x$, and we must solve the equation

$$x(10 - x) = 40.$$

Using methods we will describe in Chapter 2, Cardano obtained the roots

$$5 + \sqrt{-15} \qquad \text{and} \qquad 5 - \sqrt{-15}$$

This frustrated him terribly since he had obtained an answer that was "nonsense." He wrote, "So progresses arithmetic subtlety the end of which, as is said, is as refined as it is useless." (See Morris Kline, *Mathematical Thought from Ancient to Modern Times,* Oxford Press, New York, 1972.)

Other famous mathematicians at that time also rejected complex numbers as worthless and fictitious objects, and in fact it was the philosopher and scientist René Descartes who called them "imaginary." It was not until the 1700s that these numbers began to be understood and used. Those earlier mathematicians would be very surprised to learn that complex numbers have been used in thousands of applications ranging from problems in aerodynamics to explanations of the inner workings of the atom.

EXAMPLE 14   Addition and Subtraction of Complex Numbers
Perform the indicated operations.
(a) $(7 - 2i) + (4 - 3i)$ (b) $14 - (3 - 8i)$

SOLUTION
(a) $(7 - 2i) + (4 - 3i) = (7 + 4) + (-2 - 3)i = 11 - 5i$
(b) $14 - (3 - 8i) = (14 - 3) + 8i = 11 + 8i$

∎

PROGRESS CHECK
Perform the indicated operations.
(a) $(-9 + 3i) + (6 - 2i)$ (b) $7i - (3 + 9i)$

ANSWERS
(a) $-3 + i$ (b) $-3 - 2i$

Assuming that the usual laws of algebra apply to complex numbers, we can write

$$(a + bi)(c + di) = a(c + di) + bi(c + di)$$
$$= ac + adi + bci + bdi^2$$
$$= ac + (ad + bc)i + bd(-1)$$
$$= (ac - bd) + (ad + bc)i$$

This entices us into defining multiplication of complex numbers by

$$(a + bi)(c + di) = (ac - bd) + (ad + bc)i$$

This result is significant since it demonstrates that the product of two complex numbers is again a complex number. It need not be memorized; simply use the FOIL method to form all the products and the substitution $i^2 = -1$ to simplify.

EXAMPLE 15 Multiplication of Complex Numbers
Find the product of $(2 - 3i)$ and $(7 + 5i)$.

SOLUTION

$$
\begin{aligned}
(2 - 3i)(7 + 5i) &= 2(7 + 5i) - 3i(7 + 5i) \\
&= 14 + 10i - 21i - 15i^2 \\
&= 14 - 11i - 15(-1) \\
&= 29 - 11i
\end{aligned}
$$

PROGRESS CHECK
Find the product.
(a) $(-3 - i)(4 - 2i)$ (b) $(-4 - 2i)(2 - 3i)$

ANSWERS
(a) $-14 + 2i$ (b) $-14 + 8i$

EXERCISE SET 1.3
In Exercises 1–23 simplify, using the rules for exponents. Write the answers using only positive exponents.

1. $(y^4)^{2n}$

2. $\dfrac{(-4)^6}{(-4)^{10}}$

3. $-\left(\dfrac{x}{y}\right)^3$

4. $-3r^3 r^3$

5. $\dfrac{(r^2)^4}{(r^4)^2}$

6. $[(3b + 1)^5]^5$

7. $\left(\dfrac{3}{2}x^2 y^3\right)^n$

8. $\dfrac{(-2a^2 b)^4}{(-3ab^2)^3}$

9. $2^0 + 3^{-1}$

10. $(xy)^0 - 2^{-1}$

11. $\dfrac{3}{(2x^2 + 1)^0}$

12. $(-3)^{-3}$

13. $\dfrac{1}{3^{-4}}$

14. x^{-5}

15. $\dfrac{3a^5 b^{-2}}{9a^{-4} b^2}$

16. $\left(\dfrac{x^3}{x^{-2}}\right)^2$

17. $\left(\dfrac{2a^2 b^{-4}}{a^{-3} c^{-3}}\right)^2$

18. $\dfrac{2x^{-3} y^2}{x^{-3} y^{-3}}$

19. $(a - 2b^2)^{-1}$

20. $\left(\dfrac{y^{-2}}{y^{-3}}\right)^{-1}$

21. $\dfrac{a^{-1} + b^{-1}}{a^{-1} - b^{-1}}$

22. $\left(\dfrac{a}{b}\right)^{-1} + \left(\dfrac{b}{a}\right)^{-1}$

23. Show that $\left(\dfrac{a}{b}\right)^{-n} = \left(\dfrac{b}{a}\right)^n$

In Exercises 24–26 evaluate each expression.

 24. $[(1.20)^2]^{-1}$

25. $[(-3.67)^2]^{-1}$

26. $\left[\dfrac{(7.65)^{-1}}{7.65^2}\right]^2$

In Exercises 27–32 simplify and write answers using only positive exponents.

27. $16^{3/4}$

28. $(-125)^{-1/3}$

29. $\left(\dfrac{x^{3/2}}{x^{2/3}}\right)^{1/6}$

30. $\dfrac{125^{4/3}}{125^{2/3}}$

31. $(x^{1/3}y^2)^6$

32. $(x^6y^4)^{-1/2}$

In Exercises 33–36 write in radical form.

33. $\left(\dfrac{1}{4}\right)^{2/5}$

34. $x^{2/3}$

35. $a^{3/4}$

36. $(-8x^2)^{2/5}$

In Exercises 37–40 write in exponent form.

37. $\sqrt[4]{8^3}$

38. $\sqrt[5]{3^2}$

39. $\dfrac{1}{\sqrt[5]{(-8)^2}}$

40. $\dfrac{1}{\sqrt[3]{x^7}}$

In Exercises 41–46 evaluate.

41. $\sqrt[4]{-81}$

42. $\sqrt[3]{\dfrac{1}{27}}$

43. $\sqrt{(-5)^2}$

44. $\sqrt{\left(-\dfrac{1}{3}\right)^2}$

45. $\sqrt{\left(\dfrac{5}{4}\right)^2}$

46. $(14.43)^{3/2}$

In Exercises 47 and 48 provide real values for x and y to demonstrate the result.

47. $\sqrt{x^2+y^2} \neq x+y$

48. $\sqrt{x+y} \neq \sqrt{x}+\sqrt{y}$

In Exercises 49–60 simplify and write the answer in simplified form. Every variable represents a positive real number.

49. $\sqrt{48}$

50. $\sqrt{200}$

51. $\sqrt[3]{54}$

52. $\sqrt{x^8}$

53. $\sqrt[3]{y^7}$

54. $\sqrt[4]{b^{14}}$

55. $\sqrt[4]{96x^{10}}$

56. $\sqrt{x^5y^4}$

57. $\sqrt{\dfrac{1}{5}}$

58. $\dfrac{4}{3\sqrt{11}}$

59. $\dfrac{1}{\sqrt{3y}}$

60. $\sqrt{\dfrac{2}{y}}$

In Exercises 61–67 rationalize the denominator.

61. $\dfrac{3}{\sqrt{2}+3}$

62. $\dfrac{-3}{\sqrt{7}-9}$

63. $\dfrac{-3}{3\sqrt{a}+1}$

64. $\dfrac{4}{2-\sqrt{2y}}$

65. $\dfrac{\sqrt{2}+1}{\sqrt{2}-1}$

66. $\dfrac{\sqrt{5}+\sqrt{3}}{\sqrt{5}-\sqrt{3}}$

67. $\dfrac{\sqrt{6}+\sqrt{2}}{\sqrt{3}-\sqrt{2}}$

In Exercises 68–73 rationalize the numerator.

68. $\dfrac{\sqrt{x}-\sqrt{2}}{x-2}$

69. $\dfrac{\sqrt{x}+\sqrt{3}}{x-3}$

70. $\dfrac{\sqrt{x}-\sqrt{y}}{x-y}$

71. $\dfrac{\sqrt{x}-5}{x-25}$

72. $\dfrac{\sqrt{x+h}-\sqrt{x}}{h}$

73. $\dfrac{\sqrt{x+2}+\sqrt{2}}{2x}$

74. Prove that $|ab| = |a||b|$. (*Hint:* Begin with $|ab| = \sqrt{(ab)^2}$.)

In Exercises 75–80 factor and simplify the expression.

75. $(2x-1)^{1/2} - (2x-1)^{3/2}$

76. $(2x+1)^{-2/3} + (2x+1)^{-5/3}$

77. $(x^2 - 1)^{-2/3} - (x^2 - 1)^{-5/3}$

78. $(x^2 + 1)^{3/2} + (x^2 + 1)^{5/2}$

79. $(x - 1)^{1/2} + (x - 1)^{3/2} + (x - 1)^{5/2}$

80. $(x + 1)^{-1/2} + (x + 1)^{-3/2} + (x + 1)^{-5/2}$

In Exercises 81–84 simplify.

81. i^{60} 82. i^{27} 83. i^{-84} 84. $-i^{39}$

In Exercises 85–92 write as a complex number in the form $a + bi$.

85. $-\dfrac{1}{2}$ 86. -0.3 87. $\sqrt{-25}$ 88. $-\sqrt{-5}$

89. $3 - \sqrt{-49}$ 90. $-\dfrac{3}{2} - \sqrt{-72}$ 91. $0.3 - \sqrt{-98}$ 92. $-0.5 + \sqrt{-32}$

In Exercises 93–96 solve for x and for y.

93. $(3x - 1) + (y + 5)i = 1 - 3i$

94. $\left(\dfrac{1}{2}x + 2\right) + (3y - 2)i = 4 - 7i$

95. $(2y + 1) - (2x - 1)i = -8 + 3i$

96. $(y - 2) + (5x - 3)i = 5$

In Exercises 97–104 compute the answer and write it in the form $a + bi$.

97. $2i + (3 - i)$ 98. $-3i + (2 - 5i)$ 99. $2 + 3i + (3 - 2i)$ 100. $(3 - 2i) - \left(2 + \dfrac{1}{2}i\right)$

101. $i\left(-\dfrac{1}{2} + i\right)$ 102. $\dfrac{i}{2}\left(\dfrac{4 - i}{2}\right)$ 103. $(2 - i)(2 + i)$ 104. $(5 + i)(2 - 3i)$

In Exercises 105–108 evaluate the polynomial $x^2 - 2x + 5$ for the given complex value of x.

105. $1 + 2i$ 106. $2 - i$ 107. $1 - i$ 108. $1 - 2i$

1.4
LINEAR EQUATIONS AND INEQUALITIES IN ONE UNKNOWN

Expressions of the form

$$x - 2 = 0 \qquad x^2 - 9 = 0 \qquad 3(2x - 5) = 3$$

$$2x + 5 = \sqrt{x - 7} \qquad \frac{1}{2x + 3} = 5 \qquad x^3 - 3x^2 = 32$$

are examples of equations in the unknown x. An **equation** states that two algebraic expressions are equal. We refer to these expressions as the **left-hand** and **right-hand sides** of the equation.

Our task is to find values of the unknown for which the equation holds true. These values are called **solutions** or **roots** of the equation, and the set of all solutions is called the **solution set.** For example, 2 is a solution of the equation $3x - 1 - 5$ since $3(2) - 1 = 5$. However, -2 is *not* a solution since $3(-2) - 1 \neq 5$.

The solutions of an equation depend upon the number system we are using. For example, the equation $2x = 5$ has no integer solutions but does have a solution among the rational numbers, namely $\frac{5}{2}$. Similarly, the equation $x^2 = -4$ has no solutions among the real numbers but does have solutions if we consider complex numbers, namely $2i$ and $-2i$.

We say that an equation is an **identity** if it is true for every real number for which both sides of the equation are defined. For example,

$$x^2 - 1 = (x + 1)(x - 1)$$

is an identity since it is true for all real numbers; that is, every real number is a solution of the equation. The equation

$$x - 5 = 3$$

is a false statement for all values of x except 8. This is an example of a **conditional equation,** which means that there are real values of the variable for which both sides of the equation are defined but unequal.

When we say that we want to "solve an equation," we mean that we want to find *all* solutions or roots. If we can replace an equation by another, simpler equation that has the same solutions, we will have an approach to solving equations. Equations having the same solutions are called **equivalent equations.** For example, $3x - 1 = 5$ and $3x = 6$ are equivalent equations since it can be shown that $\{2\}$ is the solution set of both equations.

There are two important rules that allow us to replace an equation with an equivalent equation.

Equivalent Equations   The solutions of a given equation are not affected by the following operations.

(**1**) Addition or subtraction of the same number or expression on both sides of the equation.

(**2**) Multiplication or division of both sides of the equation by a number other than 0.

EXAMPLE 1   Solving an Equation
Solve $3x + 4 = 13$.

SOLUTION
We apply the preceding rules to this equation. The strategy is to isolate x, so we *subtract 4 from both sides* of the equation.

$$3x + 4 - 4 = 13 - 4$$
$$3x = 9$$

Dividing both sides by 3, we obtain the solution

$$x = 3$$

■

To be technically accurate, the *solution* of the equation in Example 1 is 3, while $x = 3$ is an equation that is *equivalent* to the original equation. Now that this distinction is understood, we will join in the common usage that says that the equation $3x + 4 = 13$ "has the solution $x = 3$."

LINEAR EQUATIONS   An equation in which the variable appears only in the first degree is called a **first-degree equation in one unknown,** or more simply, a **linear equation.** The general form of such an equation is

$$ax + b = 0$$

where a and b are real numbers and $a \neq 0$. Solving this equation for x produces the following result.

Roots of a Linear Equation	The linear equation $ax + b = 0$, $a \neq 0$, has exactly one solution: $-b/a$.

LINEAR INEQUALITIES

Much of the terminology of equations carries over to inequalities. A **solution of an inequality** is a value of the unknown that satisfies the inequality, and the **solution set** is composed of all solutions. The properties of inequalities listed in Section 1.1 enable us to use the same procedures in solving inequalities as those used in solving equations *with one exception.*

> Multiplication or division of an inequality by a negative number reverses the sense of the inequality.

We will concentrate for now on solving a **linear inequality,** that is, an inequality in which the unknown appears only in the first degree.

EXAMPLE 2   Solving a Linear Inequality
Solve the inequality.

$$2x + 11 \geq 5x - 1$$

SOLUTION
We perform addition and subtraction to collect terms in x just as we did for equations.

$$2x + 11 \geq 5x - 1$$
$$2x \geq 5x - 12$$
$$-3x \geq -12$$

We now divide both sides of the inequality by -3, a negative number, and therefore *reverse* the sense of the inequality.

$$\frac{-3x}{-3} \leq \frac{-12}{-3}$$
$$x \leq 4$$

∎

 PROGRESS CHECK
Solve the inequality $3x - 2 \geq 5x + 4$.

ANSWER
$x \leq -3$

There are three methods commonly used to describe subsets of the real numbers: graphs on a real number line, interval notation, and set-builder notation. Since there will be occasions when we want to use each of these schemes, this is a convenient time to introduce them and to apply them to inequalities.

The **graph of an inequality** is the set of all points satisfying the inequality. The graph of the inequality $a \leq x < b$ is shown in Figure 7. The portion of the real number line that is in color is the solution set of the inequality. The circle at point a has been filled in to indicate that a is also a solution of the inequality; the circle at point b has been left open to indicate that b is not a member of the solution set.

FIGURE 7 **Graph of $a \leq x < b$**

An **interval** is a set of numbers on the real number line that forms a line segment, a half-line, or the entire real number line. The subset shown in Figure 7 would be written in **interval notation** as $[a, b)$, where a and b are the **endpoints** of the interval. A bracket, [or], indicates that the endpoint is included, while a parenthesis, (or), indicates that the endpoint is not included. The interval $[a, b]$ is called a **closed interval** because both endpoints are included. The interval (a, b) is called an **open interval** because neither endpoint is included. Finally, the intervals $[a, b)$ and $(a, b]$ are called **half-open intervals.**

The set of all real numbers satisfying a given property P is written as

$$\{x \mid x \text{ satisfies property } P\}$$

which is read as "the set of all x such that x satisfies property P." This form, called **set-builder notation,** provides a third means of designating subsets of the real number line. Thus, the interval $[a, b)$ shown in Figure 7 is written as

$$\{x \mid a \leq x < b\}$$

which indicates that x must satisfy the inequalities $x \geq a$ and $x < b$.

EXAMPLE 3 **Intervals on a Real Number Line**
Graph each of the given intervals on a real number line and indicate the same subset of the real number line in set-builder notation.
(a) $(-3, 2]$ **(b)** $(1, 4)$ **(c)** $[-4, -1]$

SOLUTION

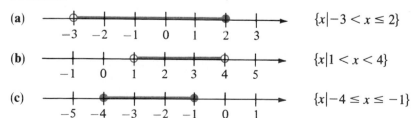

FIGURE 8 **Graphs of Inequalities**

To describe the inequalities $x > 2$ and $x \le 3$ in interval notation, we need to introduce the symbols ∞ and $-\infty$ (read "infinity" and "minus infinity," respectively). The inequalities $x > 2$ and $x \le 3$ are then written as $(2, \infty)$ and $(-\infty, 3]$, respectively, in interval notation and would be graphed on a real number line as shown in Figure 9. Note that ∞ and $-\infty$ are symbols (not numbers) indicating that the intervals extend indefinitely. An interval using one of these symbols is called an **infinite interval.** The interval $(-\infty, \infty)$ designates the entire real number line. Square brackets must never be used around ∞ and $-\infty$, since they are not real numbers.

FIGURE 9   **Graphs of Infinite Intervals**

EXAMPLE 4   Graphing Linear Inequalities
Graph each inequality and write the solution set in interval notation.
(a) $x \le -2$ **(b)** $x \ge -1$ **(c)** $x < 3$

SOLUTION

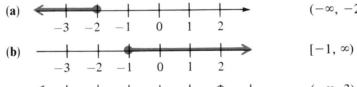

FIGURE 10   **Graphs of Infinite Intervals**

EXAMPLE 5   Linear Inequalities: Solving and Graphing
Solve the inequality.

$$\frac{x}{2} - 9 < \frac{1 - 2x}{3}$$

Graph the solution set and write the solution set in both interval notation and set-builder notation.

SOLUTION
To clear the inequality of fractions, we multiply both sides by the least common denominator (L.C.D.) of all fractions, which is 6.

$$3x - 54 < 2(1 - 2x)$$
$$3x - 54 < 2 - 4x$$
$$7x < 56$$
$$x < 8$$

We may write the solution set as $\{x \mid x < 8\}$ or as the infinite interval $(-\infty, 8)$. The graph of the solution set is shown in Figure 11.

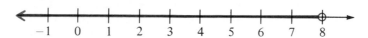

FIGURE 11   Graph of the Solution of a Linear Inequality

∎

EXAMPLE 6   Linear Inequality with No Solution
Solve the inequality.

$$\frac{2(x + 1)}{3} < \frac{2x}{3} - \frac{1}{6}$$

SOLUTION
The L.C.D. of all fractions is 6. Multiplying both sides of the inequality by 6, we obtain

$$4(x + 1) < 4x - 1$$
$$4x + 4 < 4x - 1$$
$$4 < -1$$

Our procedure has led to a contradiction, indicating that there is no solution to the inequality.

∎

PROGRESS CHECK
Solve, and write the answers in interval notation.

(a) $\dfrac{3x - 1}{4} + 1 > 2 + \dfrac{x}{3}$ **(b)** $\dfrac{2x - 3}{2} \geq x + \dfrac{2}{5}$

ANSWERS
(a) $(3, \infty)$ (b) no solution

We can solve double inequalities such as

$$1 < 3x - 2 \leq 7$$

by operating on both inequalities at the same time.

$$3 < 3x \leq 9 \quad \text{Add } +2 \text{ to each member.}$$
$$1 < x \leq 3 \quad \text{Divide each member by 3.}$$

The solution set is the half-open interval $(1, 3]$.

EXAMPLE 7   Solving Double Inequalities
Solve the inequality $-3 \leq 1 - 2x < 6$ and write the answer in interval notation.

SOLUTION

Operating on both inequalities we have

$$-4 \leq -2x < 5 \quad \text{Add } (-1) \text{ to each member.}$$

$$2 \geq x > -\frac{5}{2} \quad \text{Divide each member by } -2.$$

The solution set is the half-open interval $\left(-\frac{5}{2}, 2 \right]$.

■

PROGRESS CHECK

Solve the inequality $-5 < 2 - 3x < -1$ and write the answer in interval notation.

ANSWER

$\left(1, \frac{7}{3} \right)$

EXAMPLE 8   Inequalities in Word Problems

A taxpayer may choose to pay a 20% tax on the gross income or a 25% tax on the gross income less $4000. Above what income level should the taxpayer elect to pay at the 20% rate?

SOLUTION

If we let x = gross income, then the choice available to the taxpayer is
(a) pay at the 20% rate on the gross income, that is, pay $0.20x$, or

(b) pay at the 25% rate on the gross income less $4000, that is, pay $0.25(x - 4000)$.
To determine when (a) produces a lower tax than (b), we must solve

$$0.20x \leq 0.25(x - 4000)$$
$$0.20x \leq 0.25x - 1000$$
$$-0.05x \leq -1000$$
$$x \geq \frac{1000}{0.05} = 20{,}000$$

The taxpayer should choose to pay at the 20% rate if the gross income is $20,000 or more.

■

PROGRESS CHECK

A customer is offered the following choice of telephone services: unlimited local calls at a fixed $20 monthly charge, or a base rate of $8 per month plus 6¢ per message unit. At what level of usage does it cost less to choose the unlimited service?

ANSWER

When the anticipated use exceeds 200 message units.

EXERCISE SET 1.4

In Exercises 1–10 solve the given linear equation and check your answer.

1. $-2x + 6 = -5x - 4$

2. $6x + 4 = -3x - 5$

3. $2(3b + 1) = 3b - 4$

4. $-3(2x + 1) = -8x + 1$

5. $4(x - 1) = 2(x + 3)$

6. $-3(x - 2) = 2(x + 4)$

7. $2(x + 4) - 1 = 0$

8. $3a + 2 - 2(a - 1) = 3(2a + 3)$

9. $-4(2x + 1) - (x - 2) = -11$

10. $3(a + 2) - 2(a - 3) - 0$

In Exercises 11–14 solve for x.

11. $kx + 8 = 5x$

12. $8 - 2kx = -3x$

13. $2 - k + 5(x - 1) = 3$

14. $3(2 + 3k) + 4(x - 2) = 5$

In Exercises 15–18 indicate whether the equation is an identity (I) or a conditional equation (C).

15. $x^2 + x - 2 = (x + 2)(x - 1)$

16. $(x - 2)^2 = x^2 - 4x + 2$

17. $2x + 1 = 3x - 1$

18. $3x - 5 = 4x - x - 2 - 3$

In Exercises 19–36 solve the inequality and graph the result.

19. $x + 4 < 8$

20. $x + 5 < 4$

21. $x + 3 < -3$

22. $x - 2 \le 5$

23. $x - 3 \ge 2$

24. $x + 5 \ge -1$

25. $2 < a + 3$

26. $-5 > b - 3$

27. $2y < -1$

28. $3x < 6$

29. $2x \ge 0$

30. $-\dfrac{1}{2}y \ge 4$

31. $2r + 3 < 9$

32. $3x - 2 > 4$

33. $3x - 1 \ge 2$

34. $\dfrac{-1}{2x + 3} > 0$

35. $\dfrac{4}{5 - 3x} < 0$

36. $\dfrac{3}{3x - 1} > 0$

In Exercises 37–60 solve the given inequality and write the solution set in interval notation.

37. $4x + 3 \le 11$

38. $\dfrac{1}{2}y - 2 \le 2$

39. $\dfrac{3}{2}x + 1 \ge 4$

40. $-5x + 2 > -8$

41. $4(2x + 1) < 16$

42. $3(3r - 4) \ge 15$

43. $2(x - 3) < 3(x + 2)$

44. $4(x - 3) \ge 3(x - 2)$

45. $3(2a - 1) > 4(2a - 3)$

46. $2(3x - 1) + 4 < 3(x + 2) - 8$

47. $\dfrac{2}{3}(x + 1) + \dfrac{5}{6} \ge \dfrac{1}{2}(2x - 1) + 4$

48. $\dfrac{1}{4}(3x + 2) - 1 \le -\dfrac{1}{2}(x - 3) + \dfrac{3}{4}$

49. $\dfrac{x - 1}{3} + \dfrac{1}{5} < \dfrac{x + 2}{5} - \dfrac{1}{3}$

50. $\dfrac{x}{5} - \dfrac{1 - x}{2} > \dfrac{x}{2} - 3$

51. $3(x + 1) + 6 \ge 2(2x - 1) + 4$

52. $4(3x + 2) - 1 \le -2(x - 3) + 15$

53. $-2 < 4x \le 5$

54. $3 \le 6x < 12$

55. $-4 \le 2x + 2 \le -2$

56. $5 \le 3x - 1 \le 11$

57. $3 \le 1 - 2x < 7$

58. $5 < 2 - 3x \le 11$

59. $-8 < 2 - 5x \le 7$

60. $-10 < 5 - 2x < -5$

61. A student has grades of 42 and 70 in the first two tests of the semester. If an average of 70 is required to obtain a C grade, what is the minimum score the student must achieve on the third exam to obtain a C?

62. A compact car can be rented from firm A for $160 per week with no charge for mileage, or from firm B for $100 per week plus 20 cents for each mile driven. If the car is driven m miles, for what values of m does it cost less to rent from firm A?

63. An appliance salesperson is paid $30 per day plus $25 for each appliance sold. How many appliances must be sold for the salesperson's income to exceed $130 per day?

64. A pension trust invests $6000 in a bond that pays 5% simple interest per year. Additional funds are to be invested in a more speculative bond paying 9% simple interest per year, so that the return on the total investment will be at least 6%. What is the minimum amount that must be invested in the more speculative bond?

65. A manufacturer finds that the revenue R from sales of x units of its product is given by

$$R = 22x$$

and the cost C of producing these units is

$$C = 16x + 900$$

For what levels will production of this product provide a profit?

66. If the area of a right triangle is not to exceed 80 square inches and the base is 10 inches, what values may be assigned to the altitude h?

67. A total of 70 meters of fencing material is available with which to enclose a rectangular area. If the width of the rectangle is 15 meters, what values can be assigned to the length L?

68. The relationship

$$C = \frac{5}{9}(F - 32)$$

is used to convert from temperature in degrees Fahrenheit to degrees Celsius. If $50 \leq F \leq 77$, what is the corresponding range for C?

1.5
ABSOLUTE VALUE IN EQUATIONS AND INEQUALITIES

In Section 1.1 we discussed the use of absolute value notation to indicate distance and provided this formal definition.

$$|x| = \begin{cases} x & \text{when} & x \geq 0 \\ -x & \text{when} & x < 0 \end{cases}$$

Note the definition applies to the *expression* that is found between the bars that indicate absolute value. The following example illustrates the application of this definition to the solution of equations involving absolute value.

EXAMPLE 1   Absolute Value in Equations
Solve the equation $|2x - 7| = 11$.

SOLUTION
We apply the definition of absolute value to the two cases.
Case 1. $2x - 7 \geq 0$, that is, the *expression* $2x - 7$ is assumed to be nonnegative. We therefore apply the first part of the definition to obtain

$$|2x - 7| = 2x - 7 = 11$$
$$2x = 18$$
$$x = 9$$

Case 2. $2x - 7 < 0$, that is, the *expression* $2x - 7$ is assumed to be negative. We therefore apply the second part of the definition to obtain

$$|2x - 7| = -(2x - 7) = 11$$
$$-2x + 7 = 11$$
$$x = -2$$

Conclusion: the values $x = -2$ and $x = 9$ are solutions of the equation. (You are urged to substitute these values into the original equation.)

■

PROGRESS CHECK
Solve each equation and check the solution(s).
(a) $|x + 8| = 9$ **(b)** $|3x - 4| = 7$

ANSWERS

(a) $1, -17$ (b) $\dfrac{11}{3}, -1$

When used in inequalities, absolute value notation plays an important and frequently used role in higher mathematics. To solve inequalities involving absolute value, first recall that $|x|$ is the distance between the origin and the point on the real number line corresponding to x. For $a > 0$, the solution set of the inequality $|x| < a$ is then seen to consist of all real numbers whose distance from the origin is less than a, that is, all real numbers in the open interval $(-a, a)$ shown in Figure 12. Similarly, if $|x| > a > 0$, the solution set consists of all real numbers whose distance from the origin is greater than a, that is, all points in the infinite intervals $(-\infty, -a)$ and (a, ∞) as shown in Figure 13. Of course, $|x| \le a$ and $|x| \ge a$ would include the endpoints a and $-a$, and the circles would be filled in.

We can summarize the results in this way.

If $a > 0$, then the solution set to the inequality

$$|x| < a$$

consists of all real numbers in the interval $(-a, a)$ and the solution set to the inequality

$$|x| > a$$

consists of all real numbers in the intervals $(-\infty, a)$ and (a, ∞).

FIGURE 12 Solution to the Inequality $|x| < a$

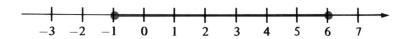

FIGURE 13   Solution to the Inequality $|x| > a$

The following examples illustrate how we apply this result to the *expression* between the bars indicating absolute value.

EXAMPLE 2   Absolute Value in Inequalities

Solve $|2x - 5| \le 7$, graph the solution set, and write the solution set in interval notation.

SOLUTION

We must solve the equivalent double inequality

$$-7 \le 2x - 5 \le 7$$
$$-2 \le 2x \le 12 \qquad \text{Add } +5 \text{ to each member.}$$
$$-1 \le x \le 6 \qquad \text{Divide each member by 2.}$$

The graph of the solution set is then
Thus, the solution set is the closed interval $[-1, 6]$.

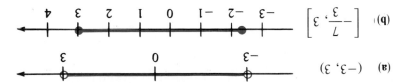

☑ **PROGRESS CHECK**

Solve each inequality, graph the solution set, and write the solution set in interval notation.

(a) $|x| < 3$ **(b)** $|3x - 1| \le 8$ **(c)** $|x| < -2$

ANSWERS

(a) $(-3, 3)$

(b) $\left[-\dfrac{7}{3}, 3 \right]$

(c) No solution. Since $|x|$ is always nonnegative, $|x|$ cannot be less than -2.

EXAMPLE 3   Absolute Value in Inequalities

Solve the inequality $\left| 3 - \dfrac{t}{2} \right| > 1$, write the solution set in interval notation, and graph the solution.

SOLUTION

We must solve the equivalent inequalities

$$3 - \frac{t}{2} > 1 \qquad 3 - \frac{t}{2} < -1$$
$$6 - t > 2 \qquad 6 - t < -2$$
$$-t > -4 \qquad -t < -8$$
$$t < 4 \qquad t > 8$$

The solution set consists of the real numbers in the infinite intervals $(-\infty, 4)$ and $(8, \infty)$. The graph of the solution set is then

■

 PROGRESS CHECK

Solve each inequality, write the solution set in interval notation, and graph the solution.

(a) $|5x - 6| > 9$ **(b)** $|2x - 2| \geq 8$

ANSWERS

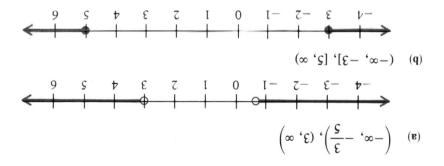

$$(-\infty, -3], [5, \infty) \qquad \textbf{(b)}$$

$$\left(-\infty, -\frac{3}{5}\right), (3, \infty) \qquad \textbf{(a)}$$

 WARNING

Students sometimes write

$$1 > x > 5$$

This is a misuse of the inequality notation since it states that x is simultaneously less than 1 *and* greater than 5, which is impossible. What is usually intended is the pair of infinite intervals $(-\infty, 1)$ and $(5, \infty)$, and the inequalities that must be written are

$$x < 1 \qquad x > 5$$

EXERCISE SET 1.5

In Exercises 1–9 solve and check.

1. $|x + 2| = 3$
2. $|r - 5| = \dfrac{1}{2}$
3. $|2x - 4| = 2$
4. $|5y + 1| = 11$

5. $|-3x + 1| = 5$
6. $|2t + 2| = 0$
7. $3|-4x - 3| = 27$
8. $\dfrac{1}{|x|} = 5$

9. $\dfrac{1}{|s - 1|} = \dfrac{1}{3}$

In Exercises 10–15 solve and graph the solution set.

10. $|x + 3| < 5$
11. $|x + 1| > 3$
12. $|3x + 6| \le 12$
13. $|4x - 1| > 3$

14. $|3x + 2| \ge -1$
15. $\left|\dfrac{1}{3} - x\right| < \dfrac{2}{3}$

In Exercises 16–24 solve and write the solution set using interval notation.

16. $|x - 2| \le 4$
17. $|x - 3| \ge 4$
18. $|2x + 1| < 5$
19. $\dfrac{|2x - 1|}{4} < 2$

20. $\dfrac{|3x + 2|}{2} \le 4$
21. $\dfrac{|2x + 1|}{3} < 0$
22. $\left|\dfrac{4}{3x - 2}\right| < 1$
23. $\left|\dfrac{5 - x}{3}\right| > 4$

24. $\left|\dfrac{2x + 1}{3}\right| \le 5$

In Exercises 25 and 26, x and y are real numbers.

25. Prove that $\left|\dfrac{x}{y}\right| = \dfrac{|x|}{|y|}$. (*Hint:* Treat as four cases.)

26. Prove that $|x|^2 = x^2$.

27. The number of units u produced daily by a factory is given by

$$|u - 190{,}000| < 10{,}000$$

What is the daily production range?

28. A company policy states that the hours h a full-time employee may work in a week is given by

$$\left|\dfrac{2h - 81}{3}\right| < 4$$

Express the acceptable range of hours as an inequality.

29. A machine that packages 100 vitamin pills per bottle can make an error of 2 pills per bottle. If x is the number of pills in a bottle, write an inequality, using absolute value, that indicates a maximum error of 2 pills per bottle. Solve the inequality.

30. The weekly income of a worker in a manufacturing plant differs from \$300 by no more than \$50. If x is the weekly income, write an inequality, using absolute value, that expresses this relationship. Solve the inequality.

1.6 SECOND-DEGREE INEQUALITIES AND CRITICAL VALUES

To solve a second-degree inequality, such as

$$x^2 - 2x > 15$$

we rewrite the inequality in the form

$$x^2 - 2x - 15 > 0$$

or, after factoring,

$$(x + 3)(x - 5) > 0$$

With the right-hand side equal to 0, this inequality requires that the product of the two factors, which represent real numbers, be positive. That means that both factors must have the same sign. We must therefore analyze the *signs* of $(x + 3)$ and $(x - 5)$.

In any situation like this we are interested in knowing all values of x for which the general expression $ax + b$ will be positive and those values for which it will be negative. Since $ax + b = 0$ when $x = -b/a$, we see that

> The linear factor $ax + b$ equals 0 at the **critical value** $x = -b/a$ and has opposite signs to the left and right of the critical value on a number line.

A practical means for solving such problems as the current example is illustrated in Figure 14. Since the critical values occur where $x + 3 = 0$ and $x - 5 = 0$, the values -3 and $+5$ are displayed on a real number line. The rows above the real number line display the *signs* of the factors $x + 3$ and $x - 5$ for all real values of x. The row below the real number line displays the *signs* of the product $(x + 3)(x - 5)$. The product is positive when the factors have the same sign, is negative when the factors are of opposite sign, and is zero when either factor is zero. The row below the real number line shows the solution set of the inequality $(x + 3)(x - 5) > 0$ to be

$$\{x \mid x < -3 \text{ or } x > 5\}$$

which consists of the real numbers in the open intervals $(-\infty, -3)$ and $(5, \infty)$. The solution set is shown in Figure 15.

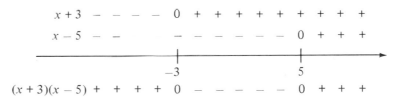

FIGURE 14 Critical Value Diagram

FIGURE 15 Solution Set

EXAMPLE 1   Solving a Second-Degree Inequality
Solve the inequality $x^2 \leq -3x + 4$ and graph the solution set on a real number line.

SOLUTION
We rewrite the inequality and factor.

$$x^2 + 3x - 4 \leq 0$$
$$(x - 1)(x + 4) \leq 0$$

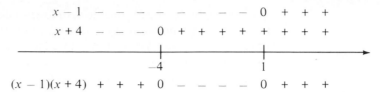

FIGURE 16 Critical Value Diagram for Example 1

We seek values of x for which the factors $(x - 1)$ and $(x + 4)$ have opposite signs or are zero. The critical values occur where $x - 1 = 0$ and where $x + 4 = 0$, that is, at $+1$ and -4. Figure 16 gives an analysis of the signs of the factors $x - 1$ and $x + 4$ as well as the signs of their product, $(x - 1)(x + 4)$. We see that the solution set consists of all real numbers.

$$\{x \mid -4 \le x \le 1\}$$

which is the closed interval $[-4, 1]$, shown in Figure 17.

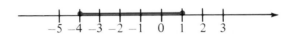

FIGURE 17 Solution Set for Example 1

■

☑ **PROGRESS CHECK**

Solve the inequality $2x^2 \ge 5x + 3$ and graph the solution set on a real number line.

ANSWERS

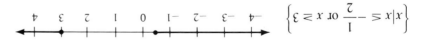

$$\left\{x \mid x \le -\frac{1}{2} \text{ or } x \ge 3\right\}$$

Although

$$\frac{ax + b}{cx + d} < 0$$

is not a second-degree inequality, the solution of this inequality can be found by using the methods developed to solve the inequality

$$(ax + b)(cx + d) < 0$$

since both inequalities require that the two expressions composing them have different signs.

EXAMPLE 2 Solving Inequalities

Solve the inequality $\dfrac{y + 1}{2 - y} \le 0$.

SOLUTION

Figure 18 gives an analysis of the signs of $y + 1$ and $2 - y$. The critical values occur where $y + 1 = 0$ and where $2 - y = 0$, that is, at -1 and $+2$. The bottom row shows the signs of the quotient $(y + 1)/(2 - y)$, from which we see that the solution set is $\{y \mid y \leq -1 \text{ or } y > 2\}$ or all real numbers in the intervals $(-\infty, -1], (2, \infty)$. Note that $y = 2$ would result in division by 0 and must be excluded.

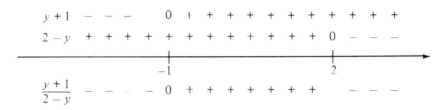

FIGURE 18   Critical Value Diagram for Example 2

PROGRESS CHECK

Solve the inequality $\dfrac{2x - 3}{1 - 2x} \geq 0$.

ANSWER

$\left\{x \,\middle|\, \dfrac{1}{2} > x > \dfrac{3}{2}\right\}$ or $\left[\dfrac{1}{2}, \dfrac{3}{2}\right)$

EXAMPLE 3   Solving Inequalities

Solve the inequality $\dfrac{4y - 5}{2 - y} \leq -3$.

SOLUTION

To use the method of sign analysis, we must first manipulate the expression algebraically into the form $(ax + b)/(cx + d)$.

$$\frac{4y - 5}{2 - y} + 3 \leq 0$$

$$\frac{4y - 5 + 3(2 - y)}{2 - y} \leq 0$$

$$\frac{y + 1}{2 - y} \leq 0$$

In this form, we can analyze the signs of the expressions $y + 1$ and $2 - y$. Note that the problem is now identical to that of Example 2 and that the same solution method and answers apply.

PROGRESS CHECK

Solve the inequality $\dfrac{2}{x+1} < 1$.

ANSWER

$$\{x \mid x > 1 \text{ or } x < -1\} \qquad \text{or} \qquad (-\infty, -1), (1, \infty)$$

EXAMPLE 4 Solving Inequalities

Solve the inequality $(x - 2)(2x + 5)(3 - x) < 0$.

SOLUTION

Although this is a third-degree inequality, the same approach will work. Figure 19 gives an analysis of the signs of $x - 2$, $2x + 5$, and $3 - x$. The product of three factors is negative when there are an odd number of negative factors. The solution set is then

$$\left\{x \mid -\frac{5}{2} < x < 2 \text{ or } x > 3\right\} \qquad \text{or} \qquad \left(\frac{5}{2}, 2\right), (3, \infty)$$

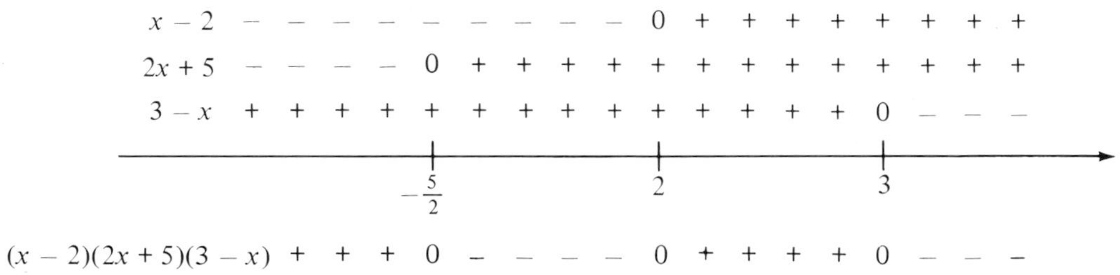

$$
\begin{array}{lccccccccccccccccc}
x - 2 & - & - & - & - & - & - & - & - & - & 0 & + & + & + & + & + & + & + & + \\
2x + 5 & - & - & - & - & 0 & + & + & + & + & + & + & + & + & + & + & + & + & + \\
3 - x & + & + & + & + & + & + & + & + & + & + & + & + & + & + & 0 & - & - & - \\
\end{array}
$$

$$-\frac{5}{2} \qquad\qquad 2 \qquad\qquad 3$$

$$(x - 2)(2x + 5)(3 - x) \quad + \quad + \quad + \quad 0 \quad - \quad - \quad - \quad - \quad 0 \quad + \quad + \quad + \quad + \quad 0 \quad - \quad - \quad -$$

FIGURE 19 Diagram for Example 4

∎

PROGRESS CHECK

Solve the inequality $(2y - 9)(6 - y)(y + 5) \geq 0$.

ANSWER

$$\left\{y \mid y \leq -5 \text{ or } \frac{9}{2} \leq y \leq 6\right\} \qquad \text{or} \qquad (-\infty, -5], \left[\frac{9}{2}, 6\right]$$

EXERCISE SET 1.6

In Exercises 1–14 determine the solution set of each inequality.

1. $x^2 + 5x + 6 > 0$ 2. $x^2 + 3x - 4 \leq 0$ 3. $2x^2 - x - 1 < 0$ 4. $3x^2 - 4x - 4 \geq 0$

5. $4x - 2x^2 < 0$ 6. $r^2 + 4r + 4 \leq 0$ 7. $\dfrac{x + 5}{x + 3} \leq 0$ 8. $\dfrac{x - 6}{x + 4} \geq 0$

9. $\dfrac{2r + 1}{r - 3} \leq 0$ 10. $\dfrac{x - 1}{2x - 3} \geq 0$ 11. $\dfrac{3s + 2}{2s - 1} \geq 0$ 12. $\dfrac{4x + 5}{x^2} \leq 0$

13. $(x + 2)(3x - 2)(x - 1) > 0$ 14. $(x - 4)(2x + 5)(2 - x) \le 0$

In Exercises 15–28 indicate the solution set of each inequality on a real number line.

15. $x^2 + x - 6 > 0$ 16. $x^2 - 3x - 10 \ge 0$ 17. $2x^2 - 3x - 5 < 0$ 18. $3x^2 - 4x - 4 \le 0$

19. $\dfrac{2r + 3}{2r - 1} < 0$ 20. $\dfrac{3x + 2}{2x - 3} \ge 0$ 21. $\dfrac{x - 1}{x + 1} \ge 0$ 22. $\dfrac{2x - 1}{x + 2} \le 0$

23. $6x^2 + 8x + 2 \ge 0$ 24. $2x^2 + 5x + 2 \le 0$

25. $(y - 3)(2 - y)(2y + 4) \ge 0$ 26. $(2x + 5)(3x - 2)(x + 1) < 0$

27. $(x - 3)(1 + 2x)(3x + 5) > 0$ 28. $(1 - 2x)(2x + 1)(x - 3) \le 0$

In Exercises 29–36 find the values of x for which the given expression has real values.

29. $\sqrt{(x - 2)(x + 1)}$ 30. $\sqrt{(2x + 1)(x - 3)}$ 31. $\sqrt{x^2 - 4}$ 32. $\sqrt{1 - x^2}$

33. $\sqrt{x^2 + 1}$ 34. $\sqrt{a^2 - x^2}$ 35. $\sqrt{2x^2 + 7x + 6}$ 36. $\sqrt{2x^2 + 3x + 1}$

37. A photographer wants to mount a 6-inch by 8-inch photo with a mat that has the same width on each side. If the total area of the mounted photo is not to exceed 168 square inches, what is the maximum size of the mat?

38. If the cost C of producing x units of a product is

$$C = 1.3x + 558$$

and the revenue R from sale of these units is

$$R = 2.2x$$

for what production levels will the company show a profit?

39. A manufacturer of solar heaters finds that when x units are made and sold, the profit (in thousands of dollars) is given by $x^2 - 50x - 5000$. For what values of x will the firm show a loss?

40. A ball thrown directly upward from level ground at an initial velocity of 40 feet per second attains a height d given by $d = 40t - 16t^2$ after t seconds. During what time interval is the ball at a height of at least 16 feet?

TERMS AND SYMBOLS

KEY IDEAS FOR REVIEW

Topic	Page	Key Idea
Set	3	A set is simply a collection of objects or numbers.

Real number system
  rational numbers
  irrational numbers
	3	The real number system is composed of the rational and irrational numbers. The rational numbers are those that can be written as the ratio of two integers, p/q, with $q \neq 0$; the irrational numbers cannot be written as a ratio of integers.

Real number line	4	There is a one-to-one correspondence between the set of all real numbers and the set of all points on the real number line. That is, for every point on the line there is a real number and for every real number there is a point on the line.
Inequality symbols	4	Algebraic statements using inequality symbols have straightforward geometric interpretations using the real number line. For example, $a < b$ says that a lies to the left of b on the real number line.
Absolute value	6	Absolute value specifies distance independent of the direction. Three important properties of absolute value are $$\lvert a \rvert \geq 0 \qquad \lvert a \rvert = \lvert -a \rvert \qquad \lvert a - b \rvert = \lvert b - a \rvert$$
distance on a line	7	The distance between points A and B whose coordinates are a and b, respectively, is given by $$\overline{AB} = \lvert b - a \rvert$$
Polynomials	10	Algebraic expressions of the form $$P = a_n x^n + a_{n-1} x^{n-1} + \cdots + a_1 x + a_0$$ are called polynomials.
operations	12	To add (subtract) polynomials, simply add (subtract) like terms. To multiply polynomials, form all possible products using the rule for exponents: $a^m a^n = a^{m+n}$.
factored	13	A polynomial is said to be factored when it is written as a product of polynomials of lower degree.
Exponents	20	The rules for positive integer exponents also apply to zero and negative integer exponents and to rational exponents.
Radical notation	24	Radical notation is simply another way of writing a rational exponent. That is, $\sqrt[n]{b} = b^{1/n}$.
principal root	24	If n is even and b is positive, there are two real numbers a such that $b^{1/n} = a$. Under these circumstances, we insist that the nth root be positive. That is, $\sqrt[n]{b}$ is a positive number if n is even and b is positive. Thus, $\sqrt{16} = 4$. We must write $\sqrt{x^2} = \lvert x \rvert$ to insure that the result is a positive number.
Complex numbers	30	Complex numbers were created because there are no real numbers that satisfy such simple polynomial equations as $x^2 + 5 = 0$.

Topic	Page	Key Idea
imaginary unit	30	Using the imaginary unit $i = \sqrt{-1}$, a complex number is of the form $a + bi$, where a and b are real numbers.
real number system	32	The real number system is a subset of the complex number system.
Linear equations	37	The linear equation $ax + b = 0$, $a \neq 0$, has the solution $-b/a$.
Inequalities	38	Inequalities can be operated upon in the same manner as statements involving an equals sign, with one important exception. When an inequality is multiplied or divided by a negative number, the sense of the inequality is reversed.
graphing a linear inequality	39	The solution set of a linear inequality can be indicated by graphing on a real number line, by set-builder notation, or by interval notation.
absolute value in inequalities	44	$\|x\| < a$ is equivalent to $-a < x < a$; $\|x\| > a$ is equivalent to $x > a$ or $x < -a$
solving quadratic inequalities	48	If a second-degree inequality can be written in the factored form $$(ax + b)(cx + d) < 0$$ or $$(ax + b)(cx + d) > 0$$ then the solution set is easily found. First, on the real number line, determine the intervals in which each factor is positive and the intervals in which each is negative. If the product of the factors is negative (<0), then the solution set consists of the intervals in which the factors are opposite in sign; if the product is positive (>0), the solution set consists of the intervals in which the factors are of the same sign.

REVIEW EXERCISES

Solutions to exercises whose numbers are in color are in the Solutions section in the back of the book.

1.1 For Exercises 1–4 determine whether the statement is true (T) or false (F).

1. $\sqrt{7}$ is a real number.

2. -35 is a natural number.

3. -14 is not an integer.

4. 0 is an irrational number.

In Exercises 5–8, provide a counterexample to the given statement.

5. The sum of two irrational numbers is an irrational number.

6. The product of two irrational numbers is an irrational number.

7. If a is a nonnegative real number, then $\sqrt{a}$ is irrational.

8. If a and b are real numbers such that $|a| = |b|$, then $a = b$.

In Exercises 9–11 sketch the given set of numbers on a real number line.

9. The negative real numbers.

10. The real numbers x such that $x > 4$.

11. The real numbers x such that $-1 \leq x < 1$.

12. Find the value of $|-3| - |1 - 5|$.

13. Find $\overline{PQ}$ if the coordinates of P and Q are $\frac{9}{2}$ and 6, respectively.

1.2 14. A salesperson receives $3.25x + 0.15y$ dollars, where x is the number of hours worked and y is the number of miles of automobile usage. Find the amount due the salesperson if $x = 12$ hours and $y = 80$ miles.

15. Which of the following expressions are not polynomials?

(a) $-2xy^2 + x^2y$ (b) $3b^2 + 2b - 6$

(c) $x^{-1/2} + 5x^2 - x$ (d) $7.5x^2 + 3x - \frac{1}{2}x^0$

In Exercises 16 and 17 indicate the leading coefficient and the degree of each polynomial.

16. $-0.5x^7 + 6x^3 - 5$ 17. $2x^2 + 3x^4 - 7x^5$

In Exercises 18–20 perform the indicated operations.

18. $(3a^2b^2 - a^2b + 2b - a) - (2a^2b^2 + 2a^2b - 2b - a)$

19. $x(2x - 1)(x + 2)$ 20. $3x(2x + 1)^2$

In Exercises 21–26 factor each expression.

21. $2x^2 - 2$ 22. $x^2 - 25y^2$

23. $2a^2 + 3ab + 6a + 9b$ 24. $4x^2 + 19x - 5$

25. $x^8 - 1$ 26. $27r^6 + 8s^6$

1.3   In Exercises 27–32 simplify, using only positive exponents to express the answers. All variables are positive numbers.

27. $(2a^2b^{-3})^{-3}$ 28. $2(a^2 - 1)^0$

29. $\left(\dfrac{x^3}{y^{-6}}\right)^{-4/3}$ 30. $\dfrac{x^{3+n}}{x^n}$

31. $\sqrt{80}$ 32. $\dfrac{2}{\sqrt{12}}$

33. Rationalize the denominator: $\dfrac{\sqrt{x}}{\sqrt{x} + \sqrt{y}}$

34. Rationalize the numerator: $\dfrac{\sqrt{3} - \sqrt{x}}{x - 3}$

35. Factor and simplify:
$$(2x + 1)^{2/3} - (2x + 1)^{5/3}$$

36. Solve for x and for y:
$$(x - 2) + (2y - 1)i = -4 + 7i$$

37. Simplify i^{47}.

In Exercises 38–40 perform the indicated operations and write all answers in the form $a + bi$.

38. $2 + (6 - i)$ 39. $(2 + i)^2$

40. $(4 - 3i)(2 + 3i)$

1.4   In Exercises 41–44 solve for x.

41. $3x - 5 = 3$

42. $2(2x - 3) - 3(x + 1) = -9$

43. $\dfrac{2 - x}{3 - x} = 4$ 44. $k - 2x = 4kx$

45. Indicate whether the statement is true (T) or false (F): The equation $3x^2 = 9$ is an identity.

46. Indicate whether the statement is true (T) or false (F): $x = 3$ is a solution of the equation $3x - 1 = 10$.

47. Solve and graph $3 \le 2x + 1$.

48. Solve and graph $-4 < -2x + 1 \le 10$.

In Exercises 49–51 solve, and express the solution set in interval notation.

49. $2(a + 5) > 3a + 2$ 50. $\dfrac{-1}{2x - 5} < 0$

51. $\dfrac{2x}{3} + \dfrac{1}{2} \ge \dfrac{x}{2} - 1$

1.5   52. Solve $|3x + 2| = 7$ for x.

53. Solve and graph $|4x - 1| = 5$.

54. Solve and graph $|2x + 1| > 7$.

55. Solve $|2 - 5x| < 1$ and write the solution in interval notation.

56. Solve $|3x - 2| \ge 6$ and write the solution in interval notation.

1.6   57. Write the solution set of the inequality $x^2 + 4x - 5 \ge 0$ in interval notation.

58. Write the solution set of $\dfrac{2x + 1}{x + 5} \ge 0$ in interval notation.

59. Write the solution set of
$$(3 - x)(2x + 3)(x + 2) < 0$$
in interval notation.

PROGRESS TEST 1A

In Problems 1–4 determine whether the statement is true (T) or false (F).

1. -1.36 is an irrational number.

2. π is equal to $\frac{22}{7}$.

3. $\sqrt{4}$ is a real number.

4. $\sqrt{x^2} = x$ for all real numbers x.

In Problems 5 and 6 sketch the given set of numbers on a real number line.

5. The integers that are greater than -3 and less than or equal to 3.

6. The real numbers x such that $-2 \le x < \frac{1}{2}$.

7. Find the value of $|2 - 3| - |4 - 2|$.

8. Find $\overline{AB}$ if the coordinates of A and B are -6 and -4, respectively.

9. The area of a region is given by the expression $3x^2 - xy$. Find the area when $x = 5$ meters and $y = 10$ meters.

10. Evaluate the expression $\dfrac{-|y - 2x|}{|xy|}$ when $x = 3$ and $y = -1$.

11. Which of the following expressions are not polynomials?
    (a) x^5 (b) $5x^{-4}y + 3x^2 - y$
    (c) $4x^3 + x$ (d) $2x^2 + 3x^0$

In Problems 12 and 13 indicate the leading coefficients and the degree of each polynomial.

12. $-2.2x^5 + 3x^3 - 2x$ 13. $14x^6 - 2x + 1$

In Problems 14 and 15 perform the indicated operations.

14. $3xy + 2x + 3y + 2 - (1 - y - x + xy)$

15. $(a + 2)(3a^2 - a + 5)$

In Problems 16 and 17 factor each expression.

16. $8a^3b^5 - 12a^5b^2 + 16a^2b$

17. $4 - 9x^2$

In Problems 18–21 simplify, and use only positive exponents to express the answers.

18. $\left(\dfrac{x^{7/2}}{x^{2/3}}\right)^{-6}$ 19. $\dfrac{y^{2n}}{y^{n-1}}$

20. $\dfrac{-1}{(x - 1)^0}$ 21. $(2a^2b^{-1})^2$

22. Rationalize the denominator: $\dfrac{2}{\sqrt{x} + 2}$

23. Factor and simplify:
$$(2x - 1)^{-2/3} + (2x - 1)^{-5/3}$$

24. For what values of x is $\dfrac{1}{\sqrt{x - 2}}$ a real number?

In Problems 25 and 26 perform the indicated operations and write all answers in the form $a + bi$.

25. $(2 - i) + (-3 + i)$ 26. $(5 + 2i)(2 - 3i)$

In Problems 27 and 28 solve for y.

27. $5 - 4y = 2$ 28. $\dfrac{2 + 5y}{3y - 1} = 6$

29. Indicate whether the statement is true (T) or false (F): The equation $(2x - 1)^2 = 4x^2 - 4x + 1$ is an identity.

30. Solve $-1 \le 2x + 3 < 5$ and graph the solution set.

In Problems 31 and 32 solve, and express the solution set in interval notation.

31. $3(2a - 1) - 4(a + 2) \le 4$

32. $-2 < 2 - x \le 6$ 33. Solve $|4x - 1| = 9$.

34. Solve $|2x - 1| \le 5$ and graph the solution set.

35. Solve $|1 - 3x| < 5$ and write the solution in interval notation.

In Problems 36 and 37 write the solution set in interval notation.

36. $-2x^2 + 3x - 1 \le 0$

37. $(x - 1)(2 - 3x)(x + 2) \le 0$

PROGRESS TEST 1B

In Problems 1–4 determine whether the statement is true (T) or false (F).

1. 19.6 is a real number.

2. π is equal to 3.14.

3. $\sqrt{5}$ is a rational number.

4. If a and b are real numbers, then $|a - b| = |b - a|$.

In Problems 5 and 6 sketch the given set of numbers on a real number line.

5. The natural numbers that are less than 5.

6. The real numbers x such that $\frac{3}{2} < x < 3$.

7. Find the value of $\dfrac{|2 - 5| + |1 - 5|}{|-7|}$.

8. Find $\overline{AB}$ if the coordinates of A and B are -2 and 5, respectively.

9. The area of a trapezoid is given by the formula $A = \frac{1}{2}h(b + b')$. Find the area if $h = 4$ meters, $b = 3$ meters, and $b' = 4$ meters.

10. Evaluate the expression $|x|/|x - y|$ when $x = -2$ and $y = -3$.

11. Which of the following expressions are not polynomials?
    (a) $3x^2 + x^{-1} - 2$ (b) $2x^3 - xy^2 + x$
    (c) $2x^2y^2 + xy - 4$ (d) $x^2y + x^{1/2}y + 2$

In Problems 12 and 13 indicate the leading coefficient and the degree of each polynomial.

12. $-3x^3 + 4x^5$ 13. $1.5x^{10} - x^9 + 17x^8$

In Problems 14 and 15 perform the indicated operations.

14. $(2s^2t^3 - st^2 + st - s + t)$
$$- (3s^2t^2 - 2s^2t - 4st^2 - t + 3)$$

15. $(b + 3)(-3b^2 + 2b + 4)$

In Problems 16 and 17 factor each expression.

16. $5r^3s^4 - 40r^4s^3t$ 17. $2x^2 + 7x - 4$

In Problems 18–21 simplify, using only positive exponents to express the answers.

18. $\dfrac{4x^{-3}}{x^{-2}}$ 19. $(b^2)^5(b^3)^6$

20. $\left(\dfrac{x^8}{y^{12}}\right)^{3/4}$ 21. $\dfrac{2(x + 2)^0}{-2}$

22. Rationalize the numerator: $\dfrac{\sqrt{a} + \sqrt{b}}{a - b}$

23. Factor and simplify:
$$(x^2 - 1)^{-1/2} + (x^2 - 1)^{-3/2}$$

24. For what values of x is $\sqrt{2 - x}$ a real number?

In Problems 25 and 26 perform the indicated operations and write all answers in the form $a + bi$.

25. $(4 - 2i) - \left(2 - \dfrac{1}{2}i\right)$ 26. $(3 - 2i)(2 - i)$

In Problems 27 and 28 solve for x.

27. $3(2x + 5) = 5 - (3x - 1)$

28. $3x - k^2 = -kx$

29. Indicate whether the statement is true (T) or false (F):
$x = -1$ is a solution of the equation $\dfrac{x - 1}{x + 1} = 0$.

30. Solve $-9 \leq 1 - 5x \leq -4$ and graph the solution set.

In Problems 31 and 32 solve, and express the solution set in interval notation.

31. $\dfrac{x}{4} - \dfrac{1}{2} \leq \dfrac{1}{2} - x$ 32. $\dfrac{-2}{3 - x} \geq 0$

33. Solve $|1 - 3x| = 7$.

34. Solve $\dfrac{|x - 4|}{2} \geq 1$ and graph the solution set.

35. Solve $|5x + 2| > 3$ and write the solution set in interval notation.

In Problems 36 and 37 write the solution set in interval notation.

36. $\dfrac{x^2}{x + 5} < 0$

37. $(3x - 2)(x + 4)(1 - x) > 0$

FUNCTIONS AND GRAPHS

What is the effect of increased fertilization on the growth of an azalea? If the minimum wage is increased, what will be the impact on the number of unemployed workers? When a submarine dives, can we calculate the water pressure against the hull at a given depth?

Each of the questions posed above seeks a relationship between phenomena. The search for relationships, or correspondence, is a central activity in our attempts to understand the universe; it is used in mathematics, engineering, the physical and biological sciences, the social sciences, and business and economics.

The concept of a function has been developed as a means of organizing and assisting in the study of relationships. Since graphs are powerful means of exhibiting relationships, we begin with a study of the Cartesian, or rectangular, coordinate system. We will then formally define a function and will provide a number of ways of viewing the function concept. Function notation will be introduced to provide a convenient means of writing functions.

Much of the material in this chapter focuses on the graphs of functions. The information available at a glance from a graph is so impressive that it is vital for a student planning to study advanced mathematics to be familiar with techniques for quickly sketching the graphs of those functions that occur most frequently.

2.1
THE RECTANGULAR COORDINATE SYSTEM

In Chapter 1 we associated the system of real numbers with points on the real number line. That is, we saw that there is a one-to-one correspondence between the system of real numbers and points on the real number line.

We will now develop an analogous way to handle points in a plane. We begin by drawing a pair of perpendicular lines intersecting at a point O called the **origin.** One of the lines, called the **x-axis,** is usually drawn in a horizontal position. The other line, called the **y-axis,** is usually drawn vertically. The coordinate axes divide the plane into four **quadrants,** which we label I, II, III, and IV as in Figure 1.

If we think of the x-axis as a real number line, we may mark off some convenient unit of length, with positive numbers to the right of the origin and negative numbers to the left of the origin. Similarly, we may think of the y-axis as a real number line. Again, we may mark off a convenient unit of length (usually the same as the unit of length on the x-axis) with the upward direction representing positive numbers and the downward direction negative numbers. The x- and y-axes are called **coordinate axes,** and together they constitute a **rectangular** or **Cartesian coordinate system.** The term rectangular stems from the fact that the axes are perpendicular to each other. The term Cartesian honors the "father of modern mathematics," René Descartes (1596–1650).

By using the coordinate axes, we can outline a procedure for labeling a point P in the plane. From P, draw a perpendicular to the x-axis and note that it meets the x-axis at $x = a$. Now draw a perpendicular from P to the y-axis and note that it meets the y-axis at $y = b$. We say that the **coordinates** of P are given by the **ordered pair** (a, b). The term "ordered pair" means that the order is significant; that is, the ordered pair (a, b) is different from the ordered pair (b, a).

The first number of the ordered pair (a, b) is sometimes called the **abscissa** or **x-coordinate** of P. The second number is called the **ordinate** or **y-coordinate** of P.

We have now developed a procedure for associating with each point P in the plane a unique ordered pair of real numbers (a, b). We usually write the point P as $P(a, b)$. Conversely, every ordered pair of real numbers (a, b) determines a unique point P in the plane. The point P is located at the intersection of the lines perpendicular to the x-axis

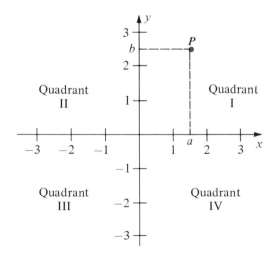

FIGURE 1 Rectangular Coordinate System

and to the y-axis at the points on the axes having coordinates a and b, respectively. We have thus established a one-to-one correspondence between the set of all points in the plane and the set of all ordered pairs of real numbers.

We have indicated a number of points in Figure 2. Note that all points on the x-axis have a y-coordinate of 0 and all points on the y-axis have an x-coordinate of 0. Since we always measure distance from a point to a line along the perpendicular, we have this important observation:

> The x-coordinate of a point P is the directed distance of P from the y-axis; the y-coordinate is its directed distance from the x-axis.

The point $(2, 3)$ in Figure 2 is 2 units from the y-axis and 3 units from the x-axis.

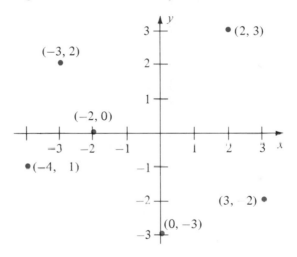

FIGURE 2 Plotting Points

THE DISTANCE FORMULA

Our first use of the rectangular coordinate system deals with finding a formula that gives the distance $\overline{PQ}$ between two points $P(x_1, y_1)$ and $Q(x_2, y_2)$. In Figure 3a we have shown the x-coordinate of a point in quadrant I as the distance of the point from the y-axis, and the y-coordinate as its distance from the x-axis. Thus we labeled the horizontal segments x_1 and x_2 and the vertical segments y_1 and y_2. In Figure 3b we use the lengths from Figure 3a to indicate that $\overline{PR} = x_2 - x_1$ and $\overline{QR} = y_2 - y_1$. Since triangle PRQ is a right triangle, we can apply the Pythagorean theorem.

$$d^2 = (x_2 - x_1)^2 + (y_2 - y_1)^2$$

Although the points in Figure 3 are both in quadrant I, the same result will be obtained for any two points. Since distance cannot be negative, we have

The Distance Formula

The distance $\overline{PQ}$ between the points $P(x_1, y_1)$ and $Q(x_2, y_2)$ in the plane is

$$\overline{PQ} = \sqrt{(x_2 - x_1)^2 + (y_2 - y_1)^2}$$

It is also clear from the distance formula that $\overline{PQ} = \overline{QP}$.

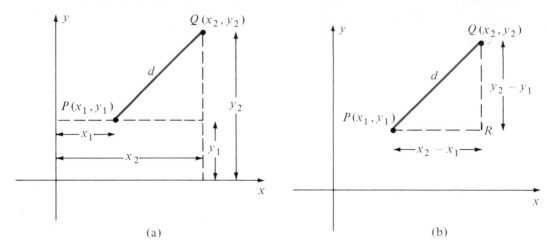

(a)　　　　　　　　　　　　(b)

FIGURE 3　Deriving the Distance Formula

EXAMPLE 1　The Distance Formula

Find the distance between the points $P(-2, -3)$ and $Q(1, 2)$.

SOLUTION

Using the distance formula, we have
$$\overline{PQ} = \sqrt{[1 - (-2)]^2 + [2 - (-3)]^2} = \sqrt{3^2 + 5^2} = \sqrt{34}$$

■

　PROGRESS CHECK

Find the distance between the points $P(-3, 2)$ and $Q(4, -2)$.

ANSWER

$\sqrt{65}$

EXAMPLE 2　Applying the Distance Formula

Show that the triangle with vertices $A(-2, 3)$, $B(3, -2)$, and $C(6, 1)$ is a right triangle.

SOLUTION

It is a good idea to draw a diagram as in Figure 4. We compute the lengths of the three sides.
$$\overline{AB} = \sqrt{(3 + 2)^2 + (-2 - 3)^2} = \sqrt{50}$$
$$\overline{BC} = \sqrt{(6 - 3)^2 + (1 + 2)^2} = \sqrt{18}$$
$$\overline{AC} = \sqrt{(6 + 2)^2 + (1 - 3)^2} = \sqrt{68}$$

If the Pythagorean theorem holds, then triangle ABC is a right triangle. We see that
$$(\overline{AC})^2 = (\overline{AB})^2 + (\overline{BC})^2 \quad \text{since } 68 = 50 + 18$$

and we conclude that triangle ABC is a right triangle whose hypotenuse is AC.

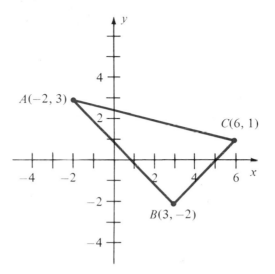

FIGURE 4   Diagram for Example 2

GRAPHS OF EQUATIONS
A graph is an "information multiplier." Graphs have achieved enormous popularity because they display the relationship between variables in a concise yet comprehensive manner. Financial journals display graphs of the Dow-Jones Industrial Average, the price of gold, and the consumer price index over a span of time. Graphics programs are used by businesses to summarize production and sales figures. Every "spreadsheet" program has the ability to graph data that has been entered and results that have been computed.

We often use an equation to express the relationship between two quantities or variables and would like to see a graphic display of this relationship. In mathematics, this has a very specific meaning.

> The **graph of an equation in two variables** consists of the set of all points $P(a, b)$ whose coordinates satisfy the equation. The ordered pair (a, b) is called a **solution** of the equation.

Let's apply these definitions to the task of graphing the equation

$$y = x^2 - 4$$

We want to find ordered pairs (a, b) such that substituting $x = a$ and $y = b$ satisfies the equation. Then the points $P(a, b)$ lie on the graph of the equation and enable us to see the shape of the curve. To obtain the ordered pairs (a, b), we can assign arbitrary values to x and compute corresponding values of y. For example,

$$x = -3 \qquad\qquad x = 1$$
$$y = x^2 - 4 - (-3)^2 \quad 4 - 5 \qquad y = x^2 - 4 = 1^2 - 4 = -3$$

which shows us that $(-3, 5)$ and $(1, -3)$ are solutions of the equation and that the

TABLE 1 $y = x^2 - 4$

x	-3	-2	-1	0	1	2	$5/2$
y	5	0	-3	-4	-3	0	$9/4$

points $P(-3, 5)$ and $P(1, -3)$ lie on the graph of the equation. Table 1 displays a number of additional solutions. We next plot the points corresponding to these ordered pairs. Since the equation has an infinite number of solutions, the plotted points represent only a portion of the graph. We assume that the curve behaves nicely between the plotted points and connect these points by a smooth curve (Figure 5). We must plot enough points to feel reasonably certain of the outline of the curve.

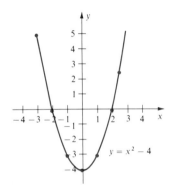

FIGURE 5 Graph of $y = x^2 - 4$

INTERCEPTS

We are going to spend much of our time in this chapter in sketching graphs. Accordingly, it is very helpful to know the **intercepts,** those points where the graph meets the axes. The abscissa (x-coordinate) of a point at which a graph meets the x-axis is called an **x-intercept.** Since the graph in Figure 5 meets the x-axis at the points $(2, 0)$ and $(-2, 0)$, we see that 2 and -2 are the x-intercepts. Similarly, we define the **y-intercept** as the ordinate (y-coordinate) of a point at which the graph meets the y-axis. In Figure 5 the y-intercept is -4.

Since the y-coordinate of every point on the x-axis is 0 and the x-coordinate of every point on the y-axis is 0, we have this simple rule for finding intercepts.

> To find the x-intercepts, set y to 0 and solve the equation for x.
> To find the y-intercepts, set x to 0 and solve the equation for y.

EXAMPLE 3 Finding Intercepts
(a) Sketch the graph of the equation $y = 2x + 1$. From the graph, determine the x- and y-intercepts, if any.
(b) Find the intercepts of the graph of the equation $y = -3x + 5$ algebraically.

SOLUTION

(a) We form a short table of values and sketch the graph in Figure 6. The graph appears to be a straight line that intersects the x-axis at $(-\frac{1}{2}, 0)$ and the y-axis at $(0, 1)$. The x-intercept is $-\frac{1}{2}$ and the y-intercept is 1.

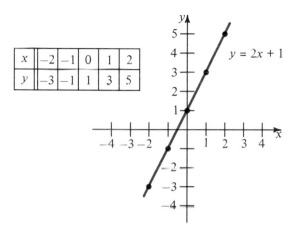

x	-2	-1	0	1	2
y	-3	-1	1	3	5

$y = 2x + 1$

FIGURE 6 Graph of $y = 2x + 1$

(b) We find the y-intercept algebraically by letting $x = 0$ so that

$$y = -3x + 5 = -3(0) + 5 = 5$$

and the x-intercept by letting $y = 0$ so that

$$y = -3x + 5$$
$$0 = -3x + 5$$
$$x = \frac{5}{3}$$

■

SYMMETRY

In sketching a graph, we can halve the effort if we can determine that the graph exhibits one or more types of symmetry. For example, if we folded the graph of Figure 7a along the x-axis, the top and bottom portions would exactly match, which is what we intuitively mean when we speak of symmetry about the x-axis. We would like to develop a means of testing for symmetry that doesn't rely upon examining a graph. We can then use information about symmetry to help in sketching the graph.

In Figure 7a, we see that every point (x_1, y_1) on the portion of the curve above the x-axis is reflected in a point $(x_1, -y_1)$ that lies on the portion of the curve below the x-axis. Similarly, using the graph of Figure 7b, we can argue that symmetry about the y-axis occurs if, for every point (x_1, y_1) on the curve, $(-x_1, y_1)$ also lies on the curve. Finally, using the graph sketched in Figure 7c, we see that symmetry about the origin occurs if, for every point (x_1, y_1) on the curve, $(-x_1, -y_1)$ also lies on the curve. We now summarize these results.

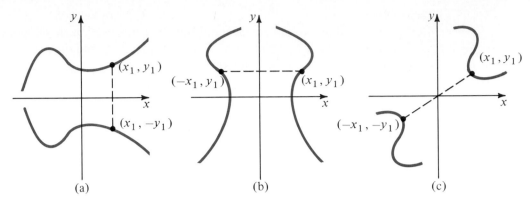

FIGURE 7 Symmetry with Respect to (a) *x*-axis (b) *y*-axis (c) origin

Tests for Symmetry	The graph of an equation is **symmetric with respect to the**

(i) **x-axis** if replacing y with $-y$ results in an equivalent equation;
(ii) **y-axis** if replacing x with $-x$ results in an equivalent equation;
(iii) **origin** if replacing x with $-x$ and y with $-y$ results in an equivalent equation.

EXAMPLE 4 **Applying Intercepts and Symmetry**
Use intercepts and symmetry to assist in graphing the equations.
(a) $y = 1 - x^2$ **(b)** $x = y^2 + 1$

SOLUTION
(a) To determine the intercepts, set $x = 0$ to yield $y = 1$ as the y-intercept. Setting $y = 0$, we have $x^2 = 1$ or $x = \pm 1$ as the x-intercepts.

To test for symmetry, replace x with $-x$ in the equation $y = 1 - x^2$ to obtain

$$y = 1 - (-x)^2 = 1 - x^2$$

Since the equation is unaltered, the curve is symmetric with respect to the y-axis. Now, replacing y with $-y$, we have

$$-y = 1 - x^2$$

which is *not* equivalent to the original equation. The curve is therefore not symmetric with respect to the x-axis. Finally, replacing x with $-x$ and y with $-y$ repeats the last result and shows that the curve is not symmetric with respect to the origin.

We can now form a table of values for $x \geq 0$ and use symmetry with respect to the y-axis to help sketch the graph of the equation (see Figure 8a).

(b) The y-intercepts occur where $x = 0$. Since this leads to the equation $y^2 = -1$, which has no real roots, there are no y-intercepts. Setting $y = 0$, we have $x = 1$ as the x-intercept.

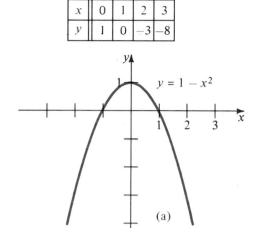

x	0	1	2	3
y	1	0	−3	−8

y	0	1	2	3
x	1	2	5	10

FIGURE 8   Graphs for Example 4

Replacing x with $-x$ in the equation $x = y^2 + 1$ gives us

$$-x = y^2 + 1$$

which is *not* an equivalent equation. The curve is therefore not symmetric with respect to the y-axis. Replacing y with $-y$, we find that

$$x = (-y)^2 + 1 = y^2 + 1$$

which is the same as the original equation. Thus, the curve is symmetric with respect to the x-axis. Replacing x with $-x$ and y with $-y$ also results in the equation

$$-x = y^2 + 1$$

and demonstrates that the curve is not symmetric with respect to the origin. We next form the table of values shown in Figure 8b by assigning nonnegative values to y and calculating the corresponding values of x from the equation; symmetry enables us to sketch the lower half of the graph without plotting points.

Solving the given equation for y yields $y = \pm\sqrt{x - 1}$, which confirms the symmetry about the x-axis. We can think of the upper half of Figure 8b as the graph of the equation $y = \sqrt{x - 1}$ and the lower half as the graph of the equation $y = -\sqrt{x - 1}$.

∎

EXAMPLE 5   Determining Symmetry

Without sketching the graph, determine symmetry with respect to the x-axis, the y-axis, and the origin.

(a) $x^2 + 4y^2 - y = 1$ (b) $xy = 5$ (c) $y^2 = \dfrac{x^2 + 1}{x^2 - 1}$

SOLUTION

(a) Replacing x with $-x$ in the equation, we have

$$(-x)^2 + 4y^2 - y = 1$$
$$x^2 + 4y^2 - y = 1$$

Since the equation is unaltered, the curve is symmetric with respect to the y-axis. Next, replacing y with $-y$, we have

$$x^2 + 4(-y)^2 - (-y) = 1$$
$$x^2 + 4y^2 + y = 1$$

which is *not* an equivalent equation. Replacing x with $-x$ and y with $-y$ repeats the last result. The curve is therefore not symmetric with respect to either the x-axis or the origin.

(b) Replacing x with $-x$, we have $-xy = 5$, which is *not* an equivalent equation. Replacing y with $-y$, we again have $-xy = 5$. Thus the curve is not symmetric with respect to either axis. However, replacing x with $-x$ and y with $-y$ gives us

$$(-x)(-y) = 5$$

which is equivalent to $xy = 5$. We conclude that the curve is symmetric with respect to the origin.

(c) Since x and y both appear to the second power only, all tests will lead to an equivalent equation. The curve is therefore symmetric with respect to both axes and the origin.

■

PROGRESS CHECK

Without graphing, determine symmetry with respect to the coordinate axes and the origin.

(a) $x^2 - y^2 = 1$ (b) $x + y = 10$ (c) $y = x + \dfrac{1}{x}$

ANSWERS

(c) Symmetric with respect to the origin only.

(b) Not symmetric with respect to either axis or the origin.

(a) Symmetric with respect to the x-axis, the y-axis, and the origin.

Note that in Example 5c and in (a) of the last Progress Check, the curves are symmetric with respect to both the x- and y-axes, as well as the origin. In fact, we have the following general rule.

> A curve that is symmetric with respect to both coordinate axes is also symmetric with respect to the origin. However, a curve that is symmetric with respect to the origin need not be symmetric with respect to the coordinate axes.

The curve in Figure 7c illustrates the last point. The curve is symmetric with respect to the origin but not with respect to the coordinate axes.

EXERCISE SET 2.1

In each of Exercises 1 and 2 plot the given points on the same coordinate axes.

1. $(2, 3), (-3, -2), \left(-\frac{1}{2}, \frac{1}{2}\right), \left(0, \frac{1}{4}\right), \left(-\frac{1}{2}, 0\right), (3, -2)$

2. $(-3, 4), (5, -2), (-1, -3), \left(-1, \frac{3}{2}\right), (0, 1.5)$

In Exercises 3–8 find the distance between each pair of points.

3. $(5, 4), (2, 1)$

4. $(-4, 5), (-2, 3)$

5. $(-1, -5), (-5, -1)$

6. $(-3, 0), (2, -4)$

7. $\left(\frac{2}{3}, \frac{3}{2}\right), (-2, -4)$

8. $\left(-\frac{1}{2}, 3\right), \left(-1, -\frac{3}{4}\right)$

In Exercises 9–12 find the length of the shortest side of the triangle determined by the three given points.

9. $A(6, 2), B(-1, 4), C(0, -2)$

10. $P(2, -3), Q(4, 4), R(-1, -1)$

11. $R\left(-1, \frac{1}{2}\right), S\left(-\frac{3}{2}, 1\right), T(2, -1)$

12. $F(-5, -1), G(0, 2), H(1, -2)$

In Exercises 13–16 determine if the given points form a right triangle. (*Hint:* A triangle is a right triangle if and only if the lengths of the sides satisfy the Pythagorean theorem.)

13. $(1, -2), (5, 2), (2, 1)$

14. $(2, -3), (-1, -1), (3, 4)$

15. $(-4, 1), (1, 4), (4, -1)$

16. $(1, -1), (-6, 1), (1, 2)$

In Exercises 17–20 show that the points lie on the same line. (*Hint:* Three points are collinear if and only if the sum of the lengths of two sides equals the length of the third side.)

17. $(-1, 2), (1, 1), (5, -1)$

18. $(-1, -4), (1, 10), (0, 3)$

19. $(-1, 2), (1, 5), \left(-2, \frac{1}{2}\right)$

20. $(-1, -5), (1, 1), (-2, -8)$

21. Find the perimeter of the quadrilateral whose vertices are $(-2, -1), (-4, 5), (3, 5), (4, -2)$.

22. Show that the points $(-2, -1), (2, 2), (5, -2)$ are the vertices of an isosceles triangle.

23. Show that the points $(9, 2), (11, 6), (3, 5)$, and $(1, 1)$ are the vertices of a parallelogram.

24. Show that the point $(-1, 1)$ is the midpoint of the line segment whose endpoints are $(-5, -1)$ and $(3, 3)$.

25. The points $A(2, 7), B(4, 3)$, and $C(x, y)$ determine a right triangle whose hypotenuse is AB. Find x and y. (*Hint:* There is more than one answer.)

26. The points $A(2, 6), B(4, 6), C(4, 8)$, and $D(x, y)$ form a rectangle. Find x and y.

In Exercises 27–32 determine the intercepts and sketch the graph of the given equation.

27. $y = 2x + 4$

28. $y = -2x + 5$

29. $y = \sqrt{x}$

30. $y = \sqrt{x - 1}$

31. $y = |x + 3|$

32. $y = 2 - |x|$

In Exercises 33–38 determine the intercepts and use symmetry to assist in sketching the graph of the given equation.

33. $y = 3 - x^2$

34. $y = 3x - x^2$

35. $y = x^3 + 1$

36. $x = y^3 - 1$

37. $x = y^2 - 1$

38. $y = 3x$

In Exercises 39–54 use the tests for symmetry to determine whether each curve is symmetric with respect to the x-axis, the y-axis, the origin, or none of these.

39. $3x + 2y = 5$

40. $y = 4x^2$

41. $y^2 = x - 4$

42. $x^2 - y = 2$

43. $y^2 = 1 + x^3$

44. $y - (x - 2)^2$

45. $y^2 - (x - 2)^2$

46. $y^2 x + 2x = 4$

47. $y^2x + 2x^2 = 4x^2y$

48. $y^3 = x^2 - 9$

49. $y = \dfrac{x^2 + 4}{x^2 - 4}$

50. $y = \dfrac{1}{x^2 + 1}$

51. $y^2 = \dfrac{x^2 + 1}{x^2 - 1}$

52. $4x^2 + 9y^2 = 36$

53. $xy = 4$

54. $y = \dfrac{1}{x}$

55. Sketch the graph of the equation

$$y = \frac{x^2 - 4}{x - 2}$$

(*Hint:* y is not defined when $x = 2$. However, for $x \neq 2$, the expression can be simplified. Your graph should be that of the simplified expression with a "hole" at $x = 2$.)

2.2
FUNCTIONS AND
FUNCTION NOTATION

The equation

$$y = 2x + 3$$

assigns a value to y for every value of x. If we let X denote the set of values that we can assign to x, and let Y denote the set of values that the equation assigns to y, we can show the correspondence schematically as in Figure 9. The equation can be thought of as a rule defining the correspondence between the sets X and Y.

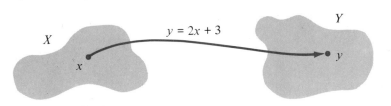

FIGURE 9 Correspondence Defined by $y = 2x + 3$

We are particularly interested in the situation where, for each element x in X, there corresponds one and only one element y in Y; that is, the rule assigns exactly one y for a given x. This type of correspondence plays a fundamental role in mathematics and is given a special name.

Function, Domain, Image, and Range	A **function** is a rule that, for each x in a set X, assigns exactly one y in a set Y. The element y is called the **image** of x. The set X is called the **domain** of the function and the set of all images is called the **range** of the function.

We can think of the rule defined by the equation $y = 2x + 3$ as a function machine (see Figure 10). Each time we drop a value of x from the domain into the input hopper, exactly one value of y falls out of the output chute. If we drop in $x = 5$, the function

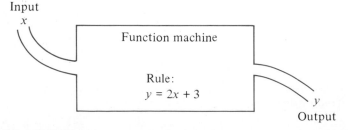

FIGURE 10 A Function Machine

machine follows the rule and produces $y = 13$. Since we are free to choose the values of x that we drop into the machine, we call x the **independent variable;** the value of y that drops out depends upon the choice of x, so y is called the **dependent variable.** We say that the dependent variable is a function of the independent variable; that is, *the output is a function of the input.*

Let's look at a few schematic presentations. The correspondence in Figure 11a is a function; for each x in X there is exactly one corresponding value of y in Y. The fact that y_1 is the image of both x_1 and x_2 does not violate the definition of a function. However, the correspondence in Figure 11b is not a function, since x_1 has two images, y_1 and y_2, assigned to it, thus violating the definition of a function.

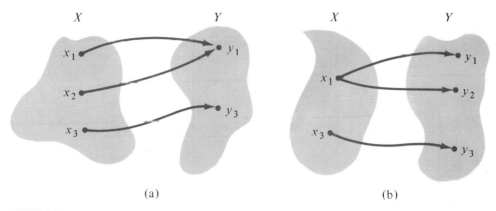

(a)   (b)

FIGURE 11 **Which Correspondence Is a Function?**

VERTICAL LINE TEST

There is a simple graphic way to test whether a correspondence determines a function. When we draw vertical lines on the graph of Figure 12a, we see that no vertical line intersects the graph at more than one point. This means that the correspondence used in sketching the graph assigns exactly one y-value for each x-value and therefore

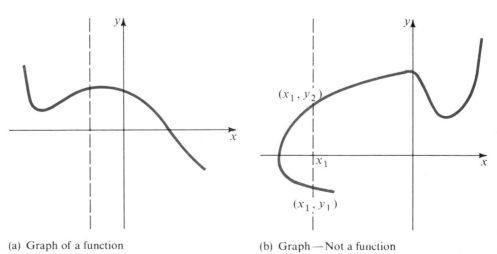

(a) Graph of a function   (b) Graph—Not a function

FIGURE 12 **Vertical Line Test**

determines y as a function of x. When we draw vertical lines on the graph of Figure 12b, however, some vertical lines intersect the graph at two points. Since the correspondence graphed in Figure 12b assigns the values y_1 and y_2 to x_1, it does not determine y as a function of x. Thus, *not every equation or correspondence in the variables x and y determines y as a function of x.*

Vertical Line Test	A graph determines y as a function of x if and only if no vertical line meets the graph at more than one point.

EXAMPLE 1   Applying the Vertical Line Test

Which of the following graphs determine y as a function of x?

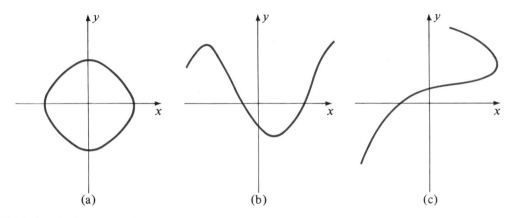

(a)   (b)   (c)

FIGURE 13   **Which Graphs Determine a Function?**

SOLUTION

(**a**) Not a function. Some vertical lines meet the graph in more than one point.

(**b**) A function. Passes the vertical line test.

(**c**) Not a function. Fails the vertical the test.

■

DOMAIN AND RANGE

We have defined the domain of a function as the set of values assumed by the independent variable. In more advanced courses in mathematics, the domain may include complex numbers. In this book we will restrict the domain of a function to those real numbers for which the image is also a real number, and we say that the function **is defined at** such values. When a function is defined by an equation, we must always be alert to two potential problems.

(**a**) *Division by zero.* For example, the domain of the function

$$y = \frac{2}{x - 1}$$

is the set of all real numbers for which the denominator is not zero. If we set the denominator equal to zero and solve, we have

$$x - 1 = 0$$
$$x = 1$$

and we can conclude that the domain of the function is the set of all real numbers other than $x = 1$.

(b) *Even roots of negative numbers.* For example, the function

$$y = \sqrt{x} - 1$$

is defined only for $x - 1 \geq 0$ since we exclude the square root of negative numbers. Solving the inequality,

$$x - 1 \geq 0$$
$$x \geq 1$$

we see that the domain of the function consists of all real numbers $x \geq 1$.

The range of a function is, in general, not as easily determined as is the domain. The range is the set of all y-values that occur in the correspondence; that is, it is the set of all outputs of the function. For our purposes, it will suffice to determine the range by examining the graph.

x	0	1	4	9
y	0	1	2	3

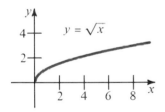

FIGURE 14 **Graph of** $y = \sqrt{x}$

EXAMPLE 2 Applying the Vertical Line Test
Graph the equation $y = \sqrt{x}$. If the correspondence determines a function, find the domain and range.

SOLUTION
We obtain the graph of the equation by plotting points and connecting them to form a smooth curve. Applying the vertical line test to the graph as shown in Figure 14, we see that the equation determines a function.

Since the square root function is not defined for negative arguments, the domain of the function is the set $\{x \mid x \geq 0\}$. By examining the graph, we see that the range is the set $\{y \mid y \geq 0\}$. ∎

☑ **PROGRESS CHECK**
Graph the equation $y = x^2 - 4, -3 \leq x \leq 3$. If the correspondence determines a function, find the domain and range.

ANSWER

The graph is that portion of the curve shown in Figure 5 that lies between $x = -3$ and $x = 3$.
The domain is $\{x \mid -3 \leq x \leq 3\}$; the range is $\{y \mid -4 \leq y \leq 5\}$.

FUNCTION NOTATION

It is customary to designate a function by a letter of the alphabet, such as f, g, F, or C. We then denote the output corresponding to x by $f(x)$, which is read "f of x." Thus,

$$f(x) = 2x + 3$$

specifies a rule f for determining an output $f(x)$ for a given value of x. To find $f(x)$ when $x = 5$, we simply substitute 5 for x and obtain

$$f(5) = 2(5) + 3 = 13$$

The notation $f(5)$ is a convenient way of specifying "the value of the function f that corresponds to $x = 5$." The symbol f represents the function or rule; the notation $f(x)$ represents the output produced by the rule. For convenience, however, we will at times join in the common practice of designating the function f by $f(x)$.

EXAMPLE 3   Evaluating a Function

The function f is given as $f(x) = 2x^2 - x - 1$. Find
(a) $f(-3)$ **(b)** $f(3t)$ **(c)** $f(t - 1)$ **(d)** $f(t - 1) - f(t)$

SOLUTION
(a) Substituting -3 for *each occurrence of* x,

$$f(-3) = 2(-3)^2 - (-3) - 1 = 20$$

(b) Substituting $3t$ for x,

$$f(3t) = 2(3t)^2 - 3t - 1 = 18t^2 - 3t - 1$$

(c) Substituting $t - 1$ for x,

$$f(t - 1) = 2(t - 1)^2 - (t - 1) - 1 = 2t^2 - 5t + 2$$

(d) Using the result from (c) above,

$$f(t - 1) - f(t) = 2t^2 - 5t + 2 - (2t^2 - t - 1)$$
$$= -4t + 3$$

■

PROGRESS CHECK
(a) If $f(u) = u^3 + 3u - 4$, find $f(-2)$.
(b) If $f(t) = t^2 + 1$, find $f(t - 1)$.

ANSWERS
(a) -18 (b) $t^2 - 2t + 2$

EXAMPLE 4   Functions and Word Problems

A newspaper makes this offer to its advertisers: The first column inch will cost $40, and each subsequent column inch will cost $30. If T is the total cost of running an ad whose length is n column inches, and the minimum space is 1 column inch,
(a) express T as a function of n;
(b) find T when $n = 4$.

SOLUTION

(a) The equation

$$T = 40 + 30(n - 1)$$
$$= 10 + 30n$$

gives the correspondence between n and T. In function notation,

$$T(n) = 10 + 30n, \qquad n \geq 1$$

(b) When $n = 4$,

$$T(4) = 10 + 30(4) = 130$$

∎

In calculus, you will be working with expressions of the form

$$\frac{f(x + h) - f(x)}{h}$$

and

$$\frac{f(x) - f(a)}{x - a}$$

The following examples will show you how to deal with these **difference quotients.**

EXAMPLE 5   Working with Function Notation

Given $f(x) = x^2 - x + 2$, compute $\dfrac{f(x + h) - f(x)}{h}$.

SOLUTION

$$f(x + h) = (x + h)^2 - (x + h) + 2$$
$$= x^2 + 2hx + h^2 - x - h + 2$$
$$f(x + h) - f(x) = x^2 + 2hx + h^2 - x - h + 2 - (x^2 - x + 2)$$
$$= 2hx + h^2 - h$$
$$\frac{f(x + h) - f(x)}{h} = \frac{2hx + h^2 - h}{h} = 2x + h - 1, h \neq 0$$

∎

PROGRESS CHECK

Given $f(x) = 2x^2 + 3x - 1$, compute $\dfrac{f(x + h) - f(x)}{h}$.

ANSWER

$4x + 2h + 3, h \neq 0$

EXAMPLE 6 **Working with Function Notation**

Given $f(x) = \dfrac{1}{x}$, find $\dfrac{f(x) - f(3)}{x - 3}$.

SOLUTION

$$\frac{f(x) - f(3)}{x - 3} = \frac{\dfrac{1}{x} - \dfrac{1}{3}}{x - 3} = \frac{3 - x}{3x(x - 3)} \qquad \text{[multiplying both numerator and denominator by } 3x]$$

$$= -\frac{1}{3x}, \ x \neq 3 \qquad \text{[since } 3 - x = -(x - 3)]$$

■

EXERCISE SET 2.2

In Exercises 1–6 graph the equation. If the graph determines y as a function of x, find the domain and use the graph to determine the range of the function.

1. $y = 2x - 3$

2. $y = x^2 + x, \quad -2 \leq x \leq 1$

3. $x = y + 1$

4. $x = y^2 - 1$

5. $y = \sqrt{x - 1}$

6. $y = |x|$

In Exercises 7–14 determine the domain of the function defined by the given rule.

7. $f(x) = \sqrt{2x - 3}$

8. $f(x) = \sqrt{5 - x}$

9. $f(x) = \dfrac{1}{\sqrt{x - 2}}$

10. $f(x) = \dfrac{-2}{x^2 + 2x - 3}$

11. $f(x) = \dfrac{\sqrt{x - 1}}{x - 2}$

12. $f(x) = \dfrac{x}{x^2 - 4}$

13. $f(x) = \dfrac{x - 1}{x + 1}$

14. $f(x) = \dfrac{x - 1}{|x|}$

In Exercises 15–18 find the number (or numbers) whose image is 2.

15. $f(x) = 2x - 5$

16. $f(x) = x^2$

17. $f(x) = \dfrac{1}{x - 1}$

18. $f(x) = \sqrt{x - 1}$

In Exercises 19–26, given the function f defined by $f(x) = 2x^2 + 5$, find:

19. $f(0)$

20. $f(-2)$

21. $f(a)$

22. $f(3x)$

23. $3f(x)$

24. $-f(x)$

25. $-f(-1)$

26. $f(\sqrt{2})$

In Exercises 27–32, given the function g defined by $g(x) = x^2 + 2x$, find:

27. $g(-3)$

28. $g\left(\dfrac{1}{x}\right)$

29. $\dfrac{1}{g(x)}$

30. $g(-x)$

31. $g(a + h)$

32. $g(x + a) - g(a)$

In Exercises 33–38, given the function F defined by $F(x) = \dfrac{x^2 + 1}{3x - 1}$, find:

 33. $F(-2.73)$

 34. $F(16.11)$

35. $\dfrac{1}{F(x)}$

36. $F(-x)$

37. $2F(2x)$

38. $F(x^2)$

In Exercises 39–44, given the function r defined by $r(t) = \dfrac{t - 2}{t^2 + 2t - 3}$, find:

 39. $r(-8.27)$

40. $r(2.04)$

41. $r(2a)$

42. $2r(a)$

43. $r(a + 1)$

44. $r(1 + h)$

In Exercises 45–50 compute $\dfrac{f(x+h)-f(x)}{h}$.

45. $f(x) = -2x + 3$

46. $f(x) = x^2 - 2$

47. $f(x) = 2x^2 + x - 3$

48. $f(x) = 5$

49. $f(x) = \dfrac{1}{2x}$

50. $f(x) = x^3$

In Exercises 51–54 compute $\dfrac{f(x) - f(2)}{x - 2}$.

51. $f(x) - 3x - 2$

52. $f(x) = x^2 - 2x + 2$

53. $f(x) = \dfrac{1}{x-1}$

54. $f(x) = x^3$

55. If x dollars are borrowed at 7% simple annual interest, express the interest I at the end of 4 years as a function of x.

56. Express the area A of an equilateral triangle as a function of the length s of its side.

57. Express the diameter d of a circle as a function of its circumference C.

58. Express the perimeter P of a square as a function of its area A.

2.3 GRAPHS OF FUNCTIONS

We have used the graph of an equation to help us find out whether or not the equation determines a function. It is not surprising, therefore, that the **graph of a function** f is defined as the graph of the equation $y = f(x)$. For example, the graph of the function f defined by the rule $f(x) = \sqrt{x}$ is the graph of the equation $y = \sqrt{x}$, which was sketched in Figure 14.

There is an important element of notation that we need to clarify before proceeding to graphs of functions. Let's focus on the point $P(x_1, y_1)$ on the graph of the function f in Figure 15a. Since the graph of the function f is that of $y = f(x)$, we must have

$$y_1 = f(x_1)$$

That is, y_1 is the value of f at $x = x_1$. This says that the ordinate y_1 can also be designated as $f(x_1)$ as in Figure 15b. In summary,

(x_1, y_1) and $(x_1, f(x_1))$ are alternative ways of designating the same point P on the graph of the function $f(x)$. The directed distance from P to the x-axis can be designated as y_1 or as $f(x_1)$.

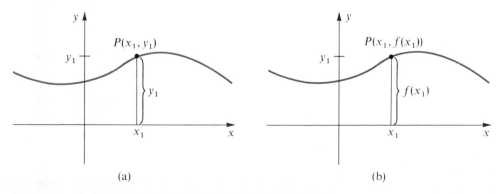

(a) (b)

FIGURE 15 Alternative Notations for the Coordinates of a Point

"SPECIAL" FUNCTIONS AND THEIR GRAPHS

x	-2	-1	0	1	2
y	-2	-1	0	1	2

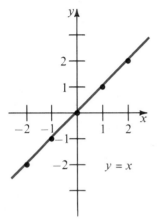

FIGURE 16   Graph of $y = x$

There are a number of "special" functions that a calculus instructor is likely to use to demonstrate a point. The instructor will sketch the graph of the function, since the graph shows at a glance many characteristics of the function. For example, information about symmetry, domain, and range is available from a graph. In fact, we have already used the graphs of some of these functions to illustrate these characteristics.

You should become thoroughly acquainted with the following functions and their graphs. For each function we will form a table of values, sketch the graph of the function, and discuss symmetry, domain, and range.

$f(x) = x$ Identity function

The domain of f is the set of all real numbers. We form a table of values and use it to sketch the graph of $y = x$ in Figure 16. The graph is symmetric with respect to the origin (note that $-y = -x$ is equivalent to $y = x$). The range of f is seen to be the set of all real numbers.

$f(x) = -x$ Negation function

The domain of f is the set of all real numbers. A table of values is used to sketch the graph of $y = -x$ in Figure 17. The graph is symmetric with respect to the origin (note that $-y = x$ is equivalent to $y = -x$). The range of f is seen to be the set of all real numbers.

$f(x) = |x|$ Absolute value function

The domain of f is the set of all real numbers. A table of values allows us to sketch the graph in Figure 18. The graph is symmetric with respect to the y-axis. Since the graph always lies above the x-axis, the range of f is the set of all nonnegative real numbers.

x	-2	-1	0	1	2
y	2	1	0	-1	-2

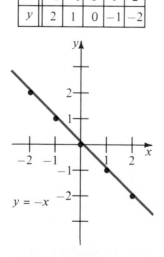

FIGURE 17   Graph of $y = -x$

x	-2	-1	0	1	2
y	2	1	0	1	2

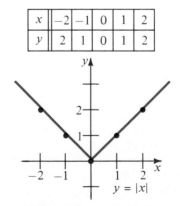

FIGURE 18   Graph of $y = |x|$

x	-2	-1	0	1	2
y	c	c	c	c	c

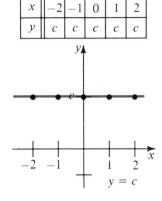

FIGURE 19   Graph of $y = c$

$f(x) = c$ Constant function

The domain of f is the set of all real numbers. In fact, the value of f is the same for all values of x (see Figure 19). The range of f is the set $\{c\}$. The graph is symmetric with respect to the y-axis (note that $y = c$ is unaltered when x is replaced by $-x$).

$f(x) = x^2$ Squaring function

The domain of f is the set of all real numbers. The graph in Figure 20 is called a **parabola** and illustrates the general shape of all second-degree polynomials. The graph of f is symmetric with respect to the y-axis (note that $y = (-x)^2 = x^2$). Since $y \geq 0$ for all values of x, the range is the set of all nonnegative real numbers.

$f(x) = \sqrt{x}$ Square root function

Since $\sqrt{x}$ is not defined for $x < 0$, the domain is the set of nonnegative real numbers. The graph in Figure 21 always lies above the x-axis, so the range of f is $\{y \mid y \geq 0\}$, that is, the set of all nonnegative real numbers. The graph is not symmetric with respect to either axis or the origin.

x	-2	-1	0	1	2
y	4	1	0	1	4

x	-2	-1	0	1	2
y	-8	-1	0	1	8

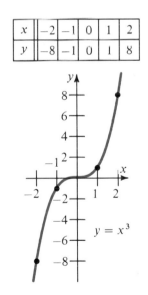

FIGURE 22   Graph of $y = x^3$

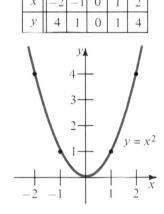

FIGURE 20   Graph of $y = x^2$

x	0	1	4	9
y	0	1	2	3

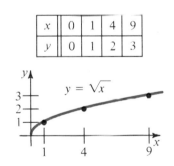

FIGURE 21   Graph of $y = \sqrt{x}$

$f(x) = x^3$ Cubing function

The domain is the set of all real numbers. Since the graph in Figure 22 extends indefinitely both upward and downward with no gaps, the range is also the set of all real numbers. The graph is symmetric with respect to the origin (note that $-y = (-x)^3 = -x^3$ is equivalent to $y = x^3$).

PIECEWISE-DEFINED FUNCTIONS

Thus far we have defined each function by means of an equation. A function can also be defined by a table, by a graph, or by several equations. When a function is defined in different ways over different parts of its domain, it is said to be a **piecewise-defined function.** We illustrate this idea by several examples.

EXAMPLE 1 Graphing a Piecewise-Defined Function

Sketch the graph of the function f defined by

$$f(x) = \begin{cases} -1 & \text{when} & x < -2 \\ x^2 & \text{when} & -2 \le x \le 2 \\ 2x + 1 & \text{when} & x > 2 \end{cases}$$

SOLUTION

Since this is our first exposure to a piecewise-defined function, we will form a table of points to be plotted. We must exercise extreme care in forming Table 2 to use the first part of the definition when $x < -2$, the second part when $-2 \le x \le 2$, and the third part when $x > 2$. (It's a good idea to separate the intervals in the table, as shown.)

TABLE 2 Table for Example 1

x	-4	-3	-2	-1	0	1	2	3	4	5
y	-1	-1	4	1	0	1	4	7	9	11

The graph in Figure 23 is a horizontal line when $x < -2$ and a parabola when $-2 \le x \le 2$, which is what we would expect since $f(x) = -1$ is a constant function and $f(x) = x^2$ is a squaring function. Note that the graph has gaps, a frequent occurrence when we deal with piecewise-defined functions. Also note that the points $(-2, -1)$ and

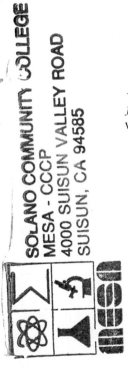

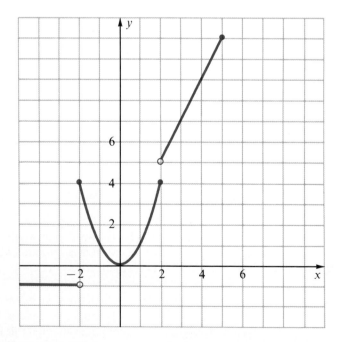

FIGURE 23 Graph for Example 1

(2, 5) have been marked with open circles to indicate that they are not on the graph of the function.

■

EXAMPLE 2   Graphing a Function Involving Absolute Value

Sketch the graph of the function $f(x) = |x + 1|$.

SOLUTION

We apply the definition of absolute value to obtain

$$y = |x + 1| = \begin{cases} x + 1 & \text{when} & x + 1 \geq 0 \\ -(x + 1) & \text{when} & x + 1 < 0 \end{cases}$$

or

$$y = \begin{cases} x + 1 & \text{when} & x \geq -1 \\ -x - 1 & \text{when} & x < -1 \end{cases}$$

From this example it is easy to see that a function involving absolute value will usually be a piecewise-defined function. As usual, we form a table of values (see Table 3), being careful to use $y = x + 1$ when $x \geq -1$ and $y = -x - 1$ when $x < -1$. It is a good idea to include the value $x = -1$ in the table.

TABLE 3   Table for Example 2

x	-3	-2	-1	0	1	2	3
y	2	1	0	1	2	3	4

The points are joined by a smooth curve (Figure 24), which consists of two rays or half-lines intersecting at $(-1, 0)$. In Section 2.4 we will show you how to obtain this graph directly from your knowledge of the graph of the absolute value function $f(x) = |x|$.

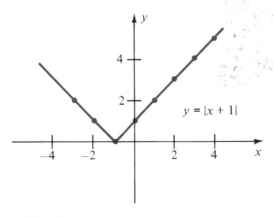

FIGURE 24   Graph of $y = |x + 1|$

■

EXAMPLE 3   Piecewise-Defined Functions in Word Problems

The commission earned by a door-to-door cosmetics salesperson is determined as shown in Table 4.

TABLE 4   Table for Example 3

Weekly sales s	Commission
less than $300	20% of sales
$300 or more but less than $400	$60 + 35% of sales over $300
$400 or more	$95 + 60% of sales over $400

(a) Express the commission C as a function of sales s.
(b) Find the commission if the weekly sales are $425.
(c) Sketch the graph of the function.

SOLUTION

(a) The function C can be described by three equations.

$$C(s) = \begin{cases} 0.20s & \text{when} & s < 300 \\ 60 + 0.35(s - 300) & \text{when} & 300 \le s < 400 \\ 95 + 0.60(s - 400) & \text{when} & s \ge 400 \end{cases}$$

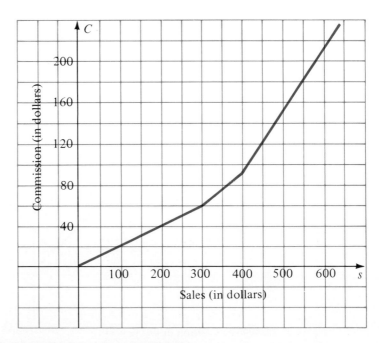

FIGURE 25   Commission as a Function of Sales

(b) When $s = 425$, we must use the third equation and substitute to determine $C(425)$.

$$C(425) = 95 + 0.60(425 - 400)$$
$$= 95 + 0.60(25)$$
$$= 110$$

The commission on sales of $425 is $110.

(c) The graph of the function C consists of three line segments (Figure 25). ∎

INCREASING AND DECREASING FUNCTIONS   When we apply the terms *increasing* and *decreasing* to the graph of a function, we assume that we are viewing the graph from left to right. The straight line of Figure 26a is increasing, since the values of y increase as we move from left to right; similarly, the graph in Figure 26b is decreasing, since the values of y decrease as we move from left to right.

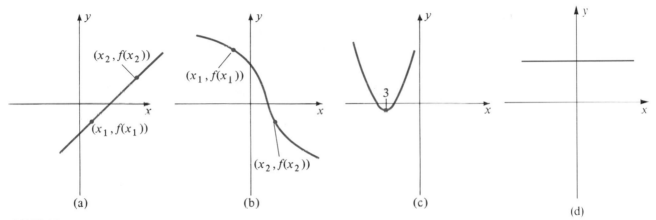

FIGURE 26   Increasing and Decreasing Functions

One portion of the graph pictured in Figure 26c is decreasing and another is increasing. Since this is the most common situation, we define increasing and decreasing on an interval.

> Assume a function f is defined on an interval I and that x_1 and x_2 are any two numbers in I. Then
>
> f is **increasing** on I if $x_1 < x_2$ implies $f(x_1) < f(x_2)$ [Figure 26a]
> f is **decreasing** on I if $x_1 < x_2$ implies $f(x_1) > f(x_2)$ [Figure 26b]
> f is **constant** on I if $f(x_1) = f(x_2)$ [Figure 26d]

Returning to Figure 26c, note that the function is decreasing when $x \leq -3$ and increasing when $x \geq -3$; that is, the function is decreasing on the interval $(-\infty, -3]$ and increasing on the interval $[-3, \infty)$. The graph shows that the function has a minimum value at the point $x = -3$. Finding such points is very useful in sketching graphs and is an important technique taught in calculus courses.

It is important to become accustomed to the notation used in Figure 26, where the y-coordinate at the point $x = x_1$ is denoted by $f(x_1)$.

EXAMPLE 4   Increasing and Decreasing on an Interval
Use the graph of the function $f(x) = x^3 - 3x + 2$, shown in Figure 27, to determine where the function is increasing and where it is decreasing.

SOLUTION
From the graph we see that there are turning points at $(-1, 4)$ and at $(1, 0)$. We conclude that

$$f \text{ is increasing on the intervals } (-\infty, -1] \text{ and } [1, \infty)$$
$$f \text{ is decreasing on the interval } [-1, 1]$$

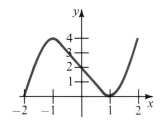

FIGURE 27
$f(x) = x^3 - 3x + 2$

EXAMPLE 5   Increasing, Decreasing, and Constant on an Interval
The function f is defined by

$$f(x) = \begin{cases} |x| & \text{when} & x \le 2 \\ -3 & \text{when} & x > 2 \end{cases}$$

Use a graph to find the values of x for which the function is increasing, decreasing, and constant.

SOLUTION
Note that the piecewise-defined function f is composed of the absolute value function when $x \le 2$ and a constant function when $x > 2$. We can therefore sketch the graph of

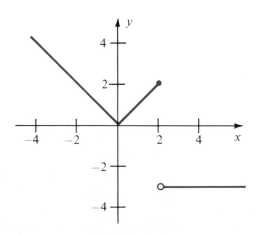

FIGURE 28   Piecewise-Defined Function of Example 5

f immediately as shown in Figure 28. From the graph in Figure 28 we determine that

f is decreasing on the interval $(-\infty, 0]$

f is increasing on the interval $[0, 2]$

f is constant and has value -3 on the interval $(2, \infty)$

■

 PROGRESS CHECK

The function f is defined by

$$f(x) = \begin{cases} 2x + 1 & \text{when} & x < -1 \\ 0 & \text{when} & -1 \le x \le 3 \\ -2x + 1 & \text{when} & x > 3 \end{cases}$$

Use a graph to find the values of x for which the function is increasing, decreasing, and constant.

ANSWER

Increasing on the interval $(-\infty, -1)$. Constant on $[-1, 3]$. Decreasing on $(3, \infty)$.

EXERCISE SET 2.3

In Exercises 1–20 sketch the graph of the function and state where it is increasing, decreasing, and constant.

1. $f(x) = 3x + 1$
2. $f(x) = 3 - 2x$
3. $f(x) = x^2 + 1$
4. $f(x) = 9 - x^2$

5. $f(x) = 4x - x^2$
6. $f(x) = |x| - 3$
7. $f(x) = |2x + 1|$
8. $f(x) = |1 - x|$

9. $f(x) = \sqrt{2x}$
10. $f(x) = 2x^3$
11. $f(x) = -3$
12. $f(x) = x^3 + 1$

13. $f(x) = \begin{cases} 2x, & x > -1 \\ -x - 1, & x \le -1 \end{cases}$

14. $f(x) = \begin{cases} x + 1, & x > 2 \\ 1, & -1 \le x \le 2 \\ -x + 1, & x < -1 \end{cases}$

15. $f(x) = \begin{cases} x, & x < 2 \\ 2, & x \ge 2 \end{cases}$

16. $f(x) = \begin{cases} -x, & x \le -2 \\ x^2, & -2 < x \le 2 \\ -x, & 3 \le x \le 4 \end{cases}$

17. $f(x) = \begin{cases} -x^2, & -3 < x < 1 \\ 0, & 1 \le x \le 2 \\ -3x, & x > 2 \end{cases}$

18. $f(x) = \begin{cases} 2 & \text{when } x \text{ is an integer} \\ -1 & \text{when } x \text{ is not an integer} \end{cases}$

19. $f(x) = \begin{cases} -2, & x < -2 \\ -1, & -2 \le x \le -1 \\ 1, & x > -1 \end{cases}$

20. $f(x) = \begin{cases} \dfrac{x^2 - 1}{x - 1}, & x \ne 1 \\ 3, & x = 1 \end{cases}$

21. The telephone company charges a fee of $6.50 per month for the first 100 message units and an additional fee of 6 cents for each of the next 100 message units. A reduced rate of 5 cents is charged for each message unit after the first 200 units. Express the monthly charge C as a function of the number of message units u.

22. The annual dues of a union are as shown in the table.

Employee's annual salary	Annual dues
less than $8000	$60
$8000 or more but less than $15,000	$60 + 1% of the salary in excess of $8000
$15,000 or more	$130 + 2% of the salary in excess of $15,000

Express the annual dues d as a function of the salary S.

23. A tour operator who runs charter flights to Rome has established the following pricing schedule. For a group of no more than 100 people, the round trip fare per person is $300, with a minimum rental of $30,000 for the plane. For a group that has more than 100 but fewer than 150 people, the fare per person for all passengers will be reduced by $1 for each passenger in excess of 100. Write the tour operator's total revenue R as a function of the number of people x in the group.

24. A firm packages and ships 1-pound jars of instant coffee. The cost C of shipping is 40 cents for the first pound and 25 cents for each additional pound.
   (a) Write C as a function of the weight w (in pounds) for $0 < w \le 30$.
   (b) What is the cost of shipping a package containing 24 jars of instant coffee?

25. The daily rates of a car rental firm are $14 plus 8 cents per mile.
   (a) Express the cost C of renting a car as a function of the number of miles m traveled.
   (b) What is the domain of the function?
   (c) How much would it cost to rent a car for a 100-mile trip?

26. In a wildlife preserve, the population P of eagles depends on the population x of its basic food supply, rodents. Suppose that P is given by

$$P(x) = 0.002x + 0.004x^2$$

Find the eagle population when the rodent population is
(a) 500   (b) 2000

2.4 ASSISTS IN GRAPHING

You now know the shape of the graph of the function $f(x) = |x|$. Can this knowledge be used to sketch the graph of the function $f(x) = |x| + 1$? Or, knowing the graph of $f(x) = x^2$, can you quickly sketch the graph of $f(x) = (x - 1)^2$? Of $f(x) = 3x^2$?

This section discusses techniques that will enable you to answer yes to each of the above questions. You will see that we can expand upon our knowledge of the graphs of the basic functions, the "special" functions listed in Section 2.3, to quickly sketch the graphs of a wide variety of functions. Since these functions are frequently encountered in calculus courses, the ability to sketch their graphs without using a table of values will give you a big boost toward succeeding in the study of calculus.

REFLECTIONS

Let's compare the graph of the functions $f(x)$ and $-f(x)$. For any value of x, say $x = x_0$, the point $(x_0, f(x_0))$ lies on the graph of $f(x)$ and the point $(x_0, -f(x_0))$ lies on the graph of $-f(x)$. We see that, for every value of x, the y-coordinate on the graph of $-f(x)$ is the negative of the y-coordinate on the graph of $f(x)$. The graph of $-f(x)$ is called a **reflection about the x-axis** of the graph of $f(x)$.

Graphing $-f(x)$	The graph of $-f(x)$ is the graph of $f(x)$ reflected about the x-axis.

EXAMPLE 1   Reflections of Known Graphs
Sketch the graphs of the functions.
   (a) $f(x) = -|x|$ (b) $f(x) = -x^2$ (c) $f(x) = -\sqrt{x}$ (d) $f(x) = -x^3$

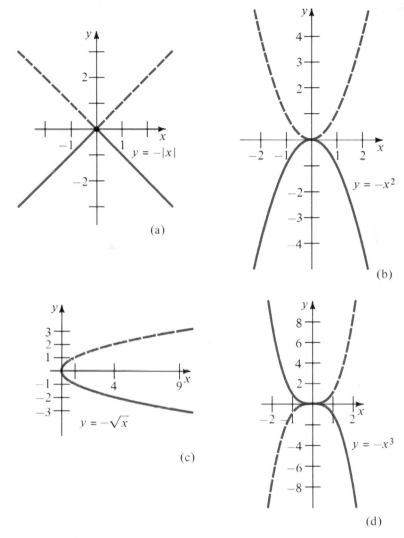

FIGURE 29   Reflections of Known Graphs (Example 1)

SOLUTION
See Figure 29.

■

VERTICAL SHIFTS The value of the function $f(x) + c$, where c is a constant, is obtained by adding c to the value of $f(x)$ for all x. This leads to the following result.

Graphing $f(x) + c$	The graph of $f(x) + c$ is the graph of $f(x)$ shifted vertically c units. If $c > 0$, the shift is upward; if $c < 0$, the shift is downward.

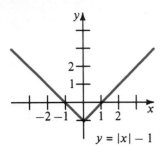

$$y = |x| - 1$$

FIGURE 30 **Vertical Shift**

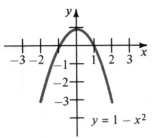

$$y = 1 - x^2$$

FIGURE 31 **Vertical Shift**

EXAMPLE 2 **Vertical Shift of a Known Graph**

Sketch the graph of the function $f(x) = |x| - 1$.

SOLUTION

The graph is that of $f(x) = |x|$ shifted downward 1 unit as in Figure 30.

■

EXAMPLE 3 **Vertical Shift of a Known Graph**

Sketch the graph of the function $f(x) = 1 - x^2$.

SOLUTION

If we rewrite the function in the form

$$f(x) = -x^2 + 1$$

we see that the graph is that of $f(x) = -x^2$ shifted upward 1 unit. However, the graph of $f(x) = -x^2$ is a reflection of the graph of $f(x) = x^2$. We therefore combine the methods of reflecting and shifting to obtain the graph of Figure 31.

■

HORIZONTAL SHIFTS

Graphing $f(x - c)$	The graph of $f(x - c)$, where c is a constant, is the graph of $f(x)$ shifted horizontally c units. If $c > 0$, the shift is to the right; if $c < 0$, the shift is to the left.

Assuming that we can quickly sketch the graph of

$$y = f(x) \tag{1}$$

we would like to use this knowledge to sketch the graph of

$$y = f(x - c) \tag{2}$$

where c is a constant. Focusing on the value $x = x_0$, we see that the point $(x_0, f(x_0))$ lies on the graph of Equation (1). Similarly, the point $(x_0 + c, f(x_0))$ lies on the graph of Equation (2). This shows that Equation (2) attains the y-coordinate $f(x_0)$ at $x_0 + c$, whereas Equation (1) attains this y-coordinate at x_0. Thus, the graph of Equation (2) is that of Equation (1) shifted horizontally c units. In summary,

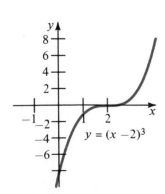

$$y = (x - 2)^3$$

FIGURE 32 **Horizontal Shift Right**

EXAMPLE 4 **Horizontal Shift of a Known Graph**

Sketch the graph of the function $f(x) = (x - 2)^3$.

SOLUTION

The graph is that of $y = x^3$ shifted right 2 units as in Figure 32.

■

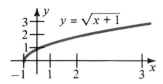

FIGURE 33 Horizontal Shift Left

EXAMPLE 5 Horizontal Shift of a Known Graph

Sketch the graph of the function $f(x) = \sqrt{x + 1}$.

SOLUTION

If we write the function in the form

$$f(x) = \sqrt{x + 1} = \sqrt{x - (-1)}$$

we see that the graph of $f(x)$ is that of the square root function shifted left 1 unit as in Figure 33.

∎

VERTICAL STRETCHING AND SHRINKING

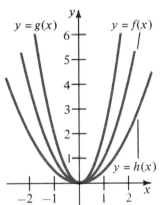

FIGURE 34 Stretching and Shrinking

We now want to tackle the problem of graphing $c \cdot f(x)$, where c is a constant, assuming we already know the appearance of the graph of $f(x)$. If $c < 0$, then we may treat this as the graph of $-f(x)$ multiplied by $|c|$. For this reason we will restrict our results to the case where $c > 0$.

EXAMPLE 6 Stretching and Shrinking

Sketch the graphs of the functions $f(x) = x^2$, $g(x) = 2x^2$, and $h(x) = \frac{1}{2}x^2$ on the same coordinate axes.

SOLUTION

See Figure 34. The graph of $g(x) = 2x^2$ rises more rapidly than that of $f(x) = x^2$ while the graph of $h(x) = \frac{1}{2}x^2$ rises less rapidly.

∎

For every value of x, the y-coordinate on the graph of $c \cdot f(x)$ is precisely c times the y-coordinate on the graph of $f(x)$. This permits us to summarize in this way.

Graphing $c \cdot f(x)$	When $c > 1$, the graph of $c \cdot f(x)$ is "stretched" vertically away from the x-axis, whereas for $0 < c < 1$ the graph is flattened, or "shrunk" vertically, toward the x-axis.

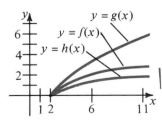

FIGURE 35 Stretching and Shrinking

EXAMPLE 7 Stretching and Shrinking

Sketch the graphs of $f(x) = \sqrt{x - 2}$, $g(x) = 2\sqrt{x - 2}$, and $h(x) = \frac{1}{2}\sqrt{x - 2}$ on the same coordinate axes.

SOLUTION

The graph of $f(x) = \sqrt{x - 2}$ is the same as that of $y = \sqrt{x}$ shifted right 2 units. The graph of $g(x)$ rises more rapidly and the graph of $h(x)$ rises less rapidly. See Figure 35.

∎

EXERCISE SET 2.4

In Exercises 1–12 sketch the graph of the function by shifting or reflecting a known graph. Where needed, perform both operations.

1. $f(x) = x + 2$ 2. $f(x) = \sqrt{x} - 2$ 3. $f(x) = 1 - x^3$ 4. $f(x) = 2 - |x|$

5. $f(x) = (x + 2)^2$ 6. $f(x) = |x - 1|$ 7. $f(x) = \sqrt{x - 2}$ 8. $f(x) = (x + 1)^3$

9. $f(x) = (x - 1)^2 + 2$ 10. $f(x) = \sqrt{x + 2} - 1$ 11. $f(x) = (x + 2)^3 - 1$ 12. $f(x) = \sqrt{x - 2} + 2$

In Exercises 13–20 sketch the graphs of the given functions on the same coordinate axes.

13. $f(x) = \sqrt{x}, \; g(x) = \frac{1}{2}\sqrt{x}, \; h(x) = \frac{1}{4}\sqrt{x}$ 14. $f(x) = \frac{1}{2}x^2, \; g(x) = \frac{1}{3}x^2, \; h(x) = \frac{1}{4}x^2$

15. $f(x) = 2x^2, \; g(x) = -2x^2$ 16. $f(x) = 2x^2 + 1, \; g(x) = 1 - 2x^2$

17. $f(x) = x^3, \; g(x) = 2x^3$ 18. $f(x) = -x^3, \; g(x) = -2x^3$

19. $f(x) = \sqrt{x - 2}, \; g(x) = 2\sqrt{x - 2}$ 20. $f(x) = \sqrt{x}, \; g(x) = \sqrt{x - 2}, \; h(x) = \sqrt{x + 2}$

2.5
LINEAR FUNCTIONS

The polynomial function of the first degree

$$f(x) = ax + b$$

is called a **linear function.** In this section we will show that the graph of a linear function is a straight line. We will also show that there is a property unique to the straight line that differentiates it from all other curves.

SLOPE OF THE STRAIGHT LINE

In Figure 36 we have drawn a straight line L that is not vertical. We have indicated the distinct points $P_1(x_1, y_1)$ and $P_2(x_2, y_2)$ on L. The increments or changes $x_2 - x_1$ and $y_2 - y_1$ in the x- and y-coordinates, respectively, from P_1 to P_2 are also indicated. Note that the increment $x_2 - x_1$ cannot be zero, since L is not vertical.

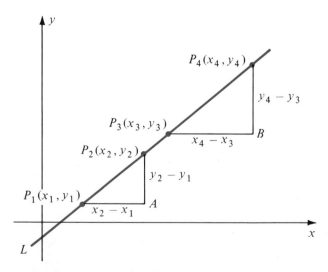

FIGURE 36 Slope of a Line Is Independent of Points

If $P_3(x_3, y_3)$ and $P_4(x_4, y_4)$ are another pair of points on L, the increments $x_4 - x_3$ and $y_4 - y_3$ will, in general, be different from the increments obtained by using P_1 and P_2. However, since triangles P_1AP_2 and P_3BP_4 are similar, the corresponding sides are in proportion; that is, the ratios

$$\frac{y_4 - y_3}{x_4 - x_3} \quad \text{and} \quad \frac{y_2 - y_1}{x_2 - x_1}$$

are the same. This ratio is called the **slope of the line** L and is denoted by m.

Slope of a Line	The slope m of a line that is not vertical is given by $$m = \frac{y_2 - y_1}{x_2 - x_1}$$ where $P_1(x_1, y_1)$ and $P_2(x_2, y_2)$ are any two distinct points on the line.

For a vertical line, $x_1 = x_2$ so that $x_2 - x_1 = 0$. Since we cannot divide by 0, we say that the slope of a vertical line is **undefined.**

The property of constant slope characterizes the straight line; that is, no other curve has this property. In fact, to define slope for a curve other than a straight line is not a trivial task; it requires use of the concept of limit, which is fundamental to calculus. (See Appendix B.)

EXAMPLE 1 Determining Slope

Find the slope of the line that passes through the points $(4, 2)$, $(1, -2)$.

SOLUTION

We may choose either point as (x_1, y_1) and the other as (x_2, y_2). Our choice is

$$(x_1, y_1) = (4, 2) \quad \text{and} \quad (x_2, y_2) = (1, -2)$$

Then

$$m = \frac{y_2 - y_1}{x_2 - x_1} = \frac{-2 - 2}{1 - 4} = \frac{-4}{-3} = \frac{4}{3}$$

The student should verify that reversing the choice of P_1 and P_2 produces the same result for the slope m. We may choose either point as P_1 and the other as P_2, but we must use this choice consistently for both the x- and y-coordinates.

∎

Slope is a means of measuring the steepness of a line. That is, slope specifies the number of units we must move up or down to reach the line after moving 1 unit to the left or right of the line. In Figure 37 we have displayed several lines with positive and negative slopes. We can summarize this way.

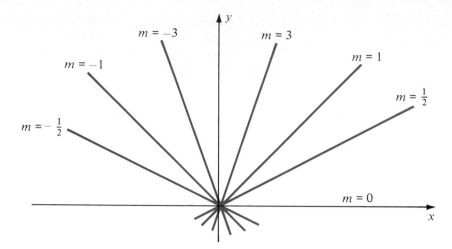

FIGURE 37 Slope and Steepness of a Line

(a) When $m > 0$, the line is the graph of an increasing function.
(b) When $m < 0$, the line is the graph of a decreasing function.
(c) When $m = 0$, the line is the graph of a constant function.
(d) Slope is not defined for a vertical line, and a vertical line is not the graph of a function.

EQUATIONS OF THE STRAIGHT LINE

We can apply the concept of slope to develop two important forms of the equation of a straight line. In Figure 38 the point $P_1(x_1, y_1)$ lies on a line L whose slope is m. If $P(x, y)$

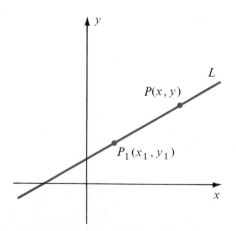

FIGURE 38 Point-Slope Form

THE PIRATE TREASURE (PART I)

Five pirates traveling with a slave found a chest of gold coins. The pirates agreed to divide the coins among themselves the following morning.

During the night Pirate 1 awoke and, not trusting his fellow pirates, decided to remove his share of the coins. After dividing the coins into five equal lots, he found that one coin remained. The pirate took his lot and gave the remaining coin to the slave to ensure his silence.

Later that night Pirate 2 awoke and decided to remove his share of the coins. After dividing the remaining coins into five equal lots, he found one coin left over. The pirate took his lot and gave the extra coin to the slave.

That same night the process was repeated by Pirates, 3, 4, and 5. Each time there remained one coin, which was given to the slave.

In the morning these five compatible pirates divided the remaining coins into five equal lots. Once again a single coin remained.

Question: What is the minimum number of coins there could have been in the chest? (For help, see Part II on page 99.)

is any other point on L, then we may use P and P_1 to compute m; that is,

$$m = \frac{y - y_1}{x - x_1}$$

This can be written in the form

$$y - y_1 = m(x - x_1)$$

Since (x_1, y_1) satisfies this equation, every point on L satisfies this equation. Conversely, any point satisfying this equation must lie on the line L, since there is only one line through $P_1(x_1, y_1)$ with slope m. This equation is called the **point-slope form** of a line.

Point-Slope Form
$$y - y_1 = m(x - x_1)$$
is an equation of the line with slope m that passes through the point (x_1, y_1).

EXAMPLE 2   Determining an Equation of a Line
(a) Find an equation of the line that passes through the points $(6, -2)$ and $(-4, 3)$.

(b) If f is a linear function and $f(1) = -1$ and $f(-2) = -7$, find an equation for f.

SOLUTION

(a) We first find the slope. Letting $(x_1, y_1) = (6, -2)$ and $(x_2, y_2) = (-4, 3)$, then

$$m = \frac{y_2 - y_1}{x_2 - x_1} = \frac{3 - (-2)}{-4 - 6} = \frac{5}{-10} = -\frac{1}{2}$$

Next, the point-slope form is used with $m = -\frac{1}{2}$ and $(x_1, y_1) = (6, -2)$.

$$y - y_1 = m(x - x_1)$$

$$y - (-2) = -\frac{1}{2}(x - 6)$$

$$y = -\frac{1}{2}x + 1$$

The student should verify that using the point $(-4, 3)$ and $m = -\frac{1}{2}$ in the point-slope form will yield the same equation.

(b) We have been provided with the coordinates of two points but in a disguised manner that requires us to work with function notation. If you think in terms of the form $y = f(x)$, then you see that $(1, -1)$ and $(-2, -7)$ are two points on the graph of the line. The slope of the line is then

$$m = \frac{y_2 - y_1}{x_2 - x_1} = \frac{-7 - (-1)}{-2 - 1} = \frac{-6}{-3} = 2$$

The point-slope form with $m = 2$ and $(x_1, y_1) = (1, -1)$ yields

$$y - y_1 = m(x - x_1)$$
$$y - (-1) = 2(x - 1)$$
$$y = 2x - 3$$

∎

PROGRESS CHECK
(a) Find an equation of the line that passes through the points $(-5, 0)$ and $(2, -5)$.

(b) If f is a linear function and $f(0) = 2$ and $f(2) = 0$, find an equation for f.

ANSWERS
(a) $y = -5x - 25$ (b) $y = -x + 2$

There is another form of the equation of the straight line that is very helpful. In Figure 39 the line L meets the y-axis at the point $(0, b)$, and is assumed to have slope m. Then we can let $(x_1, y_1) = (0, b)$ and use the point-slope form.

$$y - y_1 = m(x - x_1)$$
$$y - b = m(x - 0)$$
$$y = mx + b$$

Recalling that b is the y-intercept, we call this equation the **slope-intercept form** of the line.

Slope-Intercept Form   The graph of the equation

$$y = mx + b$$

is a straight line with slope m and y-intercept b.

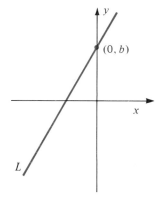

FIGURE 39 Slope-Intercept

The last result leads to the important conclusion mentioned in the introduction to this section. Since the graph of $y = mx + b$ is the graph of the function $f(x) = mx + b$, we have shown that the *graph of a linear function is a straight line*.

EXAMPLE 3
Find the slope and y-intercept of the line $y - 3x + 1 = 0$.

SOLUTION
The equation must be placed in the form $y = mx + b$. Solving for y gives

$$y = 3x - 1$$

and we find that $m = 3$ is the slope and $b = -1$ is the y-intercept.

$\blacksquare$

✓ **PROGRESS CHECK**
Find the slope and y-intercept of the line $2y + x - 3 = 0$.

ANSWER

slope $= m = -\dfrac{1}{2}$; y-intercept $= b = \dfrac{3}{2}$

EXAMPLE 4 Linear Functions in Word Problems
A small business firm buys a computer for $6000 and depreciates it linearly. At the end of 5 years the salvage value is $2000.
(a) Find a formula for the value V of the computer after t years.

(b) After how many years will the computer have been completely written off, that is, have lost all of its value?

SOLUTION
(a) Since we are dealing with *linear* depreciation, we may write the formula as

$$V = mt + b$$

where m is the rate of depreciation and b is the initial value (see Figure 40). Since the initial value V of the computer at time $t = 0$ is $6000, the point $(0, 6000)$ lies on the graph of V. Similarly, the point $(5, 2000)$ also lies on the graph. We use these points to determine the slope m:

$$m = \frac{V_2 - V_1}{t_2 - t_1} = \frac{2000 - 6000}{5 - 0} = \frac{-4000}{5} = -800$$

Then

$$V = -800t + 6000$$

(b) Setting $V = 0$,

$$0 = -800t + 6000$$

$$t = \frac{6000}{800} = \frac{15}{2} = 7.5 \text{ years}$$

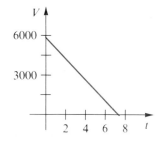

FIGURE 40 Diagram for Example 4

$\blacksquare$

HORIZONTAL AND VERTICAL LINES

In Figure 41 we have drawn a horizontal line through the point (a, b). Every point on this line has the form (x, b) since the y-coordinate remains constant. If $P(x_1, b)$ and $Q(x_2, b)$ are any two distinct points on the line, then the slope is

$$m = \frac{b - b}{x_2 - x_1} = 0$$

We have established the following.

Horizontal Lines

The equation of the horizontal line through the point (a, b) is

$$y = b$$

The slope of a horizontal line is 0.

This result corresponds to our prior observation that the graph of the constant function $f(x) = c$ is a horizontal line.

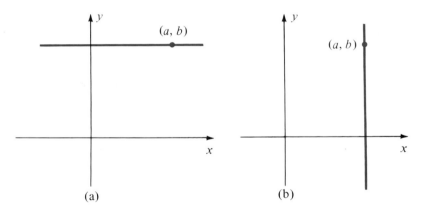

(a)   (b)

FIGURE 41   Horizontal and Vertical Lines

In Figure 41b, every point on the vertical line through the point (a, b) has the form (a, y) since the x-coordinate remains constant. The slope computation using any two points $P(a, y_1)$ and $Q(a, y_2)$ on the line produces

$$m = \frac{y_2 - y_1}{a - a} = \frac{y_2 - y_1}{0}$$

Since we cannot divide by 0, slope is not defined for a vertical line.

Vertical Lines

The equation of the vertical line through the point (a, b) is

$$x = a$$

The slope of a vertical line is undefined.

Note, too, that a vertical line fails the vertical line test and is not the graph of a function.

EXAMPLE 5 Horizontal and Vertical Lines

Find the equations of the horizontal and vertical lines through $(-4, 7)$.

SOLUTION

The horizontal line has the equation $y = 7$. The vertical line has the equation $x = -4$.

∎

WARNING

Don't confuse "undefined slope" and "zero slope." A horizontal line has zero slope. Slope is undefined for a vertical line.

PARALLEL AND PERPENDICULAR LINES

The concept of slope of a line can be used to determine when two lines are parallel or perpendicular. Since parallel lines have the same "steepness," we intuitively recognize that they must have the same slope.

> Two lines with slopes m_1 and m_2 are parallel if and only if
>
> $$m_1 = m_2$$

The criterion for perpendicular lines can be stated in this way.

> Two lines with slopes m_1 and m_2 are perpendicular if and only if
>
> $$m_2 = -\frac{1}{m_1}$$

These two theorems do not apply to vertical lines, since the slope of a vertical line is undefined. The proofs of these theorems are geometric in nature and are outlined in Exercises 66 and 68.

EXAMPLE 6 Parallel and Perpendicular Lines

Given the line $y = 3x - 2$, find an equation of the line passing through the point $(-5, 4)$ that is (a) parallel to the given line and (b) perpendicular to the given line.

SOLUTION

We first note that the line $y = 3x - 2$ has slope $m_1 = 3$.

(a) Every line parallel to the line $y = 3x - 2$ must have slope $m_2 = m_1 = 3$. We therefore seek a line with slope 3 that passes through the point $(-5, 4)$. Using the point-slope formula,

$$y - y_1 = m(x - x_1)$$
$$y - 4 = 3(x + 5)$$
$$y = 3x + 19$$

(b) Every line perpendicular to the line $y = 3x - 2$ has slope $m_2 = -1/m_1 = -\frac{1}{3}$. The line we seek has slope $-\frac{1}{3}$ and passes through the point $(-5, 4)$. We can again apply the point-slope formula to obtain

$$y - y_1 = m(x - x_1)$$

$$y - 4 = -\frac{1}{3}(x + 5)$$

$$y = -\frac{1}{3}x + \frac{7}{3}$$

The three lines are shown in Figure 42.

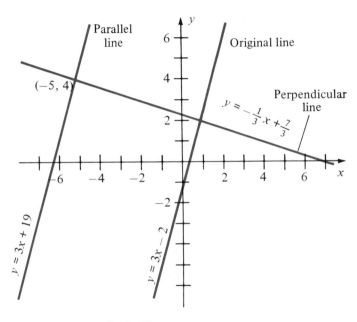

FIGURE 42   Parallel and Perpendicular Lines

GENERAL FIRST-DEGREE EQUATION

The **general first-degree equation** in x and y can always be written in the form

$$Ax + By + C = 0$$

where A, B, and C are constants and A and B are not both zero. We can rewrite this equation as

$$By = -Ax - C$$

If $B \neq 0$, the equation becomes

$$y = -\frac{A}{B}x - \frac{C}{B}$$

which we recognize as having a straight line graph with slope $-A/B$ and y-intercept $-C/B$. If $B = 0$, the original equation becomes $Ax + C = 0$, whose graph is a vertical line.

THE PIRATE TREASURE (PART II)

First, note that any number that is a multiple of 5 can be written in the form $5n$, where n is an integer. Since the number of coins found in the chest by Pirate 1 was one more than a multiple of 5, we can write the original number of coins C in the form $C = 5n + 1$, where n is a positive integer. Now, Pirate 1 removed his lot of n coins and gave one to the slave. The remaining coins can be calculated as

$$5n + 1 - (n + 1) = 4n$$

and since this is also one more than a multiple of 5, we can write $4n = 5p + 1$, where p is a positive integer. Repeating the process, we have the following sequence of equations.

$$
\begin{aligned}
C &= 5n + 1 & &\text{found by Pirate 1}\\
4n &= 5p + 1 & &\text{found by Pirate 2}\\
4p &= 5q + 1 & &\text{found by Pirate 3}\\
4q &= 5r + 1 & &\text{found by Pirate 4}\\
4r &= 5s + 1 & &\text{found by Pirate 5}\\
4s &= 5t + 1 & &\text{found next morning}
\end{aligned}
$$

Solving for s in the last equation and substituting successively in the preceding equations leads to the requirement that

$$1024n - 3125t = 2101 \qquad (1)$$

where n and t are positive integers. Equations, such as this, that require integer solutions are called Diophantine equations, and there is an established procedure for solving them that is studied in courses in number theory.

You might want to try to solve Equation (1) using a computer program. Since

$$n = \frac{3125t + 2101}{1024}$$

you can substitute successive integer values for t until you produce an integer result for n. The accompanying BASIC program does just that.

```
10 FOR K = 1 TO 3200
20 X = (3125*K
   + 2101)/1024
30 I = INT(X)
40 IF X = I THEN GO
   TO 60
50 NEXT K
60 PRINT "MINIMUM
   NUMBER OF
   COINS = ";
   5*I + 1
70 END
```

**The General
First-Degree Equation**

- The graph of the general first-degree equation

$$Ax + By + C = 0$$

  is a straight line.
- If $B = 0$, the graph is a vertical line.
- If $A = 0$, the graph is a horizontal line.

SUMMARY

The slope m of a line tells us a great deal about its graph; the slopes m_1 and m_2 of a pair of lines are useful in determining special relationships of their graphs.

TABLE 5   Slope of a Line

Slope(s)	Graph(s)	Example
$m > 0$	rising	
$m < 0$	falling	
$m = 0$	horizontal	
m undefined	vertical	
$m_1 = m_2$	parallel	
$m_1 = -1/m_2$	perpendicular	

EXERCISE SET 2.5

In Exercises 1–6 determine the slope of the line through the given points. State whether the line is the graph of an increasing function, a decreasing function, or a constant function.

1. $(2, 3), (-1, -3)$

2. $(1, 2), (-2, 5)$

3. $(-2, 3), (0, 0)$

4. $(2, 4), (-3, 4)$

5. $\left(\frac{1}{2}, 2\right), \left(\frac{3}{2}, 1\right)$

6. $(-4, 1), (-1, -2)$

7. Use slopes to show that the points $A(-1, -5)$, $B(1, -1)$, and $C(3, 3)$ are collinear (lie on the same line).

8. Use slopes to show that the points $A(-3, 2)$, $B(3, 4)$, $C(5, -2)$, and $D(-1, -4)$ are the vertices of a parallelogram.

In Exercises 9–12 determine an equation of the line with the given slope m that passes through the given point.

9. $m = 2, (-1, 3)$

10. $m = -\frac{1}{2}, (1, -2)$

11. $m = 3, (0, 0)$

12. $m = 0, (-1, 3)$

In Exercises 13–18 determine an equation of the line through the given points.

13. $(2, 4), (-3, -6)$

14. $(-3, 5), (1, 7)$

15. $(0, 0), (3, 2)$

16. $(-2, 4), (3, 4)$

17. $\left(-\frac{1}{2}, -1\right), \left(\frac{1}{2}, 1\right)$

18. $(-8, -4), (3, -1)$

In Exercises 19–24 find an equation for the linear function f satisfying the given conditions.

19. $f(2) = -5, f(5) = -2$

20. $f(-2) = 2, f(4) = 5$

21. $f(-1) = 1/2, f(1) = -3/2$

22. $f(0) = -5, f(3) = -4$

23. $f(-1) = -9, f(1) = 5$

24. $f(1/2) = 1/2, f(1/4) = 1$

In Exercises 25–30 determine an equation of the line with the given slope m and the given y-intercept b.

25. $m = 3, b = 2$

26. $m = -3, b = -3$

27. $m = 0, b = 2$

28. $m = -\frac{1}{2}, b = \frac{1}{2}$

29. $m = \frac{1}{3}, b = -5$

30. $m = -2, b = -\frac{1}{2}$

In Exercises 31–36 determine an equation of the linear function f with the given slope m that passes through the given point.

31. $m = -1, f(0) = 2$

32. $m = 2, (-1, 3)$

33. $m = 1/2, f(0) = -1/2$

34. $m = -2, f(2) = 0$

35. $m = 0, f(0) = 5$

36. $m = -1, f(1) = 2$

In Exercises 37–42 determine the slope m and y-intercept b of the given line.

37. $3x + 4y = 5$

38. $2x - 5y + 3 = 0$

39. $y - 4 = 0$

40. $x = -5$

41. $3x + 4y + 2 = 0$

42. $x = -\frac{1}{2}y + 3$

In Exercises 43–48 write an equation of (a) the horizontal line passing through the given point and (b) the vertical line passing through the given point.

43. $(-6, 3)$

44. $(-5, -2)$

45. $(-7, 0)$

46. $(0, 5)$

47. $(9, -9)$ 48. $\left(-\dfrac{3}{2}, 1\right)$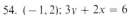

In Exercises 49–52 determine the slope of (a) every line that is parallel to the given line and (b) every line that is perpendicular to the given line.

49. $y = -3x + 2$ 50. $2y - 5x + 4 = 0$ 51. $3y = 4x - 1$ 52. $5y + 4x = -1$

In Exercises 53–56 determine an equation of the line through the given point that (a) is parallel to the given line; (b) is perpendicular to the given line.

53. $(1, 3)$; $y = -3x + 2$

54. $(-1, 2)$; $3y + 2x = 6$

55. $(-3, 2)$; $3x + 5y = 2$

56. $(-1, -3)$; $3y + 4x - 5 = 0$

57. The Celsius (C) and Fahrenheit (F) temperature scales are related by a linear equation. Water boils at $212°$F or $100°$C, and freezes at $32°$F or $0°$C.
    (a) Write a linear equation expressing F in terms of C.
    (b) What is the Fahrenheit temperature when the Celsius temperature is $20°$?

58. The college bookstore sells a textbook that costs $10 for $13.50, and a textbook that costs $12 for $15.90. If the markup policy of the bookstore in linear, write a linear function that relates sale price S and cost C. What is the cost of a book that sells for $22?

59. An appliance manufacturer finds that it had sales of $200,000 five years ago and sales of $600,000 this year. If the growth in sales is assumed to be linear, what will the sales amount be five years from now?

60. A product that cost $2.50 three years ago sells for $3 this year. If price increases are assumed to be linear, how much will the product cost six years from now?

61. Find a real number c such that $P(-2, 2)$ is on the line $3x + cy = 4$.

62. Find a real number c such that the line
    $$cx - 5y + 8 = 0$$
    has x-intercept 4.

63. If the point $(-2, -3)$ and $(-1, 5)$ are on the graph of a linear function f, find $f(x)$.

64. If $f(1) = 4$ and $f(-1) = 3$ and the function f is linear, find $f(x)$.

65. Prove that the linear function $f(x) = ax + b$ is an increasing function if $a > 0$ and is a decreasing function if $a < 0$.

66. In the accompanying figure, lines L_1 and L_2 are parallel. Points A and D are selected on lines L_1 and L_2, respectively. Lines parallel to the x-axis are constructed through A and D that intersect the y-axis at points B and

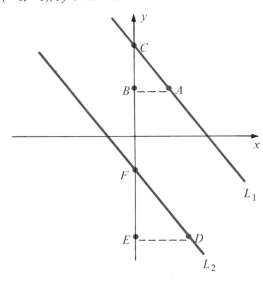

E. Supply a reason for each of the steps in the following proof.

(a) Angles ABC and DEF are equal.

(b) Angles ACB and DFE are equal.

(c) Triangles ABC and DEF are similar.

(d) $\dfrac{\overline{CB}}{\overline{BA}} = \dfrac{\overline{FE}}{\overline{ED}}$

(e) $m_1 = \dfrac{\overline{CB}}{\overline{BA}}$, $m_2 = \dfrac{\overline{FE}}{\overline{ED}}$

(f) $m_1 = m_2$

(g) Parallel lines have the same slope.

67. Prove that if two lines have the same slope, they are parallel.

68. In the accompanying figure, lines perpendicular to each other, with slopes m_1 and m_2, intersect at a point Q. A

perpendicular from Q to the x-axis intersects the x-axis at the point C. Supply a reason for each of the steps in the following proof.

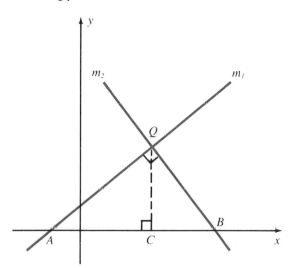

(a) Angles CAQ and BQC are equal.

(b) Triangles ACQ and BCQ are similar.

(c) $\dfrac{\overline{CQ}}{\overline{AC}} = \dfrac{\overline{CB}}{\overline{CQ}}$

(d) $m_1 = \dfrac{\overline{CQ}}{\overline{AC}}, \quad m_2 = \dfrac{\overline{CQ}}{\overline{CB}}$

(e) $m_2 = -\dfrac{1}{m_1}$

69. Prove that if two lines have slopes m_1 and m_2 such that $m_2 = -1/m_1$, the lines are perpendicular.

70. If x_1 and x_2 are the abscissas of two points on the graph of the function $y = f(x)$, show that the slope m of the line connecting the two points can be written as

$$m = \frac{f(x_2) - f(x_1)}{x_2 - x_1}$$

2.6 QUADRATIC EQUATIONS

The second-degree polynomial equation of the form

$$ax^2 + bx + c = 0$$

where a, b, and c are real numbers and $a \neq 0$, is known as a **quadratic equation.** This type of equation occurs frequently, and we will devote this section to studying methods for finding solutions, or **roots,** of quadratic equations, methods for analyzing the nature of the roots, and applications of quadratics to practical problems. In the next section we will relate our findings to quadratic functions and their graphs.

SOLVING BY FACTORING

If we can factor the left-hand side of the quadratic equation

$$ax^2 + bx + c = 0, \qquad a \neq 0$$

into two linear factors, then we can solve the equation quickly. For example, the quadratic equation

$$3x^2 + 5x - 2 = 0$$

can be written as

$$(3x - 1)(x + 2) = 0$$

To solve this equation we make use of a property of the real numbers.

> If the product of two real numbers is zero, then at least one of the factors must be zero.

We can therefore set each factor equal to zero to find the roots.

$$3x - 1 = 0 \qquad \text{or} \qquad x + 2 = 0$$

$$x = \frac{1}{3} \qquad x = -2$$

The solutions of the given quadratic equation are $\frac{1}{3}$ and -2.

PROGRESS CHECK

Solve each of the given equations by factoring.

(a) $3x^2 - 4x = 0$ **(b)** $2x^2 - 3x - 2 = 0$

ANSWERS

(a) $0, \dfrac{4}{3}$ (b) $-\dfrac{1}{2}, 2$

COMPLETING THE SQUARE

A quadratic equation of the form

$$a(x - h)^2 + k = 0 \tag{1}$$

where a, h, and k are constants, can be easily solved. For example, from

$$2(x - 5)^2 - 8 = 0$$

we conclude that

$$(x - 5)^2 = 4$$

$$x - 5 = \pm\sqrt{4} = \pm 2$$

$$x = 7 \qquad x = 3$$

The solutions of the given equations are the numbers 7 and 3. (Verify!)

A technique known as **completing the square** permits us to rewrite *any* quadratic equation in the form of Equation (1). Beginning with the expression $x^2 + dx$, we seek a constant h^2 to complete the square so that

$$x^2 + dx + h^2 = (x - h)^2$$

Expanding and solving, we have

$$x^2 + dx + h^2 = x^2 - 2hx + h^2$$

$$dx = -2hx$$

$$h = -\frac{d}{2}$$

so h^2 is the square of half the coefficient of x.

EXAMPLE 1 Completing the Square

Complete the square for each of the following.

(a) $x^2 - 6x$ **(b)** $x^2 + 3x$

SOLUTION

(a) The coefficient of x is -6 so that $h = 3$ and $h^2 = 9$. Then

$$(x^2 - 6x + 9) = (x - 3)^2$$

(b) The coefficient of x is 3 and $h^2 = \left(-\dfrac{3}{2}\right)^2 = \dfrac{9}{4}$.

Then

$$x^2 + 3x + \frac{9}{4} = \left(x + \frac{3}{2}\right)^2$$

We are now in a position to use this method to solve a quadratic equation.

EXAMPLE 2 Completing the Square

Solve the quadratic equation $2x^2 - 10x + 1 = 0$ by completing the square.

SOLUTION

We now outline and explain each step of the process.

Completing the Square	
Step 1. Rewrite the equation with the constant term on the right-hand side.	*Step 1.* $2x^2 - 10x = -1$
Step 2. Divide through by the coefficient of x^2.	$x^2 - 5x = -\dfrac{1}{2}$
Step 3. Complete the square, $$x^2 + dx + h^2 = (x - h)^2$$ where $h^2 = (-d/2)^2$. Balance the equation by adding h^2 to the right-hand side. Simplify.	*Step 3.* $d = -5$ $$h^2 = \left(\frac{-d}{2}\right)^2 = \left(\frac{5}{2}\right)^2 = \frac{25}{4}$$ $$x^2 - 5x + \frac{25}{4} = -\frac{1}{2} + \frac{25}{4}$$ $$\left(x - \frac{5}{2}\right)^2 = \frac{23}{4}$$
Step 4. Solve for x.	*Step 4.* $$x - \frac{5}{2} = \pm\frac{\sqrt{23}}{2}$$ $$x = \frac{5 \pm \sqrt{23}}{2}$$

PROGRESS CHECK

Solve by completing the square.

(a) $x^2 - 3x + 2 = 0$ (b) $3x^2 - 4x + 2 = 0$

ANSWERS

(a) 1, 2   (b) $\dfrac{2 \pm i\sqrt{2}}{3}$

THE QUADRATIC FORMULA

If we apply the method of completing the square to the general quadratic equation

$$ax^2 + bx + c = 0, \qquad a \neq 0$$

we arrive at the following *formula* that gives us the solutions for *any* quadratic equation.

Quadratic Formula	The equation

$$ax^2 + bx + c = 0, \qquad a \neq 0$$

has the solutions

$$x = \frac{-b \pm \sqrt{b^2 - 4ac}}{2a}$$

The derivation of the quadratic formula is outlined in Exercise 106 at the end of this section.

EXAMPLE 3   The Quadratic Formula

Solve $-5x^2 + 3x = 2$ by the quadratic formula.

SOLUTION

We first rewrite the given equation as $5x^2 - 3x + 2 = 0$ so that $a > 0$ and the right-hand side equals 0. Then $a = 5$, $b = -3$, and $c = 2$. Substituting in the quadratic formula, we have

$$x = \frac{-b \pm \sqrt{b^2 - 4ac}}{2a}$$

$$= \frac{-(-3) \pm \sqrt{(-3)^2 - 4(5)(2)}}{2(5)}$$

$$= \frac{3 \pm \sqrt{-31}}{10} = \frac{3 \pm i\sqrt{31}}{10}$$

∎

PROGRESS CHECK

Solve by use of the quadratic formula.

(a) $x^2 - 8x = -10$ (b) $4x^2 - 2x + 1 = 0$

ANSWERS

WARNING

There are a number of errors that students make in using the quadratic formula.
(a) To solve $x^2 - 3x = -4$, you must write the equation in the form $x^2 - 3x + 4 = 0$ to properly identify a, b, and c.

(b) The quadratic formula is

$$x = \frac{-b \pm \sqrt{b^2 - 4ac}}{2a}$$

Note that

$$x \neq -b \pm \frac{\sqrt{b^2 - 4ac}}{2a}$$

since the term $-b$ must also be divided by $2a$.

Now that you have a formula that works for any quadratic equation, you may be tempted to use it all the time. However, if you see an equation of the form

$$x^2 - 15$$

it is certainly easier to immediately supply the answer: $x = \pm\sqrt{15}$. Similarly, if you are faced with

$$x^2 + 3x + 2 = 0$$

it is faster to solve if you see that

$$x^2 + 3x + 2 = (x + 1)(x + 2)$$

The method of completing the square is generally not used for solving quadratic equations once you have learned the quadratic formula. The *technique* of completing the square is helpful in a variety of applications, and we will use it in a later chapter when we graph second-degree equations.

THE DISCRIMINANT

By analyzing the quadratic formula

$$x = \frac{-b \pm \sqrt{b^2 - 4ac}}{2a}$$

we can learn a great deal about the roots of the quadratic equation

$$ax^2 + bx + c = 0, \qquad a > 0$$

The key to the analysis is the **discriminant** $b^2 - 4ac$ found under the radical.
(a) If $b^2 - 4ac$ is negative, we have the square root of a negative number, and the roots of the quadratic equation are complex numbers.

(b) If $b^2 - 4ac$ is positive, we have the square root of a positive number, and the roots of the quadratic equation will be real numbers.

(c) If $b^2 - 4ac = 0$, then $x = -b/2a$, which we call a **double root** or **repeated root** of the quadratic equation. For example, if $x^2 - 10x + 25 = 0$, then the discriminant is 0 and $x = 5$. But

$$x^2 - 10x + 25 = (x - 5)(x - 5) = 0$$

We call $x = 5$ a double root because the factor $(x - 5)$ is a double factor of $x^2 - 10x + 25 = 0$. This hints at the importance of the relationship between roots and factors, a relationship that we will explore in a later chapter on roots of polynomial equations.

If the roots of the quadratic equation are real and a, b, and c are rational numbers, the discriminant enables us to determine whether the roots are rational or irrational. Since $\sqrt{k}$ is a rational number only if k is rational and a perfect square, we see that the quadratic formula produces a rational result only if $b^2 - 4ac$ is a perfect square. We summarize as follows.

The quadratic equation $ax^2 + bx + c = 0$, $a > 0$, has exactly two roots, the nature of which is determined by the discriminant $b^2 - 4ac$.

Discriminant	Roots
Negative	Two complex roots
0	A double root
Positive	Two real roots
a, b, c $\begin{cases} \text{A perfect square} \\ \text{rational} \begin{cases} \end{cases} \text{Not a perfect square} \end{cases}$	Rational roots
	Irrational roots

EXAMPLE 4   Using the Discriminant

Without solving, determine the nature of the roots of the quadratic equation $3x^2 - 4x + 6 = 0$.

SOLUTION

We evaluate $b^2 - 4ac$ using $a = 3$, $b = -4$, and $c = 6$.

$$b^2 - 4ac = (-4)^2 - 4(3)(6) = 16 - 72 = -56$$

The discriminant is negative and the equation has two complex roots.

■

EXAMPLE 5   Using the Discriminant

Without solving, determine the nature of the roots of the equation

$$2x^2 - 7x = -1$$

SOLUTION

We rewrite the equation in the standard form

$$2x^2 - 7x + 1 = 0$$

DETERMINING THE GOLDEN RATIO

When a point divides a line segment of length L into two parts of lengths a and b so that

$$\frac{L}{a} = \frac{a}{b}$$

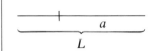

then a/b is called the **golden ratio.** To determine the golden ratio ϕ, assume that $b = 1$. Then we see from the accompanying figure that

$$L = a + 1$$

In addition, the above proportion will simplify to

$$L = a^2$$

Equating the expressions for L, we have

$$a^2 = a + 1$$
$$a^2 - a - 1 = 0$$

The length a satisfies this quadratic equation. Using the quadratic formula, we have

$$a = \frac{1 \pm \sqrt{5}}{2}$$

You can use a calculator to verify that $(1 + \sqrt{5})/2$ yields an approximate value for the golden ratio ϕ of 1.61803. The negative number $(1 - \sqrt{5})/2$ is rejected because the segment has a positive length.

and then substitute $a = 2$, $b = -7$, and $c = 1$ in the discriminant. Thus,

$$b^2 - 4ac = (-7)^2 - 4(2)(1) = 49 - 8 = 41$$

The discriminant is positive and is not a perfect square; thus, the roots are real, unequal, and irrational.

■

 PROGRESS CHECK

Without solving, determine the nature of the roots of the quadratic equation by using the discriminant.

(a) $4x^2 - 20x + 25 = 0$ (b) $5x^2 - 6x = -2$

(c) $10x^2 = x + 2$ (d) $x^2 + x - 1 = 0$

ANSWERS

(d) 2 real, irrational roots   (c) 2 real, rational roots

(b) 2 complex roots   (a) a real, double root

FORMS LEADING TO QUADRATICS

Certain types of equations can be transformed into quadratic equations, which can be solved by the methods discussed in this section. One form that leads to a quadratic equation is the **radical equation,** such as

$$x - \sqrt{x - 2} = 4$$

which is solved in Example 6. To solve the equation we isolate the radical and raise both sides to suitable powers. The following is the key to the solution of such equations.

> If P and Q are algebraic expressions, then the solution set of the equation
>
> $$P = Q$$
>
> is a subset of the solution set of the equation
>
> $$P^n = Q^n$$
>
> where n is a natural number.

This suggests that we can solve radical equations if we observe a precaution.

> If both sides of an equation are raised to the same power, the solutions of the resulting equation must be checked to see that they satisfy the original equation.

EXAMPLE 6 Solving a Radical Equation
Solve $x - \sqrt{x - 2} = 4$.

SOLUTION

Solving Radical Equations	
Step 1. When possible, isolate the radical on one side of the equation.	*Step 1.* $x - 4 = \sqrt{x - 2}$
Step 2. Raise both sides of the equation to a suitable power to eliminate the radical.	*Step 2.* Squaring both sides, $$x^2 - 8x + 16 = x - 2$$
Step 3. Solve for the unknown.	*Step 3.* $$x^2 - 9x + 18 = 0$$ $$(x - 3)(x - 6) = 0$$ $$x = 3 \qquad x = 6$$
Step 4. Check each solution by substituting in the *original* equation.	*Step 4.* $$\begin{array}{cc} \text{checking } x = 3 & \text{checking } x = 6 \\ 3 - \sqrt{3 - 2} \overset{?}{=} 4 & 6 - \sqrt{6 - 2} \overset{?}{=} 4 \\ 3 - 1 \overset{?}{=} 4 & 6 - \sqrt{4} \overset{?}{=} 4 \\ 2 \neq 4 & 4 = 4 \end{array}$$

We conclude that 6 is a solution of the original equation and 3 is not a solution of the original equation. We say that 3 is an **extraneous solution** that was introduced when we raised each side of the original equation to the second power.

∎

PROGRESS CHECK

Solve $x - \sqrt{1 - x} = -5$.

ANSWER

-3

Although the equation

$$x^4 - x^2 - 2 = 0$$

is not a quadratic in the unknown x, it is a quadratic in the unknown x^2; that is,

$$(x^2)^2 - (x^2) - 2 = 0$$

This may be seen more clearly by replacing x^2 by a new unknown u such that $u = x^2$. Substituting, we have

$$u^2 - u - 2 = 0$$

which is a quadratic equation in the unknown u. Solving,

$$(u + 1)(u - 2) = 0$$

$$u = -1 \qquad \text{or} \qquad u = 2$$

Since $x^2 = u$, we must next solve the equations

$$x^2 = -1 \qquad \text{and} \qquad x^2 = 2$$

$$x = \pm i \qquad\qquad x = \pm\sqrt{2}$$

The original equation has four solutions: i, $-i$, $\sqrt{2}$, and $-\sqrt{2}$.

The technique we have used is called a **substitution of variable.** Although simple in concept, this is a powerful device that is commonly used in calculus.

PROGRESS CHECK

Indicate an appropriate substitution of variable and solve each of the following equations.

(a) $3x^4 - 10x^2 - 8 = 0$ (b) $4x^{2/3} + 7x^{1/3} - 2 = 0$

(c) $\dfrac{2}{x^2} + \dfrac{1}{x} - 10 = 0$ (d) $\left(1 + \dfrac{2}{x}\right)^2 - 8\left(1 + \dfrac{2}{x}\right) + 15 = 0$

ANSWERS

(a) $u = x^2$; ± 2, $\mp \dfrac{\sqrt{6}}{3}$

(b) $u = x^{1/3}$; $\dfrac{1}{64}$, -8

(c) $u = \dfrac{1}{x}$, $\dfrac{5}{2}$, $-\dfrac{1}{2}$

(d) $u = 1 + \dfrac{2}{x}$; 1, $\dfrac{2}{1}$

APPLICATIONS

As your knowledge of mathematical techniques and ideas grows, you will be capable of tackling an ever-wider variety of applications. For example, the two practical problems that follow can be expressed as quadratic equations, which you now know how to solve.

One word of caution: It is possible to arrive at a solution that is meaningless. For example, a negative solution that represents hours worked or the age of an individual is meaningless and must be rejected.

EXAMPLE 7 Quadratic Equations and Word Problems

A number of students rented a car for $160 for a one-week camping trip. If another student had joined the original group, each person's share of expenses would have been reduced by $8. How many students were there in the original group?

SOLUTION

Let $n =$ the number of students in the original group. Then the cost per student was $160/n$ and the cost per student for the augmented group would have been $160/(n + 1)$. So

$$\text{original cost per student} = \text{reduced cost per student} + 8$$

$$\frac{160}{n} = \frac{160}{n + 1} + 8$$

$$160(n + 1) = 160n + 8n(n + 1)$$

$$8n^2 + 8n - 160 = 0$$

$$n^2 + n - 20 = 0$$

$$(n - 4)(n + 5) = 0$$

$$n = 4 \qquad n = -5$$

We reject the solution $n = -5$ and conclude that there were 4 students in the original group.

■

EXAMPLE 8 Quadratic Equations and Word Problems

A piece of wire 100 inches in length is cut into two pieces, each of which is then bent into the shape of a square. If the sum of the areas of the two squares is 425 square inches, what is the length of each of the two pieces of wire?

SOLUTION

A diagram such as Figure 43 is helpful in solving geometric problems. Since the original length of wire is 100 inches, we can designate the two pieces as having lengths x and $(100 - x)$ as in Figure 43a. The sides of the squares are then of length $x/4$ and $(100 - x)/4$ (Figures 43b and 43c) and we can express the sum of the areas by

$$\left(\frac{x}{4}\right)^2 + \left(\frac{100 - x}{4}\right)^2 = 425$$

Solving, we have

$$x^2 + (100 - x)^2 = 6800$$
$$x^2 + 10{,}000 - 200x + x^2 - 6800 = 0$$
$$2x^2 - 200x + 3200 = 0$$
$$x^2 - 100x + 1600 = 0$$
$$(x - 20)(x - 80) = 0$$
$$x = 20 \qquad x = 80$$

Either solution is acceptable (why?) and the two pieces have lengths of 20 inches and 80 inches.

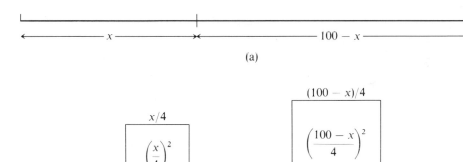

(a)

(b)   (c)

FIGURE 43   Diagram for Example 8

EXERCISE SET 2.6

In Exercises 1–8 solve the given equation.

1. $3x^2 - 27 = 0$
2. $4x^2 - 64 = 0$
3. $9x^2 + 64 = 0$
4. $81x^2 + 25 = 0$
5. $(2r + 5)^2 = 8$
6. $(3x - 4)^2 = -6$
7. $(3x - 5)^2 - 8 = 0$
8. $(4t + 1)^2 - 3 = 0$

In Exercises 9–18 solve by factoring.

9. $x^2 - 3x + 2 = 0$
10. $x^2 - 6x + 8 = 0$
11. $x^2 + x - 2 = 0$
12. $3r^2 - 4r + 1 = 0$
13. $x^2 + 6x = -8$
14. $x^2 + 6x + 5 = 0$
15. $y^2 - 4y = 0$
16. $2x^2 - x = 0$
17. $2x^2 - 5x = -2$
18. $2s^2 - 5s - 3 = 0$

In Exercises 19–30 solve by completing the square.

19. $x^2 - 2x = 8$
20. $t^2 - 2t = 15$
21. $2r^2 - 7r = 4$
22. $9x^2 + 3x = 2$
23. $3x^2 + 8x = 3$
24. $2y^2 + 4y = 5$
25. $2y^2 + 2y = -1$
26. $3x^2 - 4x = -3$
27. $4x^2 - x = 3$
28. $2x^2 + x = 2$
29. $3x^2 + 2x = -1$
30. $3u^2 - 3u = -1$

For Exercises 31–42 solve Exercises 19–30 by using the quadratic formula. In Exercises 43–48 solve by any method.

43. $2x^2 + 2x - 5 = 0$
44. $2t^2 + 2t + 3 = 0$
45. $3x^2 + 4x - 4 = 0$
46. $4u^2 - 1 = 0$
47. $x^2 + 2 = 0$
48. $4x^3 + 2x^2 + 3x = 0$

In Exercises 49–54 solve for the indicated variable in terms of the remaining variables.

49. $a^2 + b^2 = c^2$, for b

50. $s = \dfrac{1}{2}gt^2$, for t

51. $V = \dfrac{1}{3}\pi r^2 h$, for r

52. $A = \pi r^2$, for r

53. $s = \dfrac{1}{2}gt^2 + vt$, for t

54. $F = g\dfrac{m_1 m_2}{d^2}$, for d

In Exercises 55–62, without solving, determine the nature of the roots of each quadratic equation.

55. $x^2 - 2x + 3 = 0$

56. $3x^2 + 2x - 5 = 0$

57. $4x^2 - 12x + 9 = 0$

58. $2x^2 + x + 5 = 0$

59. $-3x^2 + 2x + 5 = 0$

60. $-3y^2 + 2y - 5 = 0$

61. $3x^2 + 2x = 0$

62. $4x^2 + 20x + 25 = 0$

In Exercises 63–66 determine a value or values for k that will result in the quadratic having a double root.

63. $kx^2 - 4x + 1 = 0$

64. $2x^2 + 3x + k = 0$

65. $x^2 - kx - 2k = 0$

66. $kx^2 - 4x + k = 0$

In Exercises 67–74 find the solution set.

67. $x + \sqrt{x + 5} = 7$

68. $x - \sqrt{13 - x} = 1$

69. $2x + \sqrt{x + 1} = 8$

70. $3x - \sqrt{1 + 3x} = 1$

71. $\sqrt{3x + 4} - \sqrt{2x + 1} = 1$

72. $\sqrt{4 - 4x} - \sqrt{x + 4} = 3$

73. $\sqrt{2x - 1} + \sqrt{x - 4} = 4$

74. $\sqrt{5x + 1} + \sqrt{4x - 3} = 7$

In Exercises 75–82 indicate an appropriate substitution of variable and solve each of the equations.

75. $3x^4 + 5x^2 - 2 = 0$

76. $2x^6 + 15x^3 - 8 = 0$

77. $\dfrac{6}{x^2} + \dfrac{1}{x} - 2 = 0$

78. $\dfrac{2}{x^4} - \dfrac{3}{x^2} - 9 = 0$

79. $2x^{2/5} + 5x^{1/5} + 2 = 0$

80. $3x^{4/3} - 4x^{2/3} - 4 = 0$

81. $2\left(\dfrac{1}{x} + 1\right)^2 - 3\left(\dfrac{1}{x} + 1\right) - 20 = 0$

82. $3\left(\dfrac{1}{x} - 2\right)^2 + 2\left(\dfrac{1}{x} - 2\right) - 1 = 0$

83. Find the width of a strip that has been mowed around a rectangular field 60 feet by 80 feet, if one-half of the lawn has not yet been mowed.

84. A 16-by-20-inch mounting board is used to mount a photograph. How wide a uniform border is there if the photograph occupies $\frac{3}{5}$ of the area of the mounting board?

85. The length of a rectangle exceeds twice its width by 4 feet. If the area of the rectangle is 48 square feet, find the dimensions.

86. The length of a rectangle is 4 centimeters less than twice its width. Find the dimensions if the area of the rectangle is 96 square centimeters.

87. The area of a rectangle is 48 square centimeters. If the length and width are each increased by 4 centimeters, the area of the newly formed rectangle is 120 square centimeters. Find the dimensions of the original rectangle.

88. The base of a triangle is 2 feet more than twice its altitude. If the area is 12 square feet, find the dimensions.

89. The smaller of two numbers is 4 less than the larger. If the sum of their squares is 58, find the numbers.

90. The sum of the reciprocals of two consecutive odd numbers is $\frac{8}{15}$. Find the numbers.

91. An investor placed an order totaling $1200 for a certain number of shares of a stock. If the price of each share of stock were $2 more, the investor would get 30 shares less for the same amount of money. How many shares did the investor buy?

92. A fraternity charters a bus for a ski trip at a cost of $360. When 6 more students join the trip, each person's cost decreases by $2. How many students were in the original group of travelers?

93. A salesman worked a certain number of days to earn $192. If he had been paid $8 more per day, he would have earned the same amount of money in two fewer days. How many days did he work?

94. A freelance photographer worked a certain number of days for a newspaper to earn $480. If she had been paid $8 less per day, she would have earned the same amount in two more days. What was her daily rate of pay?

95. A piece of wire 20 cm long is cut into two pieces, each of which is then bent into the shape of a circle. If the total area of the circles is 20 square cm, what is the length of the shorter piece of wire? (Round the answer to two decimal places.)

96. A piece of wire 20 cm long is cut into two pieces, which are then bent to form a circle and a square. If the total area of the circle and square is 20 square cm, what is the length of the shorter piece of wire? (Round the answer to two decimal places.)

In Exercises 97 and 98, provide a proof of the stated theorem.

97. If r_1 and r_2 are the roots of the equation

$$ax^2 + bx + c = 0,$$

then (a) $r_1 r_2 = c/a$ and (b) $r_1 + r_2 = -b/a$.

98. If a, b, and c are rational numbers, and the discriminant of the equation $ax^2 + bx + c = 0$ is positive, then the quadratic has either two rational roots or two irrational roots.

In Exercises 99–105 use the theorems of Exercise 97 to find a value or values of k so that the indicated condition is satisfied.

99. $kx^2 + 3x + 5 = 0$; sum of the roots is 6.

100. $2x^2 - 3kx - 2 = 0$; sum of the roots is -3.

101. $3x^2 - 10x + 2k = 0$; product of the roots is -4.

102. $2kx^2 + 5x - 1 = 0$; product of the roots is $\frac{1}{2}$.

103. $2x^2 - kx + 9 = 0$; one root is double the other.

104. $3x^2 - 4x + k = 0$; one root is triple the other.

105. $6x^2 - 13x + k = 0$; one root is the reciprocal of the other.

106. The quadratic formula can be derived by applying the method of completing the square to the quadratic equation

$$ax^2 + bx + c = 0, \qquad a \neq 0$$

Supply a reason for each of the following steps and complete the derivation.

$$ax^2 + bx = -c$$

$$x^2 + \frac{b}{a}x = -\frac{c}{a}$$

$$x^2 + \frac{b}{a}x + \left(\frac{b}{2a}\right)^2 = \left(\frac{b}{2a}\right)^2 - \frac{c}{a}$$

$$a\left(x + \frac{b}{2a}\right)^2 = \frac{b^2}{4a} - c$$

$$\left(x + \frac{b}{2a}\right)^2 = \frac{b^2}{4a^2} - \frac{c}{a} = \frac{b^2 - 4ac}{4a^2}$$

2.7 GRAPHING QUADRATIC FUNCTIONS

A function of the form

$$f(x) = ax^2 + bx + c \tag{1}$$

where a, b, and c are real numbers and $a \neq 0$, is called a **quadratic function.** The results we have obtained earlier in this chapter will be very helpful in studying the graph of quadratic functions.

By completing the square, it is always possible to rewrite Equation (1) in the form

$$f(x) = a(x - h)^2 + k \tag{2}$$

where h and k are constants. Here is an example.

EXAMPLE 1   Completing the Square
Write the quadratic function

$$f(x) = 2x^2 - 4x - 1$$

in the form of Equation (2).

SOLUTION
We complete the square in a manner analogous to that used in the previous section for solving quadratic equations. Here, we will factor out the coefficient a of x^2, complete the square, and balance the equation as follows.

$$
\begin{aligned}
f(x) &= 2(x^2 - 2x \quad) - 1 & \text{Factor out } a. \\
&= 2(x^2 - 2x + 1) - 1 - 2 & \text{Complete the square and balance.} \\
&= 2(x - 1)^2 - 3
\end{aligned}
$$

which is in the form of Equation (2) with $a = 2$, $h = 1$, and $k = -3$.

∎

PROGRESS CHECK
Write each quadratic function f in the form $f(x) = a(x - h)^2 + k$.
(a) $f(x) = -3x^2 - 12x - 13$ **(b)** $f(x) = 2x^2 - 2x + 3$

ANSWERS
(a) $f(x) = -3(x + 2)^2 - 1$ (b) $f(x) = 2\left(x - \dfrac{1}{2}\right)^2 + \dfrac{5}{2}$

WARNING
Be careful to balance the equation properly when completing the square. In Example 1, we wrote

$$
\begin{aligned}
f(x) &= 2(x^2 - 2x \quad) - 1 \\
&= 2(x^2 - 2x + 1) - 1 - 2
\end{aligned}
$$

We added $+1$ to the expression in parentheses which, due to the factor 2 in front of the parentheses, adds $+2$ to $f(x)$. We must balance by subtracting 2 as shown.

In Example 1 we completed the square to show that

$$f(x) = 2x^2 - 4x - 1 = 2(x - 1)^2 - 3$$

From our prior work (see Section 2.4), we know that the graph is that of the parabola $f(x) = 2x^2$ shifted 3 units downward and 1 unit to the right. In general, the graph of Equation (2) is that of the parabola $f(x) = ax^2$ shifted vertically k units and horizontally h units. Thus the graph of Equation (2) is a parabola opening from the point (h, k), which is called the **vertex** of the parabola. If $a > 0$, the parabola opens upward from the vertex; if $a < 0$, the parabola opens downward. We can summarize in this way.

Graph of
$f(x) = ax^2 + bx + c$

The quadratic function

$$f(x) = ax^2 + bx + c, \qquad a \neq 0$$

can be written in the form

$$f(x) = a(x - h)^2 + k$$

where h and k are constants. The graph is a parabola with vertex at (h, k), opening upward if $a > 0$ and downward if $a < 0$.

EXAMPLE 2   Graphing Quadratic Functions
Sketch the graphs of the following functions.
(a) $f(x) = 2x^2 + 4x - 1$ (b) $f(x) = -2x^2 + 4x$

SOLUTION
(a) Completing the square in x, we have

$$\begin{aligned}
f(x) &= 2x^2 + 4x - 1 \\
&= 2(x^2 + 2x \qquad) - 1 \\
&= 2(x^2 + 2x + 1) - 1 - 2 \\
&= 2(x + 1)^2 - 3
\end{aligned}$$

which is in the form $f(x) = a(x - h)^2 + k$ with $a = 2$, $h = -1$, and $k = -3$. The vertex of the parabola is at $(-1, -3)$ and the graph opens upward as shown in Figure 44a.

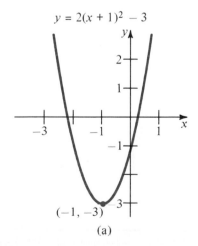

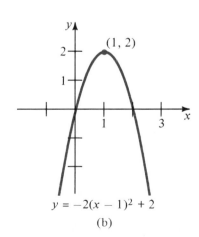

(a)   (b)

FIGURE 44   Graphs of Quadratic Functions

(b) Completing the square,

$$f(x) = -2x^2 + 4x$$
$$= -2(x^2 - 2x \qquad)$$
$$= -2(x^2 - 2x + 1) + 2$$
$$= -2(x - 1)^2 + 2$$

Here, $a = -2$, $h = 1$, and $k = 2$. The vertex of the parabola is at $(1, 2)$ and the parabola opens downward as in Figure 44b.

■

INTERCEPTS AND ROOTS

Since the graph of the quadratic function of Equation (1) is the graph of the equation

$$y = ax^2 + bx + c, \qquad a \neq 0$$

the graph intersects the x-axis at those points where $y = 0$. The x-intercepts are, then, those points where

$$y = 0 = ax^2 + bx + c \qquad (3)$$

which, of course, is a quadratic equation.

---

x-Intercepts of the Parabola

The x-intercepts of the parabola

$$f(x) = ax^2 + bx + c, \qquad a \neq 0$$

are the real roots of the quadratic equation

$$ax^2 + bx + c = 0$$

and are given by the quadratic formula.

The discriminant of the quadratic equation, Equation (3), tells us the number of real roots and, therefore, the number of x-intercepts of the parabola. The possibilities are two real roots (Figure 45a), a double root (Figure 45b), and two complex roots (Figure 45c).

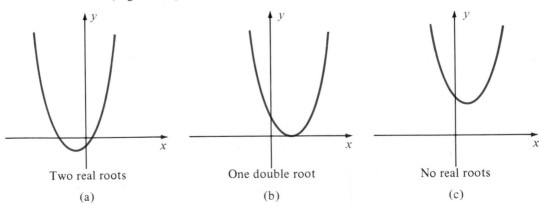

Two real roots

(a)

One double root

(b)

No real roots

(c)

FIGURE 45   Roots, Intercepts, and Graphs of Quadratic Functions

EXAMPLE 3 Graphing Parabolas

Find the vertex and all intercepts of each of the following parabolas. Sketch the graph.

(a) $f(x) = -x^2 + 3x - 2$ **(b)** $f(x) = x^2 + 1$ **(c)** $f(x) = x^2 + 2x + 1$

SOLUTION

(a) To find the vertex we must complete the square

$$f(x) = -x^2 + 3x - 2 = -\left(x - \frac{3}{2}\right)^2 + \frac{1}{4}$$

The vertex is at $(\frac{3}{2}, \frac{1}{4})$ and the parabola opens downward. Setting $y = f(x) = 0$, we see that

$$x^2 - 3x + 2 = 0$$
$$(x - 1)(x - 2) = 0$$

and the x-intercepts occur at $x = 1$ and $x = 2$. Finally, the y-intercept occurs where $x = 0$ and is found by evaluating $y = f(0) = -2$. The graph is shown in Figure 46a.

(b) Since

$$f(x) = x^2 + 1 = (x - 0)^2 + 1$$

the vertex is at $(0, 1)$ and the parabola opens upward. To find the x-intercepts, we set $y = f(x) = 0$ so that

$$x^2 + 1 = 0$$

Since this quadratic equation has no real roots, there are no x-intercepts. To find the y-intercept we set $x = 0$ and find that $y = f(0) = 1$. See Figure 46b for the graph.

(c) Completing the square,

$$f(x) = x^2 + 2x + 1 = (x + 1)^2 + 0$$

The vertex is at the point $(-1, 0)$ and the parabola opens upward. Setting $f(x) = 0$,

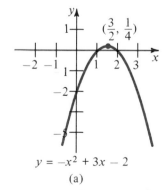

$y = -x^2 + 3x - 2$

(a)

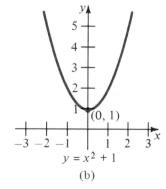

$y = x^2 + 1$

(b)

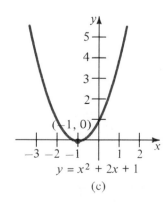

$y = x^2 + 2x + 1$

(c)

FIGURE 46 Graphs of Quadratic Functions

we see that

$$x^2 + 2x + 1 = (x + 1)^2 = 0$$

so that $x = -1$ is the only x-intercept. Finally, we set $x = 0$ and find that $y = f(0) = 1$ is the y-intercept. See Figure 46c.

■

EXERCISE SET 2.7

In Exercises 1–10 write the quadratic function f in the form $f(x) = a(x - h)^2 + k$.

1. $f(x) = x^2 - 6x + 10$
2. $f(x) = -x^2 - 2x - 3$
3. $f(x) = -2x^2 + 4x - 5$
4. $f(x) = 3x^2 + 12x + 14$

5. $f(x) = 2x^2 + 6x + 5$
6. $f(x) = -4x^2 + 4x$
7. $f(x) = -x^2 - x$
8. $f(x) = 3x^2 + 18x + 9$

9. $f(x) = -2x^2 + 5$
10. $f(x) = -\frac{1}{2}x^2 + 2x + 2$

In Exercises 11–18 find the vertex and all intercepts of the parabola. Sketch the graph.

11. $f(x) = 2x^2 - 4x$
12. $f(x) = -x^2 - 2x + 3$
13. $f(x) = -4x^2 + 4x - 1$
14. $f(x) = x^2 + 4x + 4$

15. $f(x) = \frac{1}{2}x^2 + 2x + 4$
16. $f(x) = -2x^2 + 2x - \frac{5}{2}$
17. $f(x) = -\frac{1}{2}x^2 + 3x - 4$
18. $f(x) = x^2 - x + 1$

2.8
MAXIMA AND MINIMA

There are many processes for which we seek the "best" solution in the sense of minimizing costs or maximizing the utilization of raw materials. Some of these problems simply require that we find where the largest or smallest value of a function occurs. The methods of calculus provide powerful tools for solving these types of **optimization** problems. We will restrict ourselves here to an approach that works when the function is a quadratic and will provide an introduction to applied "max/min" problems.

We know that the graph of the function

$$f(x) = ax^2 + bx + c, \qquad a \neq 0$$

is a parabola that opens from its vertex. If $a > 0$, the graph opens upward and the function takes on its *minimum* value at the vertex (see Figure 47a); if $a < 0$, the graph opens downward and the function takes on its *maximum* value at the vertex (see Figure 47b). You already know how to determine the coordinates (h, k) of the vertex by completing the square. However, by doing this task in a general manner, we can obtain

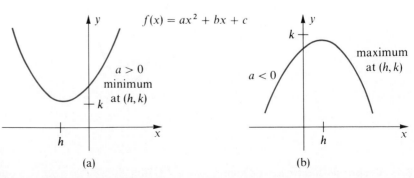

FIGURE 47 Maximum and Minimum

a simple expression for the x-coordinate of the vertex. We can write

$$f(x) = a\left(x^2 + \frac{b}{a}x + \right) + c$$

$$= a\left[x^2 + \frac{b}{a}x + \left(\frac{b}{2a}\right)^2\right] + c - a\left(\frac{b}{2a}\right)^2$$

$$= a\left(x + \frac{b}{2a}\right)^2 + \left(c - \frac{b^2}{4a}\right)$$

Since this last expression is of the form

$$f(x) = a(x - h)^2 + k$$

we see that the vertex occurs at $x = -b/2a$. Although we had to perform some messy algebraic manipulation, we arrive at a simple expression for the x-coordinate of the vertex of a parabola.

The maximum or minimum value of the quadratic function

$$f(x) = ax^2 + bx + c, \quad a \neq 0$$

occurs at the vertex where

$$x = -\frac{b}{2a}$$

It is a maximum value if $a < 0$ and a minimum value if $a > 0$; the maximum or minimum value is $f(-b/2a)$.

EXAMPLE 1 Maximum and Minimum Values of Quadratic Functions
Find the maximum or minimum value of the function $f(x) = -x^2 + 3x - 2$.

SOLUTION
For this quadratic function, $a = -1$, $b = 3$, and the vertex occurs at $x = -b/2a = \frac{3}{2}$. Since $a < 0$, the curve opens downward and the function attains a maximum value at the vertex. We find this maximum value by evaluating f when $x = \frac{3}{2}$ and find that $f(\frac{3}{2}) = \frac{1}{4}$. Conclusion: the maximum value of the function is $\frac{1}{4}$ and occurs when $x = \frac{3}{2}$. (This example is the same as Example 3a in Section 2.7 and the graph is shown in Figure 46a.)

■

PROGRESS CHECK
Given the function $f(x) = 2x^2 + x - 1$. Determine
(a) if f has a maximum or minimum value;
(b) the value of x at which the maximum or minimum occurs;
(c) the maximum or minimum value of f.

ANSWERS
(a) minimum   (b) $-\frac{1}{4}$ (c) $-\frac{9}{8}$

APPLIED MAXIMA AND MINIMA

In earlier courses in algebra you were faced with "word problems," that is, applied problems where you had to translate from English to algebraic symbols. Your objective was to set up an equation or inequality whose solution (or solutions) provided an answer to the problem. If the process of going from words to algebra gave you troubles or if you would like to review an approach to applied problems along with several worked-out examples, this is a good time to make use of Appendix A, "Applications: From Words to Algebra."

We are going to concentrate here on word problems that demand that you find a maximum or minimum value of a function. You will be surprised at the variety of problems of this type.

EXAMPLE 2   Maximum and Minimum in Word Problems
A rectangular region with one side against an existing building is to be fenced in. If 100 feet of fencing material is available, what dimensions will maximize the area of the region? What is the maximum area?

SOLUTION
In Figure 48 we indicate the lengths of the two parallel sides by x. Since we have 100 feet of material available, the length of the remaining side must be $(100 - 2x)$. The area A is a function of x and is given by

$$A(x) = x(100 - 2x) = -2x^2 + 100x$$

This quadratic function has a maximum (why?) which is assumed when $x = -b/2a = 25$. The dimensions of the three sides are 25, 25, and 50 feet and the maximum area is 1250 square feet.

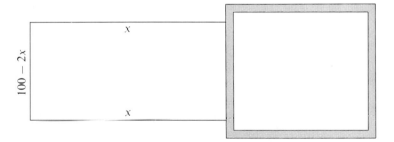

FIGURE 48   Diagram for Example 2

■

EXAMPLE 3   Maximum and Minimum in Word Problems
Find two numbers whose sum is 18 and whose product is a maximum.

SOLUTION
You could approach this problem by listing possible integer solutions: 1 and 17, 2 and 16, and so on. Although this is tempting, it is not appropriate. Let's see what we can do by using algebraic methods.

If one of the numbers is x, then the other must be $(18 - x)$. The product P is a function of x such that

$$P(x) = x(18 - x) = -x^2 + 18x$$

The function has a maximum (why?) at $x = -b/2a = 9$. The two numbers are 9 and 9 and the product is 81.

■

EXAMPLE 4   Applied Maximum and Minimum

Which point on the graph of the function $f(x) = \sqrt{x}$ is closest to the point $(2, 0)$?

SOLUTION

In Figure 49 we use the symbol d to denote the distance from the point $(2, 0)$ to a point $P(x, y)$ on the graph of $y = \sqrt{x}$. By the distance formula, we can write

$$d = \sqrt{(x - 2)^2 + (y - 0)^2}$$

Since the point lies on the graph of $y = \sqrt{x}$, we can substitute $\sqrt{x}$ for y and write d as a function of the variable x:

$$d(x) = \sqrt{x^2 - 4x + 4 + x} = \sqrt{x^2 - 3x + 4}$$

Since the square root function is an increasing function, the same value of x that provides a minimum for d will provide a minimum for $d^2 = x^2 - 3x + 4$. The minimum occurs where $x = -b/2a = \frac{3}{2}$. The corresponding value of y is

$$y = \sqrt{x} = \sqrt{\frac{3}{2}} = \frac{\sqrt{6}}{2}$$

The point we seek has coordinates $(\frac{3}{2}, \sqrt{6}/2)$.

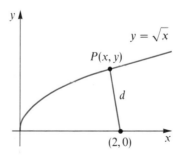

FIGURE 49   Diagram for Example 4

■

EXAMPLE 5   Applied Maximum and Minimum

A corporate vice-president finds that the cost C (in dollars) to produce and sell x units of the company's product is approximated by the function

$$C(x) = 3x^2 - 30x + 3000$$

If each unit sells for $300, find the number of units manufactured and sold that will maximize profit.

SOLUTION
Since each unit sells for $300, the revenue R can be expressed as a function of x in this way:

$$R(x) = 300x$$

The profit P is the difference between revenue and cost, that is,

$$\begin{aligned} P(x) &= R(x) - C(x) \\ &= 300x - (3x^2 - 30x + 3000) \\ &= -3x^2 + 330x - 3000 \end{aligned}$$

Since P is a quadratic function and its leading coefficient is negative, P attains a maximum where $x = -b/2a = -330/-6 = 55$. The company will maximize its profit if it manufactures and sells 55 units. (You should verify that the maximum profit is $6075.)

EXAMPLE 6   Applied Maximum and Minimum
An open box is to be made from a sheet of tin measuring 8 inches by 10 inches by cutting a square out of each corner and folding up the sides. If we let V represent the volume of the box and x the side of the square (see Figure 50), write an expression for the function V in terms of the variable x.

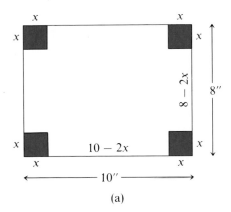

(a)

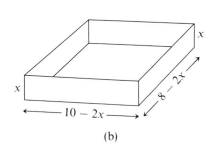

(b)

FIGURE 50   Diagram for Example 6

SOLUTION
The volume of the box is given by the formula

$$\text{volume} = \text{length} \times \text{width} \times \text{height}$$

In Figure 50b we see that the side of the square of length x becomes the height of the box and, accordingly, the length and width of the box have dimensions $(8 - 2x)$ and $(10 - 2x)$. We can then write

$$V(x) = x(8 - 2x)(10 - 2x) = 4x^3 - 36x^2 + 80x$$

Note that the function in Example 6 is not a quadratic. For that reason, we didn't ask you to find the value of x that will maximize the volume of the box. This is, however, typical of max/min problems that are handled in a calculus course.

■

EXERCISE SET 2.8

In Exercises 1 8 you are given a quadratic function f. Determine

(a) if f has a maximum value or a minimum value;

(b) the value of x at which the maximum or minimum occurs;

(c) the maximum or minimum value of f.

1. $f(x) = 3x^2 - 2x + 4$

2. $f(x) = -x^2 - x - 4$

3. $f(x) = -2x^2 - 5$

4. $f(x) = 2x^2 + 3x + 2$

5. $f(x) = x^2 + 5x$

6. $f(x) = -4x^2 + 3$

7. $f(x) = 2x^2 - \frac{1}{2}x - \frac{3}{2}$

8. $f(x) = \frac{2}{3}x^2 + x - 1$

9. Find two positive numbers such that their sum is 20 and their product is a maximum.

10. Find two positive numbers such that their sum is 100 and their product is a maximum.

11. Find two numbers whose sum is 50 and the sum of the squares of the numbers is a minimum.

12. Find two numbers such that their sum is 20 and the sum of their squares is a minimum.

13. Which point on the graph of the function $f(x) = \sqrt{x}$ is closest to the point $(1, 0)$?

14. A rectangle has a perimeter of 100 meters. What are the dimensions of the sides if the area is a maximum?

15. A farmer has 1000 feet of fencing material with which to enclose a rectangular field that borders on a straight stream. If the farmer does not enclose the side bordering on the stream, what are the dimensions of the field of maximum area?

16. Suppose that 320 feet of fencing is available to enclose a rectangular field and that one side of the field must be given double fencing. What are the dimensions of the rectangle that will yield the largest possible area?

17. A rectangle has a perimeter of 40 feet. What are the dimensions of the sides if the square of its diagonal is a minimum?

18. A piece of wire 20 inches long is to be cut into two pieces, one of which will be bent into a circle and the other into a square. How long should each piece be to minimize the sum of the areas? (Round your answer accurate to two decimal places.)

19. A ball is thrown straight up from the ground with an initial velocity of 80 feet per second. The height s (in feet) can be expressed as a function of time t (in seconds) by

$$s(t) = 80t - 16t^2$$

When does the ball reach its maximum height and what is the maximum height?

20. At a rate of $40 per room, a 100-room motel is fully occupied each night. For each $1 increase in the room rate, 2 fewer rooms are rented. What increase in room rate will maximize revenue?

21. A movie theater finds that 200 people attend each performance at the current rate of $4 and that attendance decreases by 10 persons for each 25 cent increase in price. What increase yields the greatest gross revenue?

22. An apple grower finds that the average yield is 60 bushels per tree when 80 or fewer trees are planted in an orchard. Each additional tree decreases the average yield per tree by 2 bushels. How many trees will maximize the total yield of the orchard?

23. Show that the coordinates of the point on the line $y = mx + b$ that is closest to the origin are

$$\left(\frac{-bm}{1 + m^2}, \frac{b}{1 + m^2} \right)$$

24. A manufacturer believes that the net profit P is related to the advertising expenditure x (both in thousands of dollars) by

$$P(x) = 10 + 46x - \frac{1}{2}x^2$$

What advertising expenditure will maximize the profit?

25. A manufacturer is limited by the firm's production facilities to an output of at most 125 units per day. The daily cost C and revenue R from the production and sale of x units are given as

$$C(x) = 3x^2 - 750x + 100$$

$$R(x) = 50x - x^2$$

If all units produced are sold, how many units should be manufactured daily to maximize profit?

26. An athletic field is in the shape of a rectangle with semicircular ends of radius r. If the perimeter of the field is p units, show that the maximum area is attained when $r = p/2\pi$.

27. A piece of wire 40 cm in length is to be cut into two pieces which will be used to form a circle and a square. How should the wire be cut to minimize the sum of the areas?

In Exercises 28–31 write an expression for the function to be optimized.

28. A rectangular sheet of cardboard 16 × 30 centimeters will be used to make a box by cutting a square from each corner and folding up the sides. If the side of the square is of length s, what value of s should be used to maximize the volume V of the box?

29. The United States Postal Service has this restriction on mailing a fourth-class parcel in the form of a rectangular box: The perimeter of one end plus the length of the box must be no more than 100 inches. What is the largest volume V of a permissible rectangular parcel whose ends are squares of length x?

30. A box with a square base is to have a volume of 50 cubic inches. Find the dimensions of the box that require the least material if (a) the box has no top; (b) the box has a top.

31. A container in the shape of a right circular cylinder is to have a volume of 50 cubic inches. Find the radius of the base that minimizes the surface area if (a) the container is open; (b) the container has a top.

TERMS AND SYMBOLS

KEY IDEAS FOR REVIEW

Topic	Page	Key Idea
Rectangular coordinate system	60	In a rectangular coordinate system, every ordered pair of real numbers (a, b) corresponds to a point in the plane, and every point in the plane corresponds to an ordered pair of real numbers.

Topic	Page	Key Idea
Distance formula	61	The distance $\overline{PQ}$ between points $P(x_1, y_1)$ and $Q(x_2, y_2)$ is given by the distance formula $$\overline{PQ} = \sqrt{(x_2 - x_1)^2 + (y_2 - y_1)^2}$$
Graph of an equation	63	The graph of an equation in two variables consists of all points whose coordinates satisfy the equation.
plotting	63	An equation in two variables can be graphed by plotting points that satisfy the equation and joining the points to form a smooth curve.
symmetry	65	There are simple algebraic means for testing whether the graph of an equation is symmetric with respect to the x-axis, the y-axis, and the origin.
Function	70	A function is a rule that assigns exactly one element y of a set Y to each element x of a set X. The domain is the set of inputs and the range is the set of outputs.
domain	72	The domain of a function is the set of all real numbers for which the function is defined. Beware of division by zero and even roots of negative numbers.
graph of a function	77	To graph $f(x)$, simply graph $y = f(x)$.
vertical line test	71	A graph represents a function if no vertical line meets the graph in more than one point.
piecewise-defined	79	The graph of a function can have holes or gaps, and can be defined in "pieces."
function notation	73	Function notation gives the definition of the function and also the value or expression at which to evaluate the function. If the function f is defined by $f(x) = x^2 + 2x$, then the notation $f(3)$ denotes the result of replacing the independent variable x with 3 wherever it appears. $$f(x) = x^2 + 2x$$ $$f(3) = 3^2 + 2(3) = 15$$
point on a graph	77	The point (x_1, y_1) on the graph of the function f can also be denoted as $(x_1, f(x_1))$.
tables and charts	71	An equation is not the only way to define a function. Sometimes a function is defined by a table or chart, or by several equations. Moreover, not every equation determines a function.
increasing, decreasing, constant	83	As we move from left to right, the graph of an increasing function rises and the graph of a decreasing function falls. The graph of a constant function neither rises nor falls; it is horizontal.
assists in graphing	86	If we know the graph of $f(x)$, then the graph of (a) $-f(x)$ is the reflection of the graph of $f(x)$ about the x-axis. (b) $f(x) + c$ is the graph of $f(x)$ shifted vertically c units. (c) $f(x - c)$ is the graph of $f(x)$ shifted horizontally c units. (d) $cf(x)$ is the graph of $f(x)$ "stretched" or "shrunk" vertically.
Slope of a line	91	Any two points on a line can be used to find its slope m. $$m = \frac{y_2 - y_1}{x_2 - x_1}$$
positive and negative slope	92	Positive slope indicates that a line is rising; negative slope indicates that a line is falling
horizontal and vertical lines	96	The slope of a horizontal line is 0; the slope of a vertical line is undefined.

Topic	Page	Key Idea
parallel lines	97	Parallel lines have the same slope.
perpendicular lines	97	The slopes of perpendicular lines are negative reciprocals of each other.
Equations of a line	98	The equation of a line can always be written in the form $Ax + By + C = 0$, where A, B, and C are constants and A and B are not both zero.
point-slope form	93	The point-slope form of a line is $y - y_1 = m(x - x_1)$. The slope is m and the point (x_1, y_1) lies on the line.
slope-intercept form	94	The slope-intercept form of a line is $y = mx + b$. The slope is m and the y-intercept is b.
horizontal and vertical lines	96	The equation of the horizontal line through the point (a, b) is $y = b$; the equation of the vertical line through the point (a, b) is $x = a$.
Linear function	99	The graphs of the linear function $f(x) = ax + b$ and of the general first-degree equation $Ax + By = C$ are always straight lines.
Quadratic equations quadratic formula	106	The quadratic equation $ax^2 + bx + c = 0$ always has two solutions, which may be found by using the quadratic formula: $$x = \frac{-b \pm \sqrt{b^2 - 4ac}}{2a}$$ If $b = 0$ or if the quadratic can be factored, then faster solution methods are available.
discriminant	109	The solutions or roots of a quadratic equation may be complex numbers. The expression $b^2 - 4ac$, called the discriminant, which appears under the radical of the quadratic formula, permits the nature of the roots to be analyzed without solving the equation.
Radical equations	110	Radical equations often can be transformed into quadratic equations. Since the process involves raising both sides of an equation to a power, the answers must be checked to see that they satisfy the original equation.
Substitution of variable	111	The method called "substitution of a variable" can be used to transform certain equations into quadratics. This technique is a handy tool and will be used in other chapters of this book.

REVIEW EXERCISES

Solutions to exercises whose numbers are in color are in the Solutions section in the back of the book.

2.1   **1.** Find the distance between the points $(-4, -6)$ and $(2, -1)$.

  2. Find the length of the longest side of the triangle whose vertices are $A(3, -4)$, $B(-2, -6)$ and $C(-1, 2)$.

In Exercises 3 and 4 sketch the graph of the given equation by forming a table of values.

  3. $y = 1 - |x|$ **4.** $y = \sqrt{x - 2}$

In Exercises 5 and 6 analyze the given equation for symmetry with respect to the x-axis, y-axis, and origin.

  5. $y^2 = 1 - x^3$ **6.** $y^2 = \dfrac{x^2}{x^2 - 5}$

2.2   In Exercises 7 and 8 state if the graph determines y to be a function of x.

  7.

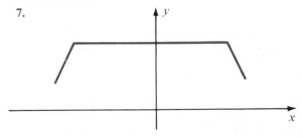

8.

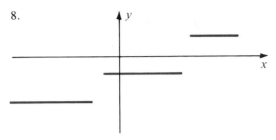

In Exercises 9 and 10 determine the domain of the given function.

9. $f(x) = \sqrt{3x - 5}$ 10. $f(x) = \dfrac{x}{x^2 + 2x + 1}$

11. In $f(x) = \sqrt{x - 1}$, find a real number whose image is 15.

12. If $f(t) = t^2 + 1$, find a real number whose image is 10.

In Exercises 13–15, $f(x) = x^2 - x$. Evaluate the following

13. $f(-3)$ 14. $f(y - 1)$

15. $\dfrac{f(2 + h) - f(2)}{h}$, $\quad h \neq 0$

2.3   Exercises 16–19 refer to the function f defined by

$$f(x) = \begin{cases} x - 1, & x \leq -1 \\ x^2, & -1 < x \leq 2 \\ -2, & x > 2 \end{cases}$$

16. Sketch the graph of the function f.

17. Determine where the function f is increasing, decreasing, and constant.

18. Evaluate $f(-4)$.   19. Evaluate $f(4)$.

2.4   In Exercises 20 and 21 sketch the graph of the function by shifting or reflecting a known graph, or by doing both.

20. $f(x) = |x| + 2$ 21. $f(x) = -\sqrt{x + 1}$

2.5   In Exercises 22–27 the points A and B have coordinates $(-4, -6)$ and $(-1, 3)$, respectively.

22. Find the slope of the line through A and B.

23. Find an equation of the line through the points A and B.

24. Find an equation of the line through A that is parallel to the y-axis.

25. Find an equation of the horizontal line through B.

26. Find an equation of the line through A that is parallel to the line $2x - y - 3 = 0$.

27. Find an equation of the line through B that is perpendicular to the line $2y + x - 5 = 0$.

2.6   28. Solve $x^2 - x - 20 = 0$ by factoring.

29. Solve $6x^2 - 11x + 4 = 0$ by factoring.

30. Solve $x^2 - 2x + 6 = 0$ by completing the square.

31. Solve $2x^2 - 4x + 3 = 0$ by the quadratic formula.

32. Solve $3x^2 + 2x - 1 = 0$ by the quadratic formula.

In Exercises 33–35 solve for x.

33. $49x^2 - 9 = 0$ 34. $kx^2 - 3\pi = 0$

35. $x^2 + x = 12$

In Exercises 36–38 determine the nature of the roots of the quadratic equation without solving.

36. $3r^2 = 2r + 5$ 37. $4x^2 + 20x + 25 = 0$

38. $6y^2 - 2y = -7$

In Exercises 39–41 solve the given equation.

39. $\sqrt{x + 2} = x$ 40. $3x^4 + 5x^2 - 5 = 0$

41. $\left(1 - \dfrac{2}{x}\right)^2 - 8\left(1 - \dfrac{2}{x}\right) + 15 = 0$

42. A charitable organization rented an auditorium at a cost of \$420 and split the cost among the attendees. If 10 additional persons had attended the meeting, the cost per person would have decreased by \$1. How many attendees were there in the original group?

2.7   In Exercises 43 and 44 find the vertex and all intercepts of the parabola.

43. $f(x) = -x^2 - 4x$ 44. $f(x) = x^2 - 5x + 7$

2.8   In Exercises 45 and 46 determine (a) if f has a maximum or minimum value; (b) the value of x at which the maximum or minimum occurs; (c) the maximum or minimum value of f.

45. $f(x) = 2x^2 - x + 1$ 46. $f(x) = -x^2 - 3x - 1$

PROGRESS TEST 2A

1. Find the perimeter of the triangle whose vertices are $(2, 5)$, $(-3, 1)$, and $(-3, 4)$.

2. Use symmetry to assist in sketching the graph of the equation $y = 2x^2 - 1$.

3. Analyze the equation $y = 1/x^3$ for symmetry with respect to the axes and origin.

4. Determine the domain of the function $\dfrac{1}{\sqrt{x - 1}}$.

5. If $f(x) = \sqrt{x - 1}$, find a real number whose image is 4.

6. If $f(x) = 2x^2 + 3$, find $f(2t)$.

Problems 7–9 refer to the function f defined by

$$f(x) = \begin{cases} 0, & x < -2 \\ |x|, & -2 \le x \le 3 \\ x^2 - x, & x > 3 \end{cases}$$

7. Determine where the function f is increasing, decreasing, and constant.

8. Evaluate $f(-5)$.   9. Evaluate $f(-2)$.

10. Sketch the graphs of the functions $f(x) = \sqrt{x}$ and $g(x) = \sqrt{x - 1}$ on the same coordinate axes.

11. Find an equation of the line through the points $(-3, 5)$ and $(-5, 2)$.

12. Find an equation of the vertical line through the point $(-3, 4)$.

13. Find the slope m and y-intercept b of the line whose equation is $2y - x = 4$.

14. Find an equation of the line through the point $(4, -1)$ that is parallel to the x-axis.

15. Find an equation of the line through the point $(-2, 3)$ that is perpendicular to the line $y - 3x - 2 = 0$.

16. Solve $x^2 - 5x = 14$ by factoring.

17. Solve $5x^2 - x + 4 = 0$ by completing the square.

18. Solve $12x^2 + 5x - 3 = 0$ by the quadratic formula.

In Problems 19 and 20 determine the nature of the roots of the quadratic equation without solving.

19. $6x^2 + x - 2 = 0$ 20. $3x^2 - 2x = -6$

In Problems 21 and 22 solve the given equation.

21. $x - \sqrt{4 - 3x} = -8$ 22. $3x^4 + 5x^2 - 2 = 0$

23. Sketch the graphs of $f(x) = (x + 1)^2$ and $g(x) = 2x^2 + 4x + 2$ on the same coordinate axes.

24. Find the vertex and intercepts of the parabola whose equation is $y = 3x^2 - 2x + 1$. Find the maximum or minimum value on the graph of the parabola.

PROGRESS TEST 2B

1. Find the length of the shorter diagonal of the parallelogram whose vertices are $(-3, 2), (-5, -4), (3, -4)$, and $(5, 2)$.

2. Use symmetry to assist in sketching the graph of the equation $y^2 = -3x + 4$.

3. Analyze the equation $x^2 - xy + 2 = 0$ for symmetry with respect to the axes and the origin.

4. Determine the domain of the function

$$f(x) = \frac{x^2}{16 - x^2}$$

5. If $f(x) = x^2 - 2x$, find a real number whose image is -1.

6. If $f(x) = \sqrt{x - 1}$, find $f(4)$.

Problems 7–9 refer to the function f defined by

$$f(x) = \begin{cases} x^2 - 1, & x \le -3 \\ 10, & -3 < x \le 3 \\ \sqrt{x}, & x > 3 \end{cases}$$

7. Determine where the function f is increasing, decreasing, and constant.

8. Evaluate $f(2)$.   9. Evaluate $f(-5)$.

10. Sketch the graphs of the functions $f(x) = |x|$ and $g(x) = |x - 2|$ on the same coordinate axes.

11. Find the slope of the line through the points $(-2, -3)$ and $(-4, 6)$.

12. Find an equation of the horizontal line through the point $(-6, -5)$.

13. Find the y-intercept of the line through the points $(4, -3)$ and $(-1, 2)$.

14. Determine the slope of every line that is perpendicular to the line $6y - 2x = 5$.

15. Determine an equation of the line through the point $(3, -2)$ that is parallel to the line $3y + x - 4 = 0$.

16. Solve $6x^2 + 13x - 5 = 0$ by factoring.

17. Solve $2x^2 - 5x + 2 = 0$ by completing the square.

18. Solve $3x^2 - x = -7$ by the quadratic formula.

In Problems 19 and 20 determine the nature of the roots of the quadratic equation without solving.

19. $6z^2 - 4z = -2$ 20. $4y^2 - 20y + 25 = 0$

In Problems 21 and 22 solve the given equation.

21. $8 + \sqrt{1 - x} = 10$ 22. $\dfrac{8}{x^{4/3}} + \dfrac{9}{x^{2/3}} + 1 = 0$

23. Sketch the graphs of $f(x) = x^2$ and $g(x) = -2(x - 1)^2 + 1$ on the same coordinate axes.

24. Find the vertex and intercepts of the parabola whose equation is $y = -2x^2 + 3x - 1$. Find the maximum or minimum value on the graph of the parabola.

CHAPTER 3

POLYNOMIAL AND RATIONAL FUNCTIONS

In Chapter 2 we observed that the polynomial function

$$f(x) = ax + b \qquad (1)$$

is called a linear function and that the polynomial function

$$g(x) = ax^2 + bx + c, \qquad a \neq 0 \qquad (2)$$

is called a quadratic function. To facilitate the study of polynomial functions in general, we will use the notation

$$P(x) = a_n x^n + a_{n-1} x^{n-1} + \cdots + a_1 x + a_0, \qquad a_n \neq 0 \qquad (3)$$

to represent a **polynomial function of degree n.** Note that the subscript k of the coefficient a_k is the same as the exponent of x in x^k. In general, the coefficients a_k may be real or complex numbers; we will restrict our work in this chapter to real values for a_k.

If $a \neq 0$ in Equation (1), we set the polynomial function equal to zero and obtain the linear equation

$$ax + b = 0$$

which has precisely one solution, $-b/a$. If we set the polynomial function in Equation (2) equal to zero, we have the quadratic equation

$$ax^2 + bx + c = 0$$

which has the two solutions given by the quadratic formula. If we set the polynomial function in Equation (3) equal to zero, we have the **polynomial equation of degree n**

$$a_n x^n + a_{n-1} x^{n-1} + \cdots + a_1 x + a_0 = 0 \qquad (4)$$

Our attention in this chapter will turn to finding the solutions of Equation (4). These solutions are also known as the **roots** of the polynomial equation $P(x) = 0$ or as **zeros** of the polynomial function P.

3.1
SYNTHETIC DIVISION;
THE REMAINDER AND
FACTOR THEOREMS

POLYNOMIAL DIVISION

To find the roots of a polynomial, it will be necessary to divide the polynomial by a second polynomial. There is a procedure for polynomial division that parallels the long division process of arithmetic. In arithmetic, if we divide an integer p by an integer $d \neq 0$, we obtain a quotient q and a remainder r, so we can write

$$\frac{p}{d} = q + \frac{r}{d} \tag{1}$$

where

$$0 \leq r < d \tag{2}$$

This result can also be written in the form

$$p = qd + r, \qquad 0 \leq r < d \tag{3}$$

For example,

$$\frac{7284}{13} = 560 + \frac{4}{13}$$

or

$$7284 = (560)(13) + 4$$

In the long division process for polynomials, we divide the dividend $P(x)$ by the divisor $D(x) \neq 0$ to obtain a quotient $Q(x)$ and a remainder $R(x)$. We then have

$$\frac{P(x)}{D(x)} = Q(x) + \frac{R(x)}{D(x)} \tag{4}$$

where $R(x) = 0$ or where

$$\text{degree of } R(x) < \text{degree of } D(x) \tag{5}$$

This result can also be written as

$$P(x) = Q(x)D(x) + R(x) \tag{6}$$

Note that Equations (1) and (4) have the same form and that Equation (6) has the same form as Equation (3). Equation (2) requires that the remainder be less than the divisor, and the parallel requirement for polynomials in Equation (5) is that the *degree* of the remainder be less than that of the divisor.

We illustrate the long division process for polynomials by an example.

EXAMPLE 1 Polynomial Division
Divide $3x^3 - 7x^2 + 1$ by $x - 2$.

SOLUTION

Polynomial Division	
Step 1. Arrange the terms of both polynomials by descending powers of x. If a power is missing, write the term with a zero coefficient.	*Step 1.* $$x - 2\overline{)3x^3 - 7x^2 + 0x + 1}$$
Step 2. Divide the first term of the dividend by the first term of the divisor. The answer is written above the first term of the dividend.	*Step 2.* $$\begin{array}{r} 3x^2 \\ x - 2\overline{)3x^3 - 7x^2 + 0x + 1} \end{array}$$
Step 3. Multiply the divisor by the quotient obtained in Step 2 and then subtract the product.	*Step 3.* $$\begin{array}{r} 3x^2 \\ x - 2\overline{)3x^3 - 7x^2 + 0x + 1} \\ \underline{3x^3 - 6x^2} \\ -x^2 + 0x + 1 \end{array}$$
Step 4. Repeat Steps 2 and 3 until the remainder is zero or the degree of the remainder is less than the degree of the divisor.	*Step 4.* $$\begin{array}{r} 3x^2 - x - 2 = Q(x) \\ x - 2\overline{)3x^3 - 7x^2 + 0x + 1} \\ \underline{3x^3 - 6x^2} \\ -x^2 + 0x + 1 \\ \underline{-x^2 + 2x} \\ -2x + 1 \\ \underline{-2x + 4} \\ -3 = R(x) \end{array}$$
Step 5. Write the answer in the form of Equation (4) or Equation (6).	*Step 5.* $$\begin{aligned} P(x) &= 3x^3 - 7x^2 + 1 \\ &= \underbrace{(3x^2 - x - 2)}_{Q(x)}\underbrace{(x - 2)}_{D(x)} + \underbrace{-3}_{R(x)} \end{aligned}$$

■

✓ **PROGRESS CHECK**
Divide $4x^2 - 3x + 6$ by $x + 2$.

ANSWER
$4x - 11 + \dfrac{28}{x + 2}$

SYNTHETIC DIVISION   Our work in this chapter will frequently require division of a polynomial by a first-degree polynomial $x - r$, where r is a constant. Fortunately, there is a shortcut called

synthetic division that simplifies this task. To demonstrate synthetic division we will do Example 1 again, writing only the coefficients.

$$
\begin{array}{r}
\mathbf{3} \quad -\mathbf{1} \quad -\mathbf{2} \\
-2\overline{)3 \quad -7 \quad -0 \quad\;\; 1} \\
\mathbf{3} \quad -6 \\
\hline
-\mathbf{1} \quad\;\; 0 \quad\;\; 1 \\
-1 \quad\;\; 2 \\
\hline
-\mathbf{2} \quad\;\; 1 \\
-2 \quad\;\; 4 \\
\hline
-3
\end{array}
$$

Note that the boldface numerals are duplicated. We can use this to our advantage and simplify the process as follows.

$$
\begin{array}{r|rrrr}
-2 & 3 & -7 & 0 & 1 \\
& & -6 & 2 & 4 \\
\hline
& 3 & -1 & -2 & -3
\end{array}
$$

$$\underbrace{3 \quad -1 \quad -2}_{\substack{\text{coefficients} \\ \text{of the} \\ \text{quotient}}} \;\Big|\; \underset{\text{remainder}}{-3}$$

In the third row we copied the leading coefficient (3) of the dividend, multiplied it by the divisor (-2), and wrote the result (-6) in the second row under the next coefficient. The numbers in the second column were subtracted to obtain $-7 - (-6) = -1$. The procedure is repeated until the third row is of the same length as the first row.

Since subtraction is more apt to produce errors than is addition, we can modify this process slightly. If the divisor is $x - r$, we will write r instead of $-r$ in the box and use addition in each step instead of subtraction. Repeating our example, we have

$$
\begin{array}{r|rrrr}
2 & 3 & -7 & 0 & 1 \\
& & 6 & -2 & -4 \\
\hline
& 3 & -1 & -2 & -3
\end{array}
$$

EXAMPLE 2 Synthetic Division
Divide $4x^3 - 2x + 5$ by $x + 2$ using synthetic division.

SOLUTION

PROGRESS CHECK
Use synthetic division to obtain the quotient $Q(x)$ and the constant remainder R when $2x^4 - 10x^2 - 23x + 6$ is divided by $x - 3$.

ANSWER
$Q(x) = 2x^3 + 6x^2 + 8x + 1; \; R = 9$

Synthetic Division	
Step 1. If the divisor is $x - r$, write r in the box. Arrange the coefficients of the dividend by descending powers of x, supplying a zero coefficient for every missing power.	*Step 1.* $\underline{-2}$ | 4 0 -2 5
Step 2. Copy the leading coefficient in the third row.	*Step 2.* $\underline{-2}$ | 4 0 -2 5 ——————————— 4
Step 3. Multiply the last entry in the third row by the number in the box and write the result in the second row under the next coefficient. Add the numbers in that column.	*Step 3.* $\underline{-2}$ | 4 0 -2 5 -8 ——————————— 4 -8
Step 4. Repeat Step 3 until there is an entry in the third row for each entry in the first row. The last number in the third row is the remainder, the other numbers are the coefficients of the quotient in descending order.	*Step 4.* $\underline{-2}$ | 4 0 -2 5 -8 16 -28 ——————————————— 4 -8 14 | -23 $\dfrac{4x^3 - 2x + 5}{x + 2}$ $= 4x^2 - 8x + 14 - \dfrac{23}{x + 2}$

∎

WARNING

(a) Synthetic division can be used only when the divisor is a linear factor. Don't forget to write a zero for the coefficient of each missing term.

(b) When dividing by $x - r$, place r in the box. For example, when the divisor is $x + 3$, place -3 in the box since $x + 3 = x - (-3)$. Similarly, when the divisor is $x - 3$, place $+3$ in the box since $x - 3 = x - (+3)$.

THE REMAINDER THEOREM   From our work with the division process we may surmise that division of a polynomial $P(x)$ by $x - r$ results in a quotient $Q(x)$ and a constant remainder R such that

$$P(x) = (x - r) \cdot Q(x) + R$$

Since this identity holds for all real values of x, it must hold when $x = r$. Consequently,

$$P(r) = (r - r) \cdot Q(r) + R$$
$$P(r) = 0 \cdot Q(r) + R$$

or

$$P(r) = R$$

We have proved the Remainder Theorem.

Remainder Theorem	If a polynomial $P(x)$ is divided by $x - r$, the remainder is $P(r)$.

EXAMPLE 3   Applying the Remainder Theorem

Determine the remainder when $P(x) = 2x^3 - 3x^2 - 2x + 1$ is divided by $x - 3$.

SOLUTION

By the Remainder Theorem, the remainder is $R = P(3)$. We then have

$$R = P(3) = 2(3)^3 - 3(3)^2 - 2(3) + 1 = 22$$

We may verify this result by using synthetic division.

$$
\begin{array}{r|rrrr}
3 & 2 & -3 & -2 & 1 \\
  &   & 6 & 9 & 21 \\
\hline
  & 2 & 3 & 7 & \mathbf{22} \\
\end{array}
$$

The numeral in boldface is the remainder, so we have verified that $R = 22$.

■

PROGRESS CHECK

Determine the remainder when $3x^2 - 2x - 6$ is divided by $x + 2$.

ANSWER

0I

FACTOR THEOREM

Let's assume that a polynomial $P(x)$ can be written as a product of polynomials, that is,

$$P(x) = D_1(x)D_2(x)\ldots D_n(x)$$

where $D_i(x)$ is a polynomial of degree greater than zero. Then $D_i(x)$ is called a **factor** of $P(x)$. If we focus on $D_1(x)$ and let

$$Q(x) = D_2(x)D_3(x)\ldots D_n(x)$$

then

$$P(x) = D_1(x)Q(x)$$

which suggests the following formal definition.

The polynomial $D(x)$ is a factor of a polynomial $P(x)$ if division of $P(x)$ by $D(x)$ results in a remainder of zero.

We can now combine this rule and the Remainder Theorem to prove the Factor Theorem.

Factor Theorem	A polynomial $P(x)$ has a factor $x - r$ if and only if $P(r) = 0$.

If $x - r$ is a factor of $P(x)$, then division of $P(x)$ by $x - r$ must result in a remainder of 0. By the Remainder Theorem, the remainder is $P(r)$, and hence $P(r) = 0$. Conversely, if $P(r) = 0$, then the remainder is 0 and $P(x) = (x - r)Q(x)$ for some polynomial $Q(x)$ of degree one less than that of $P(x)$. By definition, $x - r$ is then a factor of $P(x)$.

EXAMPLE 4 Applying the Factor Theorem
Show that $x + 2$ is a factor of

$$P(x) = x^3 - x^2 - 2x + 8$$

SOLUTION
By the Factor Theorem, $x + 2 = x - (-2)$ is a factor if $P(-2) = 0$. Using synthetic division to evaluate $P(-2)$,

$$
\begin{array}{r|rrrr}
-2 & 1 & -1 & -2 & 8 \\
   &   & -2 & 6 & -8 \\
\hline
   & 1 & 3 & 4 & 0
\end{array}
$$

we see that $P(-2) = 0$. Alternatively, we can evaluate

$$P(-2) = (-2)^3 - (-2)^2 - 2(-2) + 8 = 0$$

We conclude that $x + 2$ is a factor of $P(x)$.

■

 PROGRESS CHECK
Show that $x - 1$ is a factor of $P(x) = 3x^6 - 3x^5 - 4x^4 + 6x^3 - 2x^2 - x + 1$.

The student is urged to study and master the equivalent statements that follow. Understanding that these statements all say the same thing will make life in mathematics much more enjoyable!

If P is a polynomial function and a is a real number, then the following are all equivalent statements:
- $x = a$ is a *zero* of the polynomial function P.
- $x = a$ is a *root* of the polynomial equation $P(x) = 0$.
- $(x - a)$ is a *factor* of the polynomial $P(x)$.
- a is an *x-intercept* of the graph of the function P.

EXERCISE SET 3.1

In Exercises 1–10 use polynomial division to find the quotient $Q(x)$ and the remainder $R(x)$ when the first polynomial is divided by the second polynomial.

1. $x^2 - 7x + 12$, $\quad x - 5$
2. $x^2 + 3x + 3$, $\quad x + 2$
3. $2x^3 - 2x$, $\quad x^2 + 2x - 1$
4. $3x^3 - 2x^2 + 4$, $\quad x^2 - 2$
5. $3x^4 - 2x^2 + 1$, $\quad x + 3$
6. $x^5 - 1$, $\quad x^2 - 1$
7. $2x^3 - 3x^2$, $\quad x^2 + 2$
8. $3x^3 - 2x - 1$, $\quad x^2 - x$
9. $x^4 - x^3 + 2x^2 - x + 1$, $\quad x^2 + 1$
10. $2x^4 - 3x^3 - x^2 - x - 2$, $\quad x - 2$

In Exercises 11–20 use synthetic division to find the quotient $Q(x)$ and the constant remainder R when the first polynomial is divided by the second polynomial.

11. $x^3 - x^2 - 6x + 5$, $\quad x + 2$
12. $2x^3 - 3x^2 - 4$, $\quad x - 2$
13. $x^4 - 81$, $\quad x - 3$
14. $x^4 - 81$, $\quad x + 3$
15. $3x^3 - x^2 + 8$, $\quad x + 1$
16. $2x^4 - 3x^3 - 4x - 2$, $\quad x - 1$
17. $x^5 + 32$, $\quad x + 2$
18. $x^5 + 32$, $\quad x - 2$
19. $6x^4 - x^2 + 4$, $\quad x - 3$
20. $8x^3 + 4x^2 - x - 5$, $\quad x + 3$

In Exercises 21–26 use the Remainder Theorem and synthetic division to find $P(r)$.

21. $P(x) = x^3 - 4x^2 + 1$, $\quad r = 2$
22. $P(x) = x^4 - 3x^2 - 5x$, $\quad r = -1$
23. $P(x) = x^5 - 2$, $\quad r = -2$
24. $P(x) = 2x^4 - 3x^3 + 6$, $\quad r = 2$
25. $P(x) = x^6 - 3x^4 + 2x^3 + 4$, $\quad r = -1$
26. $P(x) = x^6 - 2$, $\quad r = 1$

In Exercises 27–32 use the Remainder Theorem to determine the remainder when $P(x)$ is divided by $x - r$.

27. $P(x) = x^3 - 2x^2 + x - 3$, $\quad x - 2$
28. $P(x) = 2x^3 + x^2 - 5$, $\quad x + 2$
29. $P(x) = -4x^3 + 6x - 2$, $\quad x - 1$
30. $P(x) = 6x^5 - 3x^4 + 2x^2 + 7$, $\quad x + 1$
31. $P(x) = x^5 - 30$, $\quad x + 2$
32. $P(x) = x^4 - 16$, $\quad x - 2$

In Exercises 33–40 use the Factor Theorem to decide whether or not the first polynomial is a factor of the second polynomial.

33. $x - 2$, $\quad x^3 - x^2 - 5x + 6$
34. $x - 1$, $\quad x^3 + 4x^2 - 3x + 1$
35. $x + 2$, $\quad x^4 - 3x - 5$
36. $x + 1$, $\quad 2x^3 - 3x^2 + x + 6$
37. $x + 3$, $\quad x^3 + 27$
38. $x + 2$, $\quad x^4 + 16$
39. $x + 2$, $\quad x^4 - 16$
40. $x - 3$, $\quad x^3 + 27$

In Exercises 41–46 determine whether the given value of x is a root of the given polynomial equation.

41. $x^3 - 3x + 2 = 0$; $\quad x = -2$
42. $3x^2 - x + 1 = 0$; $\quad x = 2$
43. $-2 + x + 2x^2 - x^3 = 0$; $\quad x = -1$
44. $3x^2 + 2x - 1 = 0$; $\quad x = \dfrac{1}{3}$
45. $2x^2 + 4x - 1 = 0$; $\quad x = \dfrac{3}{2}$
46. $x^3 + 27 = 0$; $\quad x = -3$

In Exercises 47–52 determine all zeros of the polynomial function f.

47. $f(x) = (x + 1)(x - 2)$
48. $f(x) = (x - 1)^2(x + 3)$

49. $f(x) = (1 - x)(2x - 1)$

50. $f(x) = (2 - x)(x + 2)(1 + x)$

51. $f(x) = (1 - 2x)^2(1 + 2x)$

52. $f(x) = x^3(x - 2)^2$

In Exercises 53–59 use synthetic division to determine the value of k or r as requested.

53. Determine the values of r for which division of $x^2 - 2x - 1$ by $x - r$ has a remainder of 2.

54. Determine the values of r for which

$$\frac{x^2 - 6x - 1}{x - r}$$

has a remainder of -9.

55. Determine the values of k for which $x - 2$ is a factor of $x^3 - 3x^2 + kx - 1$.

56. Determine the values of k for which $2k^2x^3 + 3kx^2 - 2$ is divisible by $x - 1$.

57. Use the Factor Theorem to show that $x - 2$ is a factor of $P(x) = x^8 - 256$.

58. Use the Factor Theorem to show that

$$P(x) = 2x^4 + 3x^2 + 2$$

has no factor of the form $x - r$, where r is a real number.

59. Use the Factor Theorem to show that $x - y$ is a factor of $x^n - y^n$, where n is a natural number.

3.2 GRAPHS OF POLYNOMIAL FUNCTIONS

The graph of the first-degree polynomial function

$$P(x) = ax + b \tag{1}$$

is always a straight line. The graph of the second-degree polynomial function

$$P(x) = ax^2 + bx + c, \qquad a \neq 0 \tag{2}$$

is always a parabola. For $n > 2$ the graph of the polynomial function

$$P(x) = a_nx^n + a_{n-1}x^{n-1} + \cdots + a_1x + a_0, \qquad a_n \neq 0 \tag{3}$$

does not have a name, nor is it a simple matter to describe its shape. Still, there are certain general patterns of behavior that we can explore that will enable you to see whether a graph that you have drawn makes sense for a given polynomial function. Taken together with a neat scheme for forming a table of values, these observations will aid greatly in sketching the graphs of polynomial functions of degree greater than 2.

CONTINUITY AND THE INTERMEDIATE VALUE THEOREM

We begin by exploring what the graph of a polynomial function can and cannot "do." The graphs in Figure 1 are "typical" in the sense that they illustrate these points:

Continuity and Corners

- The graph of a polynomial function is **continuous,** which means that there are no "breaks" or "holes." (The graph can be drawn without lifting the pencil from the paper.)

- The graph of a polynomial function has no corners (also called **cusps**). In particular, no segment of the graph of a polynomial function of degree greater than 1 is a straight line.

The labels accompanying the graphs in Figure 2 explain why each of these graphs cannot be that of a polynomial function.

The property of continuity is defined and studied intensively in calculus courses. It is this property that has enabled us to plot points on the graph of a polynomial function

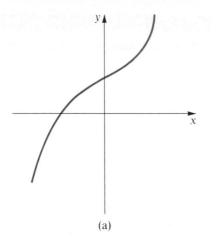

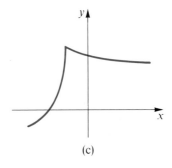

(a)   (b)

FIGURE 1   **Graphs of Polynomial Functions**

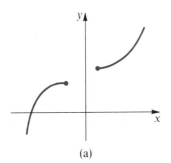

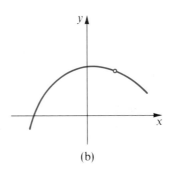

(a)   (b)   (c)

FIGURE 2   **Graphs of Non-Polynomial Functions**

P and then *connect these points by a smooth curve.* In Figure 3, to get from the point $A(a, P(a))$ on the graph of P to the point $B(b, P(b))$, the graph on the interval $[a, b]$ must pass through every ordinate between $P(a)$ and $P(b)$. This result is formally known as the **Intermediate Value Theorem for Polynomial Functions.**

Intermediate Value Theorem for Polynomial Functions	If $a < b$ and P is a polynomial function such that $P(a) \neq P(b)$, then, in the interval $[a, b]$, P takes on every value between $P(a)$ and $P(b)$.

This theorem will come in handy when we seek the roots of the polynomial equation $P(x) = 0$. If we can find values a and b such that $P(a)$ and $P(b)$ are opposite in sign, then there must be at least one value c in the interval $[a, b]$ where $P(c) = 0$, which says that there is a real root c such that $a \leq c \leq b$.

TURNING POINTS

Now, return to the graph in Figure 3. The points A and B are called **turning points** and are the points at which the graph changes "direction" from rising to falling or from falling to rising. An application of calculus that occurs early in introductory courses deals with finding these turning points. There is also a general result that is easily proven in calculus courses:

Turning Points	The graph of a polynomial function of degree n has at most $n - 1$ turning points.

The polynomial function whose graph is shown in Figure 1b has three turning points; by the last theorem, it must be of degree 4 or higher. (We'll soon see that it must also be of even degree.) Since the graph in Figure 1a has no turning points, the theorem doesn't provide any help in determining the degree of the polynomial function depicted. (We'll soon see that this polynomial function must be of odd degree.)

BEHAVIOR FOR LARGE $|x|$ By factoring out x^n we can rewrite Equation (3) as follows.

$$P(x) = a_n x^n + a_{n-1} x^{n-1} + \cdots + a_1 x + a_0$$

$$= x^n \left(a_n + \frac{a_{n-1}}{x} + \frac{a_{n-2}}{x^2} + \cdots + \frac{a_1}{x^{n-1}} + \frac{a_0}{x^n} \right), \qquad x \neq 0 \tag{4}$$

When the equation is written this way, it is easy to see how $P(x)$ behaves when $|x|$ assumes large values, that is, when x takes on large positive and large negative values. Note that the expression

$$\frac{a_{n-k}}{x^k}$$

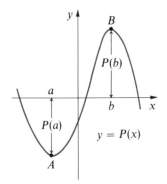

where k is a positive integer and a_{n-k} is a constant, will take on values closer and closer to zero as $|x|$ becomes larger and larger, since as the denominator of a fraction grows large, the value of the fraction becomes small. For sufficiently large values of $|x|$, we can ignore the contribution of all terms of the form a_{n-k}/x^k. What remains, then, is the term $a_n x^n$. In summary,

For large values of $|x|$, the polynomial function

$$P(x) = a_n x^n + a_{n-1} x^{n-1} + \cdots + a_1 x + a_0, \qquad a_n \neq 0$$

is dominated by its lead term $a_n x^n$.

FIGURE 3 Intermediate Value Theorem

We can make practical use of this last result. In Figures 4a and 4b we have sketched the graphs of polynomial functions of degrees 3 and 4, respectively. Note that as $|x|$ increases, the "ends" of the graph of the third-degree polynomial function extend indefinitely in opposite directions, whereas the "ends" of the graph of the fourth-degree polynomial function extend indefinitely in the same direction.

The behavior exhibited in Figure 4a will be true for all third-degree polynomial functions and, in fact, will be true for *all polynomial functions of odd degree*. To see this, simply note that the lead term $a_n x^n$ of any polynomial function of *odd* degree will assume opposite signs for positive and negative values of x. By our earlier result, this term dominates for large $|x|$, so that one end of the graph will extend upward and the other downward.

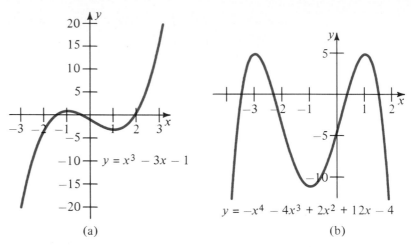

FIGURE 4   Polynomial Functions of Odd and Even Degree

We can use a similar argument to show that the behavior exhibited in Figure 4b is typical for all polynomial functions of even degree. For n that is *even*, the lead term $a_n x^n$ will have the same sign for both positive and negative values of x. Since this term dominates for large $|x|$, the ends of the curve will both extend upward or will both extend downward.

Graphs of Polynomial Functions

The graph of the polynomial function

$$P(x) = a_n x^n + a_{n-1} x^{n-1} + \cdots + a_1 x + a_0, \qquad a_n \neq 0$$

has these characteristics.

- If n is odd and $a_n > 0$, the graph extends upward for large positive values of x and downward for large negative values of x; if $a_n < 0$, the behavior is reversed.
- If n is even and $a_n > 0$, the graph extends upward for both large positive and large negative values of x; if $a_n < 0$, the graph extends downward at both ends.

EXAMPLE 1   Determining Behavior for Large $|x|$

Without sketching the graph, determine the behavior of the graph of each of the following polynomial functions for large values of $|x|$.

(a) $P(x) = -3x^5 + 26x^2 - 5$

(b) $P(x) = -\dfrac{1}{2}x^6 + 209x^3 + 16x + 2$

SOLUTION

(a) We need only concern ourselves with the lead term, $-3x^5$. Since the degree is odd and the lead coefficient is negative, the graph will extend downward for large positive values of x and upward for large negative values of x.

(b) The dominant term for large $|x|$ is $-\frac{1}{2}x^6$. Since the degree is even, the graph moves in the same direction at both ends. The negative lead coefficient tells us that this direction is downward.

■

TABLE OF VALUES

Given a polynomial function of degree n, we now know that the graph (a) is continuous; (b) has no corners; (c) has at most $n-1$ turning points; (d) behaves at the "ends" in a very predictable manner. What we need now is guidance as to what happens for intermediate values of x before the graph takes off for $+\infty$ or $-\infty$ or both.

To this end we can make use of the Remainder Theorem to find the coordinates of points on the graph. Recall that if a polynomial $P(x)$ is divided by $x-r$, the remainder is $P(r)$. Then the point $(r, P(r))$ lies on the graph of the function $P(x)$.

An efficient scheme for evaluating $P(r)$ is a streamlined form of synthetic division in which the addition is performed without writing the middle row. Given

$$P(x) = 2x^3 - 3x^2 - 2x + 1$$

we can find $P(3)$ by synthetic division in this way:

$$
\begin{array}{r|rrrr}
 & 2 & -3 & -2 & 1 \\
\hline
3\rvert & 2 & 3 & 7 & 22 = P(3)
\end{array}
$$

Then the point $(3, 22)$ lies on the graph of $P(x)$. Repeating this procedure for a number of values of r will provide a table of values for plotting.

EXAMPLE 2   Graphing P(x)

Sketching the graph of $P(x) = 2x^3 - 3x^2 - 2x + 1$.

SOLUTION

For each value r of x, the point $(r, P(r))$ lies on the graph of $y = P(x)$. We will allow x to assume integer values from -3 to $+3$ and will find $P(x)$ by using synthetic division.

	2	−3	−2	1	$(x, y) = (x, P(x))$
−3	2	−9	25	−74	$(-3, -74)$
−2	2	−7	12	−23	$(-2, -23)$
−1	2	−5	3	−2	$(-1, -2)$
0	2	−3	−2	1	$(0, 1)$
1	2	−1	−3	−2	$(1, -2)$
2	2	1	0	1	$(2, 1)$
3	2	3	7	22	$(3, 22)$

FIGURE 5a, b   Points on the Graph of P(x)

The ordered pairs shown at the right of each row in Figure 5a are the coordinates of points on the graph shown in Figure 5.

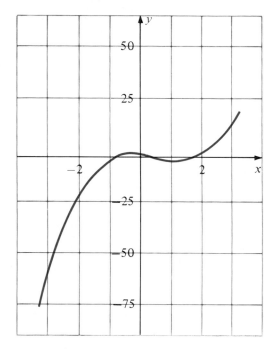

FIGURE 5   Graph for Example 2

POLYNOMIALS IN FACTORED FORM

When a polynomial can be written as a product of linear factors, it is a simple matter to find the x-intercepts and to determine where the graph of the polynomial lies above the x-axis and where it lies below the x-axis. The following example illustrates the procedure.

EXAMPLE 3   Using Intercepts in Graphing

Sketch the graph of the polynomial

$$P(x) = x^3 + x^2 - 6x$$

SOLUTION

Factoring, we find that

$$P(x) = x(x^2 + x - 6)$$
$$= x(x + 3)(x - 2)$$

Since $P(x) = 0$ when $x = 0$, when $x = -3$, and when $x = 2$, these values are the x-

intercepts. The x-intercepts divide the x-axis into the intervals

$$(-\infty, -3), (-3, 0), (0, 2), \text{ and } (2, \infty)$$

To find the signs of $P(x)$ in each of these intervals, we use the method described in Section 1.6, which requires that we analyze the sign of each factor in each interval as in Figure 6. From Figure 6 we conclude that the graph of $P(x)$ lies above the x-axis in the

x	$-$	$-$	$-$	$-$	$-$	$-$	$-$	$-$	0	$+$	$+$	$+$	$+$	$+$
$x+3$	$-$	$-$	$-$	0	$+$	$+$	$+$	$+$	$+$	$+$	$+$	$+$	$+$	$+$
$x-2$	$-$	$-$	$-$	$-$	$-$	$-$	$-$	$-$	$-$	$-$	$-$	0	$+$	$+$
				-3					0			2		
$P(x)$	$-$	$-$	$-$	0	$+$	$+$	$+$	$+$	0	$-$	$-$	0	$+$	$+$	$+$

FIGURE 6

intervals $(-3, 0)$ and $(2, \infty)$, since $P(x) > 0$ in these intervals. Similarly, the graph lies below the x-axis in the intervals $(-\infty, -3)$ and $(0, 2)$. Plotting a few points, we obtain the graph of Figure 7. Note that the graph extends indefinitely upward and downward as expected.

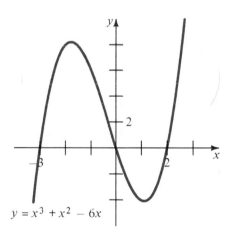

$$y = x^3 + x^2 - 6x$$

FIGURE 7 Graph for Example 3

■

EXERCISE SET 3.2

In Exercises 1 8 show that the equation has a root in the given interval.

1. $2x^4 - x^3 + 2x - 2 = 0$ $[-2, -1]$

2. $3x^3 - 2x^2 + 5x + 4 = 0$ $[-1, 0]$

3. $x^5 - 3x^3 + x^2 - 3 = 0$ $[1, 2]$

4. $-x^4 + 3x^2 + 5 = 0$ $[2, 3]$

5. $x^6 - 3x^3 + x^2 - 2 = 0$ $[1, 2]$

6. $2x^5 - x^4 + 3x^2 - 6 = 0$ $[1, 2]$

7. $-2x^3 - x^2 + 3x - 4 = 0$ $[-2, -1]$

8. $2x^4 - 3x^3 + x^2 - 1 = 0$ $[-1, 0]$

In Exercises 9–16 without sketching, determine the behavior of the graph of the given polynomial function for large values of $|x|$. Use the letters U and D to indicate whether the graph extends upward or downward.

Polynomial function	Large positive values of x	Large negative values of x
9. $P(x) = x^7 - 175x^3 + 23x^2$		
10. $P(x) = -3x^8 + 22x^4 + 3$		
11. $P(x) = -8x^3 + 17x^2 - 15$		
12. $P(x) = 14x^{12} - 5x^{11} + 3x - 1$		
13. $P(x) = -5x^{10} + 16x^7 + 5$		
14. $P(x) = 2x^5 - 11x^4 - 12x^3$		
15. $P(x) = 4x^8 - 10x^6 + x^3 - 8$		
16. $P(x) = -3x^9 + 6x^6 - 2x^5 + x$		

In Exercises 17–22 use the Remainder Theorem and synthetic division to sketch the graph of the given polynomial.

17. $P(x) = x^3 + x^2 + x + 1$

18. $P(x) = 3x^4 + 5x^3 + x^2 + 5x - 2$

19. $P(x) = 2x^3 + 3x^2 - 5x - 6$

20. $P(x) = x^3 + 3x^2 - 4x - 12$

21. $P(x) = x^4 - 3x^3 + 1$

22. $P(x) = 4x^4 + 4x^3 - 9x^2 - x + 2$

In Exercises 23–28 determine the x-intercepts and the intervals wherein $P(x) > 0$ and $P(x) < 0$. Sketch the graph of $P(x)$.

23. $P(x) = (x - 3)(2x - 1)(x + 2)$

24. $P(x) = (2 - x)(x - 4)(x + 1)$

25. $P(x) = 2x^3 + 3x^2 - 5x$

26. $P(x) = x^4 - 5x^2 + 4$

27. $P(x) = x^4 - x^3 - 6x^2$

28. $P(x) = (2x + 5)(x - 1)(x + 1)(x - 3)$

In Exercises 29–34 determine a polynomial equation whose roots include the given values.

29. $2, -4, 4$

30. $5, -5, 1, -1$

31. $-1, -2, -3$

32. $-3, \sqrt{2}, -\sqrt{2}$

33. $4, 1 \pm \sqrt{3}$

34. $1, 2, 2 \pm \sqrt{2}$

**3.3
APPROXIMATING
ROOTS OF
POLYNOMIAL
EQUATIONS (Optional)**

There are many sophisticated methods available for approximating the real roots of the polynomial equation $P(x) = 0$ that are studied in a branch of mathematics called numerical analysis. We'll discuss two methods here that are suitable for an introduction to this topic and that will require the use of a calculator. (As you will see, computer programs will prove even more suitable.)

**EVALUATING A
POLYNOMIAL**

The methods we will use here all require us to evaluate the polynomial $P(x)$ many, many times. The streamlined form of synthetic division is useful, especially when combined with a calculator. If you have access to a programmable calculator or a computer, you might just want to write a program that will evaluate a polynomial over an interval $[a, b]$ at some specifiable increment.

There is a way to rewrite a polynomial that is superbly suited for computer use. We'll illustrate this scheme by rewriting the general third-degree polynomial in

nested form:

$$P(x) = a_3 x^3 + a_2 x^2 + a_1 x + a_0 \tag{1}$$

$$= ((a_3 x + a_2)x + a_1)x + a_0 \tag{2}$$

where the innermost parentheses are evaluated first. (Verify that the expressions in (1) and (2) constitute an identity.) This is a nice scheme for use with a calculator or computer since, at each stage, there is a multiplication of the prior result by x followed by the addition of the next lower coefficient. The scheme eliminates the need to calculate powers of x and can obviously be extended to a polynomial of any degree.

EXAMPLE 1   Nested Form

Write in nested form: $P(x) = 2x^4 - x^3 + 4x^2 - 3x - 8$

SOLUTION

Following the scheme outlined and remembering to evaluate the parentheses from innermost toward outermost, we have

$$P(x) = (((2x - 1)x + 4)x - 3)x - 8$$

■

The nested form of polynomial evaluation works well in conjunction with the memory facility of a scientific calculator. To evaluate the polynomial of Example 1 for a given value of x, say $x = 3.45$, you would follow this outline:

$$3.45 \boxed{STO} \, 2 \, \boxed{\times} \, \boxed{RCL} - 1 \, \boxed{-}$$

$$\boxed{\times} \, \boxed{RCL} + 4 \, \boxed{=}$$

$$\boxed{\times} \, \boxed{RCL} - 3 \, \boxed{=}$$

$$\boxed{\times} \, \boxed{RCL} - 8 \, \boxed{=}$$

It is easy to see that this scheme can be extended for use with any polynomial and for any value of x.

APPROXIMATING ROOTS BY SUCCESSIVE DIGITS

Suppose we seek a real root of the polynomial equation $P(x) = 0$ and we can find values a and b such that $P(a)$ and $P(b)$ are opposite in sign. Then the Intermediate Value Theorem tells us that there must be at least one value c in the interval $[a, b]$ where $P(c) = 0$, which says that c is a real root and $a \le c \le b$. Our problem, then, is to get a better estimate for c.

Let's focus on a specific example. We'll deal with the equation

$$P(x) = 2x^4 + 4x^3 - x^2 - 10x - 10 = 0$$

and we'll assume that we have a reasonable means to evaluate $P(x)$ for any value of x. Suppose we are told that there is a root in the interval $[0, 3]$. Using our polynomial evaluator, we form Table 1. Since $P(1)$ is negative and $P(2)$ is positive, there is a root in the interval $(1, 2)$. To improve on our estimate, we start at $x = 1$ and increase x by tenths to form Table 2. Note that we've only recorded the *sign* of $P(x)$ and that we stopped as

TABLE 1   P(x) in the Interval [0, 3]

x	0	1	2	3
P(x)	−10	−15	30	221

TABLE 2   P(x) in the Interval [1, 2]

x	1.0	1.1	1.2	1.3	1.4	1.5	1.6	1.7	1.8	1.9	2.0
P(x)	−	−	−	−	−	−	+				

soon as we recorded a change in sign. Since $P(1.5)$ is negative and $P(1.6)$ is positive, we've isolated the root in the interval $(1.5, 1.6)$.

The process we've just illustrated can be repeated any number of times, providing yet another decimal place of accuracy at each stage. We'll take it one stage further as shown in Table 3. We can conclude that there is a root in the interval $[1.58, 1.59]$, that is, $x = 1.58^+$ is an approximate root.

TABLE 3   P(x) in the Interval [1.5, 1.6]

x	1.50	1.51	1.52	1.53	1.54	1.55	1.56	1.57	1.58	1.59	1.60
P(x)	−	−	−	−	−	−	−	−	−	+	

PROGRESS CHECK

Use the method of successive digits to find the root of the equation

$$x^3 + x^2 - 3x - 3 = 0$$

in the interval $[1, 2]$ accurate to two decimal places.

ANSWER

1.73

APPROXIMATING ROOTS BY BISECTION

Many of the sophisticated methods for approximating roots require a knowledge of calculus and are not suitable for this text. There is, however, a technique known as **bisection** that does not require calculus and can be of use in any number of circumstances.

Suppose we are told that the polynomial equation

$$P(x) = x^4 + x^3 - 5x^2 - 6x - 6 = 0$$

has a root in the interval $[2, 3]$. Let's *bisect* the interval and evaluate the polynomial at the endpoints and midpoint of the interval $[2, 3]$. We find

$$P(2) < 0 \qquad P(2.5) > 0 \qquad P(3) > 0$$

By the Intermediate Value Theorem, we've isolated the root in the interval $(2, 2.5)$.

The program shown below uses the method of bisection to find a real root of the polynomial equation $P(x) = 0$ in the interval (a, b). Should $P(a)$ and $P(b)$ be of the same sign, the program will produce an error message.

Note how few lines of code are required for the nested polynomial evaluation subroutine at the end of the listing.

```
10  '-----------------------------------------------------------
20  ' PROGRAM BISROOT finds root of polynomial by bisection
30  '-----------------------------------------------------------
40  DIM COEFF(21)
50  DEFINT I, N, D
60  DEF FNSIGN(X) = X < 0        'yields   -1=TRUE=negative or 0=FALSE=positive
70  TOLERANCE = .0001
80  INPUT ''Enter degree of polynomial: '', DEGREE
90  FOR I = DEGREE TO 0 STEP -1
100    PRINT ''Enter coefficient for term of degree ''; I;
110    INPUT '': '', COEFF(I)
120  NEXT I
130  PRINT
140  INPUT ''Enter left endpoint of interval: '', BEGINT
150  INPUT ''Enter right endpoint of interval: '', ENDINT
160  X = BEGINT
170  GOSUB 410
180  POFBEG = POFX
190  X = ENDINT
200  GOSUB 410
210  POFEND = POFX
220  ' Error if same sign at end points
230  IF ( FNSIGN(POFBEG) = FNSIGN(POFEND) ) THEN GOTO 380
240  ' Else continue to bisect the interval until a root is found
250  WHILE ( ABS(BEGINT - ENDINT) )= TOLERANCE )
260  ' Halve interval
270    MID = (BEGINT + ENDINT) / 2
280    X = MID: GOSUB 410
290    POFMID = POFX
300  ' Determine the subinterval which contains a root
310    IF ( FNSIGN(POFBEG) = FNSIGN(POFMID) )  THEN  BEGINT = MID
          ELSE  ENDINT = MID
320  WEND
330  ' A root is found
340  PRINT: PRINT ''There is a root at '';
350  PRINT USING ''###.######''; BEGINT
360  GOTO 400
370  ' There is no root
380  PRINT: PRINT ''*****Abandoning root quest because polynomial evaluation''
390  PRINT ''      at endpoints yields the same sign.''
400  END
410  '-----------------------------------------------------------
420  ' NESTED POLYNOMIAL EVALUATION (subroutine)
430  '-----------------------------------------------------------
440  POFX = 0
450  FOR N = DEGREE TO 0 STEP -1
460    POFX = POFX * X' + COEFF(N)
470  NEXT N
```

Repeating the process, we find that

$$P(2) < 0 \qquad P(2.25) < 0 \qquad P(2.5) > 0$$

which tightens the interval containing the root to $(2.25, 2.5)$. We can, of course, repeat the process until we are satisfied with the accuracy of the result.

 PROGRESS CHECK

Find a root of the equation $x^5 + x^4 + x + 2 = 0$ in the interval $[-2, -1]$ to two decimal places by the method of bisection.

ANSWER

-1.27

Which of the approximation methods is better? Clearly, neither is very sophisticated. The method of bisection is better suited for use with a computer. It is also preferable in the sense that the *technique* may be applied to finding approximate solutions in other problem situations.

EXERCISE SET 3.3

In Exercises 1–6 write the polynomial in nested form.

1. $3x^3 - 2x^2 + 5x - 1$
2. $-x^4 - 3x^2 + 4$
3. $x^5 + 2x^4 - 2x - 3$
4. $2x^4 - x^3 + x + 7$
5. $2x^4 - x^2 + x + 4$
6. $x^6 - 2x^5 - x^3 + 1$

In Exercises 7–18 find a root of the polynomial equation in the stated interval by the method of successive digits to two decimal places.

7. $2x^4 - x^3 + 2x - 2 = 0 \quad [-2, -1]$
8. $3x^2 - 2x^2 + 5x + 4 = 0 \quad [-1, 0]$
9. $x^5 - 3x^3 + x^2 - 3 = 0 \quad [1, 2]$
10. $-x^4 + 3x^2 + 5 = 0 \quad [2, 3]$
11. $x^6 - 3x^3 + x^2 - 2 = 0 \quad [1, 2]$
12. $2x^5 - x^4 + 3x^2 - 6 = 0 \quad [1, 2]$
13. $-2x^3 - x^2 + 3x - 4 = 0 \quad [-2, -1]$
14. $2x^4 - 3x^3 + x^2 - 1 = 0 \quad [-1, 0]$
15. $2x^5 - 3x^2 - 5 = 0 \quad [1, 2]$
16. $2x^3 + 3x^2 - 2x + 4 = 0 \quad [-3, -2]$
17. $2x^5 - x^4 + x^2 - 3 = 0 \quad [1, 2]$
18. $2x^3 - 2x^2 + 3x - 2 = 0 \quad [0, 1]$

In Exercises 19–30 repeat Exercises 7–18 using the method of bisection.

3.4 THE RATIONAL ROOT THEOREM AND THE DEPRESSED EQUATION

When the coefficients of a polynomial are all integers, a systematic search for the *rational* roots is possible by using the following theorem.

Rational Root Theorem	If the coefficients of the polynomial

$$P(x) = a_n x^n + a_{n-1} x^{n-1} + \cdots + a_1 x + a_0, \qquad a_n \neq 0$$

are all integers and p/q is a rational root, in lowest terms, then
(i) p is a factor of the constant term a_0, and
(ii) q is a factor of the leading coefficient a_n.

PROOF OF RATIONAL ROOT THEOREM (Optional)

Since p/q is a root of $P(x)$, then $P(p/q) = 0$. Thus,

$$a_n\left(\frac{p}{q}\right)^n + a_{n-1}\left(\frac{p}{q}\right)^{n-1} + \cdots + a_1\left(\frac{p}{q}\right) + a_0 = 0 \tag{1}$$

Multiplying Equation (1) by q^n, we have

$$a_n p^n + a_{n-1}p^{n-1}q + \cdots + a_1 pq^{n-1} + a_0 q^n = 0 \tag{2}$$

or

$$a_n p^n + a_{n-1}p^{n-1}q + \cdots + a_1 pq^{n-1} = -a_0 q^n \tag{3}$$

Taking the common factor p out of the left-hand side of Equation (3) yields

$$p(a_n p^{n-1} + a_{n-1}p^{n-2}q + \cdots + a_1 q^{n-1}) = -a_0 q^n \tag{4}$$

Since $a_1, a_2, \ldots, a_n, p$, and q are all integers, the quantity in parentheses in the left-hand side of Equation (4) is an integer. Division of the left-hand side by p results in an integer, and we conclude that p must also be a factor of the right-hand side, $-a_0 q^n$. But p and q have no common factors since, by hypothesis, p/q is in lowest terms. Hence, p must be a factor of a_0; thus we have proved part (i) of the Rational Root Theorem.

We may also rewrite Equation (2) in the form

$$q(a_{n-1}p^{n-1} + a_{n-2}p^{n-2}q + \cdots + a_1 pq^{n-2} + a_0 q^{n-1}) = -a_n p^n \tag{5}$$

An argument similar to the preceding one now establishes part (ii) of the theorem.

EXAMPLE 1 Rational Roots of a Polynomial Equation

Find the rational roots of the equation

$$8x^4 - 2x^3 + 7x^2 - 2x - 1 = 0$$

SOLUTION

If p/q is a rational root in lowest terms, then p is a factor of 1 and q is a factor of 8. We can now list the possibilities:

possible numerators: ± 1 (the factors of 1)

possible denominators: $\pm 1, \pm 2, \pm 4, \pm 8$ (the factors of 8)

possible rational roots: $\pm 1, \pm\frac{1}{2}, \pm\frac{1}{4}, \pm\frac{1}{8}$

Synthetic division can be used to test if these numbers are roots. Trying $x = 1$ and $x = -1$, we find that the remainder is not zero and they are therefore not roots. Trying $\frac{1}{2}$, we have

$$
\begin{array}{r|rrrrr}
\frac{1}{2} & 8 & -2 & 7 & -2 & -1 \\
  &   & 4 & 1 & 4 & 1 \\
\hline
  & 8 & 2 & 8 & 2 & 0 \\
\end{array}
$$

which demonstrates that $\frac{1}{2}$ is a root. Similarly,

$$-\tfrac{1}{4}\Big|\quad \begin{array}{rrrrr} 8 & -2 & 7 & -2 & -1 \\ & -2 & 1 & -2 & 1 \\ \hline 8 & -4 & 8 & -4 & \big| \ \ 0 \end{array}$$

which shows that $-\frac{1}{4}$ is also a root. The student may verify that none of the other possible rational roots will result in a zero remainder when synthetic division is employed.

∎

PROGRESS CHECK
Find the rational roots of the equation

$$9x^4 - 12x^3 + 13x^2 - 12x + 4 = 0$$

ANSWER
$\frac{2}{3}, \frac{2}{3}$

In Chapter 1 we discussed number systems and said that numbers such as $\sqrt{2}$ and $\sqrt{3}$ were irrational. The Rational Root Theorem provides a direct means of verifying that this is indeed so.

EXAMPLE 2   Applying the Rational Root Theorem
Prove that $\sqrt{3}$ is not a rational number.

SOLUTION
If we let $x = \sqrt{3}$, then $x^2 = 3$ or $x^2 - 3 = 0$. By the Rational Root Theorem, the only possible rational roots are $\pm 1, \pm 3$. Synthetic division can be used to show that none of these are roots. However, $\sqrt{3}$ is a root of $x^2 - 3 = 0$. Hence, $\sqrt{3}$ is not a rational number.

∎

THE DEPRESSED EQUATION   By using the Factor Theorem we can show that there is a close relationship between the factors and the roots of the polynomial $P(x)$. By definition, r is a root of $P(x)$ if and only if $P(r) = 0$. But the Factor Theorem tells us that $P(r) = 0$ if and only if $x - r$ is a factor of $P(x)$. This leads to the following alternative statement of the Factor Theorem.

Factor Theorem	A polynomial $P(x)$ has a root r if and only if $x - r$ is a factor of $P(x)$.

EXAMPLE 3   Applying the Factor Theorem
Find a polynomial $P(x)$ of degree 3 whose roots are $-1, 1,$ and -2.

SOLUTION
By the Factor Theorem, $x + 1, x - 1,$ and $x + 2$ are factors of $P(x)$. The product

$$P(x) = (x + 1)(x - 1)(x + 2) = x^3 + 2x^2 - x - 2$$

SOLVING POLYNOMIAL EQUATIONS

Cardan's Formula

Cardano provided this formula for one root of the cubic equation

$$x^3 + bx + c = 0$$

$$x = \sqrt[3]{\sqrt{\frac{b^3}{27} + \frac{c^2}{4}} - \frac{c}{2}}$$
$$- \sqrt[3]{\sqrt{\frac{b^3}{27} + \frac{c^2}{4}} + \frac{c}{2}}$$

Try it for the cubics

$$x^3 - x = 0$$
$$x^3 - 1 = 0$$
$$x^3 - 3x + 2 = 0$$

The quadratic formula provides us with the solutions of a polynomial equation of second degree. How about polynomial equations of third degree? of fourth degree? of fifth degree?

The search for formulas expressing the roots of polynomial equations in terms of the coefficients of the equations intrigued mathematicians for hundreds of years. A method for finding the roots of polynomial equations of degree 3 was published around 1535 and is known as Cardan's formula despite the possibility that Girolamo Cardano stole the result from his friend Nicolo Tartaglia. Shortly afterward a method that is attributed to Ferrari was published for solving polynomial equations of degree 4.

The next 250 years were spent in seeking formulas for the roots of polynomial equations of degree 5 or higher—without success. Finally, early in the nineteenth century, the Norwegian mathematician N. H. Abel and the French mathematician Evariste Galois proved that *no such formulas exist*. Galois's work on this problem was completed a year before his death in a duel at age 20. His proof, using the new concepts of group theory, was so advanced that his teachers wrote it off as being unintelligible gibberish.

is a polynomial of degree 3 with the desired roots. Note that multiplying $P(x)$ by any nonzero real number results in another polynomial that has the same roots. For example, the polynomial

$$5 \cdot P(x) = 5x^3 + 10x^2 - 5x - 10$$

also has -1, 1, and -2 as its roots. Thus, the answer is not unique.

■

PROGRESS CHECK
Find a polynomial $P(x)$ of degree 3 whose roots are 2, 4, and -3.

ANSWER
$x^3 - 3x^2 - 10x + 24$

If we know that r is a root of $P(x)$, we may write

$$P(x) = (x - r)Q(x)$$

If r_1 is a root of $Q(x)$, then $Q(r_1) = 0$ and

$$P(r_1) = (r_1 - r)Q(r_1) = (r_1 - r) \cdot 0 = 0$$

which shows that r_1 is also a root of $P(x)$. We call $Q(x) = 0$ the **depressed equation,** since $Q(x)$ is of lower degree than $P(x)$. In the next example we illustrate the use of the depressed equation in finding the roots of a polynomial.

EXAMPLE 4 Finding and Using the Depressed Equation

If 4 is a root of the polynomial $P(x) = x^3 - 8x^2 + 21x - 20$, find the other roots.

SOLUTION

Since 4 is a root of $P(x)$, $x - 4$ is a factor of $P(x)$. Therefore,

$$P(x) = (x - 4)Q(x)$$

To find the depressed equation, we compute $Q(x) = P(x)/(x - 4)$ by synthetic division

$$
\begin{array}{r|rrrr}
4 & 1 & -8 & 21 & -20 \\
  &   & 4 & -16 & 20 \\
\hline
  & 1 & -4 & 5 & 0
\end{array}
$$

$\underbrace{\qquad\qquad\qquad}$ remainder
coefficients
of $Q(x)$

The depressed equation is

$$x^2 - 4x + 5 = 0$$

Using the quadratic formula, the roots of the depressed equation are found to be $2 + i$ and $2 - i$. The roots of $P(x)$ are then seen to be 4, $2 + i$, and $2 - i$.

■

PROGRESS CHECK

If -2 is a root of the polynomial $P(x) = x^3 - 7x - 6$, find the remaining roots.

ANSWER

$-1, 3$

We can combine the Rational Root Theorem and the depressed equation to give us even greater power in seeking the roots of a polynomial.

EXAMPLE 5 Synthesizing the Theorems on Roots of Polynomials

Find the rational roots of the polynomial equation

$$8x^5 + 12x^4 + 14x^3 + 13x^2 + 6x + 1 = 0$$

SOLUTION

Since the coefficients of the polynomial are all integers, we may use the Rational Root Theorem to list the possible rational roots.

         possible numerators: ± 1 (factors of 1)

         possible denominators: $\pm 1, \pm 2, \pm 4, \pm 8$ (factors of 8)

         possible rational roots: $\pm 1, \pm \frac{1}{2}, \pm \frac{1}{4}, \pm \frac{1}{8}$

■

Trying $+1$, -1, and $+\frac{1}{2}$, we find that they are not roots. Testing $-\frac{1}{2}$ by synthetic division results in a remainder of zero.

TRANSCENDENTAL NUMBERS

Theorem: Every rational number p/q is algebraic. Proof: The number p/q is a root of the equation

$$qx - p = 0$$

since

$$q\left(\frac{p}{q}\right) - p = p - p = 0$$

Further, by definition of a rational number, q and p are integers and $q \neq 0$. So p/q is a root of a polynomial equation with integer coefficients and is therefore algebraic.

A real number that is a root of some polynomial equation with integer coefficients is said to be **algebraic.** We see that $\frac{2}{3}$ is algebraic since it is the root of the equation $3x - 2 = 0$; $\sqrt{2}$ is also algebraic, since it satisfies the equation $x^2 - 2 = 0$.

Note that every real number a satisfies the equation $x - a = 0$; that is, it satisfies a polynomial equation with *real* coefficients. But to be algebraic the number a must satisfy a polynomial equation with *integer* coefficients. To show that a real number a is *not* algebraic we must demonstrate that there is *no* polynomial equation with integer coefficients that has a as one of its roots. Although this appears to be an impossible task, it was performed in 1844 when Joseph Liouville exhibited specific examples of such numbers, called **transcendental** numbers. Consequently, Georg Cantor (1845–1918), in his brilliant work on infinite sets, provided a more general proof of the existence of transcendental numbers.

You are already familiar with a transcendental number: the number π is not a root of any polynomial equation with integer coefficients.

$$
\begin{array}{r|rrrrrr}
-\frac{1}{2} & 8 & 12 & 14 & 13 & 6 & 1 \\
& & -4 & -4 & -5 & -4 & 1 \\
\hline
& 8 & 8 & 10 & 8 & 2 & 0 \\
\end{array}
$$

coefficients of depressed equation

Rather than return to the original equation to continue the search, we will use the depressed equation

$$8x^4 + 8x^3 + 10x^2 + 8x + 2 = 0$$

The values $+1$, -1, and $+\frac{1}{2}$ have been eliminated, but the value $-\frac{1}{2}$ must be tried again.

$$
\begin{array}{r|rrrrr}
-\frac{1}{2} & 8 & 8 & 10 & 8 & 2 \\
& & -4 & -2 & -4 & -2 \\
\hline
& 8 & 4 & 8 & 4 & 0 \\
\end{array}
$$

coefficients of depressed equation

Since the remainder is zero, $-\frac{1}{2}$ is again a root. This illustrates an important point: A rational root may be a multiple root! Applying the same technique to the resulting depressed equation

$$8x^3 + 4x^2 + 8x + 4 = 0$$

we see that $-\frac{1}{2}$ is once again a root.

$$\begin{array}{r|rrrr} -\tfrac{1}{2} & 8 & 4 & 8 & 4 \\ & & -4 & 0 & -4 \\ \hline & 8 & 0 & 8 & | \ 0 \end{array}$$

$$\underbrace{}$$

coefficients of depressed equation

The final depressed equation

$$8x^2 + 8 = 0 \qquad \text{or} \qquad x^2 + 1 = 0$$

has the roots $\pm i$. Thus, the original equation has the rational roots

$$-\frac{1}{2}, \ -\frac{1}{2}, \ -\frac{1}{2}$$

PROGRESS CHECK

Find all roots of the polynomial

$$P(x) = 9x^4 - 3x^3 + 16x^2 - 6x - 4$$

ANSWER

$\frac{2}{3}, \ -\frac{1}{3}, \ \pm\sqrt{2}i$

EXERCISE SET 3.4

In Exercises 1–10 use the Rational Root Theorem to find the rational roots of the given equation.

1. $x^3 - 2x^2 - 5x + 6 = 0$
2. $3x^3 - x^2 - 3x + 1 = 0$
3. $6x^4 - 7x^3 - 13x^2 + 4x + 4 = 0$
4. $36x^4 - 15x^3 - 26x^2 + 3x + 2 = 0$
5. $5x^6 - x^5 - 5x^4 + 6x^3 - x^2 - 5x + 1 = 0$
6. $16x^4 - 16x^3 - 29x^2 + 32x - 6 = 0$
7. $4x^4 - x^3 + 5x^2 - 2x - 6 = 0$
8. $6x^4 + 2x^3 + 7x^2 + x + 2 = 0$
9. $2x^5 - 13x^4 + 26x^3 - 22x^2 + 24x - 9 = 0$
10. $8x^5 - 4x^4 + 6x^3 - 3x^2 - 2x + 1 = 0$

In Exercises 11–16 use the given root(s) to help in finding the remaining roots of the equation.

11. $x^3 - 3x - 2 = 0; \quad -1$
12. $x^3 - 7x^2 + 4x + 24 = 0; \quad 3$
13. $x^3 - 8x^2 + 18x - 15 = 0; \quad 5$
14. $x^3 - 2x^2 - 7x - 4 = 0; \quad -1$
15. $x^4 + x^3 - 12x^2 - 28x - 16 = 0; \quad -2$
16. $x^4 - 2x^2 + 1 = 0; \quad 1$ is a double root

In Exercises 17–24 use the Rational Root Theorem and the depressed equation to find all roots of the given equation.

17. $4x^4 + x^3 + x^2 + x - 3 = 0$
18. $x^4 + x^3 + x^2 + 3x - 6 = 0$
19. $5x^5 - 3x^4 - 10x^3 + 6x^2 - 40x + 24 = 0$
20. $12x^4 - 52x^3 + 75x^2 - 16x - 5 = 0$
21. $6x^4 - x^3 - 5x^2 + 2x = 0$
22. $2x^4 - \frac{3}{2}x^3 + \frac{11}{2}x^2 + \frac{23}{2}x + \frac{5}{2} = 0$
23. $2x^4 - x^3 - 28x^2 + 30x - 8 = 0$
24. $12x^4 + 4x^3 - 17x^2 + 6x = 0$

In Exercises 25–28 find the integer value(s) of k for which the given equation has rational roots, and find the roots. (*Hint:* Use synthetic division.)

25. $x^3 + kx^2 + kx + 2 = 0$
26. $x^4 - 4x^3 - kx^2 + 6kx + 9 = 0$
27. $x^4 - 3x^3 + kx^2 - 4x - 1 = 0$
28. $x^3 - 4kx^2 - k^2x + 4 = 0$

29. If $P(x)$ is a polynomial with integer coefficients and the leading coefficient is $+1$ or -1, prove that the rational roots of $P(x)$ are all integers and are factors of the constant term.

30. Prove that $\sqrt{5}$ is not a rational number.

31. If p is a prime, prove that $\sqrt{p}$ is not a rational number.

3.5 RATIONAL FUNCTIONS AND THEIR GRAPHS

We can apply our knowledge of polynomial functions to the study of a slightly more complex class of functions. A function of the form

$$f(x) = \frac{P(x)}{Q(x)} \tag{1}$$

where $P(x)$ and $Q(x)$ are polynomials and $Q(x) \neq 0$ is called a **rational function.** (You will recall that a *rational number* is the quotient of two integers; similarly, a *rational function* is the quotient of two polynomials.) We will assume that the polynomials $P(x)$ and $Q(x)$ have no common factors (other than a constant) and will call such a rational function **irreducible.** (The last example of this section will deal with a reducible rational function.) We will also assume that $Q(x)$ is of degree 1 or higher; otherwise, $Q(x)$ is a constant and $f(x)$ is simply a polynomial.

We will study the behavior of rational functions with the particular objective of sketching their graphs which we will see have characteristics very different than those of the graphs of polynomial functions.

DOMAIN AND INTERCEPTS

It's a simple task to determine the domain of the rational function f of Equation (1) above. Since $P(x)$ and $Q(x)$ are polynomials, they are both defined for all real values of x. The function f, then, can have "problems" only where the denominator is zero. Consequently, the domain of f consists of all real numbers except those for which $Q(x) = 0$.

To find the y-intercepts of the function f, we simply set x equal to 0 and evaluate $y = f(0)$. Should $Q(0) = 0$, then $f(0)$ is undefined and there are no y-intercepts; otherwise, there is precisely one y-intercept.

To find the x-intercepts, we note that $y = f(x)$ can be 0 only if the numerator $P(x)$ is zero. Therefore, the x-intercepts correspond to the roots of the polynomial equation $P(x) = 0$. This can provide an opportunity to employ all the techniques you have learned earlier in this chapter on finding the roots of a polynomial equation! (Since we have assumed that f is irreducible, if $P(r) = 0$, then $Q(r) \neq 0$.)

EXAMPLE 1 Domain and Intercepts
Determine the domain and all intercepts of each irreducible rational function.

(a) $f(x) = \dfrac{x + 1}{x - 1}$ **(b)** $g(x) = \dfrac{x^3 + 2x^2 - 3x}{x^2 - 4}$ **(c)** $h(x) = \dfrac{x^2 - 9}{x^2 + 1}$

SOLUTION
(a) The denominator is 0 when $x = 1$; thus, the domain of f is the set of all real numbers except $x = 1$.

To find the y-intercept, we set $x = 0$ and find $y = f(0) = -1$ is the only y-intercept.

To find the x-intercepts, we set the numerator to 0 and find that $y = f(x) = 0$ when $x = -1$.

(b) The denominator is 0 when $x = 2$ and when $x = -2$. The domain of g is then the set of all real numbers except $x = 2$ and $x = -2$.

To find the y-intercept, we set $x = 0$ and find $y = g(0) = 0$ is the only y-intercept of the graph of g.

To find the x-intercepts, we set the numerator to 0 and find that

$$x^3 + 2x^2 - 3x = 0$$
$$x(x^2 + 2x - 3) = 0$$
$$x(x - 1)(x + 3) = 0$$

has the solutions $x = 0$, $x = 1$, and $x = -3$ which are seen to be the x-intercepts.

(c) Since the denominator $x^2 + 1$ can never be zero (why?), the domain of h is the set of all real numbers.

To find the y-intercept, we set $x = 0$ and find $y = h(0) = -9$ is the only y-intercept of the graph of h.

To find the x-intercepts, we set the numerator $x^2 - 9 = 0$ and find that $x = 3$ and $x = -3$ are the x-intercepts.

■

PROGRESS CHECK

Determine the domain and all intercepts of each rational function.

(a) $S(x) = \dfrac{x - 3}{2x^2 - 3x - 2}$ **(b)** $T(x) = \dfrac{x + 5}{x^4 + x^2}$

ANSWERS

(b) Domain: all real numbers; no y-intercept; x-intercept: $x = -5$

y-intercept: $y = \frac{3}{2}$; x-intercept: $x = 3$

(a) Domain: all real numbers except $x = -\frac{1}{2}$, $x = 2$;

GRAPHING $\dfrac{k}{x}$ AND $\dfrac{k}{x^2}$

We'll begin the study of the graphs of rational functions with examples in which the numerator is a constant.

EXAMPLE 2 Rational Functions with Constant Numerators

Sketch the graphs of the functions

$$f(x) = \frac{1}{x} \qquad \text{and} \qquad g(x) = \frac{1}{x^2}$$

SOLUTION

Domain. Since the denominators in the expressions for both f and g are zero only when $x = 0$, the domain of both f and g is the same: the set of all real numbers other than zero.

TABLE 4 Points for Example 2

x	$\dfrac{1}{x}$	$\dfrac{1}{x^2}$
0.001	1000	1,000,000
0.01	100	10,000
0.1	10	100
1	1	1
2	0.5	0.25
4	0.25	0.06

Intercepts. It is easy to verify that neither function has any x- or y-intercepts.

Symmetry. The graph of f is symmetric with respect to the origin since the equation $y = 1/x$ remains unchanged when x and y are replaced by $-x$ and $-y$, respectively. Similarly, the graph of g is symmetric with respect to the y-axis since the equation $y = 1/x^2$ is unchanged when x is replaced by $-x$. In both cases, we therefore need plot only those points corresponding to positive values of x. We form a table of values for positive values of x (see Table 4) and sketch the branches to the right of the y-axis shown in Figure 8. The other branch is obtained from the symmetry of the graph.

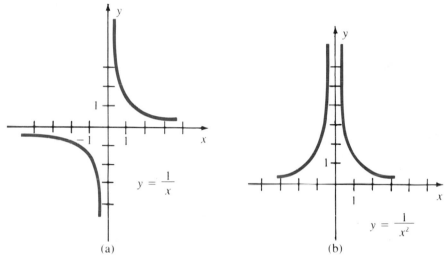

FIGURE 8 Graphs of $1/x$ and $1/x^2$

ASYMPTOTES

The graphs in Figure 8 illustrate an interesting phenomenon: they appear to approach the axes without ever touching them. Such lines play an important role in the graphs of many functions and are defined in this way.

> A line is said to be an **asymptote** of a graph if the graph gets closer and closer to the line as we move farther and farther out along the line.

Note the behavior of the graphs in Figure 8 as x gets closer and closer to 0, that is, as we approach the y-axis. We say that the line $x = 0$ (the y-axis) is a **vertical asymptote** for each of these graphs. Similarly, we note that the line $y = 0$ (the x-axis) is a **horizontal asymptote** in both cases. (We will soon show that the x-axis is a horizontal asymptote of a rational function whenever the numerator is a constant.)

EXAMPLE 3 Using Asymptotes in Graphing
Sketch the graphs of the rational functions

(a) $F(x) = \dfrac{1}{x - 1}$ (b) $G(x) = \dfrac{3}{(x + 2)^2}$

SOLUTION
If you compare these functions with those of Example 2, you see that

$$F(x) = f(x - 1) \qquad \text{and} \qquad G(x) = 3 \cdot g(x - (-2))$$

From our earlier work, this suggests that the graph of F is that of f shifted right 1 unit; similarly, the graph of G is that of g shifted left 2 units and rising more rapidly ("stretched"). The graphs are shown in Figure 9 where it is clear that the vertical asymptotes have been shifted. In both cases we say that the y-axis has been **translated.**

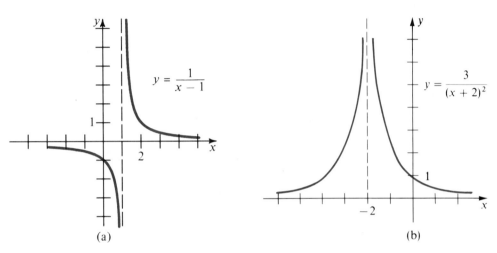

$$y = \frac{1}{x - 1}$$

$$y = \frac{3}{(x + 2)^2}$$

(a)   (b)

FIGURE 9   Graphs for Example 3

◼

Since asymptotes play a key role in graphing rational functions, we need a procedure for locating them. The graphs of Figure 9 imply that a study of horizontal asymptotes requires examining the behavior of the function as x approaches $+\infty$ and $-\infty$, that is, as $|x|$ becomes very large. Recall that the expression

$$\frac{k}{x^n}$$

where n is a positive integer and k is a constant, will become very small as $|x|$ becomes very large; that is, k/x^n approaches 0 as $|x|$ approaches $+\infty$. The procedure for determining horizontal asymptotes employs the technique used earlier of factoring out the highest power of x to determine the behavior of the function as $|x|$ becomes large.

EXAMPLE 4   Horizontal Asymptote
Determine the horizontal asymptote of the function

$$f(x) = \frac{2x^2 - 5}{3x^2 + 2x - 4}$$

SOLUTION
We illustrate the steps of the procedure.

Horizontal Asymptotes

Step 1. Factor out the highest power of x found in the numerator; factor out the highest power of x found in the denominator.	*Step 1.* $$f(x) = \frac{x^2\left(2 - \dfrac{5}{x^2}\right)}{x^2\left(3 + \dfrac{2}{x} - \dfrac{4}{x^2}\right)}$$
Step 2. Since we are interested in large values of $|x|$, we may cancel common factors in the numerator and denominator.	*Step 2.* $$f(x) = \frac{2 - \dfrac{5}{x^2}}{3 + \dfrac{2}{x} - \dfrac{4}{x^2}}, \qquad x \neq 0$$
Step 3. Let $|x|$ increase. Then all terms of the form k/x^n approach 0 and may be discarded.	*Step 3.* The terms $-\dfrac{5}{x^2}, \dfrac{2}{x}$, and $-\dfrac{4}{x^2}$ approach 0 as $|x|$ approaches $+\infty$.
Step 4. If what remains is a real number c, then $y = c$ is the horizontal asymptote. Otherwise there is no horizontal asymptote.	*Step 4.* Discarding these terms, we have $y = \frac{2}{3}$ as the horizontal asymptote.

■

EXAMPLE 5 Horizontal Asymptote

Determine the horizontal asymptote of the function

$$f(x) = \frac{2x^3 + 3x - 2}{x^2 + 5}$$

if there is one.

SOLUTION

Factoring, we have

$$f(x) = \frac{x^3\left(2 + \dfrac{3}{x^2} - \dfrac{2}{x^3}\right)}{x^2\left(1 + \dfrac{5}{x^2}\right)}$$

$$= \frac{x\left(2 + \dfrac{3}{x^2} - \dfrac{2}{x^3}\right)}{1 + \dfrac{5}{x^2}}, \qquad x \neq 0$$

As $|x|$ increases, the terms $3/x^2$, $-2/x^3$, and $5/x^2$ approach zero and can be discarded. What remains is $2x$, which becomes larger and larger as $|x|$ increases. Thus, there is no horizontal asymptote, and $|y|$ becomes larger and larger as $|x|$ approaches infinity.

■

The following theorem can be proved by utilizing the procedure of Example 4.

Horizontal Asymptote Theorem

The graph of the rational function

$$f(x) = \frac{P(x)}{Q(x)}$$

has a horizontal asymptote if the degree of $P(x)$ is less than or equal to the degree of $Q(x)$.

A more specific version of the Horizontal Asymptote Theorem can be found in Exercises 28 and 29.

 PROGRESS CHECK

Determine the horizontal asymptote of the graph of each function.

(a) $f(x) = \dfrac{x - 1}{2x^2 + 1}$

(b) $g(x) = \dfrac{4x^2 - 3x + 1}{-3x^2 + 1}$

(c) $h(x) = \dfrac{3x^3 - x + 1}{2x^2 - 1}$

ANSWERS

(a) $y = 0$ (b) $y = -\frac{4}{3}$ (c) no horizontal asymptote

---

We next turn to the examination of vertical asymptotes. In Figure 1 we see that the ordinates increase or decrease without limit as the curve approaches a vertical asymptote. We know that this occurs whenever the absolute value of the denominator of a quotient gets smaller and smaller, which leads to the following theorem.

Vertical Asymptote Theorem

The graph of the rational function

$$f(x) = \frac{P(x)}{Q(x)}$$

has a vertical asymptote at $x = r$ if r is a real root of $Q(x)$ but not of $P(x)$.

EXAMPLE 6   Vertical Asymptotes

Determine the vertical asymptotes of the graph of the function

$$T(x) = \frac{2}{x^3 - 2x^2 - 3x}$$

SOLUTION

Factoring the denominator, we have

$$T(x) = \frac{2}{x(x + 1)(x - 3)}$$

and we conclude that $x = 0$, $x = -1$, and $x = 3$ are vertical asymptotes of the graph of T.

■

Note that the graph of a rational function may have many vertical asymptotes but at most one horizontal asymptote.

There remains one more task: determining the behavior of the branches of the graph. To this end we'll make use of the function

$$S(x) = \frac{2}{x^2 - x - 2} = \frac{2}{(x + 1)(x - 2)}$$

The student is urged to verify the following:

■ x-intercepts: none

■ y-intercept: $y = -1$

■ symmetry: none

■ horizontal asymptote: $y = 0$ (x-axis)

■ vertical asymptotes: $x = -1$, $x = 2$

Although this is all very useful, we still need to know from what direction the curve approaches its asymptotes. We'll divide the x-axis into regions based upon the vertical asymptotes and analyze the behavior in each region. When x is just to the left of -1 (slightly less than -1), the terms $(x + 1)$ and $(x - 2)$ are both negative, so that $S(x)$ is positive and growing larger as x approaches -1. The same results are obtained when x is slightly larger than $+2$. Combining this with the facts that (a) the x-axis is an asymptote and (b) there are no x-intercepts, we arrive at Figure 10a.

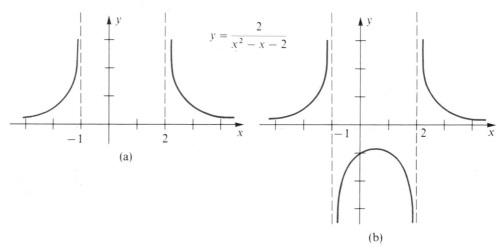

FIGURE 10   **Graph of** $y = \dfrac{2}{x^2 - x - 2}$

Next, we need to examine the behavior when x is between -1 and $+2$. If x is just to the right of -1 (slightly larger than -1), then the factor $(x + 1)$ is positive while the factor $(x - 2)$ is negative, so that $S(x)$ is negative and growing smaller as x approaches -1. The same results are obtained when x is just to the left of $+2$ (slightly less than $+2$). Since there are no x-intercepts and the y-intercept occurs at $y = -1$, this allows us to complete the graph sketched in Figure 10b. (The turning point can be estimated by plotting a few points. It's a very simple exercise in calculus to show that the turning point occurs when $x = \frac{1}{2}$.)

SUMMARY

We now summarize the information that can be gathered in preparation for sketching the graph of a rational function.

- symmetry with respect to the axes and the origin
- x- and y-intercepts
- horizontal asymptote
- vertical asymptotes
- brief table of values including points near the vertical asymptotes

EXAMPLE 7 Graphing a Rational Function

Sketch the graph of

$$f(x) = \frac{x^2}{x^2 - 1}$$

SOLUTION

Symmetry. Replacing x with $-x$ results in the same equation, establishing symmetry with respect to the y-axis.

y-intercept. Evaluating $y = f(0)$ we find that $y = 0$ is an intercept.

x-intercepts. The numerator is 0 when $x = 0$ and the denominator is defined when $x = 0$. The graph must intersect both axes at $(0, 0)$.

Vertical asymptotes. Setting the denominator equal to zero, we find that $x = 1$ and $x = -1$ are vertical asymptotes of the graph of f.

Horizontal asymptotes. We note that

$$f(x) = \frac{x^2}{x^2\left(1 - \dfrac{1}{x^2}\right)} = \frac{1}{1 - \dfrac{1}{x^2}}, \qquad x \neq 0$$

As $|x|$ gets larger and larger, $1/x^2$ approaches 0 and the values of $f(x)$ approach 1. Thus, $y = 1$ is the horizontal asymptote.

Short table of values. Choose values of x on both sides of the asymptote at $x = 1$ and use symmetry to complete the graph.

The graph is shown in Figure 11.

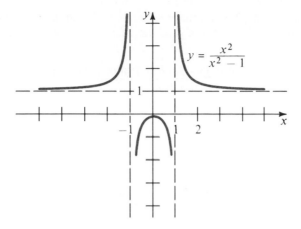

x	y
1/2	-0.33
3/4	-1.29
5/4	2.78
3/2	1.80
2	1.33

FIGURE 11   Graph for Example 7

PROGRESS CHECK
Sketch the graph of

$$f(x) = \frac{x^2 - x - 6}{x^2 - 2x}$$

ANSWER
See Figure 12.

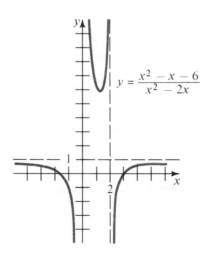

FIGURE 12   Graph of a Rational Function

REDUCIBLE RATIONAL FUNCTIONS

We conclude this section with an example of a reducible rational function, that is, one in which the numerator and denominator have a factor in common other than a constant. Reducible rational functions are often used in calculus courses to illustrate functions that have "holes" and are therefore not continuous.

EXAMPLE 8 Graphing Reducible Rational Functions

Sketch the graph of the function

$$f(x) = \frac{x^2 - 1}{x - 1}$$

SOLUTION

We observe that

$$f(x) = \frac{x^2 - 1}{x - 1} = \frac{(x + 1)(x - 1)}{x - 1} = x + 1, \qquad x \neq 1$$

Thus, the graph of the function f coincides with the straight line $y = x + 1$, with the exception that f is undefined when $x = 1$ (Figure 13).

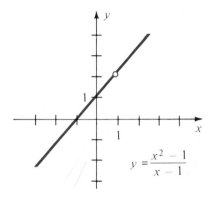

FIGURE 13 Graph for Example 8

■

 PROGRESS CHECK

Sketch the graph of the function

$$f(x) = \frac{4 - x^2}{x + 2}$$

ANSWER

See Figure 14.

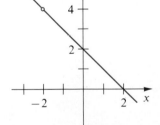

FIGURE 14 Graph of a Reducible Rational Function

EXERCISE SET 3.5

In Exercises 1–6 determine the domain and intercepts of the given function.

1. $f(x) = \dfrac{x^2}{x-1}$

2. $f(x) = \dfrac{x-1}{x^2+x-2}$

3. $g(x) = \dfrac{x^2+1}{x^2-2x}$

4. $g(x) = \dfrac{x^2+2}{x^2-2}$

5. $F(x) = \dfrac{x^2-3}{x^2+3}$

6. $T(x) = \dfrac{3x+2}{2x^3-x^2-x}$

In Exercises 7–21 determine the vertical and horizontal asymptotes of the graph of the given function. Sketch the graph.

7. $f(x) = \dfrac{1}{x-4}$

8. $f(x) = \dfrac{-2}{x-3}$

9. $f(x) = \dfrac{3}{x+2}$

10. $f(x) = \dfrac{-1}{(x-1)^2}$

11. $f(x) = \dfrac{1}{(x+1)^2}$

12. $f(x) = \dfrac{-1}{x^2+1}$

13. $f(x) = \dfrac{x+2}{x-2}$

14. $f(x) = \dfrac{x}{x+2}$

15. $f(x) = \dfrac{2x^2+1}{x^2-4}$

16. $f(x) = \dfrac{x^2+1}{x^2+2x-3}$

17. $f(x) = \dfrac{x^2+2}{2x^2-x-6}$

18. $f(x) = \dfrac{x^2-1}{x+2}$

19. $f(x) = \dfrac{x^2}{4x-4}$

20. $f(x) = \dfrac{x-1}{2x^3-2x}$

21. $f(x) = \dfrac{x^3+4x^2+3x}{x^2-25}$

In Exercise 22–27 determine the domain and sketch the graph of the reducible function.

22. $f(x) = \dfrac{x^2-25}{2x-10}$

23. $f(x) = \dfrac{2x^2-8}{x+2}$

24. $f(x) = \dfrac{2x^2+2x-12}{3x-6}$

25. $f(x) = \dfrac{x^2+2x-8}{2x^2-8x+8}$

26. $f(x) = \dfrac{x+2}{x^2-x-6}$

27. $f(x) = \dfrac{2x}{x^2+x}$

In Exercises 28–29, $f(x) = P(x)/Q(x)$ is an irreducible rational function. Provide a proof for the stated theorem.

28. If the polynomials $P(x)$ and $Q(x)$ are of the same degree, then there is a horizontal asymptote $y = k$ where k is the ratio of the leading coefficients of $P(x)$ and $Q(x)$.

29. If the degree of $P(x)$ is less than the degree of $Q(x)$, then there is a horizontal asymptote at $y = 0$.

3.6
PARTIAL FRACTIONS

Every student of algebra learns how to add fractions to arrive at

$$\frac{2}{x-1} + \frac{3}{x+2} = \frac{5x+1}{(x-1)(x+2)}$$

Strange as it might seem, there is an important application of integral calculus that requires the *reverse* procedure.

Our objective in this section, then, is to write a rational function $P(x)/Q(x)$ as a sum of fractions, each of which is called a **partial fraction.** To begin, we note that we need only consider the situation where $P(x)/Q(x)$ is a **proper fraction,** that is, where the degree of $P(x)$ is less than the degree of $Q(x)$. If it is not, we can first divide through by $Q(x)$. For example, given the **improper fraction** (one that is not a proper fraction)

$$\frac{2x^3-x^2-x+1}{x^2-1}$$

we can use long division to divide through by $x^2 - 1$ to obtain

$$\frac{2x^3 - x^2 - x + 1}{x^2 - 1} = 2x - 1 + \frac{x}{x^2 - 1}$$

We would then work with the proper fraction $x/(x^2 - 1)$.

The procedure for partial-fraction decomposition of the rational function $P(x)/Q(x)$ begins with the factorization of the denominator $Q(x)$. We will show in Chapter 8 that every polynomial with real coefficients can be written as a product of linear and quadratic factors with real coefficients such that the quadratic factors are irreducible (cannot be written as a product of linear factors). For example,

$$Q(x) = x^3 - x^2 + 2x - 2 = (x - 1)(x^2 + 2)$$

where the quadratic factor $x^2 + 2$ cannot be decomposed into linear factors with real coefficients since the equation $x^2 + 2 = 0$ has no real roots.

Having factored $Q(x)$ in this manner, we then collect the repeated factors so that $Q(x)$ is the product of distinct factors of the forms

$$(ax + b)^m \qquad \text{and} \qquad (ax^2 + bx + c)^m$$

The following rules then tell us the format of the partial-fraction decomposition of $P(x)/Q(x)$.

Rules for Partial-Fraction Decomposition

Rule 1: Linear Factors
For each distinct factor of the form $(ax + b)^m$ in the denominator $Q(x)$ introduce the sum of m partial fractions

$$\frac{A_1}{ax + b} + \frac{A_2}{(ax + b)^2} + \cdots + \frac{A_m}{(ax + b)^m}$$

where $A_1, A_2, \ldots, A_m$ are constants.

Rule 2: Quadratic Factors
For each distinct factor of the form $(ax^2 + bx + c)^m$ in the denominator $Q(x)$ introduce the sum of m partial fractions

$$\frac{A_1 x + B_1}{ax^2 + bx + c} + \frac{A_2 x + B_2}{(ax^2 + bx + c)^2} + \cdots + \frac{A_m x + B_m}{(ax^2 + bx + c)^m}$$

where $A_1, A_2, \ldots, A_m$ and $B_1, B_2, \ldots, B_m$ are constants.

What remains is the straightforward (if tedious) algebraic manipulation to determine the constants $A_1, A_2, \ldots, A_m$ and $B_1, B_2, \ldots, B_m$.

EXAMPLE 1 Partial-Fraction Decomposition: Linear Factors

Find the partial-fraction decomposition of $\dfrac{x + 5}{x^2 + x - 2}$.

SOLUTION

Factoring the denominator gives us

$$\frac{x + 5}{x^2 + x - 2} = \frac{x + 5}{(x + 2)(x - 1)}$$

The denominator consists of linear factors. By Rule 1, each linear factor introduces one term:

$$\begin{array}{ccc} \textit{Factor} & x + 2 & x - 1 \\[2mm] \textit{Term} & \dfrac{A}{x + 2} & \dfrac{B}{x - 1} \end{array}$$

The partial-fraction decomposition is then

$$\frac{x + 5}{(x + 2)(x - 1)} = \frac{A}{x + 2} + \frac{B}{x - 1}$$

To solve for the constants A and B, we clear fractions by multiplying both sides by $(x + 2)(x - 1)$ to yield

$$x + 5 = A(x - 1) + B(x + 2)$$

This last equation is easily solved by substituting values of x that will make one unknown drop out. Setting $x = 1$ gives us

$$6 = 3B \qquad \text{or} \qquad B = 2$$

Setting $x = -2$ yields

$$3 = -3A \qquad \text{or} \qquad A = -1$$

Substituting these values for A and B, we arrive at the partial-fraction decomposition

$$\frac{x + 5}{(x + 2)(x - 1)} = \frac{-1}{x + 2} + \frac{2}{x - 1}$$

The student should verify that this is indeed an *identity*.

■

EXAMPLE 2 Partial-Fraction Decomposition: Repeated Linear Factors

Find the partial-fraction decomposition of $\dfrac{x^2 - 2x - 9}{x(x + 3)^2}$.

SOLUTION

The denominator is already in factored form and consists of the linear factor x and the repeated linear factor $x + 3$. By Rule 1, the factor x introduces one term

$$\frac{A}{x}$$

and the factor $x + 3$ introduces two terms (since $m = 2$)

$$\frac{B}{x + 3} \quad \text{and} \quad \frac{C}{(x + 3)^2}$$

The partial-fraction decomposition is then

$$\frac{x^2 - 2x - 9}{x(x + 3)^2} = \frac{A}{x} + \frac{B}{x + 3} + \frac{C}{(x + 3)^2}$$

Clearing fractions by multiplying both sides by $x(x + 3)^2$ yields

$$x^2 - 2x - 9 = A(x + 3)^2 + Bx(x + 3) + Cx \qquad (1)$$

Setting $x = -3$ in Equation (1) allows us to solve for C:

$$6 = -3C \quad \text{or} \quad C = -2$$

Setting $x = 0$ in Equation (1) allows us to solve for A:

$$-9 = 9A \quad \text{or} \quad A = 1$$

Expanding the right-hand side of Equation (1) and collecting terms in the powers of x, we have

$$x^2 - 2x - 9 = (A + B)x^2 + (6A + 3B + C)x + 9A \qquad (2)$$

Substituting $A = -1$ and $C = -2$ in Equation (2) yields

$$x^2 - 2x - 9 = (B - 1)x^2 + (3B - 8)x - 9$$

Equating the coefficients of the terms in x^2, we have

$$1 = B - 1 \quad \text{or} \quad B = 2$$

(The same result is obtained by equating the coefficients of the terms in x.) The partial-fraction decomposition is

$$\frac{x^2 - 2x - 9}{x(x + 3)^2} = -\frac{1}{x} + \frac{2}{x + 3} - \frac{2}{(x + 3)^2}$$

∎

WARNING

For the rational function

$$\frac{2x - 1}{x^2(x^2 + 1)}$$

the factor x^2 in the denominator must be treated as the repeated *linear* factor $(x - 0)^2$. The partial-fraction decomposition has the structure

$$\frac{2x - 1}{x^2(x^2 + 1)} = \frac{A}{x} + \frac{B}{x^2} + \frac{Cx + D}{x^2 + 1}$$

EXAMPLE 3 Linear and Quadratic Factors

Find the partial-fraction decomposition of $\dfrac{x^2 - 5x + 1}{2x^3 - x^2 + 2x - 1}$.

SOLUTION

The first task is to factor the denominator. By the Factor Theorem, $x - r$ is a factor of

$$Q(x) = 2x^3 - x^2 + 2x - 1$$

if and only if r is a root of $Q(x)$. The Rational Root Theorem tells us that the possible rational roots are $1, -1, \frac{1}{2},$ and $-\frac{1}{2}$. Using the condensed form of synthetic division to test these possible roots

$$
\begin{array}{r|rrrr}
 & 2 & -1 & 2 & -1 \\
\hline
1 & 2 & 1 & 3 & 2 \\
\hline
-1 & 2 & -3 & 5 & -6 \\
\hline
\dfrac{1}{2} & 2 & 0 & 2 & 0 \\
\end{array}
$$

coefficients of depressed equation

we see that $Q(\frac{1}{2}) = 0$, so $x = \frac{1}{2}$ is a root of $Q(x)$. Consequently, $(x - \frac{1}{2})$, or $(2x - 1)$, is a factor of $Q(x)$, and $2x^2 + 2 = 0$ is seen to be the depressed equation.

By Rule 1, the factor $2x - 1$ will introduce a term of the form

$$\frac{A}{2x - 1}$$

and by Rule 2, the factor $x^2 + 1$ will introduce a term of the form

$$\frac{Bx + C}{x^2 + 1}$$

Then

$$\frac{x^2 - 5x + 1}{(2x - 1)(x^2 + 1)} = \frac{A}{2x - 1} + \frac{Bx + C}{x^2 + 1}$$

Multiplying by $(2x - 1)(x^2 + 1)$, we have

$$x^2 - 5x + 1 = A(x^2 + 1) + (Bx + C)(2x - 1)$$

To find A, B, and C, we first set $2x - 1 = 0$ or $x = \frac{1}{2}$:

$$-\frac{5}{4} = \frac{5}{4}A \qquad \text{or} \qquad A = -1$$

Equating the coefficients of x^2 yields

$$1 = A + 2B$$

Substituting $A = -1$,

$$1 = -1 + 2B$$
$$B = 1$$

Equating the coefficients of the constant term yields

$$1 = A - C$$
$$1 = -1 - C$$
$$C = -2$$

The partial-fraction decomposition is therefore

$$\frac{x^2 - 5x + 1}{2x^3 - x^2 + 2x - 1} = \frac{-1}{2x - 1} + \frac{x - 2}{x^2 + 1}$$

∎

EXAMPLE 4   Synthesizing the Method

Find the partial-fraction decomposition of $\dfrac{x^2 - 2x}{(x + 2)(x^2 + 4)^2}$.

SOLUTION
The linear factor $x + 2$ introduces one term of the form

$$\frac{A}{x + 2}$$

The quadratic factor $x^2 + 4$ has no real roots. By Rule 2, this irreducible quadratic factor will introduce two terms (since $m = 2$) of the form

$$\frac{Bx + C}{x^2 + 4} \qquad \text{and} \qquad \frac{Dx + E}{(x^2 + 4)^2}$$

We then have to solve

$$\frac{x^2 - 2x}{(x + 2)(x^2 + 4)^2} = \frac{A}{x + 2} + \frac{Bx + C}{x^2 + 4} + \frac{Dx + E}{(x^2 + 4)^2}$$

for values of A, B, C, D, and E that will produce an identity. Multiplying by $(x + 2)(x^2 + 4)^2$ yields

$$x^2 - 2x = A(x^2 + 4)^2 + (Bx + C)(x + 2)(x^2 + 4) + (Dx + E)(x + 2) \qquad (3)$$

Setting $x = -2$ enables us to solve for A:

$$8 = 64A \qquad \text{or} \qquad A = \frac{1}{8}$$

A methodical way to solve for B, C, D, and E is to successively equate the coefficients of the powers x^4, x^3, and x^2 in the left-hand and right-hand sides of Equation (3).

$$\text{Coefficients of } x^4\colon 0 = A + B \qquad \text{or} \qquad B = -A = -\frac{1}{8}$$

$$\text{Coefficients of } x^3\colon 0 = 2B + C \qquad \text{or} \qquad C = -2B = \frac{1}{4}$$

$$\text{Coefficients of } x^2: 1 = 8A + 4B + 2C + D$$

$$1 = 1 - \frac{1}{2} + \frac{1}{2} + D$$

$$D = 0$$

To find E, we may equate the coefficients of x or of the constant term. Choosing the latter approach,

$$0 = 16A + 8C + 2E$$

$$0 = 2 + 2 + 2E$$

$$E = -2$$

The partial-fraction decomposition is

$$\frac{x^2 - 2x}{(x + 2)(x^2 + 4)^2} = \frac{\frac{1}{8}}{x + 2} + \frac{-\frac{1}{8}x + \frac{1}{4}}{x^2 + 4} + \frac{-2}{(x^2 + 4)^2}$$

∎

EXERCISE SET 3.6

Find the partial-fraction decomposition of each of the following.

1. $\dfrac{2x - 11}{(x + 2)(x - 3)}$

2. $\dfrac{1}{x^2 + 3x + 2}$

3. $\dfrac{3x - 2}{6x^2 - 5x + 1}$

4. $\dfrac{2x + 1}{x^2 - 1}$

5. $\dfrac{x^2 + x + 2}{x^3 - x}$

6. $\dfrac{3x - 14}{(x - 3)(x^2 - 4)}$

7. $\dfrac{3x - 2}{x^3 + 2x^2}$

8. $\dfrac{4x^2 - 5x + 2}{2x^3 - x^2}$

9. $\dfrac{x^2 - x + 2}{(x - 1)(x + 1)^2}$

10. $\dfrac{3x - 1}{x(x - 1)^3}$

11. $\dfrac{1 - 2x}{x^3 + 4x}$

12. $\dfrac{x^2 - 2x + 1}{x^3 + 2x^2 + 2x}$

13. $\dfrac{2x^3 - x^2 + x}{(x^2 + 3)^2}$

14. $\dfrac{x^2 + 2x + 10}{x^3 - 2x^2 + 2x - 4}$

15. $\dfrac{-x}{x^3 - 2x^2 - 4x - 1}$

16. $\dfrac{x^3 - 2x^2 + 1}{x^4 + 2x^2 + 1}$

17. $\dfrac{x^4 - x^2 - 9}{(x + 1)(x^2 + 2)^2}$

18. $\dfrac{-x^3 - x^2 + 5x + 1}{(2x - 1)(x^2 + 1)^2}$

19. $\dfrac{x^4 + x^3 + x^2 + 3x - 2}{(x + 1)(x^2 + 1)}$

20. $\dfrac{x^3 - 5x + 5}{(x - 1)^3}$

TERMS AND SYMBOLS

KEY IDEAS FOR REVIEW

Topic	Page	Key Idea
Polynomial division	132	Polynomial division results in a quotient and a remainder, both of which are polynomials. If the remainder is not zero, then the degree of the remainder is less than the degree of the divisor.
Synthetic division	133	Synthetic division is a quick way to divide a polynomial by a first-degree polynomial $x - r$, where r is a real constant.
Remainder Theorem	135	If a polynomial $P(x)$ is divided by $x - r$, the remainder is $P(r)$.
Factor Theorem	136	A polynomial $P(x)$ has a root r if and only if $x - r$ is a factor of $P(x)$.
Lead term dominance	141	For large values of $|x|$, a polynomial function is dominated by its lead term, that is, the lead term determines the sign of the function.
Graph of a polynomial function	139	The graph of a polynomial function of degree n is continuous, has no corners, and has at most $n - 1$ turning points.
"ends" of the graph	142	For a polynomial of odd degree, the "ends" of the graph will extend indefinitely in opposite directions. For a polynomial of even degree, the "ends" of the graph will both extend upward or will both extend downward.
Intermediate Value Theorem	140	If P is a polynomial function and $P(a) \neq P(b)$, then, in the interval $[a, b]$, P assumes every value between $P(a)$ and $P(b)$.
Approximating roots of polynomials	146	The methods of successive digits and bisection can be used to approximate the roots of a polynomial equation.
nested form	147	Nested form is very convenient for automating polynomial evaluation.
Rational Root Theorem	150	If p/q is a rational root (in lowest terms) of the polynomial $P(x)$ with integer coefficients, then p is a factor of the constant term a_0 of $P(x)$ and q is a factor of the leading coefficient a_n of $P(x)$.
listing possible rational roots	151	If $P(x)$ has integer coefficients, then the Rational Root Theorem enables us to list all possible rational roots of $P(x)$. Synthetic division can then be used to test these potential rational roots, since r is a root if and only if the remainder is zero, that is, if and only if $P(r) = 0$.
Depressed equation	152	*If r is a real root of the polynomial $P(x)$, then the roots of the depressed equation are the other roots of $P(x)$. The depressed equation can be found by using synthetic division.*
Horizontal Asymptote Theorem	162	A rational function has a unique horizontal asymptote if the degree of the numerator is less than or equal to the degree of the denominator.
Vertical Asymptote Theorem	162	**An irreducible rational function has a vertical asymptote corresponding to each root of the denominator.**
Graphing rational functions	158	Always determine the intercepts, symmetry, horizontal and vertical asymptotes of a rational function before attempting to sketch its graph.
reducible rational function	165	The graph of a reducible rational function has a "hole" corresponding to each unique common factor of the numerator and denominator.
Partial fractions	167	A proper fraction can always be written as a sum of partial fractions whose denominators are of the form $$(ax + b)^m \qquad \text{and} \qquad (ax^2 + bx + c)^m$$

REVIEW EXERCISES

Solutions to exercises whose numbers are in color are in the Solutions section in the back of the book.

3.1 In Exercises 1 and 2 use synthetic division to find the quotient $Q(x)$ and the constant remainder R when the first polynomial is divided by the second polynomial.
  1. $2x^3 + 6x - 4,\ x - 1$
  2. $x^4 - 3x^3 + 2x - 5,\ x + 2$

  In Exercises 3 and 4 use synthetic division to find $P(2)$ and $P(-1)$.
  3. $7x^3 - 3x^2 + 2$ **4.** $x^5 - 4x^3 + 2x$

  In Exercises 5 and 6 use the Factor Theorem to show that the second polynomial is a factor of the first polynomial.
  5. $2x^4 + 4x^3 + 3x^2 + 5x - 2,\ x + 2$

  6. $2x^3 - 5x^2 + 6x - 2,\ x - \dfrac{1}{2}$

3.2 In Exercises 7 and 8 determine the behavior of the graph of the given polynomial function for large values of $|x|$.
  7. $P(x) = -2x^5 + 27x^2 + 100$
  8. $P(x) = 4x^3 - 10{,}000$

3.4 In Exercises 9–11 find all the rational roots of the given equation.

  9. $6x^3 - 5x^2 - 33x - 18 = 0$
  10. $6x^4 - 7x^3 - 19x^2 + 32x - 12 = 0$
  11. $x^4 + 3x^3 + 2x^2 + x - 1 = 0$

  In Exercises 12–14 use the given root to assist in finding the remaining roots of the equation.

  12. $2x^3 - x^2 - 13x - 6 = 0;\ -2$
  13. $x^3 - 2x^2 - 9x + 4 = 0;\ 4$
  14. $2x^4 - 15x^3 + 34x^2 - 19x - 20 = 0;\ -\dfrac{1}{2}$

3.5 In Exercises 15 and 16 sketch the graph of the given function.

  15. $f(x) = \dfrac{x}{x+1}$ **16.** $f(x) = \dfrac{x^2}{x+1}$

3.6 In Exercises 17–19 find the partial-fraction decomposition of the given rational function.

  17. $\dfrac{8 - x}{2x^2 + 3x - 2}$ **18.** $\dfrac{3x^3 + 5x - 1}{(x^2 + 1)^2}$

  19. $\dfrac{2x^3 - 3x^2 + 4x - 2}{(x - 1)^2}$

PROGRESS TEST 3A

1. Find the quotient and remainder when $2x^4 - x^2 + 1$ is divided by $x^2 + 2$.

2. Use synthetic division to find the quotient and remainder when $3x^4 - x^3 - 2$ is divided by $x + 2$.

3. If $P(x) = x^3 - 2x^2 + 7x + 5$, use synthetic division to find $P(-2)$.

4. Determine the remainder when $4x^5 - 2x^4 - 5$ is divided by $x + 2$.

5. Use the Factor Theorem to show that $x - 3$ is a factor of
$$2x^4 - 9x^3 + 9x^2 + x - 3$$

Problems 6–8 refer to the polynomial function
$$P(x) = -2x^9 + 3x^6 + 200$$

6. Describe the behavior of the graph of $P(x)$ for large positive values of x.

7. Describe the behavior of the graph of $P(x)$ for large negative values of x.

8. Determine the maximum number of extreme points of the graph of $P(x)$.

In Problems 9 and 10 find all rational roots of the given equation.

9. $6x^3 - 17x^2 + 14x + 3 = 0$

10. $2x^5 - x^4 - 4x^3 + 2x^2 + 2x - 1 = 0$

In Problems 11 and 12 use the given root to help in finding the remaining roots of the equation.

11. $4x^3 - 3x + 1 = 0;\ -1$

12. $x^4 - x^2 - 2x + 2 = 0;\ 1$

13. Sketch the graph of the function $f(x) = \dfrac{x^2 + 2}{x^2 - 1}$.

14. Find the partial-fraction decomposition of
$$\frac{x - 12}{x^2 + x - 6}$$

PROGRESS TEST 3B

1. Find the quotient and remainder when $3x^5 + 2x^3 - x^2 - 2$ is divided by $2x^2 - x - 1$.

2. Use synthetic division to find the quotient and remainder when $-2x^3 + 3x^2 - 1$ is divided by $x - 1$.

3. If $P(x) = 2x^4 - 2x^3 + x - 4$, use synthetic division to find $P(-1)$.

4. Determine the remainder when $3x^4 - 5x^3 + 3x^2 + 4$ is divided by $x - 2$.

5. Use the Factor Theorem to show that $x + 2$ is a factor of $x^3 - 4x^2 - 9x + 6$.

Problems 6–8 refer to the polynomial function

$$P(x) = -4x^4 + 1000x - 1$$

6. Describe the behavior of the graph of $P(x)$ for large positive values of x.

7. Describe the behavior of the graph of $P(x)$ for large negative values of x.

8. Find the maximum number of turning points of the graph of $P(x)$.

In Problems 9 and 10 find all rational roots of the given equation.

9. $3x^3 + 7x^2 - 4 = 0$

10. $4x^4 - 4x^3 + x^2 - 4x - 3 = 0$

In Problems 11 and 12 use the given root(s) to help in finding the remaining roots of the equation.

11. $x^3 - x^2 - 8x - 4 = 0$; -2

12. $x^4 - 3x^3 - 22x^2 + 68x - 40 = 0$; $2, 5$

13. Sketch the graph of the function $f(x) = \dfrac{2x}{x^2 - 1}$.

14. Find the partial-fraction decomposition of

$$\frac{5x^2 - x - 2}{x^3 + x^2}$$

CHAPTER 4

EXPONENTIAL AND LOGARITHMIC FUNCTIONS

Thus far in our study of algebra we have dealt primarily with functions that are themselves polynomials or result from some form of combining polynomials. Mathematicians make use of many other types of functions, amongst which the exponential and logarithmic functions that we will study in this chapter are of extraordinary importance.

Exponential and logarithmic functions are related in that they are a pair of inverse functions, a concept that we will explore in the opening section.

Exponential functions arise in nature and are useful in chemistry, biology, and economics, as well as in mathematics and engineering. We will look at the application of exponential functions to the fields of finance and biology. We will see that exponential functions enable us to easily calculate compound interest and to describe the growth rate of bacteria in a culture medium.

Logarithms can be viewed as another way of writing exponents. Historically, logarithms have been used to simplify calculations; in fact, the slide rule, a device long favored by engineers, is based on logarithmic scales. In today's world of sophisticated scientific calculators and readily available computers, the need for manipulating logarithms for calculating has been eliminated. However, the concept of logarithmic functions and the study of their properties is basic to entering into advanced studies in mathematics.

4.1
COMBINING
FUNCTIONS; INVERSE
FUNCTIONS

Functions such as

$$f(x) = x^2 \qquad g(x) = x - 1$$

can be combined by the usual operations of addition, subtraction, multiplication, and division. Using these functions f and g, we can form

$$(f + g)(x) = f(x) + g(x) = x^2 + x - 1$$
$$(f - g)(x) = f(x) - g(x) = x^2 - (x - 1) = x^2 - x + 1$$
$$(f \cdot g)(x) = f(x) \cdot g(x) = x^2(x - 1) = x^3 - x^2$$
$$\left(\frac{f}{g}\right)(x) = \frac{f(x)}{g(x)} = \frac{x^2}{x - 1}$$

In each case we have combined two functions f and g to form a new function. Note, however, that the domain of the new functions need not be the same as the domain of either of the original functions. The function formed by division in the above example has as its domain the set of all real numbers x except $x = 1$, since we cannot divide by 0. On the other hand, the original functions $f(x) = x^2$ and $g(x) = x - 1$ are both defined at $x = 1$.

EXAMPLE 1   Combining Functions
Given $f(x) = x - 4$, $g(x) = x^2 - 4$, find the following.
(a) $(f + g)(x)$ (b) $(f - g)(x)$ (c) $(f \cdot g)(x)$

(d) $\left(\frac{f}{g}\right)(x)$ (e) the domain of $\left(\frac{f}{g}\right)(x)$

SOLUTION
(a) $(f + g)(x) = f(x) + g(x) = x - 4 + x^2 - 4 = x^2 + x - 8$
(b) $(f - g)(x) = f(x) - g(x) = x - 4 - (x^2 - 4) = -x^2 + x$
(c) $(f \cdot g)(x) = f(x) \cdot g(x) = (x - 4)(x^2 - 4) = x^3 - 4x^2 - 4x + 16$
(d) $\left(\frac{f}{g}\right)(x) = \frac{f(x)}{g(x)} = \frac{x - 4}{x^2 - 4}$

(e) The domain of $\left(\frac{f}{g}\right)(x)$ must exclude values of x for which $x^2 - 4 = 0$.

Thus, the domain consists of the set of all real numbers except 2 and -2.

PROGRESS CHECK
Given $f(x) = 2x^2$, $g(x) = x^2 - 5x + 6$, find the following.
(a) $(f + g)(x)$ (b) $(f - g)(x)$ (c) $(f \cdot g)(x)$

(d) $\left(\frac{f}{g}\right)(x)$ (e) the domain of $\left(\frac{f}{g}\right)(x)$

ANSWERS

COMPOSITE FUNCTION

There is another, important way in which two functions f and g can be combined to form a new function. In Figure 1a the function f assigns the value y in set Y to x in set X; then, function g assigns the value z in set Z to y in Y. The net effect of this combination of f and g is a new function h, called the **composite function of f and g,** which assigns the output z in Z to the input x in X. We write the new function h as

$$h(x) = (f \circ g)(x)$$

which is read "f of g of x."

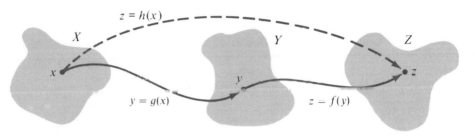

FIGURE 1a   Composite Function ($f \circ g$)(x)

We can use the "function machine" in Figure 1b to see how a composite function operates. For any input x,

(**1**) calculate $g(x)$;

(**2**) use $g(x)$ as input to f to calculate $f[g(x)]$.

Since we first evaluate $g(x)$, the input x must be in the domain of the function g. Further, since the output of g, namely, $g(x)$, is then an input to the function f, $g(x)$ must be in the domain of f. This leads us to a formal definition of the composite function.

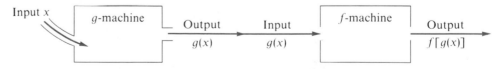

FIGURE 1b   The Function Machine for $f \circ g$

Composite Function $f \circ g$	The composite function $f \circ g$ is defined by $$(f \circ g)(x) = f[g(x)]$$ The domain of $f \circ g$ consists of all inputs x in the domain of g for which $g(x)$ is also in the domain of f.

EXAMPLE 2 Composite Function Notation

Given $f(x) = x^2$, $g(x) = x - 1$, find the following.

(a) $(f \circ g)(3)$ (b) $(g \circ f)(3)$ (c) $f[g(x)]$ (d) $g[f(x)]$

SOLUTION

(a) Noting that

$$(f \circ g)(3) = f[g(3)]$$

we begin by evaluating $g(3)$.

$$g(x) = x - 1$$
$$g(3) = 3 - 1 = 2$$

Therefore,

$$f[g(3)] = f(2)$$

Since

$$f(x) = x^2$$

then

$$f(2) = 2^2 = 4$$

Thus,

$$f[g(3)] = 4$$

(b) Noting that

$$(g \circ f)(3) = g[f(3)]$$

we begin by evaluating $f(3)$:

$$f(3) = 3^2 = 9$$

Then we find by substituting $f(3) = 9$ that

$$g[f(3)] = g(9) = 9 - 1 = 8$$

(c) Since $g(x) = x - 1$, we make the substitution

$$f[g(x)] = f(x - 1) = (x - 1)^2 = x^2 - 2x + 1$$

(d) Since $f(x) = x^2$, we make the substitution

$$g[f(x)] = g(x^2) = x^2 - 1$$

Note that $f[g(x)] \neq g[f(x)]$.

■

PROGRESS CHECK
Given $f(x) = x^2 - 2x$, $g(x) = 3x$, find the following.

(a) $f[g(-1)]$ (b) $g[f(-1)]$ (c) $f[g(x)]$

(d) $g[f(x)]$ (e) $(f \circ g)(2)$ (f) $(g \circ f)(2)$

ANSWERS

(f) 0

(e) 24

(d) $3x^2 - 6x$

(c) $9x^2 - 6x$

(b) 9

(a) 15

Although the composite function may seem like an awkward way of combining functions, it is commonly used in calculus to simplify processes. In fact, strange as it may seem, it is desirable to take a perfectly straightforward function and to write it as a composite of functions! The process is illustrated in Example 3.

EXAMPLE 3 Writing a Function as a Composite Function

Write the function $h = \sqrt{x^2 - 9}$ as a composite of two functions f and g, namely, $h(x) = (f \circ g)(x)$.

SOLUTION

For a given value of x, how would you go about evaluating h? In all likelihood, you would first evaluate $x^2 - 9$ and then take the square root of the result. Since the composite function $f \circ g$ requires that we first evaluate g, this suggests that we define g by

$$g(x) = x^2 - 9$$

Next, you would take the square root of your prior result, which suggests that we define the function f by

$$f(x) = \sqrt{x}$$

With f and g defined in this way, we see that h can be written as the composite $f \circ g$, that is, $h(x) = (f \circ g)(x)$. To verify that this is so, we evaluate

$$(f \circ g)(x) = f[g(x)]$$
$$= f(x^2 - 9)$$
$$= \sqrt{x^2 - 9}$$
$$= h(x)$$

∎

The functions f and g that we chose in Example 3 are not the only possible correct answers. You can verify that the definitions

$$f(x) = \sqrt{x - 9} \qquad \text{and} \qquad g(x) = x^2$$

would also work.

ONE-TO-ONE FUNCTIONS

An element in the range of a function may be the image of more than one element in the domain of the function. In Figure 2 we see that y in Y corresponds to both x_1 and x_2 in

FIGURE 2 $y = f(x_1) = f(x_2)$

X. If we demand that every element in the domain be assigned to a *different* element of the range, then the function is called **one-to-one.** More formally,

One-to-One Functions	A function f is one-to-one if $f(a) = f(b)$ only when $a = b$.

EXAMPLE 4 One-to-One Function Test
Prove the following:
(a) The function f defined by $f(x) = 3x - 1$ is one-to-one.

(b) The function f defined by $f(x) = x^2 - 1$ is not one-to-one.

SOLUTION
(a) We are going to use a technique known as the indirect method of proof. We'll let a and b be any two distinct numbers in the domain of f, that is, $a \neq b$, and assume that $f(a) = f(b)$. If we can show that these assumptions lead to a contradiction, we will have established that *distinct* arguments a and b lead to *distinct* function values $f(a)$ and $f(b)$, which will show that f is one-to-one.

Since

$$f(a) = 3a - 1 \qquad \text{and} \qquad f(b) = 3b - 1$$

and we assumed that $f(a) = f(b)$, we must have

$$3a - 1 = 3b - 1$$
$$a = b$$

This contradicts the assumption that $a \neq b$. Thus, $f(a) = f(b)$ only if $a = b$ and f is one-to-one.

(b) We need only find a pair of values a and b, $a \neq b$, such that $f(a) = f(b)$. If we try -3 and 3, then $f(-3) = f(3) = 8$, which is adequate to prove that f is not one-to-one.

■

There is a simple means of determining if a function f is one-to-one by examining the graph of the function. In Figure 3a we see that a horizontal line meets the graph in more than one point. Thus, $f(a) = f(b)$ although $a \neq b$; hence the function is not one-to-one. On the other hand, no horizontal line meets the graph in Figure 3b in more than

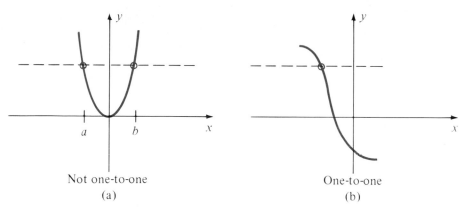

FIGURE 3 **Horizontal Line Test**

one point; the graph thus determines a one-to-one function. In summary, we have the following test.

Horizontal Line Test   If no horizontal line meets the graph of a function in more than one point, then the function is one-to-one.

EXAMPLE 5 Vertical and Horizontal Line Tests
Which of the graphs in Figure 4 are graphs of one-to-one functions?

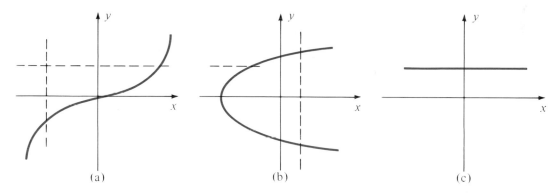

FIGURE 4

SOLUTION
(**a**) No *vertical* line meets the graph in more than one point; hence, it is the graph of a function. No *horizontal* line meets the graph in more than one point; hence, it is the graph of a one-to-one function.

(**b**) No *horizontal* line meets the graph in more than one point. But *vertical* lines do meet the graph in more than one point. It is therefore not the graph of a function and consequently cannot be the graph of a one-to-one function.

(c) No *vertical* line meets the graph in more than one point; hence, it is the graph of a function. But a *horizontal* line does meet the graph in more than one point. This is the graph of a function but not of a one-to-one function.

■

☑ **PROGRESS CHECK**
Which of the graphs in Figure 5 are graphs of one-to-one functions?

ANSWER
(q)

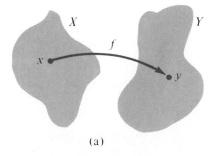

(a)   (b)   (c)

FIGURE 5

INVERSE FUNCTIONS

Suppose the function f in Figure 6a is a one-to-one function and that $y = f(x)$. Since f is one-to-one, we know that the correspondence is unique; that is, x in X is the *only* element of the domain for which $y = f(x)$. It is then possible to define a function g (Figure 6b) with domain Y and range X that reverses the correspondence, that is,

$$g(y) = x \quad \text{for every } x \text{ in } X$$

If we substitute $y = f(x)$, we have

$$g[f(x)] = x \quad \text{for every } x \text{ in } X \tag{1}$$

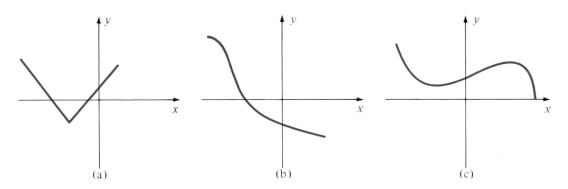

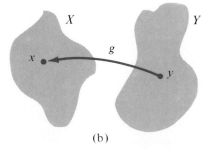

(a)   (b)

FIGURE 6   Inverse Functions

Substituting $g(y) = x$ in the equation $f(x) = y$ yields

$$f[g(y)] = y \quad \text{for every } y \text{ in } Y \tag{2}$$

The functions f and g of Figure 6 are therefore seen to satisfy the properties of Equations (1) and (2). Such functions are called inverse functions.

Inverse Functions

If f is a function with domain X and range Y, then the function g with domain Y and range X satisfying

$$g[f(x)] = x \quad \text{for every } x \text{ in } X$$
$$f[g(y)] = y \quad \text{for every } y \text{ in } Y$$

is called the **inverse function** of f.

Since the inverse (reciprocal) $1/x$ of a real number $x \neq 0$ can be written as x^{-1}, it is natural to write the inverse of a function f as f^{-1}. Thus we have

$$f^{-1}[f(x)] = x \quad \text{for every } x \text{ in } X$$
$$f[f^{-1}(y)] = y \quad \text{for every } y \text{ in } Y$$

See Figure 7 for a graphical representation.

Although we used one-to-one functions to motivate the concept of an inverse function, we made no reference to one-to-one functions in defining an inverse function. The following theorem tells you why.

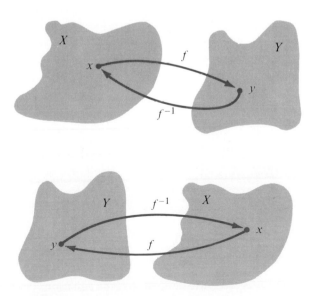

FIGURE 7   Inverse Functions

> A function has an inverse if and only if it is one-to-one.

We will omit the proof, although it is not too difficult. Instead, we will use this theorem to obtain another result:

> Increasing and decreasing functions always have an inverse.

We'll assume a function f to be an increasing function and will show that f is one-to-one. Suppose $f(a) = f(b)$ for some numbers a and b in the domain of f. It cannot be that $a < b$, for then we must have $f(a) < f(b)$; it cannot be that $a > b$, for then we must have $f(a) > f(b)$. The only possibility is that $a = b$, which establishes that f is one-to-one and therefore has an inverse. (A similar argument can be made for decreasing functions.)

It is not difficult to show that the inverse of a one-to-one function is unique (see Exercise 69).

The remaining sections of this chapter are devoted to the study of a very important class of inverse functions, the exponential and logarithmic functions. Always remember that *we can define the inverse function of f only if f is one-to-one.*

EXAMPLE 6   Verifying an Inverse Function

Let f be the function defined by

$$f(x) = x^2 - 4, \qquad x \geq 0$$

Verify that the inverse of f is given by

$$f^{-1}(x) = \sqrt{x + 4}$$

SOLUTION

We must verify that $f[f^{-1}(x)] = x$ and $f^{-1}[f(x)] = x$. Thus,

$$
\begin{aligned}
f[f^{-1}(x)] &= f(\sqrt{x + 4}) \\
&= (\sqrt{x + 4})^2 - 4 \\
&= x + 4 - 4 = x
\end{aligned}
$$

and

$$
\begin{aligned}
f^{-1}[f(x)] &= f^{-1}(x^2 - 4) \\
&= \sqrt{(x^2 - 4) + 4} \\
&= \sqrt{x^2} = |x|
\end{aligned}
$$

Since $x \geq 0$,

$$f^{-1}[f(x)] = |x| = x$$

We have verified that the equations defining inverse functions hold, and conclude that the inverse of f is as given. The student should verify that (a) the domain of f is the set of all nonnegative real numbers and the range of f is the set of all real numbers in the interval $[-4, \infty)$; (b) the domain of f^{-1} is the range of f and the range of f^{-1} is the domain of f.

■

GRAPH OF A FUNCTION AND ITS INVERSE

We may think of the function f defined by $y = f(x)$ as the set of all ordered pairs $(x, f(x))$, where x assumes all values in the domain of f. Since the inverse function reverses the correspondence, the function f^{-1} is the set of all ordered pairs $(f(x), x)$, where $f(x)$ assumes all values in the range of f.

We can illustrate this abstract idea by using the pair of inverse functions of Example 6 and list points on the graphs of the functions:

$f(x) = x^2 - 4, x \geq 0$	$f^{-1}(x) = \sqrt{x + 4}$
$(0, -4)$	$(-4, 0)$
$(3, 5)$	$(5, 3)$
$(6, 32)$	$(32, 6)$

With this approach, we see that the graphs of inverse functions are related in a distinct manner. First, note that the points (a, b) and (b, a) in Figure 8a are located symmetrically with respect to the graph of the line $y = x$. That is, if we fold the paper along the line $y = x$, the two points will coincide. And if (a, b) lies on the graph of a function f, then (b, a) must lie on the graph of f^{-1}. We conclude:

> The graphs of a pair of inverse functions are reflections of each other about the line $y = x$.

In Figure 8b we have sketched the graphs of the functions from Example 5 on the same coordinate axes to demonstrate this interesting relationship.

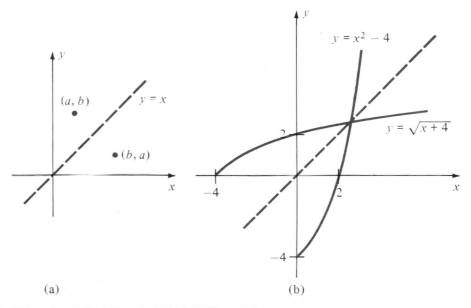

(a)   (b)

FIGURE 8 Symmetry of Inverse Functions About the Line $y = x$

FINDING THE INVERSE

It is sometimes possible to find an inverse by algebraic methods, as is shown by the following example.

EXAMPLE 7 Determining the Inverse Function

Find the inverse function of $f(x) = 2x - 3$.

SOLUTION

By definition, $f[f^{-1}(x)] = x$. Then we must have

$$f[f^{-1}(x)] = 2[f^{-1}(x)] - 3 = x$$

$$f^{-1}(x) = \frac{x + 3}{2}$$

We then verify that $f^{-1}[f(x)] = x$:

$$f^{-1}[f(x)] = \frac{2x - 3 + 3}{2} = x$$

∎

PROGRESS CHECK

Given $f(x) = 3x + 5$, find f^{-1}.

ANSWER

$f^{-1}(x) = \dfrac{x - 5}{3}$

WARNING

(a) In general, $f^{-1}(x) \neq \dfrac{1}{f(x)}$.

If $g(x) = x - 1$, then

$$g^{-1}(x) \neq \frac{1}{x - 1}$$

Use the methods of this section to show that

$$g^{-1}(x) = x + 1$$

(b) The inverse function notation is *not* to be thought of as a power.

EXAMPLE 8 Functions Defined by a Table of Values

A function h is defined in Table 1. Find the inverse of h.

TABLE 1 See Example 8

x	-3	-1	0	1	2
y	7	4	-5	-8	8

SOLUTION

Can a function really be defined by a table? Sure. After all, this table is a *rule* for obtaining $y = h(x)$ for x in the domain

$$X = \{-3, -1, 0, 1, 2\}$$

Further, for every x in X, there is just one y in the range

$$Y = \{7, 4, -5, -8, 8\}$$

Conclusion: h is indeed a function.

By definition, the inverse function h^{-1} must reverse the correspondence. To do this, we need only interchange the entries for x and for y as shown in Table 2. This table defines h^{-1}. It is then clear that (a) h and h^{-1} satisfy the definition of a pair of inverse functions; (b) the domain of h^{-1} is the range of h and the range of h^{-1} is the domain of h.

TABLE 2 The Inverse Function

x	7	4	-5	-8	8
y	-3	-1	0	1	2

■

EXERCISE SET 4.1

In Exercises 1–10 $f(x) = x^2 + 1$ and $g(x) = x - 2$. Determine the following.

1. $(f + g)(x)$ 2. $(f + g)(2)$ 3. $(f - g)(x)$ 4. $(f - g)(3)$

5. $(f \cdot g)(x)$ 6. $(f \cdot g)(-1)$ 7. $\left(\dfrac{f}{g}\right)(x)$ 8. $\left(\dfrac{f}{g}\right)(-2)$

9. the domains of f and of g 10. the domains of $\dfrac{f}{g}$ and of $\dfrac{g}{f}$

In Exercises 11–18 $f(x) = 2x + 1$ and $g(x) = 2x^2 + x$. Determine the following.

11. $(f \circ g)(x)$ 12. $(g \circ f)(x)$ 13. $(f \circ g)(2)$ 14. $(g \circ f)(3)$

15. $(f \circ g)(x + 1)$ 16. $(f \circ f)(-2)$ 17. $(g \circ f)(x - 1)$ 18. $(g \circ g)(x)$

In Exercises 19–24 $f(x) = x^2 + 4$ and $g(x) = \sqrt{x + 2}$. Determine the following.

19. $(f \circ g)(x)$ 20. $(g \circ f)(x)$

21. $(f \circ f)(-1)$ 22. the domain of $(f \circ g)(x)$

23. the domain of $(g \circ f)(x)$ 24. the domain of $(g \circ g)(x)$

In Exercises 25–28 determine $(f \circ g)(x)$ and $(g \circ f)(x)$.

25. $f(x) = x - 1$, $g(x) = x + 2$ 26. $f(x) = \sqrt{x + 1}$, $g(x) = x + 2$

27. $f(x) = \dfrac{1}{x + 1}$, $g(x) = \dfrac{1}{x - 1}$ 28. $f(x) = \dfrac{x + 1}{x - 1}$, $g(x) = x$

In Exercises 29–38 write the given function $h(x)$ as a composite of two functions f and g so that $h(x) = (f \circ g)(x)$. (There may be more than one answer.)

29. $h(x) = x^2 + 3$ 30. $h(x) = \dfrac{1}{x + 2}$ 31. $h(x) = (3x + 2)^8$ 32. $h(x) = (x^3 + 2x^2 + 1)^{13}$

33. $h(x) = (x^3 - 2x^2)^{1/3}$ **34.** $h(x) = \left(\dfrac{x^2 + 2x}{x^3 - 1}\right)^{3/2}$ **35.** $h(x) = |x^2 - 4|$ **36.** $h(x) = |x^2 + x| - 4$

37. $h(x) = \sqrt{4 - x}$ **38.** $h(x) = \sqrt{2x^2 - x + 2}$

In Exercises 39–44 verify that $g = f^{-1}$ for the given functions f and g by showing that $f[g(x)] = x$ and $g[f(x)] = x$.

39. $f(x) = 2x + 4$ $g(x) = \dfrac{1}{2}x - 2$

40. $f(x) = 3x - 2$ $g(x) = \dfrac{1}{3}x + \dfrac{2}{3}$

41. $f(x) = 2 - 3x$ $g(x) = -\dfrac{1}{3}x + \dfrac{2}{3}$

42. $f(x) = x^3$ $g(x) = \sqrt[3]{x}$

43. $f(x) = \dfrac{1}{x}$ $g(x) = \dfrac{1}{x}$

44. $f(x) = \dfrac{1}{x - 2}$ $g(x) = \dfrac{1}{x} + 2$

45. Let $F(x) = 2x - 2$. Find
   (a) $F^{-1}[F(4)]$ **(b)** $(F \circ F^{-1})(-5)$
   (c) $F^{-1}[F(y + 1)]$

46. Let $f(x) = x^3 + 2x + 3$ and assume the domain of f^{-1} to be the set of all real numbers. Find
   (a) $f[f^{-1}(-3)]$ **(b)** $(f^{-1} \circ f)(\sqrt{3})$
   (c) $f[f^{-1}(2a - 1)]$

In Exercises 47–54 find $f^{-1}(x)$. Sketch the graphs of $y = f(x)$ and $y = f^{-1}(x)$ on the same coordinate axes.

47. $f(x) = 2x + 3$ **48.** $f(x) = 3x - 4$ **49.** $f(x) = 3 - 2x$ **50.** $f(x) = \dfrac{1}{2}x + 1$

51. $f(x) = \dfrac{1}{3}x - 5$ **52.** $f(x) = 2 - \dfrac{1}{5}x$ **53.** $f(x) = x^3 + 1$ **54.** $f(x) = \dfrac{1}{x + 1}$

In Exercises 55–66 determine whether the given function is a one-to-one function.

55. $f(x) = 2x - 1$ **56.** $f(x) = 3 - 5x$ **57.** $f(x) = x^2 - 2x + 1$ **58.** $f(x) = x^2 + 4x + 4$

59. $f(x) = -x^3 + 1$ **60.** $f(x) = x^3 - 2$ **61.** $f(x) = \sqrt{x}$ **62.** $f(x) = |x|$

63. $f(x) = \dfrac{1}{x}$ **64.** $f(x) = -1$

65. $f(x) = \begin{cases} 2x, & x \le -1 \\ x^2, & -1 < x \le 0 \\ 3x - 1, & x > 0 \end{cases}$

66. $f(x) = \begin{cases} x^2 - 4x + 4, & x \le 2 \\ x, & x > 2 \end{cases}$

67. Let $f(x) = x^3 + 2x + 3$. Find $f^{-1}(3)$.

68. Let $H(x) = 2x^3 + 5x - 2$. Find $H^{-1}(-2)$.

69. Prove that a one-to-one function can have at most one inverse function. (*Hint:* Assume that the functions g and h are both inverses of the function f. Show that $g(x) = h(x)$ for all real values x in the range of f.)

70. Prove that the linear function $f(x) = ax + b$ is a one-to-one function if $a \ne 0$, and is not a one-to-one function if $a = 0$.

71. Find the inverse of the linear function $f(x) = ax + b$, $a \ne 0$.

72. A function H is defined by Table 3:
   (a) Does H have an inverse?   **(b)** Find $H(-1)$.
   (c) Find $H(1)$.

TABLE 3

x	-4	-1	0	2	3
$H(x)$	0	2	-1	0	1

73. A function G is defined by Table 4:
   (a) Is G a one-to-one function?
   (b) Find $G^{-1}(3)$.   **(c)** Find $G^{-1}(-4)$.
   (d) Find $G^{-1}(x)$.

TABLE 4

x	-10	-5	0	5	10
$G(x)$	17	20	-4	3	10

**4.2
EXPONENTIAL
FUNCTIONS**

The function $f(x) = 2^x$ is very different from any of the functions we have worked with thus far. Previously, we defined functions by using the basic algebraic operations (addition, subtraction, multiplication, division, powers, and roots). However, $f(x) = 2^x$ has a variable in the exponent and doesn't fall into the class of algebraic functions. Rather, it is our first example of an exponential function.

> An **exponential function** has the form
>
> $$f(x) = a^x, \qquad a > 0, \qquad a \neq 1$$
>
> where the real constant a is called the **base,** and the independent variable x may assume any real value.

**GRAPHS OF
EXPONENTIAL
FUNCTIONS**

The best way to become familiar with exponential functions is to sketch their graphs.

EXAMPLE 1 Graphing an Exponential Function
Sketch the graph of $f(x) = 2^x$.

SOLUTION
We let $y = 2^x$ and we form a table of values of x and y. Then we plot these points and sketch the smooth curve as in Figure 9. Note that the x-axis is a horizontal asymptote, that is, the curve gets closer and closer to the line as we move farther out along the line.

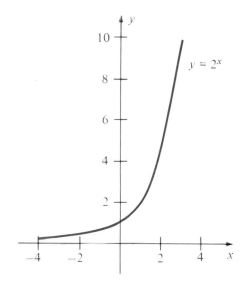

x	$y = 2^x$
-4	$1/32$
-3	$1/8$
2	$1/4$
-1	$1/2$
0	1
1	2
2	4
3	8
4	16

FIGURE 9 $y = 2^x$

■

Exponential functions are rapid growers. To see this, compare the values of 2^x and x^2 when $x = 20$. We have

$$2^{20} = 1{,}048{,}576 \qquad \text{and} \qquad 20^2 = 400$$

The exponential function 2^x far outpaces x^2. We'll see more evidence of this when we discuss exponential growth and decay.

WARNING

Don't confuse functions in which a *variable is raised to a constant power* with exponential functions where a *constant is raised to a variable power*. Here are some examples to help you distinguish:

TABLE 5 Exponential and Non-Exponential Functions

Exponential Functions	2^x	10^x	$(\sqrt{2})^x$	π^x	$\left(\dfrac{1}{2}\right)^x$
"Others"	x^2	x^{10}	$x^{\sqrt{2}}$	x^{π}	$x^{1/2}$

In a sense, we have cheated in our definition of $f(x) = 2^x$ and in sketching the graph in Figure 9. Since we have not explained the meaning of 2^x when x is irrational, we have no right to plot values such as $2^{\sqrt{2}}$. For our purposes, however, it will be adequate to think of $2^{\sqrt{2}}$ as the value we approach by taking successively closer approximations to $\sqrt{2}$, such as $2^{1.4}, 2^{1.41}, 2^{1.414}, \ldots$. A precise definition is given in more advanced mathematics courses, where it is also shown that the laws of exponents hold for irrational exponents.

We now look at $f(x) = a^x$ when $0 < a < 1$.

EXAMPLE 2 Graphing an Exponential Function

Sketch the graph of $f(x) = \left(\dfrac{1}{2}\right)^x = 2^{-x}$.

SOLUTION

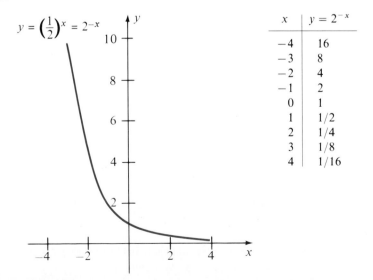

x	$y = 2^{-x}$
-4	16
-3	8
-2	4
-1	2
0	1
1	1/2
2	1/4
3	1/8
4	1/16

FIGURE 10 Graph of $y = 2^{-x}$

We form a table, plot points, and sketch the graph shown in Figure 10. Note that the graph of $y = 2^{-x}$ is a reflection about the y-axis of the graph of $y = 2^x$.

■

In Figure 11 we have sketched the graphs of

$$f(x) = 2^x \qquad g(x) = 3^x \qquad h(x) = \left(\frac{1}{2}\right)^x \qquad k(x) = \left(\frac{1}{3}\right)^x$$

on the same coordinate axes to provide additional examples of the graphs of exponential functions.

PROPERTIES OF THE EXPONENTIAL FUNCTIONS

The graphs in Figure 11 illustrate these important properties of the exponential functions. (Recall that the definition of the exponential function $f(x) = a^x$ requires that $a > 0$ and $a \neq 1$.)

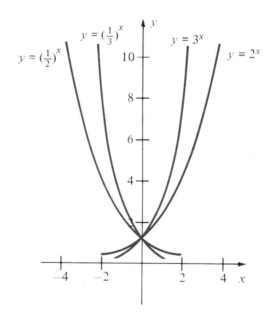

FIGURE 11 Graphs of Various Exponential Functions

Properties of the Exponential Functions	■ The graph of $f(x) = a^x$ always passes through the point $(0, 1)$ since $a^0 = 1$.

- ■ The graph of $f(x) = a^x$ always passes through the point $(0, 1)$ since $a^0 = 1$.
- ■ The domain of $f(x) = a^x$ consists of the set of all real numbers; the range is the set of all positive real numbers.
- ■ If $a > 1$, a^x is an increasing function; if $a < 1$, a^x is a decreasing function.
- ■ If $a < b$, then $a^x < b^x$ for all $x > 0$, and $a^x > b^x$ for all $x < 0$. Note in Figure 3 that $y = 3^x$ lies above $y = 2^x$ when $x > 0$ and below when $x < 0$.

Since a^x is either increasing or decreasing, it never assumes the same value twice. (Recall that $a \neq 1$.) This leads to a useful conclusion.

> If $a^u = a^v$, then $u = v$.

The graphs of a^x and b^x, $a \neq b$, intersect only at $x = 0$. This observation provides us with the following result.

> If $a^u = b^u$ for $u \neq 0$, then $a = b$.

EXAMPLE 3   Solving Exponential Equations

Solve for x.

(a) $3^{10} = 3^{5x}$ (b) $2^7 = (x - 1)^7$

SOLUTION

(a) Since $a^u = a^v$ implies $u = v$, we have

$$10 = 5x$$
$$2 = x$$

(b) Since $a^u = b^u$ implies $a = b$, we have

$$2 = x - 1$$
$$3 = x$$

∎

PROGRESS CHECK

Solve for x.

(a) $2^8 = 2^{x+1}$ (b) $4^{2x+1} = 4^{11}$

ANSWERS

(a) 7   (b) 5

THE NUMBER e

There is an irrational number that was first designated by the letter e by the Swiss mathematician Leonhard Euler (1707–1783). The number e is the value approached by the expression

$$\left(1 + \frac{1}{m}\right)^m$$

as m gets large. The procedure for studying the behavior of this expression as m gets larger and larger is developed in calculus courses. We will simply evaluate this expression for different values of m, as shown in Table 6.

TABLE 6   Approximating e

m	1	2	10	100	1000	10,000	100,000	1,000,000
$\left(1 + \dfrac{1}{m}\right)^m$	2.0	2.25	2.5937	2.7048	2.7169	2.7181	2.7182	2.71828

We admit that our definition of the number e would appear to be most unnatural. Yet, the function $f(x) = e^x$ is called the **natural exponential function.** Strange as it may seem, nature has a liking for the number e. Stranger yet, working with this function in calculus makes matters simpler rather than more complex. In fact, the natural exponential function is one of the most important functions in all of mathematics.

The graphs of $f(x) = e^x$ and $f(x) = e^{-x}$ are shown in Figure 12. Since $e \approx 2.71828$, the graph of $y = e^x$ falls between the graphs of $y = 2^x$ and $y = 3^x$. Table I in the Tables Appendix lists values for e^x and e^{-x}.

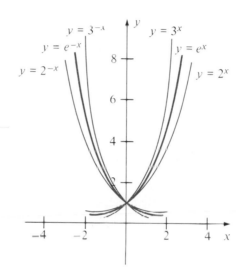

FIGURE 12   The Natural Exponential Function

APPLICATIONS

There are many fascinating applications that involve exponential functions. We'll look at the mechanism for computing the interest earned on an investment when the interest rate is compounded; problems dealing with population growth; and radioactive decay, such as determining the half-life of strontium 90.

Compound Interest

If you invest P dollars at an interest rate r that is paid to you at the end of a period of t years, then the sum S available to you is calculated by the formula for simple interest:

$$S = P + Prt = P(1 + rt)$$

Your neighborhood bank, however, advertises that it pays **compound interest,** that is, each successive payment includes interest on the previously accrued interest.

We can use Table 7 to explore the effect of annual compounding over a period of three years, both in general and with an investment of $100 compounded annually at a rate of 10%. As you would expect, the amount of interest paid increases each year, since the principal at the start of the next year has increased. Further, the amount or sum S available at the end of each year suggests the following formula.

$$S = P(1 + r)^t$$

TABLE 7   Effect of Compound Interest over Three Years

Year	Starting Amount	Ending Amount	Example: $P = \$100, r = 0.10$
1	P	$P(1 + r)$	$S = 100(1 + 0.10)$
			$= 110$
2	$P(1 + r)$	$[P(1 + r)](1 + r)$	$S = 110(1 + 0.10)$
		$= P(1 + r)^2$	$= 121$
3	$P(1 + r)^2$	$[P(1 + r)]^2(1 + r)$	$S = 121(1 + 0.10)$
		$= P(1 + r)^3$	$= 133.10$

We'll use this formula in an example and then we'll develop a variant that has more general applicability.

EXAMPLE 4   Compound Interest (Annual)

Suppose that \$5000 is invested at an interest rate of 7.5% compounded annually. Find the value of the investment after 4 years.

SOLUTION

To apply the formula for compound interest, we must express the interest rate as a decimal, that is, $7.5\% = 0.075$.

We are given $P = 5000$, $r = 0.075$, and $t = 4$. Substituting,

$$S = P(1 + r)^t$$
$$= 5000(1 + 0.075)^4$$

This is a good opportunity to use the y^x and parentheses keys on your calculator. The keystrokes are

$$5000 \;\boxed{\times}\; \boxed{(}\; 1 \;\boxed{+}\; 0.075 \;\boxed{)}\; \boxed{y^x}\; 4 \;\boxed{=}$$

and the answer displayed is \$6677.35.

∎

What happens if interest is compounded more frequently than once per year? You have surely seen banks advertise that they pay interest compounded semiannually or quarterly or even daily. We can modify the compound interest formula to handle any **compounding period,** the time period between successive additions of interest. Suppose that compounding takes place twice a year at an *annual* rate r. Then the interest rate applied to each period is $r/2$. Further, if the investment is held for t years, then there are $2t$ compounding periods. The sum available after t years is then

$$P(1 + r/2)^{2t}.$$

Generalizing from this reasoning to the situation wherein compounding takes place k times per year, the interest rate applied to each period is r/k and there are a total of kt compounding periods. The formula can then be stated this way.

Compound Interest Formula	If P dollars are invested at an interest rate r compounded k times annually, then the amount or sum S available at the end of t years is $$S = P(1 + i)^n$$ where $i = r/k$ and $n = kt$.

EXAMPLE 5 Compound Interest

Suppose that \$6000 is invested at an annual interest rate of 8%. What will the value of the investment be after 3 years if

(a) interest is compounded quarterly?

(b) interest is compounded semiannually?

SOLUTION

(a) We are given $P = 6000, r = 0.08, k = 4$, and $n = 12$ (since there are 4 compounding periods per year for 3 years). Thus,

$$i = \frac{r}{k} = \frac{0.08}{4} = 0.02$$

and

$$S = P(1 + i)^n = 6000(1 + 0.02)^{12}$$

Table IV in the Tables Appendix, with $i = r/k = 0.02 - 2\%$ and $n = 12$, yields

$$S = 6000(1.26824179) = 7609.45$$

Alternatively, we can use a calculator with the keystrokes

$$6000 \;\boxed{\times}\; \boxed{(}\; 1 \;\boxed{+}\; 0.02 \;\boxed{)}\; \boxed{y^x}\; 12 \;\boxed{=}$$

to obtain the same result. Thus, the sum at the end of the three-year period will be \$7609.45.

(b) We have $P = 6000, r = 0.08, k = 2$, and $n = 6$ (since there are 2 compounding periods per year for 3 years). Then

$$i = \frac{r}{k} = \frac{0.08}{2} = 0.04$$

$$S = P(1 + i)^n = 6000(1 + 0.04)^6$$

The appropriate keystrokes on a calculator are

$$6000 \;\boxed{\times}\; \boxed{(}\; 1 \;\boxed{+}\; 0.04 \;\boxed{)}\; \boxed{y^x}\; 6 \;\boxed{=}$$

and result in a display of 7591.91. The sum at the end of the three-year period will be \$7591.91, which is \$17.54 less than the interest earned when compounding is quarterly.

■

PROGRESS CHECK

Suppose that $5000 is invested at an annual interest rate of 6% compounded semi-annually. What is the value of the investment after 12 years?

ANSWER

$10,163.97

WARNING

Be certain that n is the total number of compounding periods and i is the interest rate per compounding period. For example, an interest rate of 18% compounded monthly for 2 years leads to $n = 24$ and $i = 0.015 = 1.5\%$.

Continuous Compounding

When P, r, and t are held fixed and the frequency of compounding is increased, the return on the investment is increased. Table 8 displays the increase in the value of an investment at the end of one year using various compounding periods.

TABLE 8 Investment of $P = \$1000$ at Rate $r = 8.0\%$ for One Year

Compounding Period	Compounding Frequency k	Sum S	Additional Interest
annual	1	$1080.00	
quarter	4	$1082.43	$2.43
month	12	$1083.00	$0.57
day	365	$1083.28	$0.28
hour	8760	$1083.28	$0.00

Although the value of the investment increases as we increase the compounding frequency, the additional benefit appears to diminish. In fact, the result in Table 8 is the same (at least to two decimal places!) for daily and hourly compounding. This observation leads us to suspect that there may a limiting value that is approached as we increase the compounding frequency, a phenomenon known as **continuous compounding**. We now have in hand all of the tools needed to pursue this idea in a formal manner.

Suppose a principal P is invested at an annual rate r, compounded k times per year. After t years, the number of conversions is $n = tk$. Then, the value of the investment after t years is

$$S = P\left(1 + \frac{r}{k}\right)^{tk}$$

Letting $m = k/r$, we can rewrite this equation as

$$S = P\left(1 + \frac{1}{m}\right)^{tmr}$$

or

$$S = P\left[\left(1 + \frac{1}{m}\right)^m\right]^{rt}$$

If the number of conversions k per year gets larger and larger, then m gets larger and larger. Since we saw in Table 1 that the expression

$$\left(1 + \frac{1}{m}\right)^m$$

gets closer and closer to e as m gets larger and larger, we have this result.

Continuous Compounding

If P dollars are invested at an interest rate r compounded continuously, then the amount or sum S available at the end of t years is

$$S = Pe^{rt} \qquad (1)$$

EXAMPLE 6 Continuous Compounding

Suppose that $20,000 is invested at an annual interest rate of 7% compounded continuously. What is the value of the investment after 4 years?

SOLUTION

We have $P = 20,000$, $r = 0.07$, and $t = 4$, and we substitute in Equation (1).

$$S = Pe^{rt}$$
$$= 20,000e^{0.07(4)} = 20,000e^{0.28}$$
$$= 20,000(1.3231) \quad \text{(from Table I, Tables Appendix, or a calculator}$$
$$= 26,462 \qquad\qquad \text{with } e \approx 2.71828)$$

The sum available after 4 years is $26,462.

■

 PROGRESS CHECK

Suppose that $10,000 is invested at an annual interest rate of 10% compounded continuously. What is the value of the investment after 6 years?

ANSWER

$18,221

By solving Equation (1) for P, we can determine the principal P that must be invested at continuous compounding to have a certain amount S at some future time. The values of e^{-x} from Table I in the Table Appendix will be used in this connection.

EXAMPLE 7 Continuous Compounding

Suppose that a principal P is to be invested at continuous compound interest of 8% per year to yield $10,000 in 5 years. Approximately how much should be invested?

SOLUTION

Using Equation (1) with $S = 10{,}000$, $r = 0.08$, and $t = 5$, we have

$$S = Pe^{rt}$$

$$10{,}000 = Pe^{0.08(5)} = Pe^{0.40}$$

$$P = \frac{10{,}000}{e^{0.40}}$$

$$= 10{,}000e^{-0.40}$$

$$= 10{,}000(0.6703) \quad \text{from Table I, Tables Appendix, or a calculator}$$

$$= 6703$$

Thus, approximately \$6703 should be invested initially.

■

PROGRESS CHECK

Approximately how much money should a 35-year-old woman invest now at continuous compound interest of 10% per year to obtain the sum of \$20,000 upon her retirement at age 65?

ANSWER

966\$

Exponential Growth

We'll say it again: nature loves the natural exponential function e^x. Biologists tell us that populations tend to grow exponentially (assuming they can avoid war or other forms of disaster) according to the **exponential growth model**

$$Q(t) = q_0 e^{kt}, \quad k > 0$$

which predicts the quantity Q that is present at time t. The independent variable in this model is t and both q_0 and k are constants. We may think of Q as the quantity of a substance available at any given time t. Note that when $t = 0$ we have

$$Q(0) = q_0 e^0 = q_0$$

which says that q_0 is the initial quantity. (It is customary to use subscript 0 to denote an initial value.) The constant k is called the **growth constant.**

EXAMPLE 8   Exponential Growth Model—Bacteria in a Culture

The number of bacteria in a culture after t hours is described by the exponential growth model

$$Q(t) = 50e^{0.7t}$$

(a) Find the initial number of bacteria, q_0, in the culture.

(b) How many bacteria are in the culture after 10 hours?

SOLUTION

(a) To find q_0 we need to evaluate $Q(t)$ at $t = 0$.

$$Q(0) = 50e^{0.7(0)} = 50e^0 = 50 = q_0$$

Thus, there are initially 50 bacteria in the culture.

(b) The number of bacteria in the culture after 10 hours is given by $Q(10)$.

$$Q(10) = 50e^{0.7(10)} = 50e^7 = 50(1096.6) = 54{,}830$$

Thus, there are 54,830 bacteria after 10 hours. (The value $e^7 \approx 1096.6$ can be found by using Table I in the Appendix; it can also be found by using a calculator with a "y^x" key, with $y = e \approx 2.71828$ and $x = 7$.)

■

PROGRESS CHECK

The number of bacteria in a culture after t minutes is described by the exponential growth model $Q(t) = q_0e^{0.005t}$. If there were 100 bacteria present initially, how many bacteria will be present after 1 hour has elapsed?

ANSWER

135

EXAMPLE 9 Exponential Growth Model—World Population

Statistics indicate that world population since World War II has been growing at the rate of 1.9% per year. Further, United Nations records indicate that the world population in 1975 was (approximately) 4 billion. Assuming an exponential growth model, what will the population of the world be in the year 2000?

SOLUTION

The growth constant $k = 0.019$ (since rate of growth is 1.9%) and the initial quantity $q_0 = 4$ (billion). Then the exponential growth model

$$Q(t) = q_0e^{kt}, \qquad k > 0$$

becomes

$$Q(t) = 4e^{0.019t}$$

For the year 2000, $t = 25$ (years) and we find that

$$Q(25) = 4e^{0.019(25)} = 4e^{0.475} \approx 6.43$$

We can project a world population of 6.43 billion persons for the year 2000.

■

Exponential Decay

Radioactive elements decay in a manner that is described by the **exponential decay model**

$$Q(t) = q_0e^{-kt}, \qquad k > 0$$

where Q is the quantity remaining after time t has elapsed. The independent variable in this model is t, q_0 is the initial quantity, and k is a constant called the **decay constant.**

We use the term **half-life** to describe the time it takes for half of the atoms of a radioactive element to break down or decay. Table 9 displays the approximate half-lives of a number of elements. Remarkably, the half-lives vary from a fraction of a second to billions of years! The long half-life of certain radioactive elements that are by-products of nuclear processes has caused environmentalists to warn us of the potential hazards of nuclear wastes.

TABLE 9   Half-Life of Radioactive Elements

Radioactive element	Half-life
Iridium 198	1 minute
Radon 222	4 days
Polonium 210	4 months
Radium 226	1620 years
Uranium 238	4.5 billion years
Thorium 232	14 billion years

EXAMPLE 10   Exponential Decay Model

A radioactive substance has a decay rate of 5% per hour. If 500 grams are present initially, how much of the substance remains after 4 hours?

SOLUTION

The general equation of an exponential decay model is

$$Q(t) = q_0 e^{-kt}, \qquad k > 0$$

In our model, $q_0 = 500$ grams (since the quantity available initially is 500 grams), and $k = 0.05$ (since the decay rate is 5% per hour). After 4 hours

$$Q(4) = 500e^{-0.05(4)} = 500e^{-0.2} \approx 500(0.8187) = 409.4$$

($e^{-0.2} \approx 0.8187$ is obtained from Table I in the Tables Appendix or a calculator with $e \approx 2.71828$.) Thus, there remain 409.4 grams of the substance.   ∎

PROGRESS CHECK

The number of grams Q of a certain radioactive substance present after t seconds is given by the exponential decay model $Q(t) = q_0 e^{-0.4t}$. If 200 grams of the substance are present initially, find how much remains after 6 seconds.

ANSWER

18.1 grams

RADIOACTIVITY IN THE NEWS

On April 25, 1986, a nuclear disaster of unprecedented proportions took place at Chernobyl in the Soviet Union. A fire occurred within a building housing a nuclear reactor, resulting in an explosion that sent radioactive material into the atmosphere. The air and soil of the surrounding farmlands was seriously contaminated.

Soil tests have indicated the presence of cesium 137, a radioactive element that has a half-life of 37 years. Scientists believe that the level of contamination in the soil must be reduced to 1/125th of its current reading before the region can again be used for farming. This will take 7 half-lives (since $2^7 = 128$), or (approximately) 259 years.

On September 13, 1988, the *New York Times* headline read

MAJOR RADON PERIL IS DECLARED BY U.S. IN CALL FOR TESTS

Radon, a colorless, invisible gas, is released by the breakdown of uranium in the earth's crust. Outdoors, the gas dissipates rapidly and is considered to be harmless. Indoors, however, it may accumulate to dangerous levels, especially in newer homes that have been constructed very tightly to conserve energy. Since the half-life of radon is only 3.8 days, homes showing high concentrations of radon may be salvaged by the installation of ventilation systems.

EXAMPLE 11 Exponential Decay Model—Half-Life of Radium 226
Radium 226 has an approximate half-life of 1620 years. Show that the quantity Q of radium 226 present after t years is given by

$$Q(t) = q_0 2^{-t/1620}$$

SOLUTION
If q_0 represents the initial quantity, then $q_0/2$ is the quantity present when $t = 1620$ (years). Substituting in the exponential decay model

$$Q(t) = q_0 e^{-kt}, \qquad k > 0$$

$$\frac{q_0}{2} = q_0 e^{-1620k}$$

We can write this last equation as

$$\frac{1}{2} = e^{-1620k}$$

and, by taking reciprocals,

$$2 = e^{1620k} = (e^k)^{1620}$$

Solving for e^k,

$$e^k = 2^{1/1620}$$

Substituting in the original equation, we have

$$Q(t) = q_0 2^{-t/1620}$$

which is the result we were asked to establish.

■

We will be able to tackle additional problems dealing with the half-life of radio-active elements after completing the next section.

EXERCISE SET 4.2

In Exercises 1–12 sketch the graph of the given function f.

1. $f(x) = 4^x$
2. $f(x) = 4^{-x}$
3. $f(x) = 10^x$
4. $f(x) = 10^{-x}$
5. $f(x) = 2^{x+1}$
6. $f(x) = 2^{x-1}$
7. $f(x) = 2^{|x|}$
8. $f(x) = 2^{-|x|}$
9. $f(x) = 2^{2x}$
10. $f(x) = 3^{-2x}$
11. $f(x) = e^{x+1}$
12. $f(x) = e^{-2x}$

In Exercises 13–20 solve for x.

13. $2^x = 2^3$
14. $2^{x-1} = 2^4$
15. $3^x = 9^{x-2}$
16. $2^x = 8^{x+2}$
17. $2^{3x} = 4^{x+1}$
18. $3^{4x} = 9^{x-1}$
19. $e^{x-1} = e^3$
20. $e^{x-1} = 1$

In Exercises 21–24 solve for a.

21. $(a + 1)^x = (2a - 1)^x$
22. $(2a + 1)^x = (a + 4)^x$
23. $(a + 1)^x = (2a)^x$
24. $(2a + 3)^x = (3a + 1)^x$

In Exercises 25–29 use Table I in the Tables Appendix, or a calculator, to evaluate e^x and e^{-x}.

25. The number of bacteria in a culture after t hours is described by the exponential growth model $Q(t) = 200e^{0.25t}$.

    (a) What is the initial number of bacteria in the culture?

    (b) Find the number of bacteria in the culture after 20 hours.

    (c) Complete the following table.

t	1	4	8	10
Q				

26. The number of bacteria in a culture after t hours is described by the exponential growth model $Q(t) = q_0 e^{0.01t}$. If there were 400 bacteria present initially, how many bacteria will be present after 2 *days*?

27. At the beginning of 1975 the world population was approximately 4 billion. Suppose that the population is described by an exponential growth model, and that the rate of growth is 2% per year. Give the approximate world population in the year 2050.

28. The number of grams of potassium-42 present after t hours is given by the exponential decay model $Q(t) = q_0 e^{-0.055t}$. If 400 grams of the substance were present initially, how much remains after 10 hours?

29. A radioactive substance has a decay rate of 4% per hour. If 1000 grams are present initially, how much of the substance remains after 10 hours?

In Exercises 30–33 use Table IV in the Tables Appendix, or a calculator, to assist in the computations.

30. An investor purchases a $12,000 savings certificate paying 10% annual interest compounded semiannually. Find the amount received when the savings certificate is redeemed at the end of 8 years.

31. The parents of a newborn infant place $10,000 in an investment that pays 8% annual interest compounded quarterly. What sum is available at the end of 18 years to finance the child's college education?

32. A widow is offered a choice of two investments. Investment A pays 8% annual interest compounded quarterly, and investment B pays 9% compounded annually. Which investment will yield a greater return?

33. A firm intends to replace its present computer in 5 years. The treasurer suggests that $25,000 be set aside in an investment paying 12% compounded monthly. What sum will be available for the purchase of the new computer?

In Exercises 34–38 use Tables I and IV in the Tables Appendix, or a calculator, to assist in the computations.

34. If $5000 is invested at an annual interest rate of 9% compounded continuously, how much is available after 5 years?

35. If $100 is invested at an annual interest rate of 5.5% compounded continuously, how much is available after 10 years?

36. A principal P is to be invested at continuous compound interest of 9% to yield $50,000 in 20 years. What is the approximate value of P to be invested?

37. A 40-year-old executive plans to retire at age 65. How much should be invested at 12% annual interest compounded continuously to provide the sum of $50,000 upon retirement?

38. Investment A offers 8% annual interest compounded semiannually, and investment B offers 8% annual interest compounded continuously. If $1000 were invested in each, what would be the approximate difference in value after 10 years?

In Exercises 39 and 40 use a calculator to determine which number is greater.

39. $2^{\pi}, \pi^{2}$

40. $3^{\pi}, \pi^{3}$

Every real number can be written as a product of a number c, $1 \le c < 10$, and an integer power of 10. For example,

$$643 = 6.43 \times 10^{2} \qquad 4629 = 4.629 \times 10^{3}$$
$$754{,}000 = 7.54 \times 10^{5} \qquad 1.76 = 1.76 \times 10^{0}$$
$$0.0423 = 4.23 \times 10^{-2} \qquad 0.0000926 = 9.26 \times 10^{-5}$$

This format is especially useful in writing very large and very small numbers and is referred to as **scientific notation.** It is the basis for a system of number representation used in computers that is called **floating point notation.**

In Exercises 41–48 write each of the numbers in scientific notation.

41. 2725

42. 493

43. 0.0084

44. 0.000914

45. 716,000

46. 527,600,000

47. 296.2

48. 32.767

4.3 LOGARITHMIC FUNCTIONS

LOGARITHMS AS EXPONENTS

We noted in the preceding section that $f(x) = a^{x}$ is an increasing function when $a > 1$ and is a decreasing function when $0 < a < 1$. It follows that exponential functions are one-to-one functions and that we can then study their inverse functions, which are called **logarithmic functions.**

The logarithmic and exponential functions are amongst the most important functions in all of mathematics. Accordingly, we use a special notation to denote a logarithmic function. If

$$f(x) = a^{x}$$

then we write the inverse using the notation

$$f^{-1}(x) = \log_{a} x$$

which is read as "log of x to the base a." Note that the base a of the exponential function ($a > 0$, $a \neq 1$) becomes the base of the corresponding logarithmic function.

Let's focus on the exponential function $f(x) = 2^{x}$ and see if we can get a better grip on the meaning of $\log_{2} x$. Since the range of f is the set of all real numbers, there is a unique point on the graph of f where $f(x) = 3.15$, that is, where the y-coordinate is

3.15. If we denote the x-coordinate at the point as k, then

$$2^k = 3.15 \tag{1}$$

where k is the *logarithm of 3.15 to the base 2* which we write as

$$k = \log_2 3.15 \tag{2}$$

Equations (1) and (2) demonstrate that a *logarithm is an exponent* and can be generalized to provide the following definition:

Logarithmic Function Base a	$y = \log_a x$ if and only if $x = a^y$

When no base is indicated, the notation $\log x$ is interpreted to mean $\log_{10} x$ which is also called the **common logarithm.**

Common Logarithm	$\log x = \log_{10} x$

The notation **ln x** is used to indicate logarithms to the base e. Since ln x is the inverse of the natural exponential function e^x, it is called the **natural logarithm.**

Natural Logarithm	$\ln x = \log_e x$

The exponential form $x = a^y$ and the logarithmic form $y = \log_a x$ are two ways of expressing the same relationship among x, y, and a. Further, it is always possible to convert from one form to the other. The natural question, then, is why bother to create a logarithmic form when we already have an equivalent exponential form. One reason is to allow us to switch an equation from the form $x = a^y$ to a form in which y is a function of x. We will also demonstrate that the logarithmic function has some very useful properties.

EXAMPLE 1   Logarithmic to Exponential Form
Write in exponential form.

(a) $\log_3 9 = 2$ (b) $\log_2 \dfrac{1}{8} = -3$ (c) $\log_{16} 4 = \dfrac{1}{2}$ (d) $\ln 7.39 = 2$

SOLUTION
We change from the logarithmic form $\log_a x = y$ to the equivalent exponential form $a^y = x$.

(a) $3^2 = 9$ (b) $2^{-3} = \dfrac{1}{8}$ (c) $16^{1/2} = 4$ (d) $e^2 = 7.39$

∎

PROGRESS CHECK
Write in exponential form.

(a) $\log_4 64 = 3$ (b) $\log \dfrac{1}{10,000} = -4$ (c) $\log_{25} 5 = \dfrac{1}{2}$ (d) $\ln 0.3679 = -1$

ANSWERS

EXAMPLE 2   Exponential to Logarithmic Form

Write in logarithmic form.

(a) $36 = 6^2$ (b) $7 = \sqrt{49}$ (c) $\dfrac{1}{100} = 10^{-2}$ (d) $0.1353 = e^{-2}$

SOLUTION

Since $y = \log_a x$ if and only if $x = a^y$, the logarithmic forms are

(a) $\log_6 36 = 2$ (b) $\log_{49} 7 = \dfrac{1}{2}$

(c) $\log \dfrac{1}{100} = -2$ (d) $\ln 0.1353 = -2$ ∎

PROGRESS CHECK

Write in logarithmic form.

(a) $1000 = 10^3$ (b) $6 = 36^{1/2}$

(c) $\dfrac{1}{7} = 7^{-1}$ (d) $20.09 = e^3$

ANSWERS

GRAPHS OF THE LOGARITHMIC FUNCTIONS

To sketch the graph of the logarithmic function $\log_a x$ we can take advantage of the fact that it is the inverse of the exponential function a^x. We know from our earlier work with inverse functions that the curves are reflections about the line $y = x$. This enables us to sketch the graphs shown in Figure 13 for $a > 1$.

The graphs of Figure 13 demonstrate that the exponential and logarithmic functions for a base $a > 1$ are both increasing functions. But what a difference in their rate of growth! To illustrate,

$$2^{16} = 65,536 \qquad \text{and} \qquad \log_2 16 = 4$$

In a sense, the exponential function is accelerating in its rate of growth while the logarithmic function is decelerating.

The next example offers an alternative means for sketching the graph of a logarithmic function.

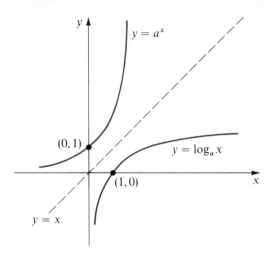

FIGURE 13   **The Logarithmic Function**

EXAMPLE 3   Graphing Logarithmic Functions

Sketch the graph of $y = \log_2 x$.

SOLUTION

To sketch the graph of a logarithmic function, we convert to the equivalent exponential form. Thus, to sketch the graph of $y = \log_2 x$, we form a table of values for the

TABLE 10   See Example 3

y	-3	-2	-1	0	1	2	3
$x = 2^y$	$\dfrac{1}{8}$	$\dfrac{1}{4}$	$\dfrac{1}{2}$	1	2	4	8

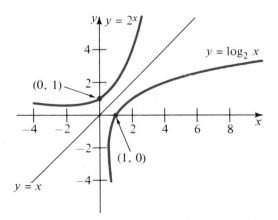

FIGURE 14 $y = \log_2 x$

equivalent exponential equation $x = 2^y$ (see Table 10). We can now plot these points and sketch a smooth curve, as in Figure 14. Note that the y-axis is a vertical asymptote. We have included the graph of $y = 2^x$ to stress that the graphs of a pair of inverse functions are reflections of each other about the line $y = x$.

∎

EXAMPLE 4   Logarithms with Base Less Than 1

Sketch the graphs of $y = \log_3 x$ and $y = \log_{1/3} x$ on the same coordinate axes.

SOLUTION

We could set up a table of values for each function and then sketch the graphs. However, by switching from the logarithmic form

$$y = \log_{1/3} x$$

to the exponential form

$$x = \left(\frac{1}{3}\right)^y = 3^{-y}$$

and then back again to logarithmic form

$$\log_3 x = -y$$

or

$$y = -\log_3 x$$

we can conclude that

$$y = \log_{1/3} x = -\log_3 x$$

This is the reason that you are unlikely to come across logarithms with a base $a < 1$. The graph will, of course, be the reflection about the x-axis of the graph of $\log_3 x$ and is shown in Figure 15.

∎

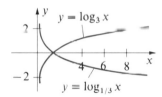

FIGURE 15   Logarithms with Base Less Than 1

LOGARITHMIC EQUATIONS AND CALCULATORS

Logarithmic equations can often be solved by changing them to equivalent exponential forms. Here are some straightforward examples; more challenging problems will be handled in Section 4.5.

EXAMPLE 5   Solving Logarithmic Equations

Solve for x.
(a) $\log_3 x = -2$ (b) $\log_x 81 = 4$ (c) $\log_5 125 = x$

SOLUTION

(a) Using the equivalent exponential form,

$$x = 3^{-2} = \frac{1}{9}$$

(b) Changing to the equivalent exponential form,

$$x^4 = 81 = 3^4$$

Since $a^u = b^u$ implies $a = b$, we have

$$x = 3$$

(c) In exponential form we have

$$5^x = 125$$

Writing 125 to the base 5, we have

$$5^x = 5^3$$

and since $a^u = a^v$ implies $u = v$, we conclude that

$$x = 3$$

∎

PROGRESS CHECK

Solve for x.

(a) $\log_x 1000 = 3$ (b) $\log_2 x = 5$ (c) $x = \log_7 \dfrac{1}{49}$

ANSWERS

(a) 10   (b) 32   (c) −2

Your calculator has keys labeled $\boxed{\log}$ and $\boxed{\ln}$ which can be used to calculate the common and natural logarithms, respectively. For example, to compute ln 5.25 you would enter

$$5.25\ \boxed{\ln}$$

and the display would read 1.6582.

Although every scientific calculator has a y^x key, many do not have an e^x key. We can take advantage of the fact that the functions ln x and e^x are a pair of inverse functions to "create" an e^x key. To do so, we will employ the $\boxed{\text{INV}}$ key on your calculator which is used to calculate the inverse of a function. For example, to calculate $e^{-0.2}$, you would enter

$$0.2\ \boxed{+/-}\ \boxed{\text{INV}}\ \boxed{\ln}$$

The answer in the display is 0.8187 and should be verified by using Table I in the Tables Appendix.

EXAMPLE 6   Calculators and Logarithms

(a) Solve for x: ln $x = -0.75$

(b) Solve for x: log $x = 1.25$

(c) What will the keystrokes 1 $\boxed{\text{INV}}$ $\boxed{\ln}$ produce in the display?

SOLUTION

(a) The equivalent exponential form is $x = e^{-0.75}$ and can be found by using the keystrokes

$$0.75 \boxed{+/-} \boxed{\text{INV}} \boxed{\ln}$$

which produces an answer of 0.4724.

(b) The equivalent exponential form is $x = 10^{1.25}$ and can be found by using the keystrokes

$$1.25 \boxed{\text{INV}} \boxed{\log}$$

which produces an answer of 17.7828. You can verify this result by trying the keystrokes

$$10 \boxed{y^x} 1.25 \boxed{=}$$

(c) The keystrokes $1 \boxed{\text{INV}} \boxed{\ln}$ will result in computing e^1 or e. This is a neat way to obtain e in your display for use in further computations.

■

LOGARITHMIC IDENTITIES

If $f(x) = a^x$, then $f^{-1}(x) = \log_a x$. Recall that inverse functions have the property that

$$f[f^{-1}(x)] = x \qquad \text{and} \qquad f^{-1}[f(x)] = x$$

Substituting $f(x) = a^x$ and $f^{-1}(x) = \log_a x$, we have

$$f[f^{-1}(x)] = x \qquad f^{-1}[f(x)] = x$$
$$f(\log_a x) = x \qquad f^{-1}(a^x) = x$$
$$a^{\log_a x} = x \qquad \log_a a^x = x$$

These two identities are useful in simplifying expressions and should be remembered.

$$a^{\log_a x} = x$$
$$\log_a a^x = x$$

The following pair of identities can be established by converting to the equivalent exponential form.

$$\log_a a = 1$$
$$\log_a 1 = 0$$

EXAMPLE 7 Working with Logarithmic Identities

Evaluate.

(a) $8^{\log_8 5}$ (b) $\log 10^{-3}$ (c) $\log_7 7$ (d) $\log_4 1$

SOLUTION

(a) 5   (b) -3 (c) 1   (d) 0

■

☑ **PROGRESS CHECK**
Evaluate.
(a) $\log_3 3^4$ **(b)** $6^{\log_6 9}$ **(c)** $\log_5 1$ **(d)** $\log_8 8$

ANSWERS
(a) 4   (b) 9   (c) 0   (d) 1

PROPERTIES OF LOGARITHMIC FUNCTIONS

The graphs in Figures 6 and 7 illustrate these important properties of logarithmic functions.

Properties of Logarithmic Functions	■ The point $(1, 0)$ lies on the graph of the function $f(x) = \log_a x$ for all real numbers $a > 0$. This is another way of saying $\log_a 1 = 0$. ■ The domain of $f(x) = \log_a x$ is the set of all positive real numbers; the range is the set of all real numbers. ■ When $a > 1$, $f(x) = \log_a x$ is an increasing function; when $0 < a < 1$, $f(x) = \log_a x$ is a decreasing function.

These results are in accord with what we anticipate for a pair of inverse functions. As expected, the domain of the logarithmic function is the range of the corresponding exponential function, and vice versa.

Since $\log_a x$ is either increasing or decreasing, the same value cannot be assumed more than once. Thus,

If $\log_a u = \log_a v$, then $u = v$.

Since the graphs of $\log_a u$ and $\log_b u$ intersect only at $u = 1$, we have:

If $\log_a u = \log_b u$ and $u \neq 1$, then $a = b$.

EXAMPLE 8   Solving Logarithmic Equations
Solve for x.
(a) $\log_5(x + 1) = \log_5 25$
(b) $\log_{x-1} 31 = \log_5 31$

SOLUTION
(a) Since $\log_a u = \log_a v$ implies $u = v$, then

$$x + 1 = 25$$
$$x = 24$$

MEASURING AN EARTHQUAKE

Richter Scale
Readings

Here's what you can anticipate from an earthquake of various Richter scale readings.

2.0 not noticed
4.5 some damage in a very limited area
6.0 hazardous; serious damage with destruction of buildings in a limited area
7.0 felt over a wide area with significant damage
8.0 great damage
8.7 maximum recorded

The great San Francisco earthquake of 1906 is estimated to have had a Richter scale reading of 8.3.

Radio and television newscasts often describe earthquakes in this way: "A minor earthquake in China registered 3.0 on the Richter scale," or, "A major earthquake in Chile registered 8.0 on the Richter scale." From statements like this we know that 3.0 is a "low" value and 8.0 is a "high" value. But just what is the Richter scale?

On the Richter scale, the magnitude R of an earthquake is defined as

$$R = \log \frac{l}{l_0}$$

where l_0 is a constant that represents a standard intensity and l is the intensity of the earthquake being measured. The Richter scale is a means of measuring a given earthquake against a "standard earthquake" of intensity l_0.

What does 3.0 on the Richter scale mean? Substituting $R = 3$ in the above equation, we have

$$3 = \log \frac{l}{l_0}$$

or, in the equivalent exponential form,

$$1000 = \frac{l}{l_0}$$

Solving for l,

$$l = 1000 \, l_0$$

which states that an earthquake with a Richter scale reading of 3.0 is 1000 times as intense as the standard! No wonder, then, that an earthquake registering 8.0 on the Richter scale is serious: it has an intensity 100,000,000 times that of the standard!

(**b**) Since $\log_a u = \log_b u$, $u \neq 1$, implies $a = b$,

$$x - 1 = 5$$
$$x = 6$$

∎

☑ **PROGRESS CHECK**
Solve for x.
(**a**) $\log_2 x^2 = \log_2 9$ (**b**) $\log_7 14 = \log_{2x} 14$

ANSWERS
(a) $3, -3$ (b) $\dfrac{7}{2}$

SUMMARY OF LOGARITHMIC IDENTITIES AND PROPERTIES	
Identities	*Properties*
$a^{\log_a x} = x$	$\log_a u = \log_a v$ implies $u = v$
$\log_a a^x = x$	$\log_a u = \log_b u,\ u \neq 1$, implies $a = b$
$\log_a a = 1$	
$\log_a 1 = 0$	

EXERCISE SET 4.3

In Exercises 1–12 write each in exponential form.

1. $\log_2 4 = 2$

2. $\log_5 125 = 3$

3. $\log_9 \dfrac{1}{81} = -2$

4. $\log_{64} 4 = \dfrac{1}{3}$

5. $\ln 20.09 = 3$

6. $\ln \dfrac{1}{7.39} = -2$

7. $\log_{10} 1000 = 3$

8. $\log_{10} \dfrac{1}{1000} = -3$

9. $\ln 1 = 0$

10. $\log_{10} 0.01 = -2$

11. $\log_3 \dfrac{1}{27} = -3$

12. $\log_{125} \dfrac{1}{5} = -\dfrac{1}{3}$

In Exercises 13–26 write each in logarithmic form.

13. $25 = 5^2$

14. $27 = 3^3$

15. $10,000 = 10^4$

16. $\dfrac{1}{100} = 10^{-2}$

17. $\dfrac{1}{8} = 2^{-3}$

18. $\dfrac{1}{27} = 3^{-3}$

19. $1 = 2^0$

20. $1 = e^0$

21. $6 = \sqrt{36}$

22. $2 = \sqrt[3]{8}$

23. $64 = 16^{3/2}$

24. $81 = 27^{4/3}$

25. $\dfrac{1}{3} = 27^{-1/3}$

26. $\dfrac{1}{2} = 16^{-1/4}$

In Exercises 27–44 solve for x.

27. $\log_5 x = 2$

28. $\log_{16} x = \dfrac{1}{2}$

29. $\log_{25} x = -\dfrac{1}{2}$

30. $\log_{1/2} x = 3$

31. $\ln x = 2$

32. $\ln x = -3$

33. $\ln x = -\dfrac{1}{2}$

34. $\log_4 64 = x$

35. $\log_5 \dfrac{1}{25} = x$

36. $\log_x 4 = \dfrac{1}{2}$

37. $\log_x \dfrac{1}{8} = -\dfrac{1}{3}$

38. $\log_3(x - 1) = 2$

39. $\log_5(x + 1) = 3$

40. $\log_2(x - 1) = \log_2 10$

41. $\log_{x+1} 24 = \log_3 24$

42. $\log_3 x^3 = \log_3 64$

43. $\log_{x+1} 17 = \log_4 17$

44. $\log_{3x} 18 = \log_4 18$

In Exercises 45–64 evaluate the expression.

45. $3^{\log_3 6}$

46. $2^{\log_2 (2/3)}$

47. $e^{\ln 2}$

48. $e^{\ln 1/2}$

49. $\log_5 5^3$

50. $\log_4 4^{-2}$

51. $\log_8 8^{1/2}$

52. $\log_{64} 64^{-1/3}$

53. $\log_7 49$

54. $\log_7 \sqrt{7}$

55. $\log_5 5$

56. $\ln e^{x^2}$

57. $\ln 1$

58. $\log_4 1$

59. $\log_2 \dfrac{1}{4}$

60. $\log_{16} 4$

61. $\log 10,000$

62. $e^{\ln(x+1)}$

63. $\ln e^2$

64. $\ln e^{-2/3}$

In Exercises 65–72 sketch the graph of each given function.

65. $f(x) = \log_4 x$

66. $f(x) = \log_{1/2} x$

67. $f(x) = \log 2x$

68. $f(x) = \frac{1}{2}\log x$

69. $f(x) = \ln\frac{x}{2}$

70. $f(x) = \ln 3x$

71. $f(x) = \log_3(x - 1)$

72. $f(x) = \log_3(x + 1)$

In Exercises 73–80 determine the domain of the given function.

73. $\ln(1 - x)$

74. $\ln(1 - x)^2$

75. $\log\dfrac{x}{x - 1}$

76. $\log\dfrac{\sqrt{x - 1}}{x}$

77. $\ln 2^x$

78. $\log e^{-x}$

79. $\log(-\sqrt{x})$

80. $\ln x^3$

4.4 FUNDAMENTAL PROPERTIES OF LOGARITHMS

There are three fundamental properties of logarithms that have made them a powerful computational aid.

$$\text{Property 1.} \quad \log_a xy = \log_a x + \log_a y$$

$$\text{Property 2.} \quad \log_a \frac{x}{y} = \log_a x - \log_a y$$

$$\text{Property 3.} \quad \log_a x^n = n\log_a x, \quad n \text{ a real number}$$

These properties can be proved by using equivalent exponential forms. To prove the first property, $\log_a xy = \log_a x + \log_a y$, we let

$$\log_a x = u \qquad \text{and} \qquad \log_a y = v$$

Then the equivalent exponential forms are

$$a^u = x \qquad \text{and} \qquad a^v = y$$

Multiplying the left-hand and right-hand sides of these equations, we have

$$a^u a^v = xy$$

or

$$a^{u+v} = xy$$

Substituting a^{u+v} for xy in $\log_a xy$ we have

$$\log_a xy = \log_a a^{u+v}$$
$$= u + v \quad \text{since } \log_a a^x = x.$$

Substituting for u and v,

$$\log_a xy = \log_a x + \log_a y$$

Properties 2 and 3 can be established in much the same way.

It is these properties of logarithms that originally made their study worthwhile. Why? Because the more complex operations of multiplication and division are converted to addition and subtraction, and exponentiation is converted to multiplication. We will next demonstrate these properties.

EXAMPLE 1 Simplifying Logarithmic Expressions

Write in terms of simpler logarithmic forms.

(a) $\log_{10}(225 \times 478) = \log_{10} 225 + \log_{10} 478$

(b) $\log_8 \dfrac{422}{735} = \log_8 422 - \log_8 735$

(c) $\log_2 2^5 = 5\log_2 2 = 5 \cdot 1 = 5$

(d) $\log_a \dfrac{xy}{z} = \log_a x + \log_a y - \log_a z$

PROGRESS CHECK

Write in terms of simpler logarithmic forms.

(a) $\log_4(1.47 \times 22.3)$ (b) $\log_a \dfrac{x-1}{\sqrt{x}}$

ANSWERS

(a) $\log_4 1.47 + \log_4 22.3$ (b) $\log_a(x-1) - \dfrac{1}{2}\log_a x$

EXAMPLE 2 Applying the Properties of Logarithms

Prove that

$$\log_a \frac{1}{x} = -\log_a x$$

SOLUTION

This result is useful in simplifying logarithmic forms and is very easy to prove. Note that

$$\log_a \frac{1}{x} = \log_a 1 - \log_a x$$

$$= -\log_a x$$

since $\log_a 1 = 0$.

SIMPLIFYING LOGARITHMS

The next example illustrates rules that speed the handling of logarithmic forms.

EXAMPLE 3 Simplifying Complex Logarithmic Expressions

Write $\log_a \dfrac{(x-1)^{-2}(y+2)^3}{\sqrt{x}}$ in terms of simpler logarithmic forms.

SOLUTION

Simplifying Logarithms	
Step 1. Rewrite the expression so that each factor has a positive exponent.	*Step 1.* $\log_a \dfrac{(x-1)^{-2}(y+2)^3}{\sqrt{x}}$ $= \log_a \dfrac{(y+2)^3}{(x-1)^2 x^{1/2}}$
Step 2. Apply Property 1 and Property 2 for multiplication and division of logarithms. Each factor in the numerator will yield a term with a plus sign. Each factor in the denominator will yield a term with a minus sign.	*Step 2.* $= \log_a(y+2)^3 - \log_a(x-1)^2 - \log_a x^{1/2}$
Step 3. Apply Property 3 to simplify.	*Step 3.* $= 3\log_a(y+2) - 2\log_a(x-1) - \dfrac{1}{2}\log_a x$

■

PROGRESS CHECK

Simplify $\log_a \dfrac{(2x-3)^{1/2}(y+2)^{2/3}}{z^4}$.

ANSWER

$\dfrac{1}{2}\log_a(2x-3) + \dfrac{2}{3}\log_a(y+2) - 4\log_a z$

EXAMPLE 4   Applying the Properties of Logarithms

If $\log_a 1.5 = r$, $\log_a 2 = s$, and $\log_a 5 = t$, find the following.

(a) $\log_a 7.5$ **(b)** $\log_a\left[(1.5)^3\ \sqrt[5]{\dfrac{2}{5}}\right]$

SOLUTION
(a) Since

$$7.5 = 1.5 \times 5$$
$$\log_a 7.5 = \log_a(1.5 \times 5)$$
$$= \log_a 1.5 + \log_a 5 \quad \text{Property 1}$$
$$= r + t \quad\quad\quad\quad\ \text{Substitution}$$

(b) Write this as

$$\log_a(1.5)^3 + \log_a\left(\frac{2}{5}\right)^{1/5} \qquad \text{Property 1}$$

$$= 3\log_a 1.5 + \frac{1}{5}\log_a\frac{2}{5} \qquad \text{Property 3}$$

$$= 3\log_a 1.5 + \frac{1}{5}[\log_a 2 - \log_a 5] \qquad \text{Property 2}$$

$$= 3r + \frac{1}{5}(s - t) \qquad \text{Substitution}$$

■

PROGRESS CHECK

If $\log_a 2 = 0.43$ and $\log_a 3 = 0.68$, find the following.

(a) $\log_a 18$ **(b)** $\log_a \sqrt[3]{\dfrac{9}{2}}$

ANSWERS

(a) 1.79   (b) 0.31

WARNING

(a) Note that

$$\log_a(x + y) \neq \log_a x + \log_a y$$

Property 1 tells us that

$$\log_a xy = \log_a x + \log_a y$$

Don't try to apply this property to $\log_a(x + y)$, which cannot be simplified.

(b) Note that

$$\log_a x^n \neq (\log_a x)^n$$

By Property 3,

$$\log_a x^n = n\log_a x$$

We can also apply the properties of logarithms to combine terms involving logarithms.

EXAMPLE 5   Combining Logarithmic Expressions

Write as a single logarithm.

$$2\log_a x - 3\log_a(x + 1) + \log_a\sqrt{x - 1}$$

SOLUTION

$$2\log_a x - 3\log_a(x+1) + \log_a\sqrt{x-1}$$

$$= \log_a x^2 - \log_a(x+1)^3 + \log_a\sqrt{x-1} \quad \text{Property 3}$$

$$= \log_a x^2\sqrt{x-1} - \log_a(x+1)^3 \quad \text{Property 1}$$

$$= \log_a \frac{x^2\sqrt{x-1}}{(x+1)^3} \quad \text{Property 2}$$

∎

PROGRESS CHECK

Write as a single logarithm.

$$\frac{1}{3}[\log_a(2x-1) - \log_a(2x-5)] + 4\log_a x$$

ANSWER

$$\log_a x^4 \sqrt[3]{\frac{2x-1}{2x-5}}$$

WARNING

(**a**) Note that

$$\frac{\log_a x}{\log_a y} \neq \log_a(x-y)$$

Property 2 tells us that

$$\log_a \frac{x}{y} = \log_a x - \log_a y$$

Don't try to apply this property to $\dfrac{\log_a x}{\log_a y}$, which cannot be simplified.

(**b**) The expressions

$$\log_a x + \log_b x$$

and

$$\log_a x - \log_b x$$

cannot be simplified. Logarithms with different bases do not readily combine except in special cases.

CHANGE OF BASE

Sometimes it is convenient to be able to write a logarithm that is given in terms of a base a in terms of another base b, that is, to convert $\log_a x$ to $\log_b x$. (As always, we must require a and b to be positive real numbers other than 1.)

To compute $\log_b x$ given $\log_a x$, let $y = \log_b x$. The equivalent exponential form is then

$$b^y = x$$

Taking logarithms to the base a of both sides of this equation, we have

$$\log_a b^y = \log_a x$$

We now apply the fundamental properties of logarithms developed earlier in this section. By Property 3,

$$y \log_a b = \log_a x$$

Solving for y,

$$y = \frac{\log_a x}{\log_a b}$$

Since $y = \log_b x$, we have

Change of Base Formula

$$\log_b x = \frac{\log_a x}{\log_a b}, \qquad a > 0, \qquad b > 0$$

EXAMPLE 6 Change of Base
Use the $\boxed{\log}$ key on your calculator to compute $\log_2 27$.

SOLUTION
We use the change of base formula

$$\log_b x = \frac{\log_a x}{\log_a b}$$

with $b = 2$, $a = 10$, and $x = 27$. Then

$$\log_2 27 = \frac{\log 27}{\log 2}$$

The appropriate key sequence is

$$27\ \boxed{\log}\ \boxed{\div}\ 2\ \boxed{\log}\ \boxed{=}$$

and the display reads 4.7549.

■

✓ **PROGRESS CHECK**
Use the $\boxed{\ln}$ key on your calculator to find $\log_5 16$.

ANSWER
1.7226

LOGARITHMIC "DO's" AND "DON'T's"

The following table summarizes the rules we have encountered for manipulating logarithms as well as the common errors that you must avoid.

DO USE	DON'T WRITE
$\log_a xy = \log_a x + \log_a y$	$\log_a(x + y) \neq \log_a x + \log_a y$
$\log_a \dfrac{x}{y} = \log_a x - \log_a y$	$\dfrac{\log_a x}{\log_a y} \neq \log_a(x - y)$
$\log_a x^n = n \log_a x$	$\log_a x^n \neq (\log_a x)^n$
$\log_a \dfrac{1}{x} = -\log_a x$	
$\log_b x = \dfrac{\log_a x}{\log_a b}$	

EXERCISE SET 4.4

In Exercises 1–20 express each in terms of simpler logarithmic forms.

1. $\log_{10}(120 \times 36)$

2. $\log_6 \dfrac{187}{39}$

3. $\log_3(3^4)$

4. $\log_3(4^3)$

5. $\log_a(2xy)$

6. $\ln(4xyz)$

7. $\log_a \dfrac{x}{yz}$

8. $\ln \dfrac{2x}{y}$

9. $\ln x^5$

10. $\log_3 y^{2/3}$

11. $\log_a(x^2 y^3)$

12. $\log_a(xy)^3$

13. $\log_a \sqrt{xy}$

14. $\log_a \sqrt[3]{xy^4}$

15. $\ln(x^2 y^3 z^4)$

16. $\log_a(xy^3 z^2)$

17. $\ln(\sqrt{x} \sqrt[3]{y})$

18. $\ln \sqrt[3]{xy^2 \sqrt[4]{z}}$

19. $\log_a\left(\dfrac{x^2 y^3}{z^4}\right)$

20. $\ln \dfrac{x^4 y^2}{z^{1/2}}$

In Exercises 21–30 if $\log 2 = 0.30$, $\log 3 = 0.47$, and $\log 5 = 0.70$, evaluate.

21. $\log 6$

22. $\log \dfrac{2}{3}$

23. $\log 9$

24. $\log \sqrt{5}$

25. $\log 12$

26. $\log \dfrac{6}{5}$

27. $\log \dfrac{15}{2}$

28. $\log 0.3$

29. $\log \sqrt{7.5}$

30. $\log \sqrt[4]{30}$

In Exercises 31–44 write each as a single logarithm.

31. $2 \log x + \dfrac{1}{2} \log y$

32. $3 \log_a x - 2 \log_a z$

33. $\dfrac{1}{3} \ln x + \dfrac{1}{3} \ln y$

34. $\dfrac{1}{3} \ln x - \dfrac{2}{3} \ln y$

35. $\dfrac{1}{3} \log_a x + 2 \log_a y - \dfrac{3}{2} \log_a z$

36. $\dfrac{2}{3} \log_a x + \log_a y - 2 \log_a z$

37. $\dfrac{1}{2}(\log_a x + \log_a y)$

38. $\dfrac{2}{3}(4 \ln x - 5 \ln y)$

39. $\dfrac{1}{3}(2\ln x + 4\ln y) - 3\ln z$

40. $\ln x - \dfrac{1}{2}(3\ln x + 5\ln y)$

41. $\dfrac{1}{2}\log_a(x-1) - 2\log_a(x+1)$

42. $2\log_a(x+2) - \dfrac{1}{2}(\log_a y + \log_a z)$

43. $3\log_a x - 2\log_a(x-1) + \dfrac{1}{2}\log_a\sqrt[3]{x+1}$

44. $4\ln(x-1) + \dfrac{1}{2}\ln(x+1) - 3\ln y$

The key labeled "ln" on a calculator is used to compute $\ln 10 = 2.3026$, $\ln 6 = 1.7918$, and $\ln 3 = 1.0986$. In Exercises 45–50 use the first value to find the required value.

45. $\ln 17 = 2.8332$; find $\log 17$

46. $\ln 22 = 3.0910$; find $\log_6 22$

47. $\ln 141 = 4.9488$; find $\log_3 141$

48. $\ln 78 = 4.3567$; find $\log_6 78$

49. $\ln 245 = 5.5013$; find $\log 245$

50. $\ln 7 = 1.9459$; find $\log_3 7$

4.5 EXPONENTIAL AND LOGARITHMIC EQUATIONS

The three properties of logarithms can be combined with the following "tips" to solve exponential and logarithmic equations.

> - To solve an exponential equation, take logarithms of both sides of the equation.
> - To solve a logarithmic equation, form a single logarithm on one side of the equation, and then convert the equation to the equivalent exponential form.

EXAMPLE 1   Solving an Exponential Equation
Solve $3^{2x-1} = 17$.

SOLUTION
Taking logarithms to the base 10 (for convenience) of both sides of the equation,

$$\log 3^{2x-1} = \log 17$$
$$(2x-1)\log 3 = \log 17 \quad \text{Property 3}$$
$$2x - 1 = \frac{\log 17}{\log 3}$$
$$2x = 1 + \frac{\log 17}{\log 3}$$
$$x = \frac{1}{2} + \frac{\log 17}{2\log 3}$$

If a numerical value is required, the key sequence

$$17 \boxed{\log} \boxed{\div} \boxed{(}\, 2 \boxed{\times} 3 \boxed{\log} \boxed{)} \boxed{=} \boxed{+} .5 \boxed{=}$$

can be used to obtain the (approximate) answer 1.7895.

$\blacksquare$

✓ PROGRESS CHECK
Solve $2^{x+1} = 3^{2x-3}$ and express your answer in terms of common logarithms (logarithms to the base 10).

ANSWER

$$\frac{\log 2 + 3 \log 3}{2 \log 3 - \log 2}$$

EXAMPLE 2   Solving an Exponential Equation

Solve the equation: $5e^{2-x} = 3$

SOLUTION

Since the equation contains the base e, it's a good idea to take logarithms to the base e of both sides of the equation.

$$\ln(5e^{2-x}) = \ln 3$$

$$\ln 5 + \ln e^{2-x} = \ln 3 \qquad \text{Property 1}$$

$$2 - x = \ln 3 - \ln 5 \qquad \log_a a^x = x$$

$$x = 2 + \ln 5 - \ln 3$$

$$x = 2 + \ln \frac{5}{3} \qquad \text{Property 2}$$

If a numerical value is required for x, the key sequence

$$5 \boxed{\div} 3 \boxed{-} \boxed{\ln} \boxed{+} 2 \boxed{=}$$

produces the approximation 2.5108. It is also instructive to redo the problem by first dividing both sides of the given equation by 5.

∎

PROGRESS CHECK

Solve the equation: $5 - 2e^{3x-1} = 0$

ANSWER

$$\frac{1}{3}\left(1 + \ln\frac{5}{2}\right)$$

EXAMPLE 3   Solving a Logarithmic Equation

Solve for x: $\log(2x + 8) = 1 + \log(x - 4)$.

SOLUTION

If we rewrite the equation in the form

$$\log(2x + 8) - \log(x - 4) = 1$$

then we can apply Property 2 to form a single logarithm.

$$\log \frac{2x + 8}{x - 4} = 1$$

Now we convert to the equivalent exponential form.

$$\frac{2x + 8}{x - 4} = 10^1 = 10$$

$$2x + 8 = 10x - 40$$

$$x = 6$$

∎

 PROGRESS CHECK
Solve for x: $\log_3(x - 8) = 2 - \log_3 x$.

ANSWER
$\dfrac{299}{101}$

EXAMPLE 4   Solving a Logarithmic Equation
Solve for x: $\log_2 x = 3 - \log_2(x + 2)$.

SOLUTION
Rewriting the equation with a single logarithm, we have

$$\log_2 x + \log_2(x + 2) = 3$$
$$\log_2[x(x + 2)] = 3 \qquad \text{Why?}$$
$$x(x + 2) = 2^3 = 8 \quad \text{Equivalent exponential form}$$
$$x^2 + 2x - 8 = 0$$
$$(x - 2)(x + 4) = 0 \qquad \text{Factor}$$
$$x = 2 \qquad \text{or} \qquad x = -4$$

The "solution" $x = -4$ must be rejected since the original equation contains $\log_2 x$, which requires that x be positive since the domain of the logarithm function does not include negative values.

∎

 PROGRESS CHECK
Solve for x: $\log_3(x - 8) = 2 - \log_3 x$.

ANSWER
$6 = x$

EXAMPLE 5   Exponential Growth Model
In a certain country, population is increasing at an annual rate of 2.5%. If we assume an exponential growth model, in how many years will the population double?

SOLUTION
The exponential growth model

$$Q(t) = q_0 e^{0.025t}$$

DATING THE LATEST ICE AGE

$$Q(t) = q_0 e^{-kt}$$
$$0.254 q_0 = q_0 e^{-0.00012t}$$
$$0.254 = e^{-0.00012t}$$
$$\ln 0.254 = \ln e^{-0.00012t}$$
$$-1.3704 = -0.00012t$$
$$t = 11{,}420$$

All organic forms of life contain radioactive carbon 14. In 1947 the chemist Willard Libby (who won the Nobel Prize in 1960) found that the percentage of carbon 14 in the atmosphere equals the percentage found in the living tissues of all organic forms of life. When an organism dies, it stops replacing carbon 14 in its living tissues. Yet the carbon 14 continues decaying at the rate of 0.012% per year. By measuring the amount of carbon 14 in the remains of an organism, it is possible to estimate fairly accurately when the organism died.

In the late 1940s radiocarbon dating was used to date the last ice sheet to cover the North American and European continents. Remains of trees in the Two Creeks Forest in northern Wisconsin were found to have lost 74.6% of their carbon 14 content. The remaining carbon 14, therefore, was 25.4% of the original quantity q_0 that was present when the descending ice sheet felled the trees. The accompanying computations use the general equation of an exponential decay model to find the age t of the wood. Conclusion: The latest ice age occurred approximately 11,420 years before the measurements were taken.

describes the population Q as a function of time t. Since the initial population is $Q(0) = q_0$, we seek the time t required for the population to double or become $2q_0$. We wish to solve the equation

$$Q(t) = 2q_0 = q_0 e^{0.025t}$$

for t. We then have

$$2q_0 = q_0 e^{0.025t}$$
$$2 = e^{0.025t} \qquad \text{Divide by } q_0$$
$$\ln 2 = \ln e^{0.025t} \qquad \text{Take natural logs of both sides}$$
$$= 0.025t \qquad \text{Since } \ln e^x = x$$
$$t = \frac{\ln 2}{0.025} \approx \frac{0.6931}{0.025} \approx 27.7$$

or approximately 28 years.

■

EXAMPLE 6 Continuous Compounding

A trust fund invests $8000 at an annual interest rate of 8% compounded continuously. How long does it take for the initial investment to grow to $12,000?

SOLUTION

We have seen (Section 4.2) that the equation

$$S = Pe^{rt}$$

gives the future value S for an investment of P dollars at a rate r for a period t (in years).

The problem statement tells us that

$$S = 12{,}000 \qquad P = 8000 \qquad r = 8\% = 0.08$$

and asks us to solve for t. Thus,

$$12{,}000 = 8000e^{0.08t}$$

$$\frac{12{,}000}{8000} = e^{0.08t}$$

$$e^{0.08t} = 1.5$$

Taking natural logarithms of both sides, we have

$$0.08t = \ln 1.5$$

$$t = \frac{\ln 1.5}{0.08} \approx \frac{0.4055}{0.08}$$

$$\approx 5.07$$

It takes approximately 5.07 years for the initial \$8000 to grow to \$12,000.

∎

EXERCISE SET 4.5

In Exercises 1–31 solve for x.

1. $5^x = 18$
2. $2^x = 24$
3. $2^{x-1} = 7$
4. $3^{x-1} = 12$
5. $3^{2x} = 46$
6. $2^{2x-1} = 56$
7. $5^{2x-5} = 564$
8. $3^{3x-2} = 23.1$
9. $3^{x-1} = 2^{2x-1}$
10. $4^{2x-1} = 3^{2x-3}$
11. $2^{-x} = 15$
12. $3^{-x+2} = 103$
13. $4^{-2x+1} = 12$
14. $3^{-3x+2} = 2^{-x}$
15. $e^x = 18$
16. $e^{x-1} = 2.3$
17. $e^{2x+3} = 30$
18. $e^{-3x+2} = 40$
19. $\log x + \log 2 = 3$
20. $\log x - \log 3 = 2$
21. $\log_x(3 - 5x) = 1$
22. $\log_x(8 - 2x) = 2$
23. $\log x + \log(x - 3) = 1$
24. $\log x + \log(x + 21) = 2$
25. $\log(3x + 1) - \log(x - 2) = 1$
26. $\log(7x - 2) - \log(x - 2) = 1$
27. $\log_2 x = 4 - \log_2(x - 6)$
28. $\log_2(x - 4) = 2 - \log_2 x$
29. $\log_2(x + 4) = 3 - \log_2(x - 2)$
30. $y = \dfrac{e^x + e^{-x}}{2}$
31. $y = \dfrac{e^x - e^{-x}}{2}$

32. Suppose that world population is increasing at an annual rate of 2%. If we assume an exponential growth model, in how many years will the population double?

33. Suppose that the population of a certain city is increasing at an annual rate of 3%. If we assume an exponential growth model, in how many years will the population triple?

34. The population P of a certain city t years from now is given by

$$P = 20{,}000e^{0.05t}$$

How many years from now will the population be 50,000?

35. Potassium 42 has a decay rate of approximately 5.5% per hour. Assuming an exponential decay model, in how many hours will the original quantity of potassium 42 have been halved?

36. Consider an exponential decay model given by

$$Q = q_0 e^{-0.4t}$$

where t is in weeks. How many weeks does it take for Q to decay to $\frac{1}{4}$ of its original amount?

37. How long does it take an amount of money to double if it is invested at a rate of 8% per year compounded semiannually?

38. At what rate of annual interest, compounded semiannually, should a certain amount of money be invested so that it will double in 8 years?

39. The number N of radios that an assembly line worker can assemble daily after t days of training is given by

$$N = 60 - 60e^{-0.04t}$$

After how many days of training does the worker assemble 40 radios daily?

40. The quantity Q (in grams) of a radioactive substance that is present after t days of decay is given by

$$Q = 400e^{-kt}$$

If $Q = 300$ when $t = 3$, find k, the decay rate.

41. A person on an assembly line produces P items per day after t days of training, where

$$P = 400(1 - e^{-t})$$

How many days of training will it take this person to be able to produce 300 items per day?

42. Suppose that the number N of mopeds sold when x thousands of dollars are spent on advertising is given by

$$N = 4000 + 1000 \ln(x + 2)$$

How much advertising money must be spent to sell 6000 mopeds?

TERMS AND SYMBOLS

base of a logarithmic function **205**
base of an exponential function **191**
common logarithm $\log_{10} x$ **206**
composite function $f \circ g$ **179**
compound interest **195**
compounding period **196**
continuous compounding **198**
decay constant **202**
Euler's constant e **194**
exponential decay model **201**
exponential function a^x **191**
exponential growth model **200**
growth constant **200**
horizontal line test **183**
inverse function f^{-1} **185**
logarithmic function $\log_a x$ **205**
natural logarithm $\ln x$ **206**
one-to-one function **181**
scientific notation **205**

KEY IDEAS FOR REVIEW

Topic	Page	Key Idea
Combining functions	178	Functions can be combined by the usual operations of addition, subtraction, multiplication, and division. However, the domain of the resulting function need not coincide with the domain of either of the original functions.
Composite function	179	A composite function is a function of a function. It is written as $(f \circ g)(x)$ or $f[g(x)]$
One-to-one function	182	We say a function is one-to-one if every element of the range corresponds to precisely one element of the domain.
horizontal line test	183	No horizontal line meets the graph of a one-to-one function in more than one point.
increasing and decreasing functions	186	An increasing (decreasing) function is always a one-to-one function.

Topic	Page	Key Idea
Inverse function	185	The inverse of a function, f^{-1}, reverses the correspondence defined by the function f. The domain of f becomes the range of f^{-1}, and the range of f becomes the domain of f^{-1}.
properties	185	A function f and its inverse satisfy the equations $$f^{-1}[f(x)] = x$$ $$f[f^{-1}(x)] = x$$
one-to-one functions	186	The inverse of a function f is defined only if f is a one-to-one function.
graphs	187	The graphs of $y = f(x)$ and $y = f^{-1}(x)$ are symmetric about the line $y = x$.
Exponential function	191	An exponential function has a variable in the exponent and has a base that is a positive constant.
domain	191	The domain of the exponential function is the set of all real numbers; the range is the set of all positive numbers.
graph	191	The graph of the exponential function $f(x) = a^x$, where $a > 0$ and $a \neq 1$, ■ passes through the points $(0, 1)$ and $(1, a)$ for any value of x; ■ is increasing if $a > 1$ and decreasing if $0 < a < 1$.
equality	194	If $a^x = a^y$, then $x = y$ (assuming $a > 0$, $a \neq 1$). If $a^x = b^x$ for all $x \neq 0$, then $a = b$ (asuming $a > 0$, $b > 0$).
Logarithmic function	205	The logarithmic function $\log_a x$ is the inverse of the function a^x.
logs as exponents	206	The logarithmic form $y = \log_a x$ and the exponential form $x = a^y$ are two ways of expressing the same relationship. In short, logarithms are exponents. Consequently, it is always possible to convert from one form to the other.
identities	211	The following identities are useful in simplifying expressions and in solving equations. $$a^{\log_a x} = x \qquad \log_a a = 1$$ $$\log_a a^x = x \qquad \log_a 1 = 0$$
domain	212	The domain of the logarithmic function is the set of all positive real numbers; the range is the set of all real numbers.
graph	212	The graph of the logarithmic function $f(x) = \log_a x$, where $x > 0$, ■ passes through the points $(1, 0)$ and $(a, 1)$ for any $a > 0$; ■ is increasing if $a > 1$ and decreasing if $0 < a < 1$.
equality	212	If $\log_a x = \log_a y$, then $x = y$. If $\log_a x = \log_b x$ and $x \neq 1$, then $a = b$.
fundamental properties	215	The fundamental properties of logarithms are as follows. *Property 1.* $\log_a(xy) = \log_a x + \log_a y$ *Property 2.* $\log_a\left(\dfrac{x}{y}\right) = \log_a x - \log_a y$ *Property 3.* $\log_a x^n = n \log_a x$
Change of base	220	The change of base formula is $$\log_b x = \frac{\log_a x}{\log_a b}$$

REVIEW EXERCISES

Solutions to exercises whose numbers are in color are in the Solutions section in the back of the book.

4.1 In Exercises 1–6 $f(x) = x + 1$ and $g(x) = x^2 - 1$. Determine the following.

1. $(f + g)(x)$

2. $(f \cdot g)(-1)$

3. $\left(\dfrac{f}{g}\right)(x)$

4. the domain of $\left(\dfrac{f}{g}\right)(x)$

5. $(g \circ f)(x)$

6. $(f \circ g)(?)$

In Exercises 7–10 $f(x) = \sqrt{x} - 2$ and $g(x) = x^2$. Determine the following.

7. $(f \circ g)(x)$

8. $(g \circ f)(x)$

9. $(f \circ g)(-2)$

10. $(g \circ f)(-2)$

In Exercises 11 and 12 $f(x) = 2x + 4$ and $g(x) = \dfrac{x}{2} - 2$.

11. Prove that f and g are inverse functions of each other.

12. Sketch the graphs of $y = f(x)$ and $y = g(x)$ on the same coordinate axes.

4.2 13. Sketch the graph of $f(x) = \left(\dfrac{1}{3}\right)^x$. Label the point $(1, f(1))$.

14. Solve $2^{2x} = 8^{x-1}$ for x.

15. Solve $(2a + 1)^x = (3a - 1)^x$ for a.

16. The sum of $8000 is invested in a certificate paying 12% annual interest compounded semiannually. What sum is available at the end of 4 years?

4.3 In Exercises 17–20 write each logarithmic form in exponential form and vice versa.

17. $27 = 9^{3/2}$

18. $\log_{64} 8 = \dfrac{1}{2}$

19. $\log_2 \dfrac{1}{8} = -3$

20. $6^0 = 1$

In Exercises 21–24 solve for x.

21. $\log_x 16 = 4$

22. $\log_5 \dfrac{1}{125} = x - 1$

23. $\ln x = -4$

24. $\log_3(x + 1) = \log_3 27$

In Exercises 25–28 evaluate the given expression.

25. $\log_3 3^5$

26. $\ln e^{-1/3}$

27. $\log_3 \left(\dfrac{1}{3}\right)$

28. $e^{\ln 3}$

29. Sketch the graph of $f(x) = \log_3 x + 1$.

4.4 In Exercises 30–33 write the given expression in terms of simpler logarithmic forms.

30. $\log_a \dfrac{\sqrt{x - 1}}{2x}$

31. $\log_a \dfrac{x(2 - x)^2}{(y + 1)^{1/2}}$

32. $\ln(x + 1)^4 (y - 1)^2$

33. $\log \sqrt[5]{\dfrac{y^2 z}{z + 3}}$

In Exercises 34–37 use the values $\log 2 = 0.30$, $\log 3 = 0.50$, and $\log 7 = 0.85$ to evaluate the given expression.

34. $\log 14$

35. $\log 3.5$

36. $\log \sqrt{6}$

37. $\log 0.7$

In Exercises 38–41 write the given expression as a single logarithm.

38. $\dfrac{1}{3}\log_a x - \dfrac{1}{2}\log_a y$

39. $\dfrac{4}{3}[\log x + \log(x - 1)]$

40. $\ln 3x + 2\left(\ln y - \dfrac{1}{2}\ln z\right)$

41. $2\log_a(x + 2) - \dfrac{3}{2}\log_a(x + 1)$

In Exercises 42 and 43 use the values $\log 32 = 1.5$, $\log 8 = 0.9$, and $\log 5 = 0.7$ to find the requested value.

42. $\log_8 32$

43. $\log_5 32$

4.5 44. A substance is known to have a decay rate of 6% per hour. Approximately how many hours are required for the remaining quantity to be half of the original quantity?

In Exercises 45–47 solve for x.

45. $2^{3x-1} = 14$

46. $2\log x - \log 5 = 3$

47. $\log(2x - 1) = 2 + \log(x - 2)$

PROGRESS TEST 4A

In Problems 1–4 $f(x) = \dfrac{1}{x - 1}$ and $g(x) = x^2$. Find the following.

1. $(f - g)(2)$

2. $\left(\dfrac{g}{f}\right)(x)$

3. $(g \circ f)(3)$

4. Prove that $f(x) - -3x + 1$ and $g(x) - -\tfrac{1}{3}(x - 1)$ are inverse functions of each other.

5. Sketch the graph of $f(x) = 2^{x+1}$. Label the point $(1, f(1))$.

6. Solve $\left(\dfrac{1}{2}\right)^x = \left(\dfrac{1}{4}\right)^{2x+1}$

In Problems 7 and 8 convert from logarithmic form to exponential form or vice versa.

7. $\log_3 \dfrac{1}{9} = -2$

8. $64 = 16^{3/2}$

In Problems 9 and 10 solve for x.

9. $\log_x 27 = 3$

10. $\log_6 \left(\dfrac{1}{36}\right) = 3x + 1$

In Problems 11 and 12 evaluate the given expression.

11. $\ln e^{5/2}$

12. $\log_5 \sqrt{5}$

In Problems 13 and 14 write the given expression in terms of simpler logarithmic forms.

13. $\log_a \dfrac{x^3}{y^2 z}$

14. $\log \dfrac{x^2 \sqrt{2y-1}}{y^3}$

In Problems 15 and 16 use the values $\log 2.5 = 0.4$ and $\log 2 = 0.3$ to evaluate the given expression.

15. $\log 5$

16. $\log 2\sqrt{2}$

In Problems 17 and 18 write the given expression as a single logarithm.

17. $2\log x - 3\log(y+1)$

18. $\dfrac{2}{3}[\log_a(x+3) - \log_a(x-3)]$

19. The number of bacteria in a culture is described by the exponential growth model

$$Q(t) = q_0 e^{0.02t}$$

Approximately how many hours are required for the number of bacteria to double?

In Problems 20 and 21 solve for x.

20. $\log x - \log 2 = 2$

21. $\log_4(x-3) = 1 - \log_4 x$

PROGRESS TEST 4B

In Problems 1–4 $f(x) = \dfrac{1}{\sqrt{x+1}}$ and $g(x) = x - 1$. Find the following.

1. $(f \cdot g)(3)$

2. $(f + g)(1)$

3. $(f \circ g)(x)$

4. Prove that $f(x) = \dfrac{1}{x}$ and $g(x) = \dfrac{1}{x}$ are inverse functions.

5. Sketch the graph of $f(x) = \left(\dfrac{1}{2}\right)^{x-1}$. Label the point $(2, f(2))$.

6. Solve $(a+3)^x = (2a-5)^x$ for a, assuming the quantities in parentheses are both positive.

In Problems 7 and 8 convert from logarithmic form to exponential form and vice versa.

7. $\dfrac{1}{1000} = 10^{-3}$

8. $\log_3 1 = 0$

In Problems 9 and 10 solve for x.

9. $\log_2(x-1) = -1$

10. $\log_{2x} 27 = \log_3 27$

In Problems 11 and 12 evaluate the given expression.

11. $\log_3 3^{10}$

12. $e^{\ln 4}$

In Problems 13 and 14 write the given expression in terms of simpler logarithmic forms.

13. $\log_a(x-1)(y+3)^{5/4}$

14. $\ln \sqrt{xy} \sqrt[4]{2z}$

In Problems 15 and 16 use the values $\log 2.5 = 0.4$, $\log 2 = 0.3$, and $\log 6 = 0.75$ to evaluate the given expression.

15. $\log 7.5$

16. $\log 36$

In Problems 17 and 18 write the given expression as a single logarithm.

17. $\dfrac{3}{5}\ln(x-1) + \dfrac{2}{5}\ln y - \dfrac{1}{5}\ln z$

18. $\log \dfrac{x}{y} - \log \dfrac{y}{x}$

19. Suppose that $500 is invested in a certificate at an annual interest rate of 12% compounded monthly. What is the value of the investment after 6 months?

In Problems 20 and 21 solve for x.

20. $\log_x(x+6) = 2$

21. $\log(x-9) = 1 - \log x$

UNIT 1 CUMULATIVE PROGRESS TEST
CHAPTERS 1–4

In Problems 1–4 simplify the expression and write the answer using only positive exponents.

1. $x^{5/2}(x^{-2/3} - 1)$

2. $\sqrt{12} - \sqrt{75} + 2\sqrt{27}$

3. $\dfrac{2}{x^2 - x} - \dfrac{2x}{x-1}$

4. $(1-i)^3$

In Problems 5 and 6 factor completely.

5. $3x^3 + x^2 - 2x$

6. $(x-1)^{-2/3} + 2(x-1)^{-5/3}$

7. Rationalize the denominator: $\dfrac{2}{\sqrt{x} - \sqrt{2}}$

In Problems 8–10 solve the inequality.

8. $3x - 2 \le x + 3$

9. $|1 - 2h| > 2$

10. $2t^2 - 5t \ge 12$

11. Write in the form $a + bi$: $(3 - 2i)(2 + 3i)$

In Problems 12 and 13 find the domain of the function.

12. $\dfrac{1}{\sqrt{2x - 1}}$

13. $\dfrac{x}{x^2 - 1}$

14. Find the length of the hypotenuse of the right triangle whose vertices are $(-2, 2)$, $(4, 2)$, and $(-2, -3)$.

15. Given $f(t) = 1 - t^2$,
    (a) find $f(2a - 1)$.
    (b) find a real number whose image is -15.
    (c) find $\dfrac{f(t + h) - f(t)}{h}$.

16. The function G is defined by

$$G(x) = \begin{cases} 1/2 & \text{when} & x < -2 \\ x & \text{when} & -2 \le x \le 2 \\ -x^2 & \text{when} & x > 2 \end{cases}$$

(a) Find $G(0)$ (b) Find $G(-3)$ (c) Find $G(3)$

In Problems 17–19 solve the given equation.

17. $2x^2 - x - 3 = 0$

18. $2x^2 - x + 19 = 15$

19. $\sqrt{x + 12} = x$

20. The hypotenuse of a right triangle has a length of 20 inches. If one leg of the triangle is three times the other leg, what is the length of the shortest side of the triangle?

21. Find the vertex and intercepts of the parabola $f(x) = -3x^2 - x + 2$

22. Given the points $A(3, -2)$ and $B(-1, 2)$, find
    (a) the slope of the line through the points A and B.
    (b) an equation of the line through the points A and B.
    (c) an equation of the line through B parallel to the y-axis.
    (d) an equation of the line through the point $C(-4, -1)$ that is perpendicular to the line AB.

23. Given $P(x) = 2x^3 + 5x^2 + 3$,
    (a) find $P(-2)$.
    (b) find the quotient and remainder when $P(x)$ is divided by $x + 1$.

24. Find all rational roots of the equation
$$5x^3 - x^2 + 20x - 4 = 0.$$

25. The graph of the function $f(x) = 2x^3 - x^2 - 6x + 3$

crosses the x-axis at $x = 1/2$. Find the other x-intercepts, if any.

26. Find the partial-fraction decomposition of $\dfrac{x + 5}{x^2 - x - 2}$.

27. Given the function $f(x) = \dfrac{2x^2 + 6x}{x^2 - 9}$
    (a) determine the domain of f.
    (b) find all asymptotes of the graph of f.
    (c) sketch the graph of f.

28. Sketch the graph of $f(x) = 2^x - 1$.

29. If $\log 2 = s$, $\log 3 = t$, and $\log 5 = u$, express the following in terms of s, t, and u:
    (a) $\log 18$ (b) $\log(\frac{5}{4})$ (c) $\log 1.5$ (d) $\log \sqrt{7.5}$

30. A certificate of deposit is purchased paying interest at the rate of 8% per year compounded continuously. If the certificate matures in 6 months at a value of $1000, how much was invested initially?

31. Simplify: $\ln \dfrac{x^2 \sqrt{x + 1}}{x - 1}$

In Problems 32–34 solve for x.

32. $27^{x-1} = 9^{3-x}$

33. $\log_x(2x + 3) = 2$

34. $\log_2(\log_3 x) = 1$

35. A radioactive isotope is known to decay exponentially. If the half-life of the isotope is 5 years, how long will it take for 9 grams to decay to 3 grams?

© Hazel Hankin/Stock, Boston

THE TRIGONOMETRIC FUNCTIONS

All nature's structuring, associating, and patterning must be based on triangles, because there is no structural validity otherwise. This is nature's basic structure, and it is modelable.

Buckminster Fuller

Philosophy . . . is written in the language of mathematics, and its characters are triangles, circles, and other geometrical figures, without which it is humanly impossible to understand a single word of it.

Galileo

The word *trigonometry* derives from the Greek, meaning *measurement of triangles*. An intense interest in astronomy motivated discoveries in this field and it is Hipparchus of Nicaea (180–125 B.C.) who is referred to as the "father of trigonometry."

The title of Chapter 5, however, provides a hint of a second approach to the study of this material. We may think of the trigonometric functions as relating the set of real numbers to points on a unit circle—which leads to the alternate name, the circular functions. The first chapter of this unit will deal with the circular functions since this is the preferred approach for students preparing for courses in calculus.

The use of trigonometry to relate the measurement of the sides and angles of a triangle is dealt with in the second chapter of this unit. Many important applications in such fields as navigation and surveying make extensive use of the results derived in this chapter.

CHAPTER 5

TRIGONOMETRY: THE CIRCULAR FUNCTIONS

We have said that trigonometry deals with the measurement of triangles. Many important applications in such fields as navigation and surveying deal with the relationship between the sides and angles of a triangle and make extensive use of results derived in trigonometry.

For the student preparing for a course in calculus and who will be taking additional courses in the sciences, there is another approach to the subject matter that has greater relevance. In this approach, the *trigonometric functions* are viewed as functions of real numbers whose ranges are related to points on a circle. For this reason, they are often referred to as *circular functions*.

We will devote this chapter to the circular functions and will deal with both angles and triangles in the next chapter.

5.0
REVIEW OF GEOMETRY

We need to recall various facts about the circle from plane geometry. A line segment joining the center of a circle to any point on the circle is called a **radius.** Since every point on the circle is the same distance from the center, the radii of a circle are all equal. Thus, in Figure 1, $\overline{OP} = \overline{OQ}$. A **chord** of a circle is a line segment joining any two points on the circle; a **diameter** of a circle is a chord that passes through the center of the circle. Note that the length of a diameter is twice that of a radius.

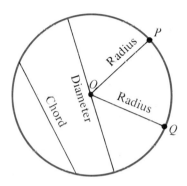

FIGURE 1 Parts of a Circle

The **circumference** C of a circle is the distance around the circle and is given by

$$C = 2\pi r$$

where r is the radius of the circle. The constant π is then seen to be the ratio of the circumference of a circle to the length of its diameter. The **area** A of a circle of radius r is given by

$$A = \pi r^2$$

An **arc** of a circle is simply a part of a circle. The are $\overset{\frown}{AB}$ of Figure 2 consists of the two endpoints A and B and the set of all points on the circle that are between A and B and are shown in color.

A **central angle** has its vertex at the center of the circle, and its sides are radii of the circle. We define the measure of a central angle to be the same as that of the arc it intercepts. Thus, in Figure 2, $\angle AOB = \overset{\frown}{AB}$. We can then show that equal arcs determine or subtend equal chords. If arc $\overset{\frown}{AB}$ = arc $\overset{\frown}{CD}$ in Figure 2, then, by definition, $\angle AOB = \angle COD$. Since $\overline{AO} = \overline{BO} = \overline{CO} = \overline{DO}$ are all radii, it follows that triangles AOB and COD are congruent. Hence, $\overline{CD} = \overline{AB}$. The converse can be proven in a similar manner. Thus,

> Equal arcs determine equal chords.
> Equal chords determine equal arcs.

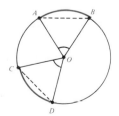

FIGURE 2 Central Angles

5.1
THE UNIT CIRCLE

We begin by discussing the **unit circle,** a circle of radius 1 whose center is at the origin of a rectangular coordinate system (Figure 3a). A point $P(x, y)$ is on the unit circle if and only if the distance $\overline{OP} = 1$. Using the distance formula,

$$\overline{OP} = \sqrt{(x - 0)^2 + (y - 0)^2} = 1$$

Squaring both sides, we conclude

> The equation of the unit circle is
> $$x^2 + y^2 = 1$$

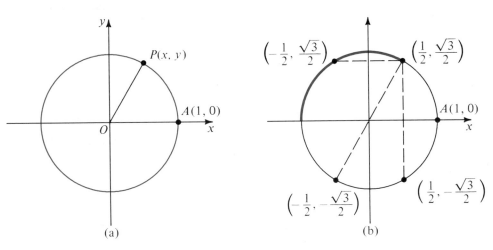

(a)   (b)

FIGURE 3   The Unit Circle

Using the methods of Section 2.1, we find that the unit circle is symmetric with respect to the x-axis, the y-axis, and the origin. These symmetries will prove to be very useful. For example, you can easily verify that the point $(1/2, \sqrt{3}/2)$ lies on the unit circle. Figure 3b shows the coordinates of various other points that can be obtained from the symmetries of the circle. (See Exercises 44 and 45.)

UNIT CIRCLE POINTS

Our objective now is to establish a correspondence between the real numbers and points on a unit circle. In Figure 4, let P be any point on the unit circle. Then the length of the arc $\overset{\frown}{AP}$ measured in the *counterclockwise* direction starting at $A(1, 0)$ is a unique nonnegative number t. (If P coincides with A we let $t = 0$.) Since the circumference of the unit circle is 2π, we see that $t < 2\pi$. Then, for *every* point P on the unit circle we have associated a real number t in the interval $[0, 2\pi)$.

It's not difficult to see that we can reverse this association. Given a real number t in the interval $[0, 2\pi)$, we begin at $A(1, 0)$ and proceed in the *counterclockwise* direction around the unit circle until we reach a point P such that arc $\overset{\frown}{AP}$ is of length t. We have

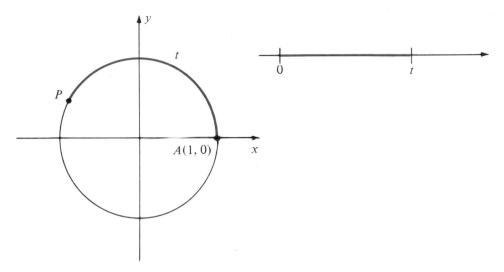

FIGURE 4   **Real Numbers t on the Unit Circle**

succeeded in establishing this result:

> There is a one-to-one correspondence between the real numbers in the interval $[0, 2\pi)$ and points on the unit circle.

Since we will be dealing with arcs on a unit circle, it is convenient to establish a standard notation and terminology. The letter A will always denote the point $A(1, 0)$ as in Figure 4. Given a real number t, if the arc that is swept out terminates at a point P, then we will say that P is the **unit circle point** corresponding to t or determined by t. We will let $P(t)$ denote the unit circle point determined by t and will write $P(t) = P(x, y)$ to indicate that the rectangular coordinates of the unit circle point are (x, y).

Our objective is to establish a correspondence between *any* real number t and its unit circle point $P(t)$. First, let's find the unit circle points that correspond to some "special" values of t.

EXAMPLE 1   Finding Coordinates of Quarter-Circle Points

Find the coordinates of the unit circle point $P(t)$ corresponding to each of the following real numbers:

(a) $t = \dfrac{\pi}{2}$ **(b)** $t = \pi$ **(c)** $t = \dfrac{3\pi}{2}$

SOLUTION

Since the unit circle has a circumference of 2π, an arc of length $\pi/2$ sweeps out $\frac{1}{4}$ of the circle. The unit circle points $P(\pi/2)$, $P(\pi)$, and $P(3\pi/2)$ are "quarter-circle points," that is, they are points at which the unit circle intersects a coordinate axis. From Figure 5 we

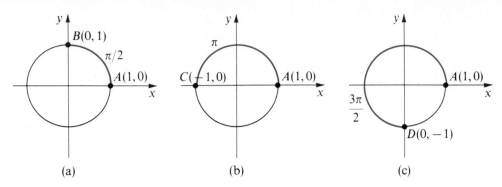

FIGURE 5 **Coordinates of Quarter-Circle Points**

see that

(a) $P\left(\dfrac{\pi}{2}\right) = P(0, 1)$ **(b)** $P(\pi) = P(-1, 0)$ **(c)** $P\left(\dfrac{3\pi}{2}\right) = P(0, -1)$

■

We will next extend the procedure so that we can determine a unit circle point for real numbers t such that $t \geq 2\pi$. We will sweep out an arc of length t in the counterclockwise direction, making as many revolutions as are needed. Since each revolution sweeps out an arc of length 2π and returns to the initial point $A(1, 0)$, we can "discard" all multiples of 2π. For example, if $t = 5\pi/2$, we can write

$$t = \frac{5\pi}{2} = 2\pi + \frac{\pi}{2}$$

Sweeping out an arc of length 2π takes us back to $A(1, 0)$. We then sweep out the arc of length $\pi/2$ to arrive at the point $P(0, 1)$. Conclusion: the unit circle point $P(0, 1)$ corresponds to both $t = 5\pi/2$ and $t = \pi/2$. Stated succinctly: $P(\pi/2) = P(5\pi/2) = P(0, 1)$. (We'll have more to say about this phenomenon in a little while.)

EXAMPLE 2 Finding Unit Circle Points
Find the unit circle point corresponding to each of the following:

(a) $t = 6\pi$ **(b)** $t = 5\pi$ **(c)** $t = \dfrac{11\pi}{2}$

SOLUTION
(a) Since $t = 6\pi = 3 \cdot 2\pi$, we must complete three revolutions, returning to the unit circle point $(1, 0)$. Thus, $P(6\pi) = P(1, 0)$.

(b) If we write

$$t = 5\pi = 2 \cdot 2\pi + \pi$$

we then need only sweep out an arc of length π to arrive at the point $(-1, 0)$. Thus, $P(5\pi) = P(-1, 0)$.

(c) We write this in the form

$$t = \frac{11\pi}{2} = 2 \cdot 2\pi + \frac{3\pi}{2}$$

Discarding the multiples of 2π, we sweep out an arc of length $3\pi/2$ that takes us to the desired unit circle point $(0, -1)$. Thus, $P(11\pi/2) = P(0, -1)$.

∎

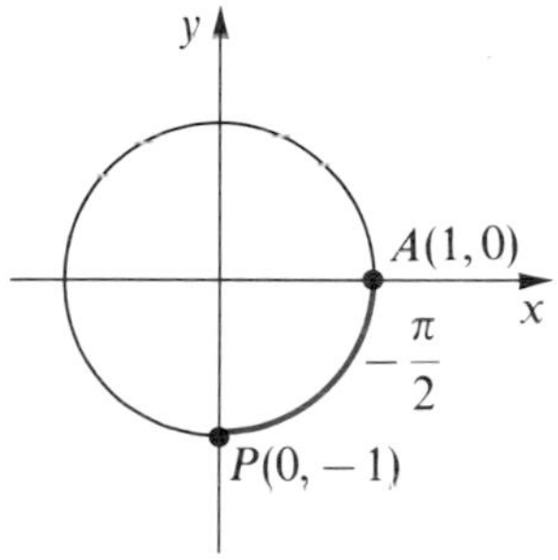

Finally, we define the unit circle point corresponding to a negative value of t. When $t < 0$, we will sweep out an arc of length $|t|$ in the *clockwise* direction. For example, when $t = -\pi/2$, sweeping out an arc of length $\pi/2$ in the clockwise direction takes us to the unit circle point $P(0, -1)$ as in Figure 6. We have established the following:

For every real number t there is a unique unit circle point $P(t)$ that is found by sweeping out an arc of length $|t|$ in the
(a) counterclockwise direction when $t \geq 0$;
(b) clockwise direction when $t < 0$.

FIGURE 6 $P(-\pi/2)$

Note that the correspondence is not one-to-one since the same unit circle point corresponds to more than one real number. For example, we noted earlier that $P(\pi/2) = P(5\pi/2) = P(0, 1)$. Since the circumference of the unit circle is 2π and every revolution *in either direction* returns to the point $A(1, 0)$, we have this important result:

For all integer values of n,

$$P(t) = P(t + 2\pi n)$$

PROGRESS CHECK
Find the unit circle point corresponding to

(a) $t = \dfrac{-3\pi}{2}$ (b) $t = -3\pi$ (c) $t = \dfrac{-\pi}{2}$

ANSWERS
(a) $P(0, 1)$ (b) $P(-1, 0)$ (c) $P(0, -1)$

"SPECIAL VALUES"

In general, finding the unit circle point $P(t)$ for a given real number t is not a trivial task. We have made use of the "quarter-circle" points that correspond to $t = \pi/2$, π, and $3\pi/2$. The remainder of this section will be dedicated to finding the unit circle points for some additional very special values.

First, let's tackle the value $t = \pi/4$. Since $\pi/4$ is one-half of $\pi/2$, the corresponding unit circle point $P(\pi/4)$ must bisect the arc between $(1, 0)$ and $(0, 1)$ as shown in Figure 7. Consequently, the point $P(\pi/4)$ must lie on the line $y = x$, that is, we may designate the coordinates of $P(\pi/4)$ as $P(a, a)$. Since the coordinates of any point on the unit circle

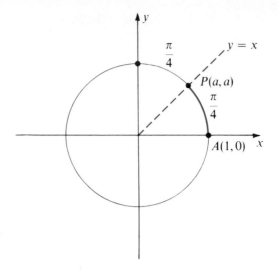

FIGURE 7 $P(\pi/4)$

satisfy $x^2 + y^2 = 1$, we have

$$a^2 + a^2 = 1$$
$$2a^2 = 1$$
$$a = \pm\sqrt{\frac{1}{2}} = \pm\frac{\sqrt{2}}{2}$$

Since P is in the first quadrant we can reject the negative result and conclude that

$$P\left(\frac{\pi}{4}\right) = \left(\frac{\sqrt{2}}{2}, \frac{\sqrt{2}}{2}\right)$$

EXAMPLE 3   Finding Unit Circle Points
Use the symmetries of the circle to find the unit circle points for $t = 3\pi/4$, $5\pi/4$, and $7\pi/4$.

SOLUTION
The points are shown in Figure 8. By using the symmetries of the circle, we determine the coordinates to be

$$P\left(\frac{3\pi}{4}\right) = \left(\frac{-\sqrt{2}}{2}, \frac{\sqrt{2}}{2}\right)$$

$$P\left(\frac{5\pi}{4}\right) = \left(\frac{-\sqrt{2}}{2}, \frac{-\sqrt{2}}{2}\right)$$

$$P\left(\frac{7\pi}{4}\right) = \left(\frac{\sqrt{2}}{2}, \frac{-\sqrt{2}}{2}\right)$$

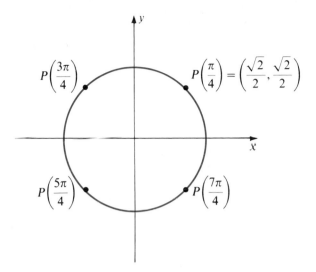

FIGURE 8   Unit Circle Points

We can employ a geometric argument to find the unit circle point for $t = \pi/6$. In Figure 9 we let $P(\pi/6) = P(a, b)$ designate the unit circle point. Then the length of arc $\overset{\frown}{PC}$ can be calculated by

$$\overset{\frown}{PC} = \overset{\frown}{AC} - \overset{\frown}{AP} = \frac{\pi}{2} - \frac{\pi}{6} = \frac{\pi}{3}$$

Locating the unit circle point D corresponding to $t = -\pi/6$, we see from the symmetries of the circle that D has coordinates $(a, -b)$ and that the length of arc $\overset{\frown}{PD}$

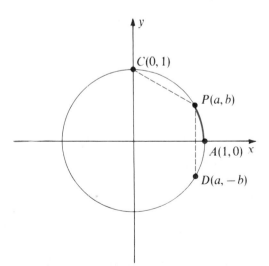

FIGURE 9 $P(\pi/6)$

can be calculated in this way:

$$\overset{\frown}{PD} = \overset{\frown}{PA} + \overset{\frown}{AD} = \frac{\pi}{6} + \frac{\pi}{6} = \frac{\pi}{3}$$

Recall from plane geometry that equal arcs subtend equal chords, so that $\overline{PC} = \overline{PD}$ and, by the distance formula,

$$\sqrt{(a-0)^2 + (b-1)^2} = \sqrt{(a-a)^2 + (b-(-b))^2} = \sqrt{(2b)^2} = 2b$$

Squaring both sides,

$$a^2 + b^2 - 2b + 1 = 4b^2$$

or

$$a^2 + b^2 = 4b^2 + 2b - 1$$

Since (a, b) are the coordinates of a point on the unit circle, they must satisfy the equation $x^2 + y^2 = 1$. Thus, $a^2 + b^2 = 1$, and we have by substitution

$$1 = 4b^2 + 2b - 1$$
$$0 = 4b^2 + 2b - 2$$
$$0 = 2(2b - 1)(b + 1)$$
$$b = \frac{1}{2} \quad \text{or} \quad b = -1$$

Since $P(\pi/6)$ is in the first quadrant, b must be positive. Thus, we have $b = 1/2$. Since $a^2 + b^2 = 1$, we find that $a = \sqrt{3}/2$. We conclude that

$$P\left(\frac{\pi}{6}\right) = \left(\frac{\sqrt{3}}{2}, \frac{1}{2}\right)$$

A similar geometric argument (see Exercise 46) can be used to show that

$$P\left(\frac{\pi}{3}\right) = \left(\frac{1}{2}, \frac{\sqrt{3}}{2}\right)$$

We will be making frequent use of the "special values" and their unit circle points throughout this chapter. Table 1 summarizes these results and should be committed to memory. (We'll provide an alternative means for remembering these results in the next chapter.)

TABLE 1   Unit Circle Points for Special Values

t	0	$\dfrac{\pi}{6}$	$\dfrac{\pi}{4}$	$\dfrac{\pi}{3}$	$\dfrac{\pi}{2}$	π	$\dfrac{3\pi}{2}$
$P(t)$	$(1, 0)$	$\left(\dfrac{\sqrt{3}}{2}, \dfrac{1}{2}\right)$	$\left(\dfrac{\sqrt{2}}{2}, \dfrac{\sqrt{2}}{2}\right)$	$\left(\dfrac{1}{2}, \dfrac{\sqrt{3}}{2}\right)$	$(0, 1)$	$(-1, 0)$	$(0, -1)$

EXAMPLE 4 Circle Symmetry and Unit Circle Points

Use the symmetries of the circle to find $P(2\pi/3)$, $P(7\pi/6)$, and $P(5\pi/3)$.

SOLUTION

The points are shown in Figure 10. By using the symmetries of the circle, we determine the coordinates to be

$$P\left(\frac{2\pi}{3}\right) = \left(\frac{-1}{2}, \frac{\sqrt{3}}{2}\right)$$

$$P\left(\frac{7\pi}{6}\right) = \left(\frac{-\sqrt{3}}{2}, \frac{-1}{2}\right)$$

$$P\left(\frac{5\pi}{3}\right) = \left(\frac{1}{2}, \frac{-\sqrt{3}}{2}\right)$$

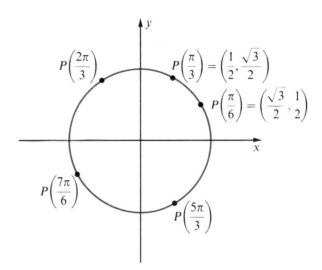

FIGURE 10 "Special Values" and Symmetry

PROGRESS CHECK

Use the symmetries of the circle to find the unit circle points corresponding to

(a) $\dfrac{4\pi}{3}$ (b) $\dfrac{5\pi}{6}$ (c) $\dfrac{11\pi}{6}$

ANSWERS

(a) $\left(-\dfrac{1}{2}, -\dfrac{\sqrt{3}}{2}\right)$ (b) $\left(-\dfrac{\sqrt{3}}{2}, \dfrac{1}{2}\right)$ (c) $\left(\dfrac{\sqrt{3}}{2}, -\dfrac{1}{2}\right)$

EXERCISE SET 5.1

In Exercises 1–12, for each given real number s, find a real number t in the interval $[0, 2\pi)$ so that $P(t) = P(s)$.

1. 4π

2. $\dfrac{13\pi}{2}$

3. $\dfrac{15\pi}{7}$

4. $-\dfrac{25\pi}{4}$

5. $-\dfrac{21\pi}{2}$

6. $-\dfrac{11\pi}{2}$

7. $\dfrac{41\pi}{6}$

8. $\dfrac{11\pi}{2}$

9. -9π

10. 7π

11. $\dfrac{27\pi}{5}$

12. $-\dfrac{22\pi}{3}$

In Exercises 13–16 plot the approximate positions of the unit circle points.

13. $P(7\pi)$, $\left(\dfrac{4\pi}{3}\right)$, $P\left(\dfrac{5\pi}{2}\right)$, $P\left(-\dfrac{7\pi}{4}\right)$, $P\left(-\dfrac{13\pi}{6}\right)$

14. $P\left(\dfrac{11\pi}{2}\right)$, $P\left(-\dfrac{3\pi}{4}\right)$, $P\left(\dfrac{11\pi}{6}\right)$, $P\left(-\dfrac{10\pi}{3}\right)$, $P\left(\dfrac{33\pi}{4}\right)$

15. $P(-10)$, $P(8)$, $P(3.3)$, $P(-4)$, $P(1.7)$

16. $P(14)$, $P(-0.5)$, $P(-8)$, $P(6)$, $P(3)$

In Exercises 17–32 find the rectangular coordinates of the given point.

17. $P(5\pi)$

18. $P\left(\dfrac{5\pi}{2}\right)$

19. $P\left(-\dfrac{\pi}{4}\right)$

20. $P\left(-\dfrac{3\pi}{2}\right)$

21. $P\left(\dfrac{5\pi}{4}\right)$

22. $P(8\pi)$

23. $P\left(\dfrac{4\pi}{3}\right)$

24. $P\left(\dfrac{2\pi}{3}\right)$

25. $P\left(-\dfrac{2\pi}{3}\right)$

26. $P\left(-\dfrac{19\pi}{3}\right)$

27. $P\left(\dfrac{19\pi}{6}\right)$

28. $P\left(\dfrac{17\pi}{6}\right)$

29. $P\left(-\dfrac{5\pi}{6}\right)$

30. $P\left(-\dfrac{11\pi}{6}\right)$

31. $P\left(\dfrac{19\pi}{3}\right)$

32. $P\left(\dfrac{25\pi}{3}\right)$

In Exercises 33–41 determine both a positive and a negative real number t, $|t| < 2\pi$, for which $P(t)$ has the following rectangular coordinates.

33. $(-1, 0)$

34. $(0, -1)$

35. $\left(\dfrac{-\sqrt{2}}{2}, \dfrac{\sqrt{2}}{2}\right)$

36. $\left(\dfrac{\sqrt{2}}{2}, \dfrac{-\sqrt{2}}{2}\right)$

37. $\left(\dfrac{-\sqrt{3}}{2}, \dfrac{1}{2}\right)$

38. $\left(\dfrac{-1}{2}, \dfrac{-\sqrt{3}}{2}\right)$

39. $\left(\dfrac{1}{2}, \dfrac{-\sqrt{3}}{2}\right)$

40. $\left(\dfrac{-\sqrt{3}}{2}, \dfrac{-1}{2}\right)$

41. $\left(\dfrac{-1}{2}, \dfrac{\sqrt{3}}{2}\right)$

42. Given $P(t) = \left(\dfrac{3}{5}, \dfrac{4}{5}\right)$, use the symmetries of the circle to find

(a) $P(t + \pi)$ (b) $P\left(t - \dfrac{\pi}{2}\right)$

(c) $P(-t)$ (d) $P(-t - \pi)$

43. Given $P(t) = \left(-\dfrac{4}{5}, -\dfrac{3}{5}\right)$, use the symmetries of the circle to find

(a) $P(t - \pi)$ (b) $P\left(t + \dfrac{\pi}{2}\right)$

(c) $P(-t)$ (d) $P(-t + \pi)$

44. If the point (a, b) is on the unit circle, show that $(a, -b)$, $(-a, b)$, and $(-a, -b)$ also lie on the unit circle.

45. If the point (a, b) is on the unit circle, show that (b, a), $(b, -a)$, $(-b, a)$, and $(-b, -a)$ also lie on the unit circle.

46. Show that $P\left(\dfrac{\pi}{3}\right) = \left(\dfrac{1}{2}, \dfrac{\sqrt{3}}{2}\right)$. (*Hint:* If $P\left(\dfrac{\pi}{3}\right) = (a, b)$, then $P\left(\dfrac{2\pi}{3}\right) = (-a, b)$. The arc from $A(1, 0)$ to $P\left(\dfrac{\pi}{3}\right)$ and the arc from $P\left(\dfrac{\pi}{3}\right)$ to $P\left(\dfrac{2\pi}{3}\right)$ are equal.)

In Exercises 47–49 the points $P(t) = P(a, b)$ and $P(t \pm \pi/2) = P'(a', b')$ as shown in Figure 11.

47. Show that $\dfrac{b'}{a'} = -\dfrac{a}{b}$. (*Hint:* Show that the lines OP and OP' are perpendicular and then determine their slopes.)

48. Show that $b' = \pm a$ and $a' = \pm b$. (*Hint:* The radii OP and OP' are equal in length. Use the distance formula combined with the result of Exercise 47 above to substitute alternately for b' and for a'.)

49. Show that either (i) $(a', b') = (-b, a)$ or (ii) $(a', b') = (b, -a)$ (*Hint:* Begin with the result of Exercise 48 and apply the result of Exercise 47.)

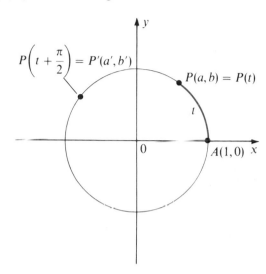

FIGURE 11

5.2
THE SINE, COSINE, AND TANGENT FUNCTIONS

The correspondence between real numbers and points on the unit circle is the key to the definitions of the six **trigonometric** or **circular functions.**

DEFINITION AND DOMAIN

In this section we will discuss the **sine, cosine,** and **tangent** functions, which are written as **sin, cos,** and **tan,** respectively. The remaining three functions are reciprocals of these and will be described in a later section. We now define the first three circular functions.

Circular Functions	If t is a real number and $P(t) = P(x, y)$, then
	$$\sin t = y$$ $$\cos t = x$$ $$\tan t = \frac{y}{x}, \qquad x \neq 0$$

It is often convenient to think of these definitions in this way (see Figure 12):

> If t is a real number and $P(t) = P(x, y)$, then
> $$P(t) = P(x, y) = (\cos t, \sin t)$$

We see immediately that the domain of the sine and cosine functions is the set of all real numbers. The tangent function, however, is not defined when $x = 0$. The unit circle points in the interval $[0, 2\pi)$ having an x-coordinate of 0 are $P(\pi/2)$ and $P(3\pi/2)$. Since

FIGURE 12 $P(t) = (\cos t, \sin t)$

$3\pi/2 = \pi/2 + \pi$ and since every revolution returns us to the same unit circle point, we see that the x-coordinate of the unit circle point is 0 whenever $t = \pi/2 + n\pi$ for all integers n.

Domain of the Circular Functions	$\sin t$: all real values of t
	$\cos t$: all real values of t
	$\tan t$: all real values of t such that $t \neq \dfrac{\pi}{2} + n\pi$ for all integers n

SPECIAL VALUES OF THE CIRCULAR FUNCTIONS

Given a real number t, if we can find the rectangular coordinates of $P(t)$, we can then easily determine the values of the circular functions. Here are some examples.

EXAMPLE 1 Finding Values of the Circular Functions
Evaluate $\sin t$, $\cos t$, and $\tan t$ for each of the following.

(a) $t = 0$ **(b)** $t = \dfrac{\pi}{6}$ **(c)** $t = \dfrac{\pi}{4}$

SOLUTION
See Figure 13.

(a) $P(0) = (1, 0)$. Then $\sin 0 = y = 0$, $\cos 0 = x = 1$, and $\tan 0 = \dfrac{y}{x} = 0$.

(b) $P\left(\dfrac{\pi}{6}\right) = \left(\dfrac{\sqrt{3}}{2}, \dfrac{1}{2}\right)$. Then

$$\sin \frac{\pi}{6} = y = \frac{1}{2}, \qquad \cos \frac{\pi}{6} = x = \frac{\sqrt{3}}{2}$$

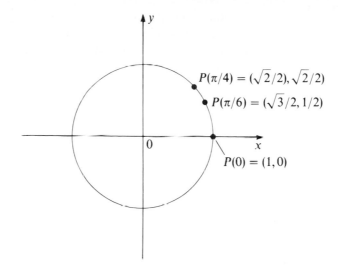

$P(\pi/4) = (\sqrt{2}/2), \sqrt{2}/2)$

$P(\pi/6) = (\sqrt{3}/2, 1/2)$

$P(0) = (1, 0)$

FIGURE 13   Special Values

$$\tan\frac{\pi}{6} = \frac{y}{x} = \frac{\dfrac{1}{2}}{\dfrac{\sqrt{3}}{2}} = \frac{1}{\sqrt{3}} = \frac{\sqrt{3}}{3}$$

(c) $P\left(\dfrac{\pi}{4}\right) = \left(\dfrac{\sqrt{2}}{2}, \dfrac{\sqrt{2}}{2}\right)$. Then

$$\sin\frac{\pi}{4} = \cos\frac{\pi}{4} = \frac{\sqrt{2}}{2}$$

$$\tan\frac{\pi}{4} = 1$$

■

PROGRESS CHECK

Evaluate $\sin t$, $\cos t$, and $\tan t$ for each of the following.

(a) $t = \dfrac{\pi}{2}$ (b) $t = \dfrac{\pi}{3}$ (c) $t = \pi$

ANSWERS

(c) $\sin \pi = 0$, $\cos \pi = -1$, $\tan \pi = 0$

(b) $\sin\dfrac{\pi}{3} = \dfrac{\sqrt{3}}{2}$, $\cos\dfrac{\pi}{3} = \dfrac{1}{2}$, $\tan\dfrac{\pi}{3} = \sqrt{3}$

(a) $\sin\dfrac{\pi}{2} = 1$, $\cos\dfrac{\pi}{2} = 0$, $\tan\dfrac{\pi}{2}$ is undefined

Since we will frequently refer to the values of the sine, cosine, and tangent functions for certain "special" real numbers t, we list these in Table 2. Note that these values of t are in the interval $[0, 2\pi)$ and that we already know $P(t)$. There are no entries for $\tan \pi/2$ and $\tan 3\pi/2$ since the tangent function is not defined for these values.

TABLE 2   "Special Values"

t	$P(t)$	$\sin t$	$\cos t$	$\tan t$
0	$(1, 0)$	0	1	0
$\dfrac{\pi}{6}$	$\left(\dfrac{\sqrt{3}}{2}, \dfrac{1}{2}\right)$	$\dfrac{1}{2}$	$\dfrac{\sqrt{3}}{2}$	$\dfrac{\sqrt{3}}{3}$
$\dfrac{\pi}{4}$	$\left(\dfrac{\sqrt{2}}{2}, \dfrac{\sqrt{2}}{2}\right)$	$\dfrac{\sqrt{2}}{2}$	$\dfrac{\sqrt{2}}{2}$	1
$\dfrac{\pi}{3}$	$\left(\dfrac{1}{2}, \dfrac{\sqrt{3}}{2}\right)$	$\dfrac{\sqrt{3}}{2}$	$\dfrac{1}{2}$	$\sqrt{3}$
$\dfrac{\pi}{2}$	$(0, 1)$	1	0	Not defined
π	$(-1, 0)$	0	-1	0
$\dfrac{3\pi}{2}$	$(0, -1)$	-1	0	Not defined

PROPERTIES OF THE CIRCULAR FUNCTIONS

In mathematics, whenever we define a new quantity or function, we then proceed to investigate its properties. We will spend the rest of this section determining some simple properties of the circular functions.

The *signs* of the circular functions in each of the four quadrants are shown in Figure 14a. These follow immediately from the definitions. For example, since both the x- and y-coordinates of any point in the third quadrant are negative, the sine and cosine

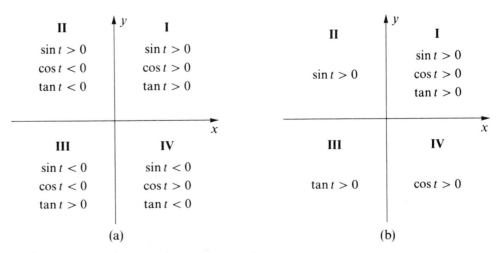

FIGURE 14   Signs of the Trigonometric Functions

functions both have negative values if $P(t)$ is in the third quadrant. The tangent will be positive in the third quadrant since it is the ratio y/x of two negative values.

Figure 14b shows where each of the circular functions is positive. It is easier to remember this and to determine the negative values by inference.

EXAMPLE 2   Finding the Quadrant of $P(t)$

Determine the quadrant in which $P(t)$ lies in each of the following.
(a) $\sin t > 0$ and $\tan t < 0$ (b) $\sin t < 0$ and $\cos t > 0$

SOLUTION

(a) $\sin t > 0$ in quadrants I and II; $\tan t < 0$ in quadrants II and IV. Both conditions therefore apply only in quadrant II.

(b) $\sin t < 0$ in quadrants III and IV; $\cos t > 0$ in quadrants I and IV. Both conditions therefore apply only in quadrant IV.

■

PROGRESS CHECK

Determine the quadrant in which $P(t)$ lies in each of the following.
(a) $\cos t < 0$ and $\tan t > 0$ (b) $\cos t < 0$ and $\sin t > 0$

ANSWERS
(a) quadrant III   (b) quadrant II

We can use the symmetries of the unit circle to find $\sin(-t)$ and $\cos(-t)$. In Figure 15 we see that $P(t)$ and $P(-t)$ correspond to points having the same x-coordinates while the y-coordinates are opposite in sign. Then $\sin t = y$ and $\sin(-t) = -y$ or $\sin(-t) = -\sin t$. Similarly, $\cos t = x = \cos(-t)$. Finally, $\tan(-t) = -y/x = -\tan t$. In summary,

$$\sin(-t) = -\sin t$$
$$\cos(-t) = \cos t$$
$$\tan(-t) = -\tan t$$

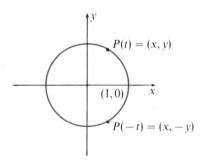

FIGURE 15 $P(t)$ and $P(-t)$ on a Unit Circle

where t is any real number in the domain of the function. A function f for which $f(-x) = f(x)$ is said to be an **even function;** if $f(-x) = -f(x)$, then f is called an **odd function.** We see that sine and tangent are odd functions and cosine is an even function. Another example of an even function is $f(x) = x^2$ while $f(x) = x^3$ is an example of an odd function. From our earlier work with graphs you can see that the graph of an even function is symmetric about the y-axis while the graph of an odd function is symmetric with respect to the origin.

EXAMPLE 3 Even and Odd Functions

Find $\sin\left(-\dfrac{\pi}{4}\right)$ and $\cos\left(-\dfrac{\pi}{3}\right)$.

SOLUTION

$$\sin\left(-\frac{\pi}{4}\right) = -\sin\left(\frac{\pi}{4}\right) = -\frac{\sqrt{2}}{2} \qquad \cos\left(-\frac{\pi}{3}\right) = \cos\left(\frac{\pi}{3}\right) = \frac{1}{2}$$

∎

IDENTITIES

Trigonometry often involves the use of **identities,** that is, equations that are true for *all* values in the domain of the variable. Identities are useful in simplifying equations and in providing alternative forms for computations. We can now establish two fundamental identities of trigonometry.

Since the coordinates (x, y) of every point on the unit circle satisfy the equation $x^2 + y^2 = 1$, we may substitute $x = \cos t$ and $y = \sin t$ to obtain

$$(\cos t)^2 + (\sin t)^2 = 1$$

Expressions of the form $(\sin t)^n$ occur so frequently that a special notation is used:

$$\sin^n t = (\sin t)^n \quad \text{when } n \neq -1$$

Using this notation and reordering the terms, the identity becomes

$$\sin^2 t + \cos^2 t = 1$$

Of course, we may also use this identity in the alternative forms

$$\sin^2 t = 1 - \cos^2 t$$
$$\cos^2 t = 1 - \sin^2 t$$

Since $\tan t = y/x$, $x \neq 0$, we may substitute $\sin t = y$ and $\cos t = x$ to obtain

$$\tan t = \frac{\sin t}{\cos t}$$

for all values of t in the domain of the tangent function.

EXAMPLE 4   Applying the Trigonometric Identities

If $\cos t = \dfrac{3}{5}$ and t is in quadrant IV, find $\sin t$ and $\tan t$.

SOLUTION

Using the identity $\sin^2 t + \cos^2 t = 1$, we have

$$\sin^2 t + \left(\frac{3}{5}\right)^2 = 1$$

$$\sin^2 t = 1 - \frac{9}{25} = \frac{16}{25}$$

$$\sin t = \pm \frac{4}{5}$$

Since t is in quadrant IV, $\sin t$ must be negative so that $\sin t -- \dfrac{4}{5}$. Then

$$\tan t = \frac{\sin t}{\cos t} = \frac{-\dfrac{4}{5}}{\dfrac{3}{5}} = -\frac{4}{3}$$

■

PROGRESS CHECK

If $\sin t = \dfrac{12}{13}$ and t is in quadrant II, find the following.

(**a**) $\cos t$ (**b**) $\tan t$

ANSWERS

(a) $-\dfrac{5}{13}$ (b) $-\dfrac{12}{5}$

EXAMPLE 5   Proving an Identity

Show that $1 + \tan^2 x = \dfrac{1}{\cos^2 x}$.

SOLUTION

We will use the trigonometric identities to transform the left-hand side of the equation into the right-hand side. Since $\tan x = \dfrac{\sin x}{\cos x}$, we have

$$1 + \tan^2 x = 1 + \frac{\sin^2 x}{\cos^2 x}$$

$$= \frac{\cos^2 x + \sin^2 x}{\cos^2 x}$$

Since $\cos^2 x + \sin^2 x = 1$,

$$1 + \tan^2 x = \frac{1}{\cos^2 x}$$

■

PROGRESS CHECK

Use identities to transform the expression $(\tan t)(\cos t) + \sin t$ to $2\sin t$.

EXERCISE SET 5.2

In Exercises 1–12 use the rectangular coordinates of $P(t)$ to find $\sin t$, $\cos t$, $\tan t$.

1. $t = \dfrac{5\pi}{3}$

2. $t = \dfrac{3\pi}{4}$

3. $t = -5\pi$

4. $t = -\dfrac{5\pi}{4}$

5. $t = \dfrac{7\pi}{4}$

6. $t = \dfrac{7\pi}{6}$

7. $t = \dfrac{2\pi}{3}$

8. $t = \dfrac{5\pi}{6}$

9. $t = -\dfrac{\pi}{3}$

10. $t = -\dfrac{5\pi}{6}$

11. $t = \dfrac{5\pi}{4}$

12. $t = -11\pi$

In Exercises 13–18 find the quadrant in which $P(t)$ lies if the following conditions hold.

13. $\sin t > 0$ and $\cos t < 0$

14. $\sin t < 0$ and $\tan t > 0$

15. $\cos t < 0$ and $\tan t > 0$

16. $\tan t < 0$ and $\sin t > 0$

17. $\sin t < 0$ and $\cos t < 0$

18. $\tan t < 0$ and $\cos t < 0$

In Exercises 19–36 find the values of t in the interval $[0, 2\pi)$ that satisfy the given equation.

19. $\cos t = 0$

20. $\sin t = 1$

21. $\tan t = 1$

22. $\cos t = -1$

23. $\sin t = \dfrac{\sqrt{2}}{2}$

24. $\cos t = \dfrac{\sqrt{3}}{2}$

25. $\cos t = -\dfrac{\sqrt{3}}{2}$

26. $\sin t = -1$

27. $\tan t = \sqrt{3}$

28. $\sin t = \dfrac{3}{2}$

29. $\sin t = \dfrac{\sqrt{3}}{2}$

30. $\tan t = \dfrac{\sqrt{3}}{3}$

31. $\sin t = -\dfrac{1}{2}$

32. $\cos t = -\dfrac{\sqrt{2}}{2}$

33. $\sin t = 2$

34. $\cos t = \dfrac{\sqrt{2}}{2}$

35. $\sin t = \dfrac{1}{2}$

36. $\cos t = \dfrac{3}{2}$

In Exercises 37–44 use trigonometric identities to find the indicated value under the given conditions.

37. $\sin t = \dfrac{3}{5}$ and $P(t)$ is in quadrant II; find $\tan t$.

38. $\tan t = -\dfrac{3}{4}$ and $P(t)$ is in quadrant II; find $\cos t$.

39. $\cos t = -\dfrac{5}{13}$ and $P(t)$ is in quadrant III; find $\sin t$.

40. $\sin t = -\dfrac{5}{13}$ and $P(t)$ is in quadrant III; find $\tan t$.

41. $\cos t = \dfrac{4}{5}$ and $\sin t < 0$; find $\sin t$.

42. $\tan t = \dfrac{12}{5}$ and $\cos t < 0$; find $\sin t$.

43. $\sin t = -\dfrac{3}{5}$ and $\tan t < 0$; find $\cos t$.

44. $\tan t = -\dfrac{5}{12}$ and $\sin t > 0$; find $\sin t$.

In Exercises 45–54 use the trigonometric identities to transform the first expression into the second.

45. $(\tan t)(\cos t), \quad \sin t$

46. $\dfrac{\cos t}{\sin t}, \quad \dfrac{1}{\tan t}$

47. $\dfrac{1 - \sin^2 t}{\sin t}, \quad \dfrac{\cos t}{\tan t}$

48. $(\tan t)(\sin t) + \cos t, \quad \dfrac{1}{\cos t}$

49. $\cos t\left(\dfrac{1}{\cos t} - \cos t\right), \quad \sin^2 t$

50. $\dfrac{1 - \cos^2 t}{\sin t}, \quad \sin t$

51. $\dfrac{1 - \cos^2 t}{\cos^2 t}, \quad \tan^2 t$

52. $\dfrac{\cos^2 t}{1 - \sin t}, \quad 1 + \sin t$

53. $(\sin t - \cos t)^2, \quad 1 - 2(\sin t)(\cos t)$

54. $\dfrac{1}{1 - \sin t} + \dfrac{1}{1 + \sin t}, \quad \dfrac{2}{\cos^2 t}$

5.3
GRAPHS OF SINE, COSINE, AND TANGENT

PERIODIC FUNCTIONS

Before we tackle the problem of graphing the sine, cosine, and tangent functions, it is helpful to observe that these functions exhibit a cyclical behavior. This characteristic makes these functions especially useful in describing cyclic phenomena in a wide variety of applications. The following definition gives a name to such functions.

> A function f is **periodic** if there exists a positive number c such that
> $$f(x + c) = f(x)$$
> for all x in the domain of f. The least number c for which f is periodic is called the **period** of f.

Since $P(t + 2\pi) = P(t)$ it follows immediately that

$$\sin(t + 2\pi) = \sin t \qquad \text{and} \qquad \cos(t + 2\pi) = \cos t$$

Thus, the sine and cosine functions are periodic functions.

It is not difficult to show that the period of the sine and cosine functions is 2π (see Exercises 11 and 12 in Exercise Set 5.3), that is, 2π is the smallest positive real value of c such that $\sin(t + c) = \sin t$ and $\cos(t + c) = \cos t$ for all real numbers t.

GRAPHS OF SINE AND COSINE

The definitions of $\sin t$ and $\cos t$ as the coordinates of points on the unit circle tell us that they must satisfy the inequalities

$$-1 \le \sin t \le 1$$
$$-1 \le \cos t \le 1$$

Focusing on the graph of $\sin t$, we can therefore indicate these maximum and minimum values on the y-axis as shown in Figure 16. Since these functions have a period of 2π, we need only sketch their graphs over the interval $[0, 2\pi]$ and then repeat the graph for every interval of length 2π. As usual, we form a table of values over the interval $[0, 2\pi]$ using the results of the last section (see Table 3). Finally, we plot these points for $y = \sin t$ on a ty coordinate system as shown in Figure 16, using the approximations

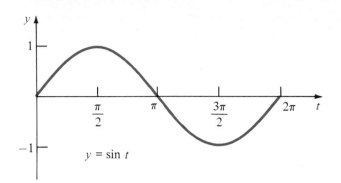

FIGURE 16   **Graph of y = sin t over [0, 2π]**

TABLE 3   Plotting sin t and cos t

t	0	$\dfrac{\pi}{6}$	$\dfrac{\pi}{4}$	$\dfrac{\pi}{3}$	$\dfrac{\pi}{2}$	$\dfrac{3\pi}{4}$	π	$\dfrac{5\pi}{4}$	$\dfrac{3\pi}{2}$	$\dfrac{7\pi}{4}$	2π
$\sin t$	0	0.50	0.71	0.87	1	0.71	0	-0.71	-1	-0.71	0
$\cos t$	1	0.87	0.71	0.50	0	-0.71	-1	-0.71	0	0.71	1

$\sqrt{2} \approx 1.414$ and $\sqrt{3} \approx 1.732$. (We use the label t for the horizontal axis since we are letting t denote the independent variable.)

We can repeat the graph of Figure 16 for adjacent intervals of width 2π to yield the graph of $y = \sin t$ in Figure 17.

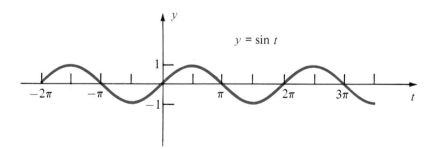

FIGURE 17   **Graph of y = sin t**

Turning to the cosine function, we can use values given in Table 2 to sketch the graph of $y = \cos t$ as in Figure 18.

To graph the tangent function, we first establish that $\tan(t + \pi) = \tan t$ for all real values of t. If $P(x, y)$ is any point on the unit circle, then $P'(-x, -y)$ also lies on the unit circle (Figure 19), and arc $\overset{\frown}{PP'}$ is of length π. If $P(t) = P(x, y)$, then

$$\tan t = \frac{y}{x}$$

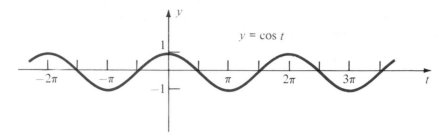

FIGURE 18   Graph of y = cos t

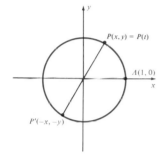

FIGURE 19   Circle Symmetry

and

$$\tan(t + \pi) = \frac{-y}{-x} = \frac{y}{x}$$

so that $\tan(t + \pi) = \tan t$. It is easy to show that there is no real number $c, 0 < c < \pi$, such that $\tan(t + c) = \tan t$ for all real numbers t (see Exercise 13). Hence, the tangent function has period π.

Table 4 provides us with some values of the tangent function for $-\pi/2 < t < \pi/2$. Since $\tan t$ is undefined at $\pi/2$ and at $-\pi/2$, we need to carefully consider the behavior of the graph *near* these values of t. As t increases from 0 toward $\pi/2$, the x-coordinate of $P(t)$ gets closer and closer to 0. Since $\tan t = y/x$, arbitrarily small values of x produce arbitrarily large values for the quotient y/x. We say that $\tan t$ **approaches positive infinity** as t approaches $\pi/2$. Similarly, as t decreases from 0 toward $-\pi/2$, $\tan t$ grows larger and larger in the negative sense. Accordingly, we say that $\tan t$ **approaches negative infinity** as t approaches $-\pi/2$. These considerations lead us to the graph of $\tan t$ shown in Figure 20. The vertical, dashed lines are vertical asymptotes.

TABLE 4   Plotting tan t

t	$-\dfrac{\pi}{2}$	$-\dfrac{\pi}{3}$	$-\dfrac{\pi}{4}$	$-\dfrac{\pi}{6}$	0	$\dfrac{\pi}{6}$	$\dfrac{\pi}{4}$	$\dfrac{\pi}{3}$	$\dfrac{\pi}{2}$
$\tan t$		-1.73	-1	-0.58	0	0.58	1	1.73	

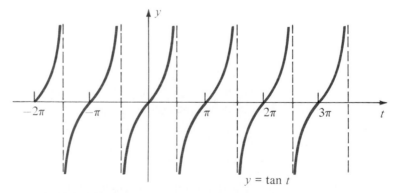

FIGURE 20   Graph of y = tan t

RANGE OF THE CIRCULAR FUNCTIONS

From the graphs of the sine, cosine, and tangent functions it is easy to determine the range of these functions. We list both the range and the period in Table 5.

TABLE 5   Range and Period

	$\sin t$	$\cos t$	$\tan t$
Range	$-1 \le y \le 1$	$-1 \le y \le 1$	all real numbers
Period	2π	2π	π

EXAMPLE 1   Graphing by "Addition"
Sketch the graph of $f(t) = 1 + \sin t$.

SOLUTION
Rather than form a table of values and plot points, we simply note that the graph of $f(t) = 1 + \sin t$ is that of $\sin t$ shifted upward one unit for each value of t. In Figure 21 we have sketched $\sin t$ with dashed lines and $f(t) = 1 + \sin t$ with a solid line.

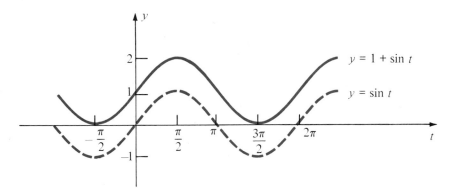

FIGURE 21   Graph of $y = 1 + \sin t$

■

EXAMPLE 2   Graphing by "Addition"
Sketch the graph of $f(t) = \sin t + \cos t$.

SOLUTION
Again, rather than plot points, we note that the y-coordinate of $f(t) = \sin t + \cos t$ is simply the sum of the y-coordinates of $\sin t$ and $\cos t$ for each value of t. In Figure 22 we have sketched the graphs of $\sin t$ and $\cos t$ with dashed lines, formed the sum of the y-coordinates geometrically, and then sketched a smooth curve through the resulting points.

■

The definition of the circular functions and their graphs can provide alternative ways of tackling problems or verifying basic results. Here is an example.

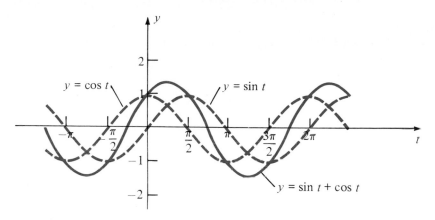

FIGURE 22 Graph of y = sin t + cos t

EXAMPLE 3 Verifying an Identity
Verify that $\tan(-t) = -\tan t$ by using the definition of $\tan t$.

SOLUTION
In Figure 23 we let $P(t) = P(x, y)$ for some real number t. We see that the coordinates of $P(-t)$ are then $(x, -y)$. By definition,

$$\tan t = \frac{y}{x} \qquad \text{and} \qquad \tan(-t) = \frac{-y}{x} = -\frac{y}{x}$$

which establishes that $\tan(-t) = -\tan t$.

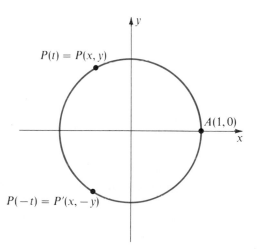

FIGURE 23 Diagram for Example 3

EXERCISE SET 5.3

In Exercises 1–10 sketch the graph of each given function.

1. $f(t) = 1 + \cos t$

2. $f(t) = -1 + \sin t$

3. $f(t) = 2 \sin t$

4. $f(t) = \dfrac{1}{2} \cos t$

5. $f(t) = \sin t + \dfrac{1}{2} \cos t$

6. $f(t) = 2 \sin t + \cos t$

7. $f(t) = \sin t - \cos t$

8. $f(t) = \sin(-t) + \cos t$

9. $f(t) = t + \sin t$

10. $f(t) = -t + \cos t$

11. Prove that the period of the sine function is 2π. (*Hint:* Assume $\sin(t + c) = \sin t, 0 < c < 2\pi$, for all t. By letting $t = 0$, show that $\sin c = 0$ and, consequently, that $c = \pi$. Finally, conclude that $\sin(t + \pi) = \sin t$ does not hold for $t = \pi/2$.)

12. Prove that the period of the cosine function is 2π.

13. Prove that the period of the tangent function is π.

14. Verify that $\sin(-t) = -\sin t$ by using the graph of the sine function.

15. Verify that $\cos(-t) = \cos t$ by using the graph of the cosine function.

16. Determine the domain and range of the functions in Exercises 1 through 10 by examining the graph of each function.

17. Use the identity

$$\tan t = \frac{\sin t}{\cos t}$$

to determine the vertical asymptotes of the graph of $\tan t$.

5.4
VALUES OF SINE, COSINE, AND TANGENT

Although we now know the shape of the curves $y = \sin t$ and $y = \cos t$, we still don't know the precise value of $\sin t$ and $\cos t$ for an arbitrary value of t. For example, what is the value of $\sin 0.34$? of $\cos 1.92$?

In this section we'll explore several ways to find values of a trigonometric function: by use of a calculator, by tables, and by polynomial approximation. Along the way we'll learn an important procedure called the Reference Number Rule.

CALCULATORS

When you turn on a scientific calculator, the display indicates one of the following modes:

DEG   RAD   GRAD

which are abbreviations for degree, radian, and grad, respectively. To insure that our arguments are treated as real numbers, *you must be sure that the display reads RAD.* To do this, find the key marked ⌑DRG⌑. Pressing this key causes the display to rotate amongst the three candidates. Simply stop when it reads RAD. (We will deal with degree measure in the next chapter. The measure known as grads is used in Europe but is virtually unknown in the United States.)

Now, with the display reading RAD, we can proceed to use the keys marked ⌑sin⌑, ⌑cos⌑, and ⌑tan⌑ to evaluate these trigonometric functions for a real valued argument. The key sequence is identical to that of any of the other functions we have used; that is, to find $\sin 1.92$, you would enter

1.92 ⌑sin⌑

and the answer in the display reads 0.9396.

PROGRESS CHECK

Use your calculator to verify the following:

(a) $\cos -3.65 \approx -0.8735$ (b) $\tan -0.5 \approx -0.5463$ (c) $\sin 24 \approx -0.9056$

TABLES OF VALUES AND THE REFERENCE NUMBER RULE

Although a calculator is the simplest and fastest means for finding $\sin t$ and $\cos t$, we can use Table V in the Tables Appendix to motivate exploration of additional properties. If you will turn to this table, you will find the values for $\sin t$ and $\cos t$ for t in the interval $[0, 1.57]$ in increments of 0.01, which corresponds (approximately) to the interval $[0, \pi/2]$.

Since the period of sine and cosine is 2π, we can agree that there is no need for the tables to go beyond 2π. But why does the table terminate at $\pi/2$? How can Table V be used to find $\sin t$ if t is in the interval $(\pi/2, 2\pi)$? The key lies in the following definition.

> The **reference number** t' associated with a real number t is the length of the shortest arc of the unit circle between $P(t)$ and the x-axis.

Figure 24 illustrates the reference number associated with values of t such that $P(t)$ lies in quadrants II, III, and IV. Table 6 summarizes these results. From the last column of

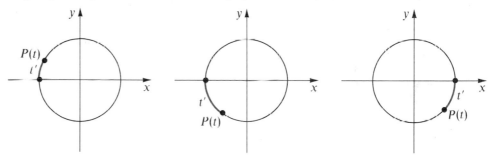

FIGURE 24 Reference Number

TABLE 6 Calculating the Reference Number t'

Interval of t	Quadrant of $P(t)$	t'	Quadrant of $P(t')$
$\left[0, \dfrac{\pi}{2}\right]$	I	t	I
$\left[\dfrac{\pi}{2}, \pi\right]$	II	$\pi - t$	I
$\left[\pi, \dfrac{3\pi}{2}\right]$	III	$t - \pi$	I
$\left[\dfrac{3\pi}{2}, 2\pi\right]$	IV	$2\pi - t$	I

this table we see that

> Given a real number t, the reference number t' lies in the interval $\left[0, \dfrac{\pi}{2}\right]$.

EXAMPLE 1 Finding the Reference Number

Find the reference number for each of the following values of t, using 3.1415 as an approximation for π where needed:

(a) $\dfrac{2\pi}{3}$ (b) $\dfrac{11\pi}{6}$ (c) $\dfrac{5\pi}{4}$ (d) 1.88

SOLUTION

(a) Since $P(2\pi/3)$ lies in quadrant II,

$$t' = \pi - t = \pi - \frac{2\pi}{3} = \frac{\pi}{3}$$

(b) Since $P(11\pi/6)$ lies in quadrant IV,

$$t' = 2\pi - t = 2\pi - \frac{11\pi}{6} = \frac{\pi}{6}$$

(c) Since $P(5\pi/4)$ lies in quadrant III,

$$t' = t - \pi = \frac{5\pi}{4} - \pi = \frac{\pi}{4}$$

(d) Noting that $\pi/2 < 1.88 < \pi$, we see that $P(1.88)$ lies in quadrant II. Therefore,

$$t' = \pi - t = \pi - 1.88 \approx 3.1415 - 1.88 \approx 1.2615$$

■

✓ PROGRESS CHECK

Find the reference number (use 3.1415 as an approximation for π).

(a) $\dfrac{5\pi}{6}$ (b) $\dfrac{7\pi}{4}$ (c) 3.55

ANSWERS

(a) $\dfrac{\pi}{6}$ (b) $\dfrac{\pi}{4}$ (c) 0.4085

We can now use the reference number concept to answer our original question: how can Table V be used to find $\sin t$ and $\cos t$ when t is in the interval $(\pi/2, 2\pi)$? In Figure 25 we illustrate the cases in which $P(t) = P(x, y)$ lies in quadrants II, III, or IV. Since the reference number t' is in the interval $[0, 2\pi]$, the unit circle point $P(t') = P(x', y')$ is in quadrant I. By the symmetries of the unit circle, in all three cases

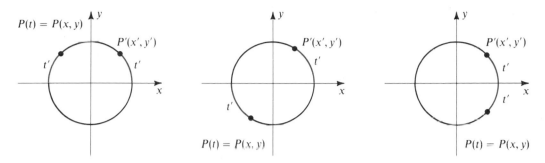

FIGURE 25 Reference Number Rule

we have

$$x' = |x| \qquad \text{and} \qquad y' = |y|$$

Since $\sin t = y$ and $\cos t = x$, we can determine $\sin t$ and $\cos t$ by first finding $\sin t' = y'$ and $\cos t' = x'$ and then adding the proper sign. The procedure we have outlined is known as the **Reference Number Rule** for finding the value of a trigonometric function of a number t, $\pi/2 < t < 2\pi$. We now list the steps involved in this procedure and illustrate with an example.

EXAMPLE 2 Using the Reference Number Rule

Find $\cos 2.55$ using the Reference Number Rule and Table V in the Tables Appendix.

SOLUTION

Reference Number Rule	
Procedure	**Example**
	Find $\cos 2.55$
Step 1. Find the reference number t' associated with t.	*Step 1.* Since $\pi/2 \approx 1.57$ and $\pi \approx 3.14$, $P(2.55)$ is in quadrant II. Thus. $$t' = \pi - t$$ $$= 3.14 - 2.55$$ $$= 0.59$$
Step 2. Obtain the (approximate) value of the required trigonometric function for the reference number t' from Table V in the Tables Appendix, or from a calculator.	*Step 2.* $$\cos 0.59 = 0.8309$$
Step 3. Append the appropriate sign according to the quadrant in which $P(t)$ lies and the trigonometric function being sought.	*Step 3.* Since $\cos t$ is negative in quadrant II, we have $$\cos 2.55 = -\cos 0.59 = -0.8309$$

PROGRESS CHECK

Find tan 5.96 using $\pi \approx 3.14$ and Table V.

ANSWER

−0.3314

POLYNOMIAL APPROXIMATIONS

A startling result that is derived in calculus texts shows that there is a relationship between the trigonometric functions and polynomials. For values of t in the interval $[0, \pi/2]$, the *approximations*

$$\sin t \approx t - \frac{t^3}{6} + \frac{t^5}{120} - \frac{t^7}{5040}$$

and

$$\cos t \approx 1 - \frac{t^2}{2} + \frac{t^4}{24} - \frac{t^6}{720} + \frac{t^8}{40,320}$$

will yield values for sine and cosine accurate to three decimal places. You can use these approximations and the method of nesting developed in Chapter 3 to write a program to produce tables of values of the trigonometric functions!

EXERCISE SET 5.4

Use Table V in the Tables Appendix to find each of the following. (Use $\pi \approx 3.14$.)

1. cos 1.12	2. sin 0.48	3. tan(−1.39)	4. sin 4.86
5. tan 3.44	6. cos(−4.79)	7. sin(−5.28)	8. tan 6.05
9. cos(−2.91)	10. sin 2.43	11. tan(−3.27)	12. cos 1.72

Use a calculator and the polynomial approximations

$$\sin t \approx t - \frac{t^3}{6} + \frac{t^5}{120} - \frac{t^7}{5040}$$

and

$$\cos t \approx 1 - \frac{t^2}{2} + \frac{t^4}{24} - \frac{t^6}{720}$$

to find each of the following.

13. sin 0.80	14. cos 1.10	15. sin(−0.20)	16. cos(−0.75)
17. tan 0.1	18. tan(−1.2)		

Use a calculator to find the following.

19. tan 0.97	20. sin 1.32	21. cos 0.15	22. sin 1.84
23. tan 2.77	24. sin(−0.65)	25. cos(−5.61)	26. cos 7.08
27. sin(−8.94)	28. tan(−6.67)	29. sin 2.62	30. cos(−3.11)

31. Using the polynomial approximation for $\sin t$, show that sine is an odd function, that is, $\sin(-t) = -\sin t$.

32. Using the polynomial approximation for $\cos t$, show that cosine is an even function, that is, $\cos(-t) = \cos t$.

**5.5
GRAPHS: AMPLITUDE,
PERIOD, AND PHASE
SHIFT**

Our objective in this section is to sketch the graph of $f(x) = A \sin(Bx + C)$, where A, B, and C are real numbers and $B > 0$. Note that we are now using the familiar symbol "x" to indicate the independent variable, rather than the symbol "t" used in earlier sections of this chapter. Of course, any symbol can be used to denote a variable; however, the symbol "x" used here is not to be confused with the x-coordinate of the point $P(t) = P(x, y)$. The results that we obtain throughout this section will also apply to the form $A \cos(Bx + C)$.

AMPLITUDE

Since the sine function has a maximum value of $+1$ and a minimum value of -1, it is clear that the function $f(x) = A \sin x$ has a maximum value of $|A|$ and a minimum value of $-|A|$. If we define the **amplitude** of a periodic function as half the difference of the maximum and minimum values, we see that the amplitude of $f(x) = A \sin x$ is $[|A| - (-|A|)]/2 = |A|$.

> The amplitude of $f(x) = A \sin x$ is $|A|$.

The multiplier A acts as a vertical "stretch" factor when $|A| > 1$, and as a vertical "shrinkage" factor when $|A| < 1$. These remarks hold for both $y = A \sin x$ and $y = A \cos x$. Here are some examples.

EXAMPLE 1 Amplitude of $A \sin x$

Sketch the graphs of $y = 2 \sin x$ and $y = \frac{1}{2} \sin x$ on the same coordinate axes.

SOLUTION

The graph of $y = 2 \sin x$ has an amplitude of 2; the maximum value of y is $+2$ and the minimum is -2. Similarly, the amplitude of $y = \frac{1}{2} \sin x$ is $\frac{1}{2}$ and the graph has a maximum value of $+\frac{1}{2}$ and a minimum of $-\frac{1}{2}$ (Figure 26).

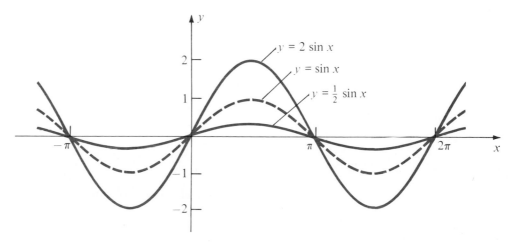

FIGURE 26   Graphs of $y = 2 \sin x$ and $y = \frac{1}{2}\sin x$

EXAMPLE 2   Graphing with a Negative "Stretch" Factor

Sketch the graph of $f(x) = -3\cos x$.

SOLUTION

The amplitude is 3, and $y = -3\cos x$ has maximum and minimum values of $+3$ and -3, respectively. Since $A = -3$, each y-coordinate will be that of $\cos x$ multiplied by -3.

The graph of $y = -3\cos x$ shown in Figure 27 is said to be a **reflection** about the x-axis of the graph of $y = 3\cos x$.

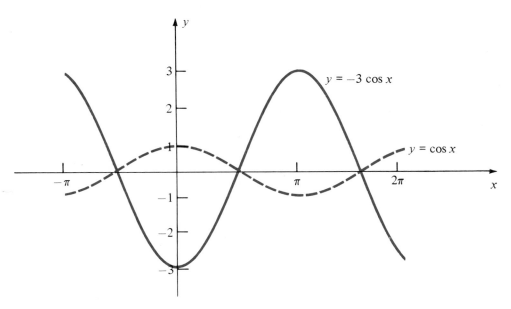

FIGURE 27   Graph of $y = -3\cos x$

PERIOD

We now seek to determine the period of the function $f(x) = \sin Bx$. Since $y = \sin x$ has period 2π, the sine function completes one cycle or wave as x varies from 0 to 2π. Then $\sin Bx$ will complete a cycle as Bx varies from 0 to 2π, which leads to the equations

$$Bx = 0 \qquad Bx = 2\pi$$

$$x = 0 \qquad x = \frac{2\pi}{B}$$

Thus, $f(x) = \sin Bx$ completes a cycle or wave as x varies from 0 to $2\pi/B$. We conclude the following:

The period of $f(x) = \sin Bx$ is $\dfrac{2\pi}{B}$.

The multiplier B acts as a horizontal "stretch" factor if $0 < B < 1$ and as a horizontal "shrinkage" factor if $B > 1$.

EXAMPLE 3   The Period of sin *Bx*
Sketch the graph of $f(x) = \sin 2x$.

SOLUTION
Since $B = 2$, the period is $2\pi/2 = \pi$. The graph will complete a cycle every π units (Figure 28).

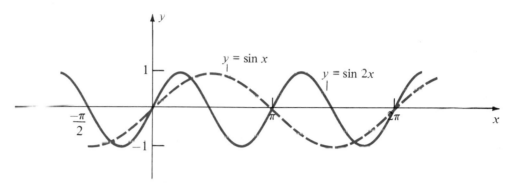

FIGURE 28   Graph of *y* = sin 2*x*

EXAMPLE 4   Graphing *A* cos *Bx*
Sketch the graph of $f(x) = 2\cos\frac{1}{2}x$.

SOLUTION
Since $B = \frac{1}{2}$, the period is $2\pi/\frac{1}{2} = 4\pi$. The graph will complete a cycle every 4π units. Note that the amplitude is 2, which provides us with maximum and minimum values of 2 and -2, respectively. The graph is shown in Figure 29.

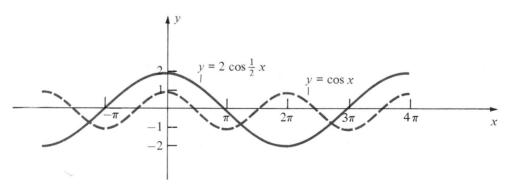

FIGURE 29   Graph of *y* = 2 cos $\frac{1}{2}$*x*

PHASE SHIFT

Let's examine the behavior of the function $f(x) = A \sin(Bx + C)$. Since $y = \sin x$ completes a cycle as x varies from 0 to 2π, the function f will complete a cycle as $Bx + C$ varies from 0 to 2π. Solving the equations

$$Bx + C = 0 \qquad Bx + C = 2\pi$$

we have

$$x = -\frac{C}{B} \qquad x = \frac{2\pi - C}{B} = \frac{2\pi}{B} - \frac{C}{B}$$

The number $-C/B$ is called the **phase shift** and indicates that the graph of the function is shifted right $-C/B$ units if $-C/B > 0$ and is shifted left if $-C/B$ is negative.

The phase shift of

$$f(x) = A \sin(Bx + C)$$

is $-C/B$.

Note that the amplitude of f is $|A|$ and the period is $2\pi/B$; that is, the introduction of a phase shift has not altered our earlier results.

EXAMPLE 5   Graphing with a Phase Shift

Sketch the graph $f(x) = 3 \sin(2x - \pi)$.

SOLUTION

We have $A = 3$, $B = 2$, and $C = -\pi$. Then

$$\text{amplitude} = |A| = 3$$

$$\text{period} = \frac{2\pi}{B} = \frac{2\pi}{2} = \pi$$

$$\text{phase shift} = -\frac{C}{B} = -\left(-\frac{\pi}{2}\right) = \frac{\pi}{2}$$

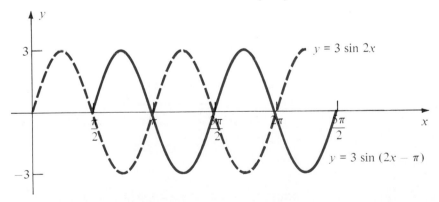

FIGURE 30   Graph of $y = 3 \sin(2x - \pi)$

PREDATOR-PREY INTERACTION

In the natural world we frequently find that two plant or animal species interact in their environment in such a manner that one species (the prey) serves as the primary food supply for the second species (the predator). Examples of such interaction are the relationships between trees (prey) and insects (predators) and between rabbits (prey) and lynxes (predators). As the population of the prey increases, the additional food supply results in an increase in the population of the predators. More predators consume more food, so the population of the prey will decrease, which, in turn, will lead to a decrease in the population of the predators. The reduction in the predator population results in an increase in the number of prey and the cycle will start all over again.

The accompanying figure, adapted from *Mathematics: Ideas and Applications*, by Daniel D. Benice, Academic Press, 1978 (used with permission), shows the interaction between lynx and rabbit populations. Both curves demonstrate periodic behavior and can be described by trigonometric functions.

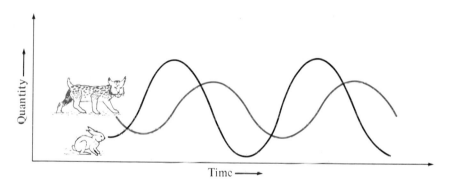

and we conclude that one cycle of $3\sin(2x - \pi)$ begins at $\pi/2$ and ends at $\pi + \pi/2 = 3\pi/2$.

The graphs of $y = 3\sin 2x$ and $y = 3\sin(2x - \pi)$ are both shown in Figure 30. To clearly demonstrate the effect of the phase shift, $y = 3\sin(2x - \pi)$ is graphed for $\pi/2 \le x \le 5\pi/2$.

■

 PROGRESS CHECK

If $f(x) = 2\cos(2x + \pi/2)$, find the amplitude, period, and phase shift of f. Sketch the graph of the function.

ANSWER

amplitude $= 2$ period $= \pi$ phase shift $= -\dfrac{\pi}{4}$ $\left(\text{or shift left } \dfrac{\pi}{4}\right)$ (Figure 31).

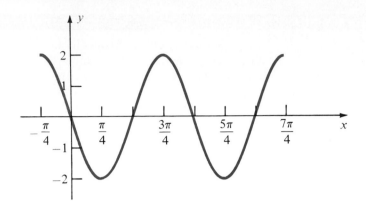

FIGURE 31   Graph of $y = 2\cos(2x + \pi/2)$

EXERCISE SET 5.5

In Exercises 1–12 determine the amplitude and period and sketch the graph of each of the following functions.

1. $f(x) = 3\sin x$

2. $f(x) = \dfrac{1}{4}\cos x$

3. $f(x) = \cos 4x$

4. $f(x) = \sin\dfrac{x}{4}$

5. $f(x) = -2\sin 4x$

6. $f(x) = -\cos\dfrac{x}{4}$

7. $f(x) = 2\cos\dfrac{x}{3}$

8. $f(x) = 4\sin 4x$

9. $f(x) = \dfrac{1}{4}\sin\dfrac{x}{4}$

10. $f(x) = \dfrac{1}{2}\cos\dfrac{x}{4}$

11. $f(x) = -3\cos 3x$

12. $f(x) = -2\sin 3x$

In Exercises 13–20 for each given function, determine the amplitude, period, and phase shift. Sketch the graph of the function.

13. $f(x) = 2\sin(x - \pi)$

14. $f(x) = \dfrac{1}{2}\cos\left(x + \dfrac{\pi}{2}\right)$

15. $f(x) = 3\cos(2x - \pi)$

16. $f(x) = 4\sin\left(x + \dfrac{\pi}{4}\right)$

17. $f(x) = \dfrac{1}{3}\sin\left(3x + \dfrac{3\pi}{4}\right)$

18. $f(x) = 2\cos\left(2x + \dfrac{\pi}{2}\right)$

19. $f(x) = 2\cos\left(\dfrac{x}{4} - \pi\right)$

20. $f(x) = 6\sin\left(\dfrac{x}{2} + \dfrac{\pi}{2}\right)$

In Exercises 21–24 use the identities $\sin(-t) = -\sin t$ and $\cos(-t) = \cos t$ to rewrite each equation as an equivalent equation with $B > 0$.

21. $y = -2\sin(-2x + \pi)$

22. $y = 4\cos\left(-\dfrac{x}{2} + \dfrac{\pi}{2}\right)$

23. $y = 3\cos\left(-\dfrac{x}{3} + \dfrac{2\pi}{3}\right)$

24. $y = -5\sin(-2x - \pi)$

5.6
SECANT, COSECANT,
AND COTANGENT

We stated earlier in this chapter that there are six trigonometric or circular functions and that the remaining three functions are reciprocals of sine, cosine, and tangent. These functions are called the **secant, cosecant,** and **cotangent** and are written as **sec, csc,** and **cot,** respectively. We now formally define these functions.

Definition of sec *t*, **csc *t*, and cot *t***	
	$\sec t = \dfrac{1}{\cos t}, \qquad \cos t \neq 0$
	$\csc t = \dfrac{1}{\sin t}, \qquad \sin t \neq 0$
	$\cot t = \dfrac{1}{\tan t}, \qquad \tan t \neq 0$

By using these definitions, we can apply the results that we have obtained for sine, cosine, and tangent to these new functions.

EXAMPLE 1   Using the Secant Function
Find $\sec \pi/3$.

SOLUTION
Since $\cos \pi/3 = \frac{1}{2}$ we see that

$$\sec \frac{\pi}{3} = \frac{1}{\cos \dfrac{\pi}{3}} = \frac{1}{\frac{1}{2}} = 2$$

■

EXAMPLE 2   Using the Cotangent Function
Find the real number t, $0 \le t \le \pi/2$, such that $\cot t = \sqrt{3}$.

SOLUTION
We seek the real number t such that

$$\tan t = \frac{1}{\cot t} = \frac{1}{\sqrt{3}} = \frac{\sqrt{3}}{3}$$

Thus, $t = \pi/6$ since $\tan \pi/6 = \sqrt{3}/3$.

■

We know that a real number and its reciprocal have the same sign; that is, if $x > 0$, then $1/x > 0$ and if $x < 0$, then $1/x < 0$. From this, we can immediately extend our conclusions (see Figure 11b) concerning the signs of the trigonometric functions in each quadrant (Figure 32). You don't have to memorize these; simply associate each function with its reciprocal.

II	**I**
$\sin t > 0$	All positive
$\csc t > 0$	
III	**IV**
$\tan t > 0$	$\cos t > 0$
$\cot t > 0$	$\sec t > 0$

FIGURE 32   Signs of the Trigonometric Functions

EXAMPLE 3   Finding the Quadrant

Find the quadrant in which $P(t)$ lies if $\sin t > 0$ and $\sec t < 0$.

SOLUTION

If $\sec t < 0$, then $\cos t < 0$. We know that sine is positive in quadrants I and II, cosine is negative in quadrants II and III (Figure 32). Both conditions are satisfied in quadrant II.

■

PROGRESS CHECK

Find the quadrant in which $P(t)$ lies if $\tan t < 0$ and $\csc t < 0$.

ANSWER

quadrant IV

EXAMPLE 4

Find t if $\sin t = \sqrt{3}/2$ and $\sec t < 0$.

SOLUTION

Since $\sec t < 0$, we have $\cos t < 0$. Then t must lie in quadrant II, since sine is positive and cosine is negative only in quadrant II (Figure 32). Finally, we know that $\sin \pi/3 = \sqrt{3}/2$. Thus $\pi/3$ is the reference number of t; that is,

$$\pi - t = \frac{\pi}{3}$$

$$t = \frac{2\pi}{3}$$

■

PROGRESS CHECK

Find t if $\cos t = -\frac{1}{2}$ and $\cot t > 0$.

ANSWER

$4\pi/3$

We can also employ the definition of cosecant to aid in sketching the graph of the function. Since $\csc t = 1/\sin t$, we compute the reciprocal of the y-coordinate of $\sin t$ at a point to determine the y-coordinate of $\csc t$ at that point. Of course, we cannot form a

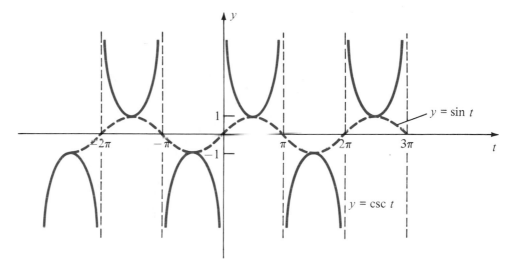

FIGURE 33   Graph of $y = \csc t$

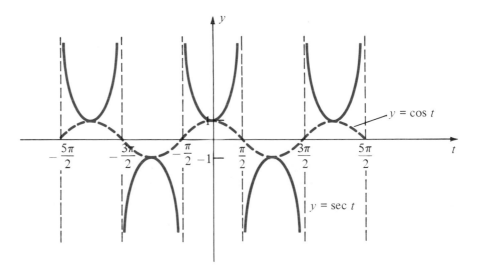

FIGURE 34   Graph of $y = \sec t$

reciprocal when $\sin t = 0$, that is, when $t = n\pi$, where n is an integer. The situation at these values of t is analogous to that of the tangent function when $t = \pi/2 + n\pi$. We conclude that the graph of $\csc t$ has vertical asymptotes when $t = n\pi$, for all integer values of n. In Figure 33 we have sketched the graph of the sine function with dashed lines, to aid in sketching the reciprocal values of the y-coordinates for the cosecant function.

A similar approach yields the graphs of $\sec t$ and $\cot t$ shown in Figures 34 and 35.

Table 7 summarizes the significant properties of the trigonometric functions.

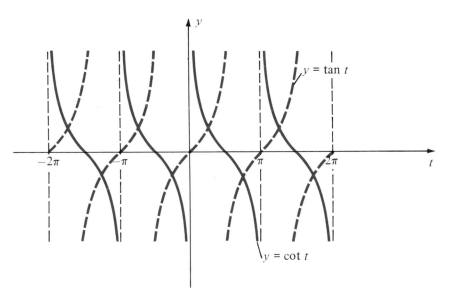

FIGURE 35   Graph of $y = \cot t$

TABLE 7   Properties of the Trigonometric Functions

	Positive in Quadrant	$-t$	Period	Domain	Range
sin	I, II	$-\sin t$	2π	all real numbers	$[-1, 1]$
cos	I, IV	$\cos t$	2π	all real numbers	$[-1, 1]$
tan	I, III	$-\tan t$	π	$t \neq \dfrac{\pi}{2} + n\pi$	$(-\infty, \infty)$
csc	I, II	$-\csc t$	2π	$t \neq n\pi$	$(-\infty, -1], [1, \infty)$
sec	I, IV	$\sec t$	2π	$t \neq \dfrac{\pi}{2} + n\pi$	$(-\infty, -1], [1, \infty)$
cot	I, III	$-\cot t$	π	$t \neq n\pi$	$(-\infty, \infty)$

EXERCISE SET 5.6

In Exercises 1–12 use the definitions of secant, cosecant, and cotangent to determine $\sec t$, $\csc t$, and $\cot t$ for each of the following values of t.

1. $\dfrac{\pi}{3}$

2. $\dfrac{\pi}{6}$

3. $\dfrac{\pi}{4}$

4. $\dfrac{\pi}{2}$

5. $\dfrac{5\pi}{6}$

6. $\dfrac{4\pi}{3}$

7. $\dfrac{3\pi}{2}$

8. $\dfrac{7\pi}{4}$

9. $\dfrac{3\pi}{4}$

10. $\dfrac{11\pi}{6}$

11. $\dfrac{5\pi}{4}$

12. $\dfrac{7\pi}{6}$

In Exercises 13–24 determine the value(s) of t, $0 \le t \le 2\pi$, that satisfy each of the following.

13. $\sec t = 1$

14. $\sec t = -1$

15. $\csc t = -2$

16. $\csc t = 0$

17. $\cot t = 1$

18. $\cot t = \sqrt{3}$

19. $\cot t = -1$

20. $\cot t = \dfrac{\sqrt{3}}{3}$

21. $\sec t = \sqrt{2}$

22. $\csc t = -\sqrt{2}$

23. $\cot t = -\sqrt{3}$

24. $\csc t = 2\dfrac{\sqrt{3}}{3}$

In Exercises 25–32 find the quadrant in which $P(t)$ lies if the following conditions hold.

25. $\sec t < 0$, $\sin t < 0$

26. $\tan t < 0$, $\sec t < 0$

27. $\csc t > 0$, $\sec t < 0$

28. $\sin t < 0$, $\cot t > 0$

29. $\sec t < 0$, $\cot t > 0$

30. $\cot t < 0$, $\sin t > 0$

31. $\sec t < 0$, $\csc t < 0$

32. $\csc t < 0$, $\cot t > 0$

In Exercises 33–40 determine the value of t, $0 \le t \le 2\pi$, that satisfies each of the following.

33. $\sin t = 1/2$, $\sec t < 0$

34. $\tan t = \sqrt{3}$, $\csc t < 0$

35. $\sec t = -2$, $\csc t > 0$

36. $\csc t = -2$, $\cot t > 0$

37. $\csc t = -\sqrt{2}$, $\sec t < 0$

38. $\sec t = \sqrt{2}$, $\cot t > 0$

39. $\cot t = -1$, $\sec t < 0$

40. $\cot t = \sqrt{3}$, $\csc t < 0$

In Exercises 41–46 use Table V in the Tables Appendix to find each of the following. (Assume $\pi \approx 3.14$ to find the reference number.)

41. $\cot 3.37$

42. $\sec 0.48$

43. $\csc(-4.68)$

44. $\csc 2.48$

45. $\sec 1.26$

46. $\cot(-1.82)$

5.7
THE INVERSE TRIGONOMETRIC FUNCTIONS

Inverse functions were introduced in Section 4.1 and were used to define logarithmic functions in Section 4.3. We have seen that if f is a one-to-one function whose domain is the set X and whose range is the set Y, then the inverse function f^{-1} reverses the correspondence; that is,

$$f^{-1}(y) = x$$

if and only if

$$f(x) = y \quad \text{for all } x \in X$$

Using this definition, we saw that the following identities characterize inverse functions.

$$f^{-1}[f(x)] = x \quad \text{for all } x \text{ in } X$$
$$f[f^{-1}(y)] = y \quad \text{for all } y \text{ in } Y$$

If we attempt to find an inverse of the sine function, we have an immediate problem. Since sine is a periodic function, it is not a one-to-one function and has no inverse. However, we can resolve this problem by defining a function that agrees with the sine function but over a restricted domain. That is, we would like to find an interval such that $y = \sin x$ is one-to-one and y assumes all values between -1 and $+1$ over this interval. If we define the function f by

$$f(x) = \sin x, \qquad -\frac{\pi}{2} \le x \le \frac{\pi}{2}$$

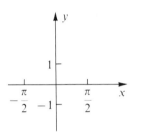

FIGURE 36 $y = \sin x, \ -\pi/2 \le x \le \pi/2$

then f takes on the same values as the sine function over the interval $[-\pi/2, \pi/2]$ and assumes all real values in the interval $[-1, 1]$. The graph of $\sin x$ over the interval $[-\pi/2, \pi/2]$ is shown in Figure 36 and demonstrates that f is an increasing function and is therefore one-to-one. Consequently, f has an inverse, and we are led to the following definition.

The inverse sine function, denoted by **arcsin** or **sin^{-1}**, is defined by

$$\sin^{-1} y = x \quad \text{if and only if} \quad \sin x = y$$

where $-\dfrac{\pi}{2} \le x \le \dfrac{\pi}{2}$.

In words, arcsin y is the value between $-\pi/2$ and $\pi/2$ whose sine is y. Note that $-1 \le y \le 1$, so the domain of the inverse sine function is the set of all real numbers in the interval $[-1, 1]$.

WARNING

When we defined $\sin^n t = (\sin t)^n$ we said that this definition does not hold when $n = -1$, allowing us to reserve the notation $\sin^{-1}$ for the inverse sine function. Therefore, $\sin^{-1} y$ is not to be confused with

$$(\sin y)^{-1} = \frac{1}{\sin y}$$

The notations arcsin and sin^{-1} are both in common use. We will therefore employ both notations to accustom you to their use. Note that if $x = \arcsin y$, then $\sin x = y$. On a unit circle, $P(x)$ determines an *arc whose sine is y*, which is the origin of the notation arcsin y.

We would like to sketch the graph of $y = \sin^{-1} x$. (Since x and y are simply symbols for variables, we have reverted to the usual practice of letting x be the independent variable.) The graph, of course, is the same as that of $\sin y = x$, with the restriction that $-\pi/2 \le y \le \pi/2$. We form a table of values and sketch the graph in Figure 37.

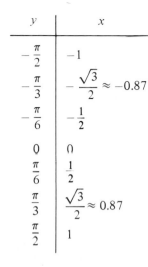

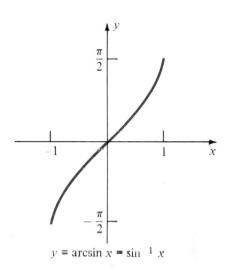

$y = \arcsin x = \sin^{-1} x$

FIGURE 37   Graph of y = arcsin x

EXAMPLE 1   Applying the Inverse Trigonometric Functions

Find   **(a)** $\arcsin \frac{1}{2}$ **(b)** $\arcsin(-1)$.

SOLUTION

(a) If $y = \arcsin \frac{1}{2}$, then $\sin y = \frac{1}{2}$ where y is restricted to the interval $[-\pi/2, \pi/2]$. Thus, $y = \pi/6$ is the *only* correct answer.

(b) If $y = \arcsin(-1)$, then $\sin y = -1$ where $-\pi/2 \le y \le \pi/2$. Thus, $-\pi/2$ is the *only* correct answer.

■

EXAMPLE 2   Applying the Inverse Trigonometric Functions

Evaluate $\sin^{-1}\left(\cos\dfrac{\pi}{4}\right)$.

SOLUTION

Since $\cos\dfrac{\pi}{4} = \dfrac{\sqrt{2}}{2}$, we have

$$\sin^{-1}\left(\cos\frac{\pi}{4}\right) = \sin^{-1}\left(\frac{\sqrt{2}}{2}\right)$$

We let

$$y = \sin^{-1}\left(\frac{\sqrt{2}}{2}\right)$$

Then

$$\sin y = \frac{\sqrt{2}}{2} \qquad \text{where} \qquad -\frac{\pi}{2} \le y \le \frac{\pi}{2}$$

$$y = \frac{\pi}{4}$$

which is the *only* solution.

■

 PROGRESS CHECK

Find   **(a)** $\sin^{-1}\left(-\frac{\sqrt{3}}{2}\right)$ **(b)** $\arcsin\left(\tan\frac{5\pi}{4}\right)$.

ANSWERS

(ʁ) $-\frac{\pi}{3}$ (q) $\frac{\pi}{2}$

We may use a similar approach to define the inverse cosine function. If we define the function f by

$$f(x) = \cos x, \qquad 0 \le x \le \pi$$

then f agrees with the cosine function over the interval $[0, \pi]$, assumes all real values in the interval $[-1, 1]$, and is a decreasing function (see Figure 38). Consequently, f is a one-to-one function and has an inverse.

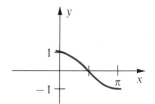

FIGURE 38   Graph of

$y = \cos x$,

$0 \le x \le \pi$

> The inverse cosine function, denoted by **arccos** or **cos⁻¹**, is defined by
>
> $$\cos^{-1} y = x \quad \text{if and only if} \quad \cos x = y$$
>
> where $0 \le x \le \pi$.

Since $-1 \le y \le 1$, the domain of the inverse cosine function is the set of all real numbers in the interval $[-1, 1]$.

To sketch the graph of $y = \cos^{-1} x$ we sketch the graph of $\cos y = x$ as in Figure 39.

EXAMPLE 3   Applying the Inverse Trigonometric Functions

Find   **(a)** $\cos^{-1}\left(-\frac{1}{2}\right)$ **(b)** $\arccos\left(\sin\frac{\pi}{2}\right)$.

SOLUTION

(a) If $y = \cos^{-1}\left(-\frac{1}{2}\right)$, then $\cos y = -\frac{1}{2}$ where y is restricted to the interval $[0, \pi]$. Consequently, $y = \frac{2\pi}{3}$ is the *only* correct answer.

y	x
0	1
$\dfrac{\pi}{6}$	$\dfrac{\sqrt{3}}{2} \approx 0.87$
$\dfrac{\pi}{3}$	$\dfrac{1}{2}$
$\dfrac{\pi}{2}$	0
$\dfrac{2\pi}{3}$	$-\dfrac{1}{2}$
$\dfrac{5\pi}{6}$	$-\dfrac{\sqrt{3}}{2} \approx -0.87$
π	-1

$$y = \arccos x = \cos^{-1} x$$

FIGURE 39 Graph of $y = \arccos x$

(b) Since $\sin \dfrac{\pi}{2} = 1$, we let $y = \arccos(1)$. Then $\cos y = 1$ where $0 \leq y \leq \pi$. Therefore, $y = 0$ is the *only* correct answer.   ∎

If we restrict the tangent function to the interval $(-\pi/2, \pi/2)$, we can define the inverse tangent function.

The inverse tangent function, denoted by **arctan** or $\tan^{-1}$, is defined by

$$\tan^{-1} y = x \quad \text{if and only if} \quad \tan x = y$$

where $-\dfrac{\pi}{2} < x < \dfrac{\pi}{2}$.

Note that the domain of the inverse tangent function is the set of all real numbers. Proceeding as before, we sketch the graph of $y = \tan^{-1} x$ in Figure 40.

EXAMPLE 4
Find   (a) $\tan^{-1}(\sqrt{3})$ (b) $\tan^{-1}(4.256)$.

SOLUTION
(a) If $y = \tan^{-1}(\sqrt{3})$, then $\tan y = \sqrt{3}$. Since $-\pi/2 < y < \pi/2$, we must have $y = \pi/3$.

(b) If $y = \tan^{-1}(4.256)$, then $\tan y = 4.256$ where $-\pi/2 < y < \pi/2$. Using Table V in the Tables Appendix or a calculator, we find $y = 1.34$.   ∎

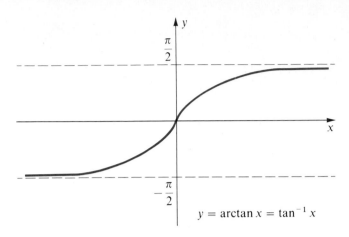

$$y = \arctan x = \tan^{-1} x$$

FIGURE 40   Graph of $y = \arctan x$

EXAMPLE 5

Find $\cos\left(\arctan\dfrac{4}{3}\right)$ without using tables or a calculator.

SOLUTION

If we let $x = \arctan\dfrac{4}{3}$, then $\tan x = \dfrac{4}{3}$ and $0 \le x < \pi/2$. Using trigonometric identities,

$$\tan x = \frac{\sin x}{\cos x} = \frac{4}{3}$$

$$3 \sin x = 4 \cos x \qquad \text{Clearing fractions}$$

$$9 \sin^2 x = 16 \cos^2 x \quad \text{Squaring both sides}$$

$$9(1 - \cos^2 x) = 16 \cos^2 x \quad \sin^2 x = 1 - \cos^2 x$$

$$\cos^2 x = \frac{9}{25}$$

$$\cos x = \pm\frac{3}{5}$$

Since $x \in [0, \pi/2]$, we conclude that $\cos x = 3/5$.   ∎

☑ **PROGRESS CHECK**

Without using tables or a calculator, find $\cot\left(\sin^{-1}-\dfrac{5}{13}\right)$.

ANSWER

$-\dfrac{5}{12}$

EXAMPLE 6
Find the solutions of the equation $5\cos^2 x - 3 = 0$ that are in the interval $[0, \pi]$.

SOLUTION
We treat the equation as a quadratic in $\cos x$. Then

$$5\cos^2 x = 3$$

$$\cos x = \pm\sqrt{\frac{3}{5}} = \pm\frac{\sqrt{15}}{5}$$

We may then write

$$x = \arccos\left(\frac{\sqrt{15}}{5}\right) \qquad \text{or} \qquad x = \arccos\left(-\frac{\sqrt{15}}{5}\right)$$

These are exact expressions for the solutions. Numerical approximations can be obtained using Table V in the Tables Appendix or a calculator. The key sequences

$$15\ \boxed{\sqrt{\ }}\ \boxed{\div}\ 5\ \boxed{=}\ \boxed{\text{INV}}\ \boxed{\cos}$$

and

$$15\ \boxed{\sqrt{\ }}\ \boxed{\div}\ 5\ \boxed{=}\ \boxed{\pm}\ \boxed{\text{INV}}\ \boxed{\cos}$$

provide the approximate solutions

$$x \approx 0.6847 \qquad \text{and} \qquad x \approx 2.4569$$

■

PROGRESS CHECK
Find the solutions of the equation $2\sin^2 x + 2\sin x - 1 = 0$ that are in the interval $[-\pi/2, \pi/2]$.

ANSWER
arcsin $\left(-\frac{1}{2} + \frac{1}{2}\sqrt{3}\right)$ or 0.3747

WARNING
It is important to remember that the range of each of the inverse trigonometric functions is a subset of the domain of the corresponding trigonometric function. Given the equation

$$t = \sin^{-1}\left(-\frac{1}{2}\right)$$

students often write $t = 7\pi/6$, which is incorrect since t must lie in the interval $[-\pi/2, \pi/2]$. The only correct answer is $-\pi/6$.

EXERCISE SET 5.7

In Exercises 1–18 evaluate the given expression.

1. $\sin^{-1}\left(-\dfrac{1}{2}\right)$
2. $\arccos\left(\dfrac{\sqrt{3}}{2}\right)$
3. $\arctan\sqrt{3}$
4. $\tan^{-1}0$

5. $\arcsin\left(-\dfrac{\sqrt{2}}{2}\right)$
6. $\cos^{-1}(-1)$
7. $\arccos\left(-\dfrac{\sqrt{3}}{2}\right)$
8. $\tan^{-1}\left(\dfrac{\sqrt{3}}{3}\right)$

9. $\sin^{-1}(-1)$
10. $\arctan 1$
11. $\cos^{-1}0$
12. $\sin^{-1}\left(-\dfrac{\sqrt{3}}{2}\right)$

13. $\cos^{-1}1$
14. $\arcsin\left(\dfrac{\sqrt{2}}{2}\right)$
15. $\arctan(-1)$
16. $\sin^{-1}0$

17. $\cos^{-1}\left(-\dfrac{1}{2}\right)$
18. $\arcsin\left(\dfrac{1}{2}\right)$

In Exercises 19–24 use Table V in the Tables Appendix to approximate the given expression.

19. $\sin^{-1}(0.3709)$
20. $\arctan(1.398)$
21. $\cos^{-1}(-0.7648)$
22. $\tan^{-1}(-3.010)$

23. $\arcsin(0.9636)$
24. $\arccos(-0.921)$

In Exercises 25–32 evaluate the given expression.

25. $\sin(\arctan 1)$
26. $\cos\left(\arcsin -\dfrac{1}{2}\right)$
27. $\tan^{-1}\left(\cos\dfrac{\pi}{2}\right)$
28. $\sin^{-1}(\sin 0.62)$

29. $\cos^{-1}\left(\sin\dfrac{9\pi}{4}\right)$
30. $\tan(\sin^{-1}0)$
31. $\cos^{-1}\left(\cos\dfrac{2\pi}{3}\right)$
32. $\sin^{-1}\left(\cos\dfrac{\pi}{6}\right)$

In Exercises 33–38 use the inverse trigonometric functions to express the solutions of the given equation exactly.

33. $7\sin^2 x - 1 = 0,\ x \in [-\pi/2, \pi/2]$
34. $6\cos^2 y - 5 = 0,\ y \in [0, \pi]$

35. $12\cos^2 x - \cos x - 1 = 0,\ x \in [0, \pi]$
36. $2\tan^2 t + 4\tan t - 3 = 0,\ t \in [-\pi/2, \pi/2]$

37. $9\sin^2 t - 12\sin t + 4 = 0,\ t \in [-\pi/2, \pi/2]$
38. $3\cos^2 x - 7\cos x - 6 = 0,\ x \in [0, \pi]$

In Exercises 39–42 provide a value for x to show that the equation is not an identity.

39. $\sin^{-1} x = \dfrac{1}{\sin x}$
40. $(\sin^{-1} x)^2 + (\cos^{-1} x)^2 = 1$

41. $\sin^{-1}(\sin x) = x$
42. $\arccos(\cos x) = x$

In Exercises 43–46 use a calculator to assist in finding all solutions in the indicated interval.

43. $2\cos^2 x + \cos x = 2,\ [0, \pi]$

45. $2\tan^2 x = 3,\ \left[0, \dfrac{\pi}{2}\right]$

44. $\sin^2 x - 2\sin x - 2 = 0,\ \left[\dfrac{-\pi}{2}, \dfrac{\pi}{2}\right]$

46. $9\cos^2 x + 3\cos x = 2,\ [0, \pi]$

TERMS AND SYMBOLS

KEY IDEAS FOR REVIEW

Topic	Page	Key Idea

Unit circle points
  $P(t)$
  rectangular coordinates
	
	237
	237
		For every real number t there is a unique unit circle point which we denote as $P(t)$. If the rectangular coordinates of the point are (x, y), we write $$P(t) = P(x, y)$$ Since $$P(t) = P(t + 2\pi n), \quad n \text{ an integer,}$$ the correspondence between points on the unit circle and real numbers is not one-to-one.

special values	239	In general, it is difficult to determine the rectangular coordinates (x, y) corresponding to the point $P(t)$ on the unit circle. When $t = 0,\ \pi/2,\ \pi,$ or $3\pi/2$, then $P(t)$ lies on a coordinate axis and the rectangular coordinates are obvious. When $t = \pi/6,\ \pi/4,$ or $\pi/3$, we can employ geometric arguments to show that $$P\left(\frac{\pi}{6}\right) = \left(\frac{\sqrt{3}}{2}, \frac{1}{2}\right)$$ $$P\left(\frac{\pi}{4}\right) = \left(\frac{\sqrt{2}}{2}, \frac{\sqrt{2}}{2}\right)$$ $$P\left(\frac{\pi}{3}\right) = \left(\frac{1}{2}, \frac{\sqrt{3}}{2}\right)$$
Circular or trigonometric functions	245	The circular or trigonometric functions sine, cosine, and tangent are defined in terms of the rectangular coordinates of a point on the unit circle. If t is a real number and $P(t) = P(x, y)$, then $$\sin t = y$$ $$\cos t = x$$ $$\tan t = \frac{y}{x}, \quad x \neq 0$$
signs	248	The signs of the trigonometric functions in each of the quadrants follow from the definitions and are displayed in Figure 32.
reciprocals	269	The secant, cosecant, and cotangent functions are defined as the reciprocals of cosine, sine, and tangent, respectively.
Odd and even functions	250	Sine and tangent are odd functions; cosine is an even function. That is $$\sin(-t) = -\sin t \qquad \cos(-t) = \cos t \qquad \tan(-t) = -\tan t$$

KEY IDEAS FOR REVIEW

Topic	Page	Key Idea
Periodic functions	253	The trigonometric functions are all periodic. The period of the sine and cosine functions is 2π; the period of the tangent function is π.
Reference Number Rule	261	Standard tables of the values of the trigonometric functions (see Table V in the Tables Appendix) display the independent variable t from 0 to $\pi/2$. If it is desired to find $\sin t'$ where $\pi/2 < t' < 2\pi$, the Reference Number Rule is used to determine a real number t, $0 \le t \le \pi/2$, such that $|\sin t| = |\sin t'|$. The appropriate sign is then added depending on the quadrant of $P(t')$.
Graph of $f(x) = A\sin(Bx + c)$	263	To sketch the graph of $f(x) = A\sin(Bx + C)$, note that (i) the amplitude is $|A|$; (ii) the period is $2\pi/B$; (iii) the phase shift is $-C/B$. The same observations hold for $f(x) = A\cos(Bx + C)$.
Inverse trigonometric functions	273	To define the inverse trigonometric functions, it is necessary to restrict the domain of the trigonometric functions to ensure that the result is a one-to-one function.

REVIEW EXERCISES

Solutions to exercises whose numbers are in color are in the Solutions section in the back of the book.

5.1 In Exercises 1–5 replace each given real number t by t', $0 \le t' < 2\pi$, so that $P(t') = P(t)$.

1. $\dfrac{9\pi}{2}$

2. $-\dfrac{15\pi}{2}$

3. -6π

4. $\dfrac{23\pi}{3}$

5. $\dfrac{16\pi}{5}$

In Exercises 6–10 find the rectangular coordinates of the given point.

6. $P\left(\dfrac{7\pi}{6}\right)$

7. $P\left(-\dfrac{8\pi}{3}\right)$

8. $P\left(\dfrac{5\pi}{6}\right)$

9. $P\left(-\dfrac{7\pi}{4}\right)$

10. $P\left(\dfrac{11\pi}{6}\right)$

In Exercises 11–15, $P(t) = (4/5, -3/5)$. Use the symmetries of the circle to find the rectangular coordinates of the given point.

11. $P(t - \pi)$

12. $P\left(t + \dfrac{\pi}{2}\right)$

13. $P(-t)$

14. $P\left(t - \dfrac{\pi}{2}\right)$

15. $P(-t - \pi)$

5.2 In Exercises 16–19 determine the value of the indicated trigonometric function, without the use of tables or a calculator.

16. $\sin\dfrac{2\pi}{3}$

17. $\sec\left(-\dfrac{5\pi}{4}\right)$

18. $\tan\dfrac{5\pi}{6}$

19. $\csc\left(-\dfrac{\pi}{6}\right)$

In Exercises 20–23 find a value of t in the interval $[0, 2\pi)$ satisfying the given conditions.

20. $\sin t = -\dfrac{\sqrt{2}}{2}$, $P(t)$ in quadrant III

21. $\cos t = \dfrac{\sqrt{3}}{2}$, $P(t)$ in quadrant IV

22. $\cot t = \dfrac{\sqrt{3}}{3}$, $P(t)$ in quadrant I

23. $\sec t = -2$, $P(t)$ in quadrant II

In Exercises 24–27 use the trigonometric identities

$$\sin^2 t + \cos^2 t = 1 \qquad \tan t = \frac{\sin t}{\cos t}$$

to find the indicated value under the given conditions.

24. $\cos t = \frac{3}{5}$ and $P(t)$ is in quadrant IV; find $\cot t$.

25. $\sin t = -\frac{4}{5}$ and $\tan t > 0$; find $\sec t$.

26. $\sin t = \frac{12}{13}$ and $\cos t < 0$; find $\tan t$.

27. $\cos t = -\frac{5}{13}$ and $\tan t < 0$; find $\csc t$.

In Exercises 28 and 29 use the trigonometric identities to transform the first expression into the second.

28. $(\sin t)(\sec t)$, $\tan t$

29. $\dfrac{\sin t}{\cos^2 t}$, $(\tan t)(\sec t)$

5.3   In Exercises 30 and 31 use Table V in the Tables Appendix to evaluate the given expression. (Assume $\pi \approx 3.14$.)

30. $\cos 3.71 - \sin 1.44$ **31.** $\tan(-2.74)$

In Exercises 32 and 33 sketch the graph of the given function.

5.4   **32.** $f(x) = 1 - \sin x$ **33.** $f(x) = 2\sin\left(\dfrac{x}{2} + \pi\right)$

5.5   In Exercises 34–36 determine the amplitude, period, and phase shift to each given function. Sketch the graph.

34. $f(x) = -\cos(2x - \pi)$

35. $f(x) = 4\sin\left(-x + \dfrac{\pi}{2}\right)$

36. $f(x) = -2\sin\left(\dfrac{x}{3} + \dfrac{\pi}{3}\right)$

5.6   In Exercises 37–39, evaluate the given expression.

37. $\arcsin\left(-\dfrac{1}{2}\right)$ **38.** $\tan(\cos^{-1} 1)$

39. $\tan(\tan^{-1} 5)$

40. Use the inverse cosine function to express the exact solutions of the equation

$$5\cos^2 x - 4 = 0$$

PROGRESS TEST 5A

In Problems 1–3 replace the given real number t by t', $0 \le t' < 2\pi$, so that $P(t') = P(t)$.

1. $\dfrac{19\pi}{3}$ **2.** -22π **3.** $\dfrac{17\pi}{4}$

In Problems 4 and 5 find the rectangular coordinates of the given point.

4. $P\left(\dfrac{29\pi}{6}\right)$ **5.** $P\left(-\dfrac{\pi}{3}\right)$

In Problems 6–8, $P(t) = (-5/13, 12/13)$. Use the symmetries of the circle to find the rectangular coordinates of the given point.

6. $P(t + \pi)$ **7.** $P\left(t - \dfrac{\pi}{2}\right)$ **8.** $P(-t)$

In Problems 9 and 10 determine the value of the indicated trigonometric function without the use of the tables or a calculator.

9. $\cos\left(\dfrac{7\pi}{3}\right)$ **10.** $\csc\left(-\dfrac{2\pi}{3}\right)$

In Problems 11 and 12 find a value of t in the interval $[0, 2\pi)$ satisfying the given conditions.

11. $\tan t = 1$, $P(t)$ in quadrant III

12. $\sec t = \sqrt{2}$, $P(t)$ in quadrant IV

In Problems 13 and 14 use the trigonometric identities

$$\sin^2 t + \cos^2 t = 1, \qquad \tan t = \frac{\sin t}{\cos t}$$

to find the indicated value under the given conditions.

13. $\cos t = -\dfrac{12}{13}$ and $\tan t > 0$; find $\sin t$.

14. $\sin t = \dfrac{3}{5}$ and $P(t)$ is in quadrant II; find $\sec t$.

15. Use the trigonometric identities given for Problems 13 and 14 to transform

$$1 - \tan x \quad \text{to} \quad \frac{\cos x - \sin x}{\cos x}$$

In Problems 16 and 17 use Table V in the Tables Appendix to evaluate the given expression. (Assume $\pi \approx 3.14$.)

16. $\tan(-3.68)$

17. $\cos 1.15 - \sin 0.72$

18. Sketch the graph of the function f defined by

$$f(x) = x + \cos x$$

In Problems 19 and 20 determine the amplitude, period, and phase shift of each given function.

19. $f(x) = -2\cos(\pi - x)$ 20. $f(x) = 2\sin\left(\dfrac{x}{2} - \dfrac{\pi}{2}\right)$

PROGRESS TEST 5B

In Problems 1–3 replace the given real number t by t', $0 \le t' < 2\pi$, so that $P(t') = P(t)$.

1. -14π 2. $\dfrac{51\pi}{5}$ 3. $-\dfrac{19\pi}{6}$

In Problems 4 and 5 find the rectangular coordinates of the given point.

4. $P\left(\dfrac{23\pi}{6}\right)$ 5. $P\left(-\dfrac{3\pi}{4}\right)$

In Problems 6–8 $P(t) = \left(-\dfrac{4}{5}, -\dfrac{3}{5}\right)$. Use the symmetries of the circle to find the rectangular coordinates of the given point.

6. $P(-t)$ 7. $P\left(t + \dfrac{\pi}{2}\right)$ 8. $P(-t + \pi)$

In Problems 9 and 10 determine the value of the indicated trigonometric function without the use of the tables or a calculator.

9. $\tan\left(\dfrac{7\pi}{4}\right)$ 10. $\sin\left(-\dfrac{3\pi}{2}\right)$

In Problems 11 and 12 find a value of t in the interval $[0, 2\pi)$ satisfying the given conditions.

11. $\sin t = \dfrac{\sqrt{3}}{2}$, $P(t)$ in quadrant I

12. $\sec t = -2$, $P(t)$ in quadrant II

In Problems 13 and 14 use the trigonometric identities

$$\sin^2 t + \cos^2 t = 1 \qquad \tan t = \dfrac{\sin t}{\cos t}$$

to find the indicated value under the given conditions.

In Problems 21 and 22 evaluate the given expression.

21. $\tan^{-1}(-\sqrt{3})$ 22. $\cos\left(\sin^{-1}\dfrac{\sqrt{3}}{2}\right)$

23. Use the inverse tangent function to express the exact solutions of the equation

$$6\tan^2 x - 13\tan x + 6 = 0$$

13. $\sin t = -\dfrac{5}{13}$ and $\tan t < 0$; find $\tan t$.

14. $\cos t = \dfrac{3}{5}$ and $\cot t < 0$; find $\cot t$.

15. Use the trigonometric identities of Problems 13 and 14 to transform $\sec^2 t \cot t$ to $\csc t$.

In Problems 16 and 17 use Table V in the Tables Appendix to evaluate the given expression. (Assume $\pi \approx 3.14$.)

16. $\sin(2.45)$ 17. $\tan(-1.25) + \cos 1.67$

18. Sketch the graph of the function f defined by

$$f(x) = \sin x + \sin\dfrac{x}{2}$$

In Problems 19 and 20 determine the amplitude, period, and phase shift of each given function.

19. $f(x) = 4\sin(3x - \pi)$

20. $f(x) = -\dfrac{1}{2}\cos\left(2x + \dfrac{\pi}{2}\right)$

In Problems 21 and 22 evaluate the given expression.

21. $\sin^{-1}\left(\cos\dfrac{\pi}{3}\right)$ 22. $\tan\left(\cos^{-1}\dfrac{\sqrt{2}}{2}\right)$

23. Use the inverse sine function to express the exact solutions of the equation $5\sin^2 x - 2\sin x - 3 = 0$.

CHAPTER 6

RIGHT TRIANGLE TRIGONOMETRY

In the previous chapter we discussed trigonometry in terms of functions of real numbers. This approach has the advantage of illustrating the centrality of the function concept in much of modern mathematical thinking.

We now turn to the more traditional approach to trigonometry, which revolves about the measurement of triangles. We will show that it is possible to define the trigonometric functions as functions of angles rather than as functions of real numbers, and that the two definitions are related in a simple manner. Our attention will then turn to the right triangle, and we will have our first opportunity to explore a wide variety of applications that clearly demonstrate the usefulness of trigonometry in such fields as surveying and navigation.

We will conclude by examining the law of sines and the law of cosines, two important rules that can be employed when dealing with an oblique triangle, that is, a triangle that does not contain a right angle.

6.1
ANGLES AND THEIR MEASUREMENT

DEFINITION OF AN ANGLE

In the study of trigonometry we view an angle as the result of a ray or half-line that rotates about its endpoint. When the ray coincides with the positive x-axis with its endpoint at the origin, the angle generated is said to be in **standard position.** In Figure 1a the x-axis, called the **initial side,** rotates in a counterclockwise direction until it coincides with the **terminal side,** forming the angle α. In this case we say that α is a **positive angle.** In Figure 1b the ray has been rotated in a clockwise direction to form the angle β. In this case, we say that β is a **negative angle.** If the terminal side coincides with a coordinate axis, the angle is called a **quadrantal angle;** otherwise, the angle is said to lie in the same quadrant as its terminal side.

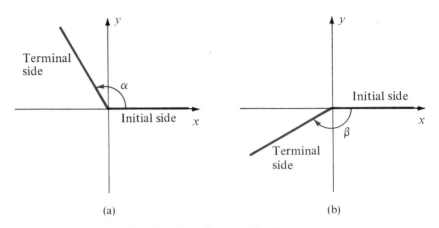

(a)   (b)

FIGURE 1   Positive and Negative Angles in Standard Position

ANGULAR MEASUREMENT: DEGREES AND RADIANS

There are two commonly used units for measuring angles. An angle is said to have a measure of one **degree** (written $1°$) if the angle is formed by rotating the initial side $\frac{1}{360}$ of a complete rotation in a counterclockwise direction. It follows that an angle obtained by a complete rotation of the initial side has a measure of $360°$, and an angle obtained by one-fourth of a complete rotation has a measure of $\frac{1}{4}(360°) = 90°$. One degree is subdivided into 60 **minutes** (written $60'$), and one minute is subdivided into 60 **seconds** (written $60''$). For example, the notation $14°24'18''$ is read *14 degrees, 24 minutes, and 18 seconds.* An angle in standard position between $0°$ and $90°$ lies in the first quadrant and is called an **acute angle;** an angle between $90°$ and $180°$ lies in the second quadrant and is called an **obtuse angle.** An angle measuring $90°$ is a quadrantal angle and is called a **right angle;** angles measuring $180°$ and $270°$ are also quadrantal angles.

To define the second unit of angular measurement, consider an angle θ in standard position and let $P(x, y)$ be the point of intersection of the terminal side of the angle with the unit circle (Figure 2). We may think of P as the point corresponding to some real number t; that is, $P(x, y) = P(t)$. We then say that θ is an **angle of t radians,** by which we mean that the measure of the angle θ is defined as the measure of the arc that θ intercepts on the unit circle. We then write $\theta = t$ radians or simply $\theta = t$. The radian measure of angles of 1 radian, -2 radians, and $3\pi/4$ radians are shown in Figure 3.

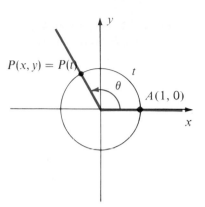

FIGURE 2   An Angle of *t* Radians

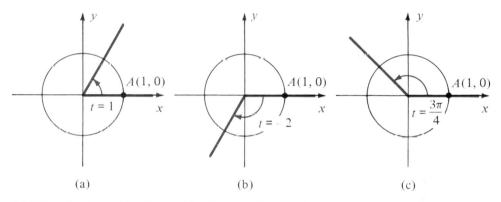

(a)   (b)   (c)

FIGURE 3   Angles of 1 Radian, −2 Radians, and 3π/4 Radians

ANGLE CONVERSION

An angle of $360°$ traces a complete revolution in the counterclockwise direction, which corresponds to an arc of length 2π on the unit circle. Thus, 2π radians $= 360°$ or

$$\pi \text{ radians} = 180° \tag{1}$$

This relationship enables us to transform angular measure from radians to degrees and vice versa. One way to handle such conversions for any angle θ is by establishing a proportion.

$$\frac{\text{radian measure of angle } \theta}{\pi \text{ radians}} = \frac{\text{degree measure of angle } \theta}{180°}$$

Alternatively, we can solve Equation (1) to provide the conversion formulas

$$1 \text{ radian} = \left(\frac{180}{\pi}\right)^{\circ} \qquad \text{and} \qquad 1° = \frac{\pi}{180} \text{ radians}$$

EXAMPLE 1 Angle Conversion
Convert $150°$ to radian measure.

SOLUTION
With θ representing the radian measure of the angle, we establish the proportion

$$\frac{\theta}{\pi} = \frac{150}{180}$$

Solving, we have

$$\theta = \frac{150\pi}{180} = \frac{5\pi}{6}$$

Alternatively, since $1° = \pi/180$ radians, we must have

$$150° = 150\left(\frac{\pi}{180}\right) = \frac{5\pi}{6} \text{ radians}$$

Thus, $\theta = 5\pi/6$ radians.

■

PROGRESS CHECK
Convert the following from degree to radian measure.
(a) $-210°$ (b) $390°$

ANSWERS
(a) $-\dfrac{7\pi}{6}$ radians   (b) $\dfrac{13\pi}{6}$ radians

---

EXAMPLE 2 Angle Conversion
Convert $2\pi/3$ radians to degree measure.

SOLUTION
With θ denoting the degree measure of the angle, we establish the proportion

$$\frac{\dfrac{2\pi}{3}}{\pi} = \frac{\theta}{180}$$

Solving, we have

$$\theta = \frac{2}{3}(180) = 120$$

Alternatively, since 1 radian $= 180/\pi$ degrees, we have

$$\frac{2\pi}{3} \text{ radians} = \frac{2\pi}{3}\left(\frac{180}{\pi}\right)° = 120°$$

Thus, $\theta = 120°$.

■

✓ **PROGRESS CHECK**
Convert the following from radian measure to degrees.

(a) $\dfrac{9\pi}{2}$ radians   **(b)** $-\dfrac{4\pi}{3}$ radians

ANSWERS
(a) 810°   (b) −240°

"SPECIAL" ANGLES

There are certain angles that we will use frequently in the examples and exercises throughout this chapter. It will prove helpful if you take the time now to verify the conversions shown in Table 1; you will then see how easy it is to switch between degree and radian measure for these values. Figure 4 displays some angles in standard position and shows both the degree measure and the radian measure.

TABLE 1   Radians and Degrees

Radians	$\dfrac{\pi}{6}$	$\dfrac{\pi}{4}$	$\dfrac{\pi}{3}$	$\dfrac{\pi}{2}$	π	$\dfrac{3\pi}{2}$	2π
Degrees	30°	45°	60°	90°	180°	270°	360°
Part of Circle	$\dfrac{1}{12}$	$\dfrac{1}{8}$	$\dfrac{1}{6}$	$\dfrac{1}{4}$	$\dfrac{1}{2}$	$\dfrac{3}{4}$	1

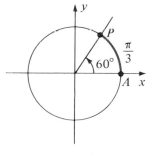 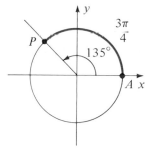 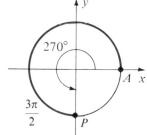

(a) $\theta = 60° = \dfrac{\pi}{3}$ radians   (b) $\theta = 135° = \dfrac{3\pi}{4}$ radians   (c) $\theta = 270° = \dfrac{3\pi}{2}$ radians

FIGURE 4   Degree and Radian Measure

THE REFERENCE ANGLE

Since a complete revolution about a circle returns to the initial position, different angles in standard position may have the same terminal side. For instance, angles of 30° and 390° in standard position have the same terminal side (Figure 5). Such angles are said to be **coterminal.** For an angle in standard position that is not a quadrantal angle, it is convenient to define an acute angle that is called the **reference angle.** The reference angle θ' associated with the angle θ is the acute angle formed by the terminal side of θ'

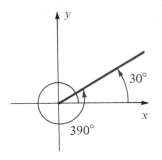

FIGURE 5 Coterminal Angles

and the x-axis. If θ lies in quadrant I, it is an acute angle and $\theta' = \theta$. The other cases are illustrated in Figure 6.

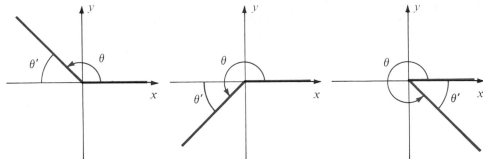

FIGURE 6 The Reference Angle

You have probably observed that the reference angle is analogous to the reference number. The reference number permits us to restrict our tables of values of the circular functions to real numbers in the interval $[0, \pi/2]$; the reference angle, when measured in radians, is also in the interval $[0, \pi/2]$ and, when measured in degrees, is in the interval $[0°, 90°]$. We will make use of the reference angle in the following section when we deal with tables of values of the trigonometric functions.

EXAMPLE 3 Reference Angle Computation
Find the reference angle θ' if

(a) $\theta = 240°$ (b) $\theta = \dfrac{5\pi}{3}$ radians

SOLUTION
(a) Since $\theta = 240°$ lies in the third quadrant, the reference angle is

$$\theta' = 240° - 180° = 60°$$

(b) Since $\theta = 5\pi/3$ radians lies in the fourth quadrant, the reference angle is

$$\theta' = 2\pi - \frac{5\pi}{3} = \frac{\pi}{3}$$

∎

PROGRESS CHECK
Find the reference angle θ' if

(a) $\theta = 160°$ (b) $\theta = \dfrac{4\pi}{3}$ radians

ANSWERS

(a) 20° (b) $\dfrac{\pi}{3}$ radians

THE CENTRAL ANGLE FORMULA

The radian measure of an angle can be found by using a circle other than a unit circle. In Figure 7 the central angle θ subtends an arc of length t on the unit circle and an arc of

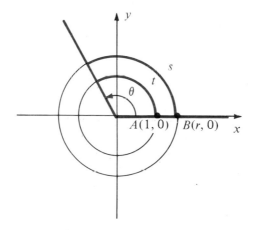

FIGURE 7 Radian Measure Using Any Circle

length s on a circle of radius r. By definition, $\theta = t$. Since the ratio of the arcs is the same as the ratio of the radii, we have

$$\frac{t}{s} = \frac{1}{r}$$

or

$$t = \frac{s}{r}$$

Since $\theta = t$, we have this useful result.

If a central angle θ subtends an arc of length s on a circle of radius r, then

$$\theta = \frac{s}{r}$$

Note that when $s = r$, $\theta = 1$ radian, which tells us that an arc of 1 radian has length equal to the radius of the circle. Also, when $\theta = 2\pi$ we see that $s = 2\pi r$ corresponds to the circumference of the circle.

EXAMPLE 4 The Central Angle Formula
A central angle θ subtends an arc of length 12 inches on a circle whose radius is 6 inches. Find the radian measure of the central angle.

SOLUTION
We have $s = 12$ and $r = 6$, so that

$$\theta = \frac{s}{r} = \frac{12}{6} = 2 \text{ radians}$$

■

WARNING
The formula

$$\theta = \frac{s}{r}$$

can only be applied if the angle θ is in radian measure.

EXAMPLE 5 **The Central Angle Formula**
A designer has to place the word ALMONDS on a can using equally spaced letters (see Figure 8a). For good visibility, the letters must cover a sector of the circle having a 90° central angle. If the base of the can is a circle of radius 2 inches (see Figure 8b), what is the maximum width of each letter?

(a)

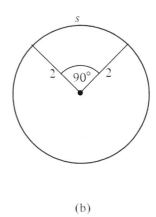

(b)

FIGURE 8 Diagram for Example 5

SOLUTION
Since $\theta = 90° = \pi/2$ radians, the arc has length

$$s = r\theta = 2\left(\frac{\pi}{2}\right) = \pi$$

Each of the seven letters can occupy 1/7 of this arc, or $\pi/7$ inches.

■

EXERCISE SET 6.1

In Exercises 1–18 the angle θ is in standard position; determine the quadrant in which the angle lies.

1. $\theta = 313°$ 2. $\theta = 182°$ 3. $\theta = 14°$ 4. $\theta = 227°$

5. $\theta = 141°$ 6. $\theta = -167°$ 7. $\theta = -345°$ 8. $\theta = 555°$

9. $\theta = 618°$ 10. $\theta = -428°$ 11. $\theta = -195°$ 12. $\theta = 730°$

13. $\theta = \dfrac{7\pi}{8}$ 14. $\theta = \dfrac{-3\pi}{5}$ 15. $\theta = \dfrac{-8\pi}{3}$ 16. $\theta = \dfrac{3\pi}{8}$

17. $\theta = \dfrac{13\pi}{3}$ 18. $\theta = \dfrac{9\pi}{5}$

In Exercises 19–32 convert from degree measure to radian measure.

19. $30°$ 20. $200°$ 21. $-150°$ 22. $-330°$

23. $75°$ 24. $570°$ 25. $-450°$ 26. $-570°$

27. $135°$ 28. $405°$ 29. $120°$ 30. $90°$

 31. $45.22°$ 32. $196.54°$

In Exercises 33–46 convert from radian measure to degree measure.

33. $\dfrac{\pi}{4}$ 34. $\dfrac{\pi}{3}$ 35. $\dfrac{3\pi}{2}$ 36. $\dfrac{5\pi}{6}$

37. $\dfrac{-\pi}{2}$ 38. $\dfrac{-7\pi}{12}$ 39. $\dfrac{4\pi}{3}$ 40. 3π

41. $\dfrac{5\pi}{2}$ 42. -5π 43. $\dfrac{-5\pi}{3}$ 44. $\dfrac{9\pi}{2}$

 45. 1.72 46. 24.98

In Exercises 47–52, for each pair of angles, write T if they are coterminal and F if they are not coterminal.

47. $30°, 390°$ 48. $50°, -310°$ 49. $45°, -45°$ 50. $120°, \dfrac{14\pi}{3}$

51. $\dfrac{\pi}{2}, \dfrac{7\pi}{2}$ 52. $-60°, 760°$

In Exercises 53–64, for each given angle, find the reference angle.

53. $130°$ 54. $\dfrac{5\pi}{6}$ 55. $-20°$ 56. $25°$

57. $-455°$ 58. $700°$ 59. $\dfrac{12\pi}{5}$ 60. $\dfrac{5\pi}{4}$

61. $72°$ 62. $\dfrac{-2\pi}{3}$ 63. $\dfrac{9\pi}{4}$ 64. $\dfrac{5\pi}{3}$

65. If a central angle θ subtends an arc of length 4 centimeters on a circle of radius 7 centimeters, find the approximate measure of θ in radians and in degrees.

66. Find the length of arc subtended by a central angle of $\pi/5$ radians on a circle of radius 6 inches.

67. Find the radius of a circle if a central angle of $2\pi/3$ radians subtends an arc of 4 meters.

68. In a circle of radius 150 centimeters, what is the length of arc subtended by a central angle of 45°?

69. A subcompact car uses a tire whose radius is 13 inches. How far has the car moved when the tire completes one rotation? How many rotations are completed when the tire has traveled one mile? (Assume $\pi \approx 3.14$.)

70. A builder intends placing 7 equally spaced homes on a semicircular plot as shown in the accompanying figure. If the circle has a diameter of 400 feet, what is the distance between any two adjacent homes? (Use $\pi \approx 3.14$.)

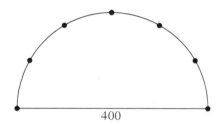

400

71. How many ribs are there in an umbrella if the length of each rib is 1.5 feet and the arc between two adjacent ribs measures $3\pi/10$ feet?

72. A microcomputer has both $5\frac{1}{4}$-inch and $3\frac{1}{2}$-inch floppy disk drives. If the disks used on both drives are divided into 8 sectors, find the ratio of the arc length of a sector of the larger disk to a sector of the smaller disk.

73. In a circle of radius r a central angle of d degrees subtends an arc of length s. Establish a formula for s in terms of the variables r and d.

74. The minute hand on a clock is 3 inches long. Through what distance does the minute hand move between the hours of 4:10 P.M. and 4:22 P.M.?

75. A steam roller whose wheel has a diameter of 2 feet has just completed a pass over a driveway that is 30 feet in length. How many rotations of the roller were required?

76. The diameter of the earth is approximately 7900 miles. What is the difference in latitude between Lexington, Kentucky, and Jacksonville, Florida, if one is 500 miles due north of the other? (*Hint:* Latitude is expressed in degrees.)

6.2 TRIGONOMETRIC FUNCTIONS OF ANGLES

DEFINITION

In Chapter 5 we thought of the trigonometric functions in terms of the correspondence between real numbers and the coordinates of points on the unit circle, which suggested the alternative name, the circular functions. We are now prepared to define the trigonometric functions as functions of angles. It will then be clear that the two approaches are completely in accord and that *all of the results of Chapter 5 hold for angular measure.*

We have previously defined the measure of angle θ as the measure of the arc of length t on the unit circle; that is, $\theta = t$. We now extend this definition to the trigonometric functions of an angle θ.

Definition of $\sin\theta$, $\cos\theta$, and $\tan\theta$	$\sin\theta = \sin t$
	$\cos\theta = \cos t$
	$\tan\theta = \tan t$

The remaining trigonometric functions are again defined as reciprocals; that is, $\csc\theta = 1/\sin\theta$, and so on.

To avoid confusion, we will continue to use small Greek letters to denote angles. Further, we will also use the degree symbol to indicate angular measure in degrees when dealing with numeric values. Thus, $\sin 2°$ and $\sin 2$ are distinct quantities: $\sin 2°$ denotes degree measure and $\sin 2$ denotes radian measure. Note that $\sin 2$ may be interpreted as the value of the sine function either for the real number 2 or for the angle of 2 radians. Since $\sin\theta = \sin t$, either interpretation is acceptable and will yield the same value.

Since we can convert from degree measure to radian measure, we can evaluate the trigonometric functions of angles.

EXAMPLE 1 Trigonometric Function of an Angle
Find sin 30°.

SOLUTION
Converting to radians, $30° = \pi/6$ radians. Then

$$\sin 30° = \sin \frac{\pi}{6} = \frac{1}{2}$$

∎

CALCULATORS

In Section 5.4 we discussed the use of scientific calculators for finding values of the trigonometric functions. You have used the key marked $\boxed{\text{DRG}}$ to ensure that the display reads RAD before entering arguments in radian measure. The $\boxed{\text{DRG}}$ key can also be used to set the display to DEG, which allows you to enter arguments in degrees. For example, with the display reading DEG, if you enter

$$27 \;\boxed{\tan}$$

the display reads 0.5095, which is the approximate value of tan 27°.

There is one other item to remember when using degree measure: you must first convert minutes to fractions of a degree. For example, to find tan 27°30′, you must enter the key sequence

$$27.5 \;\boxed{\tan}$$

that is, you must convert 30 minutes to 0.5 degrees. Your calculator expects an argument that is in degrees; it is not geared to working in minutes. Of course, you can use your calculator to convert minutes to degrees. For example, to find cos 32°51′ with the display reading DEG, you would enter

$$51 \;\boxed{\div}\; 60 \;\boxed{=}\; \boxed{+}\; 32 \;\boxed{=}\; \boxed{\cos}$$

and see 0.8401 in the display. You have effectively used the calculator to convert 51 minutes to 51/60 degrees.

THE REFERENCE ANGLE RULE

Although a scientific calculator is a wonderful tool, there are many times in a calculus course when you will be required to perform certain operations on your own. For example, it might be required to express cosine 265° in terms of the cosine of an acute angle.

In Section 5.4 we developed the Reference Number Rule to enable us to reduce the problem of finding the values of the trigonometric functions of any real number to that of finding the values for a real number in the interval $[0, \pi/2]$. Since $\pi/2 = 90°$, we would expect to have an equivalent rule that works for a non-quadrantal angle θ such that $90° < \theta < 360°$. This rule, called the **Reference Angle Rule,** can be used in conjunction with Table VI in the Tables Appendix that lists the values of the trigo-nometric functions for angles in degree measure in the interval $[0, 90°]$ in steps of 10 minutes. We illustrate the rule with an example.

EXAMPLE 2   The Reference Angle Rule
Find $\cos 150°$ without the use of a calculator.

SOLUTION
Since the angle is in quadrant II, the reference angle θ' is given by

$$\theta' = 180° - 150° = 30°$$

We know that $\cos 30° = \sqrt{3}/2$. Since cosine is negative in quadrant II,

$$\cos 150° = -\cos 30° = -\sqrt{3}/2.$$

∎

EXAMPLE 3   The Reference Angle Rule
Find $\tan 332°$ without the use of a calculator.

SOLUTION
Since the angle is in quadrant IV, the reference angle θ' is given by

$$\theta' = 360° - 332° = 28°$$

From Table VI in the Tables Appendix, $\tan 28° = 0.5317$. Since tangent is negative in quadrant IV, $\tan 332° = -\tan 28° = -0.5317$.

∎

EXAMPLE 4   The Reference Angle Rule
Find $\sin 611°20'$ without the use of a calculator.

SOLUTION
Whenever the angle exceeds $360°$ in absolute value, we must first convert to an angle in the interval $[0, 360°)$ which has the same functional values. To do this, we simply add or subtract multiples of $360°$. Thus,

$$\sin 611°20' = \sin(611°20' - 360°)$$
$$= \sin 251°20'$$

The reference angle θ' for this third-quadrant angle is given by

$$\theta' = 251°20' - 180° = 71°20'$$

From Table VI in the Tables Appendix, $\sin 71°20' = 0.9474$. Since sine is negative in quadrant III, we have

$$\sin 611°20' = -\sin 71°20' = -0.9474$$

∎

EXERCISE SET 6.2

In Exercises 1–8, without using tables or a calculator, find the values of the six trigonometric functions for each argument.

1. $135°$
2. $300°$
3. $-30°$
4. $-300°$
5. $315°$
6. $30°$
7. $270°$
8. $-180°$

In Exercises 9–20 determine the reference angle for each angle.

9. $250°$
10. $-130°$
11. $330°$
12. $125°$
13. $\dfrac{6\pi}{5}$
14. $\dfrac{9\pi}{5}$
15. $335°$
16. $-10°$
17. $-47°$
18. $110°$
19. $\dfrac{15\pi}{7}$
20. $\dfrac{-3\pi}{5}$

In Exercises 21–34 use Table VI in the Tables Appendix to determine each of the following.

21. $\cos 42°40'$
22. $\tan 65°50'$
23. $\sin 14°30'$
24. $\sec 17°40'$
25. $\cot 25°10'$
26. $\csc 26°50'$
27. $\tan 272°20'$
28. $\cos 144°40'$
29. $\sin 246°10'$
30. $\cos(-192°50')$
31. $\sin(-344°20')$
32. $\tan(-143°30')$
33. $\sin 554°$
34. $\cos(-684°)$

35. Use a calculator to verify the results of Exercises 21–34.

6.3 RIGHT TRIANGLE TRIGONOMETRY

We are now prepared to show that the trigonometric functions of an angle are related to the ratios of the sides of a right triangle. In Figure 9a we display a right triangle with sides a and b, hypotenuse r, and an acute angle θ. We can place this triangle on a Cartesian coordinate system with θ in standard position (Figure 9b). We can draw a unit circle and let $N(x, y)$ denote the point of intersection of the circle and the hypotenuse OP. If we drop the perpendicular NM as indicated, we see that the triangles OMN and OQP are similar. The corresponding sides must then be proportional so that

$$\frac{\overline{MN}}{\overline{ON}} = \frac{\overline{QP}}{\overline{OP}} \qquad \text{and} \qquad \frac{\overline{OM}}{1} = \frac{\overline{OQ}}{\overline{OP}}$$

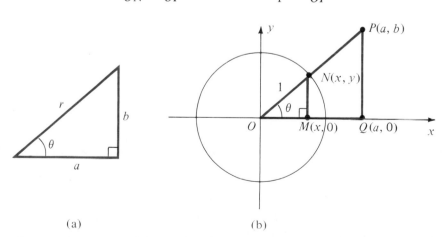

(a)   (b)

FIGURE 9 Trigonometric Functions and the Right Triangle

Since $\overline{OP} = r$, we obtain by substitution

$$\frac{y}{1} = \frac{b}{r} \quad \text{and} \quad \frac{x}{1} = \frac{a}{r}$$

By definition, $\sin \theta = y$ and $\cos \theta = x$. Substituting, we have

$$\sin \theta = y = \frac{b}{r}$$

$$\cos \theta = x = \frac{a}{r}$$

$$\tan \theta = \frac{\sin \theta}{\cos \theta} = \frac{b}{a}$$

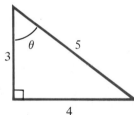

FIGURE 10   **Sides of a Right Triangle**

If we denote the sides a and b of the right triangle in Figure 9a as the adjacent and opposite sides relative to the angle θ (see Figure 10), then this last result expresses the trigonometric functions as ratios of the lengths of the sides of the right triangle.

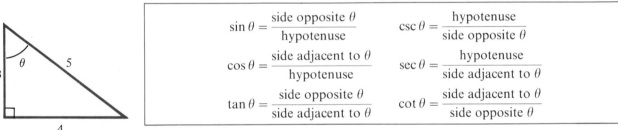

$$\sin \theta = \frac{\text{side opposite } \theta}{\text{hypotenuse}} \qquad \csc \theta = \frac{\text{hypotenuse}}{\text{side opposite } \theta}$$

$$\cos \theta = \frac{\text{side adjacent to } \theta}{\text{hypotenuse}} \qquad \sec \theta = \frac{\text{hypotenuse}}{\text{side adjacent to } \theta}$$

$$\tan \theta = \frac{\text{side opposite } \theta}{\text{side adjacent to } \theta} \qquad \cot \theta = \frac{\text{side adjacent to } \theta}{\text{side opposite } \theta}$$

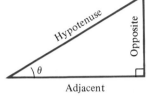

FIGURE 11   **Diagram for Example 1**

EXAMPLE 1   Right Triangle Trigonometry
Find the values of the trigonometric functions of the angle θ in Figure 11.

SOLUTION

$$\sin \theta = \frac{4}{5} \qquad \csc \theta = \frac{5}{4}$$

$$\cos \theta = \frac{3}{5} \qquad \sec \theta = \frac{5}{3}$$

$$\tan \theta = \frac{4}{3} \qquad \cot \theta = \frac{3}{4}$$

∎

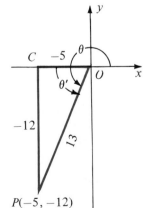

FIGURE 12   **Diagram for Example 2**

EXAMPLE 2   Right Triangle Applications
Find $\sec \theta$ if the point $P(-5, -12)$ lies on the terminal side of θ.

SOLUTION
(See Figure 12.) We construct a perpendicular from P to the x-axis to form right triangle

PCO and use the Pythagorean theorem to find $\overline{OP} = 13$. Then

$$\sec \theta' = \frac{\text{hypotenuse}}{\text{adjacent}} = \frac{13}{5}$$

Since θ is in the third quadrant, by the Reference Angle Rule,

$$\sec \theta = -\sec \theta' = -\frac{13}{5}$$

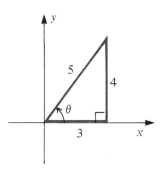

FIGURE 13 **Diagram for Example 3**

EXAMPLE 3 **Right Triangle Applications**

Find $\cos(\arctan \frac{4}{3})$ without using tables or a calculator.

SOLUTION

We have repeated Example 5 of Section 5.7 to illustrate an alternative, simpler approach using right triangle trigonometry. We let $\theta = \arctan \frac{4}{3}$ so that $\tan \theta = \frac{4}{3}$ and $0 \le \theta \le \pi/2$. The angle θ in Figure 13 satisfies these conditions. Then we see that

$$\cos\left(\arctan \frac{4}{3}\right) = \cos \theta = \frac{3}{5}$$

"SPECIAL" ANGLES

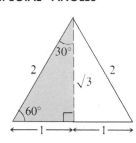

FIGURE 14a **30°-60°-90° Right Triangle**

We can use right triangle trigonometry to provide an alternate means for obtaining the values of the trigonometric functions for the "special" angles of 30°, 45°, and 60°. In Figure 14a we show an equilateral triangle having sides of length 2 and have sketched the bisector of a vertex angle. Since the angle bisector in an equilateral triangle is also a perpendicular bisector of the opposite side, each of the resulting triangles is a 30°-60°-90° right triangle and we can use the Pythagorean theorem to determine that the side opposite the 60° angle has length $\sqrt{3}$. Applying the definitions of the trigonometric functions, we have:

$$\sin 60° = \frac{\text{opp}}{\text{hyp}} = \frac{\sqrt{3}}{2} \qquad \cos 60° = \frac{\text{adj}}{\text{hyp}} = \frac{1}{2} \qquad \tan 60° = \frac{\text{opp}}{\text{adj}} = \sqrt{3}$$

$$\sin 30° = \frac{\text{opp}}{\text{hyp}} = \frac{1}{2} \qquad \cos 30° = \frac{\text{adj}}{\text{hyp}} = \frac{\sqrt{3}}{2} \qquad \tan 30° = \frac{\text{opp}}{\text{adj}} = \frac{1}{\sqrt{3}} = \frac{\sqrt{3}}{3}$$

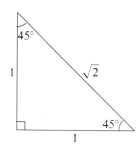

FIGURE 14b **45°-45°-90° Right Triangle**

In Figure 14b we have sketched an isosceles right triangle whose equal sides have length 1. The acute angles of the triangle both have a measure of 45° and we can use the Pythagorean Theorem to show that the hypotenuse has length $\sqrt{2}$. Then,

$$\sin 45° = \frac{\text{opp}}{\text{hyp}} = \frac{1}{\sqrt{2}} = \frac{\sqrt{2}}{2} \qquad \cos 45° = \frac{\text{adj}}{\text{hyp}} = \frac{1}{\sqrt{2}} = \frac{\sqrt{2}}{2} \qquad \tan 45° = \frac{\text{opp}}{\text{adj}} = 1$$

Of course, the results we see here using right triangle trigonometry are identical to those we obtained earlier in Chapter 5. Many students find that the 30°-60°-90° and

45°-45°-90° triangles are a good way to obtain the values of the trigonometric functions for these "special" angles. These results are summarized in Table 2 and its accompanying figures.

TABLE 2   Values of the Trigonometric Functions

θ	θ	$\sin\theta$	$\cos\theta$	$\tan\theta$
30°	$\dfrac{\pi}{6}$	$\dfrac{1}{2}$	$\dfrac{\sqrt{3}}{2}$	$\dfrac{\sqrt{3}}{3}$
45°	$\dfrac{\pi}{4}$	$\dfrac{\sqrt{2}}{2}$	$\dfrac{\sqrt{2}}{2}$	1
60°	$\dfrac{\pi}{3}$	$\dfrac{\sqrt{3}}{2}$	$\dfrac{1}{2}$	$\sqrt{3}$

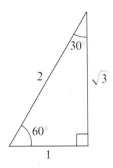 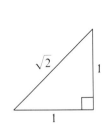

SOLVING A TRIANGLE

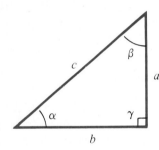

FIGURE 15   Notation for Parts of a Triangle

The expression "to solve a triangle" is used to indicate that we seek all parts of the triangle, that is, the length of each side and the measure of each angle. For any right triangle, given any two sides, or given one side and an acute angle, it is always possible to solve the triangle. We will standardize the notation as shown in Figure 15 so that (a) the acute angles are labeled α and β, the right angle is labeled γ, and (b) the sides opposite angles α, β, and γ are labeled a, b, and c, respectively.

EXAMPLE 4   Solving a Triangle

In triangle ABC, $\gamma = 90°$, $\beta = 27°$, and $b = 8.6$. Find approximate values for the remaining parts of the triangle.

SOLUTION

We begin by labeling a right triangle as in Figure 16. Since the sum of the angles of a triangle is 180°, we see that $\alpha = 63°$. Using the trigonometric functions of angle β we have

$$\sin 27° = \frac{8.6}{c} \qquad \text{and} \qquad \tan 27° = \frac{8.6}{a}$$

Solving for a and c yields

$$c = \frac{8.6}{\sin 27°} \qquad a = \frac{8.6}{\tan 27°}$$

From Table VI in the Tables Appendix, $\sin 27° = 0.4540$ and $\tan 27° = 0.5095$, so

$$c = 8.6/0.4540 \approx 18.9$$
$$a = 8.6/0.5095 \approx 16.9$$

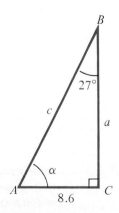

FIGURE 16   Diagram for Example 4

PROGRESS CHECK

In triangle ABC, $\gamma = 90°$, $\alpha = 64°$, and $b = 24.7$. Solve the triangle.

ANSWER

$\beta = 26°$ $a = 50.6$ $c = 56.3$

EXAMPLE 5   Solving a Triangle

In triangle ABC, $\gamma = 90°$, $a = 22.5$, and $b = 12.8$. Find approximate values for the remaining parts of the triangle.

SOLUTION

Figure 17 displays the parts of the triangle. Using angle β we have

$$\tan \beta = \frac{12.8}{22.5} = 0.5689$$

Using Table VI in the Tables Appendix, the closest entry yields $\beta \approx 29°40'$. Since the sum of the angles is 180°, we must have $\alpha \approx 60°20'$. Alternatively,

$$\tan \alpha = \frac{22.5}{12.8} = 1.7578$$

also yields $\alpha \approx 60°20'$ by use of Table VI.

    Finally, c can be found by the Pythagorean theorem or by trigonometry.

$$\sin \beta = \sin 29°40' = \frac{12.8}{c}$$

$$c = \frac{12.8}{0.4950} \approx 25.9$$

■

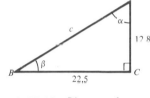

FIGURE 17   Diagram for Example 5

PROGRESS CHECK

In triangle ABC, $\gamma = 90°$, $a = 17.4$, and $b = 38.2$. Solve the triangle.

ANSWER

$\alpha = 24°30'$ $\beta = 65°30'$ $c = 42$

APPLICATIONS

Many applied problems involve right triangles. We are now prepared to use our ability in solving triangles to tackle a variety of interesting problems.

EXAMPLE 6   Applying Trigonometry

A ladder leaning against a building makes an angle of 35° with the ground. If the bottom of the ladder is 5 meters from the building, how long is the ladder? To what height does it rise along the building?

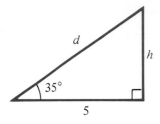

FIGURE 18 Diagram for Example 6

SOLUTION

In Figure 18 we seek the length d of the ladder and the height h along the building. Using right triangle trigonometry,

$$\cos 35° = \frac{5}{d} \qquad \text{and} \qquad \tan 35° = \frac{h}{5}$$

$$d = \frac{5}{\cos 35°} \qquad\qquad h = 5 \tan 35°$$

$$d = \frac{5}{0.8192} \qquad\qquad h = 5(0.7002)$$

$$d \approx 6.1 \text{ meters} \qquad \text{and} \qquad h \approx 3.5 \text{ meters} \qquad \blacksquare$$

PROGRESS CHECK

The string of a kite makes an angle of $32°30'$ with the ground. If 125 meters of string have been let out, how high is the kite?

ANSWER
67 meters

There are two technical terms that will occur frequently in our word problems. The **angle of elevation** is the angle between the horizontal and the line of sight. In Figure 19a, θ is the angle of elevation of the top T of a tree from a point x meters from the base of the tree.

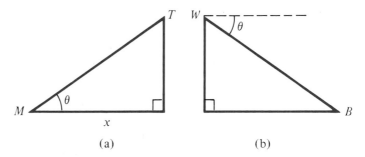

(a)   (b)

FIGURE 19 Angles of Elevation and Depression

The **angle of depression** is the angle between the horizontal and the line of sight when looking down. In Figure 19b, θ is the angle of depression of a boat B as seen from a watchtower W.

EXAMPLE 7 Angle of Elevation

A vendor of balloons inadvertently releases a balloon, which rises straight up. A child standing 50 feet from the vendor watches the balloon rise. When the angle of elevation of the balloon reaches $44°$, how high is the balloon?

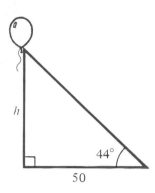

SOLUTION

We seek the height h in Figure 20. Thus,

$$\tan 44° = \frac{h}{50}$$

$$h = 50 \tan 44°$$

$$h = 50(0.9657) \approx 48$$

The balloon has risen approximately 48 feet.

■

FIGURE 20   Diagram for Example 7

EXAMPLE 8   Angle of Depression

A forest ranger is in a tower 65 feet above the ground. If the ranger spots a fire at an angle of depression of $6°40'$, how far is the fire from the base of the tower (assuming level terrain)?

SOLUTION

We need to find the distance d in Figure 21. Since $\theta + 6°40' = 90°$, $\theta = 83°20'$. Then

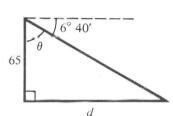

$$\tan \theta = \frac{d}{65}$$

$$d = 65 \tan 83°20'$$

$$d = 65(8.5555) \approx 556$$

The fire is approximately 556 feet from the base of the tower.

■

FIGURE 21   Diagram for Example 8

EXAMPLE 9   Applying Trigonometry

A mathematics professor walks toward the university clock tower on the way to her office, and decides to find the height of the clock above ground. She determines the angle of elevation to be $30°$ and, after proceeding an additional 60 feet toward the base of the tower, finds the angle of elevation to be $40°$. What is the height of the clock tower?

SOLUTION

This problem is somewhat more sophisticated since it involves more than one right triangle. In Figure 22 we seek to determine h. From triangle ACD,

$$\tan 30° = \frac{h}{d + 60} \qquad \text{or} \qquad h = (d + 60)(\tan 30°)$$

and from triangle ACB,

$$\tan 40° = \frac{h}{d} \qquad \text{or} \qquad h = d \tan 40°$$

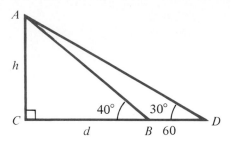

FIGURE 22 **Diagram for Example 9**

Equating the two expressions for h yields

$$(d + 60)(\tan 30°) = d \tan 40°$$

$$60 \tan 30° = d(\tan 40° - \tan 30°)$$

$$d = \frac{60 \tan 30°}{\tan 40° - \tan 30°} \approx 132 \text{ feet}$$

$$h = d \tan 40° = (132)(0.8391) \approx 110.8$$

The height of the clock tower is 110.8 feet.

■

In navigation and surveying, directions are often given by **bearings,** which specify an acute angle and its direction from the north-south line. In Figure 23a the bearing of point B from point A is N 40° E, that is, 40° east of north; in Figure 23b the bearing of point B from point A is S 60° W; and in Figure 23c it is S 20° E.

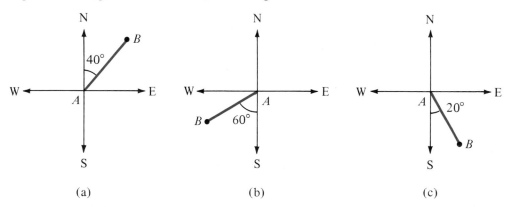

(a) (b) (c)

FIGURE 23 **Bearings**

EXAMPLE 10 Bearing

A ship leaves port at 10 A.M. and heads due east at a rate of 22 miles per hour. At 11 A.M. the course is changed to S 52° E. Find the distance and bearing of the ship from the dock at noon.

SOLUTION

The situation is depicted in Figure 24. The ship travels due east from port (point A), reaches B at 11 A.M., changes direction and arrives at E at noon. Since the ship travels at

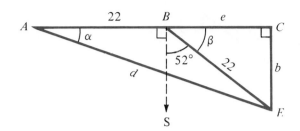

FIGURE 24   Diagram for Example 10

22 mph, $\overline{AB}$ and $\overline{BE}$ are each 22 miles in length. Further, since $\measuredangle\,EBS$ has a measure of 52°, we find angle $\beta = 38°$. From right triangle BCE,

$$\cos\beta = \frac{e}{22} \qquad \text{or} \qquad e = 22\cos 38° \approx 17.3 \text{ miles}$$

$$\sin\beta = \frac{b}{22} \qquad \text{or} \qquad b = 22\sin 38° \approx 13.5 \text{ miles}$$

We now know two sides of right triangle ACE, namely

$$\overline{AC} = 22 + e \approx 22 + 17.3 \approx 39.3$$
$$\overline{CE} = b \approx 13.5$$

We can now solve triangle ACE to obtain

$$\tan\alpha \approx \frac{13.5}{39.3} \qquad \text{or} \qquad \alpha \approx 19°$$

From triangle ACE,

$$\sin\alpha = \frac{b}{d} \approx \frac{13.5}{d}$$

$$d = \frac{13.5}{\sin 19°} \approx 41.5 \text{ miles}$$

The ship is 41.5 miles from port at a bearing of S 71° E.   ■

EXERCISE SET 6.3

In Exercises 1–8, find the values of the trigonometric functions of the angle θ in each of the following triangles.

1.

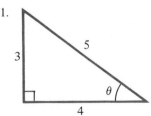

2.

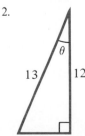

3.

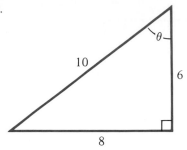

4.

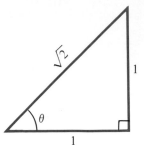

5.

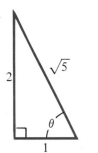

6.

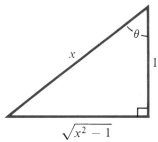

7.

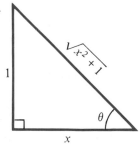

8.

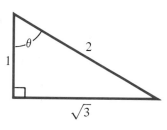

In Exercises 9–20 find the values of the trigonometric functions of the angle θ if the point P lies on the terminal side of θ.

9. $P(-5, 12)$ 10. $P(3, -4)$ 11. $P(-1, -1)$ 12. $P(1, 2)$

13. $P(-8, 6)$ 14. $P(12, 5)$ 15. $P(12, -5)$ 16. $P(-1, \sqrt{3})$

17. $P(-12, -5)$ 18. $P(-3, 4)$ 19. $P(-2, 1)$ 20. $P(-2, -1)$

In Exercises 21–26, in each of the following right triangles, express the length h as a trigonometric function of the angle θ.

21.

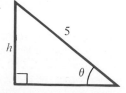

22.

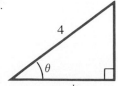

23.

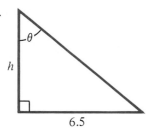

24.

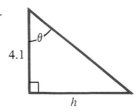

25.

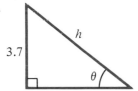

26.

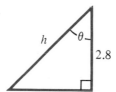

In Exercises 27–34, in triangle ABC, $\gamma = 90°$. Find the required parts of the triangle in each of the following.

27. $a = 12$, $b = 16$; find α.

28. $a = 5$, $b = 15$; find β.

29. $b = 40$, $\beta = 40°$; find c.

30. $a = 22$, $\alpha = 36°$; find b.

31. $a = 75$, $\beta = 22°$; find b.

32. $b = 60$, $\alpha = 53°$; find c.

33. $a = 25$, $\beta = 42°30'$; find c.

34. $b = 50$, $\alpha = 36°20'$; find a.

In Exercises 35–39 evaluate the given expression without using tables or a calculator.

35. $\tan\left(\sin^{-1} -\dfrac{5}{13}\right)$

36. $\sin\left(\arctan -\dfrac{12}{5}\right)$

37. $\cos\left(\sin^{-1}\dfrac{4}{5}\right)$

38. $\cos\left(\arcsin -\dfrac{2}{3}\right)$

39. $\tan\left(\cos^{-1} -\dfrac{3}{5}\right)$

40. A ladder 7 meters in length leans against a vertical wall. If the ladder makes an angle of $65°$ with the ground, find the height the ladder reaches above the ground.

41. A ladder 20 feet in length touches a wall at a point 16 feet above the ground. Find the angle the ladder makes with the ground.

42. A monument is 550 feet high. What is the length of the shadow cast by the monument when the sun is $64°$ above the horizon?

43. Find the angle of elevation of the sun when a tower 45 meters in height casts a horizontal shadow 25 meters in length.

44. A technician positioned on an oil-drilling rig 120 feet above the water spots a boat at an angle of depression of $16°$. How far is the boat from the rig?

45. A mountainside hotel is located 8000 feet above sea level. From the hotel, a trail leads farther up the mountain to an inn at an elevation of 10,400 feet. If the trail has an angle of inclination of $18°$ (that is, the angle of elevation of the inn from the hotel is $18°$), find the distance along the trail from the hotel to the inn.

46. A hill is known to be 200 meters high. A surveyor standing on the ground finds the angle of elevation of the top of the hill to be $42°50'$. Find the distance from the surveyor to a point directly below the top of the hill. (Ignore the height of the surveyor.)

47. An observer is 425 meters from a launching pad when a rocket is launched vertically. If the angle of elevation of the rocket at its apogee (highest point) is $66°20'$, how high does the rocket rise?

48. An airplane pilot wants to climb from an altitude of 6000 feet to an altitude of 16,000 feet. If the plane climbs at an angle of $9°$ with a constant speed of 22,000 feet per minute, how long will it take to reach the increased altitude?

49. A rectangle is 16 inches long and 13 inches wide. Find the measures of the angles formed by a diagonal with the sides.

50. The sides of an isosceles triangle are 15, 15, and 26 centimeters. Find the measures of the angles of the triangle. (*Hint:* The altitude of an isosceles triangle bisects the base.)

51. The side of a regular pentagon is 22 centimeters. Find the radius of the circle circumscribed about the pentagon. (*Hint:* The radii from the center of the circumscribed circle to any two adjacent vertices of the regular pentagon form an isosceles triangle. The altitude of an isosceles triangle bisects the base.)

52. To determine the width of a river, markers are placed at each side of the river in line with the base of a tower that rises 23.4 meters above the ground. From the top of the tower, the angles of depression of the markers are 58°20′ and 11°40′. Find the width of the river.

53. The angle of elevation of the top of building B from the base of building A is 29°. From the top of building A, the angle of depression of the base of building B is 15°. If building B is 110 feet high, find the height of building A.

54. A ship leaves port at 2 P.M. and heads due east at a rate of 40 kilometers per hour. At 4 P.M. the course is changed to N 32° E. Find the distance and bearing of the ship from the dock at 6 P.M.

55. An attendant in a lighthouse receives a request for aid from a stalled craft located 15 miles due east of the lighthouse. The attendant contacts a second boat located 14 miles from the lighthouse at a bearing of N 23° W. What is the distance of the rescue ship from the stalled craft?

6.4 LAW OF COSINES

In Section 6.3 we studied the trigonometry of a right triangle. In this and the next section we will examine an **oblique triangle,** a triangle that does not contain a right angle.

We can always solve an oblique triangle by dropping a perpendicular as in Figures 25a and 25b and treating the resulting right triangles *ADC* and *BDC*. It is, however, worthwhile to perform the analysis in a general way. This yields two results, known as the *law of sines* and the *law of cosines*. We shall now state and prove the law of cosines. We maintain the notation of the last section; thus, the angles of triangle *ABC* are denoted by α, β, and γ, with opposite sides a, b, and c, respectively.

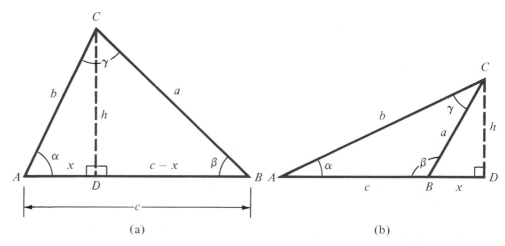

(a)   (b)

FIGURE 25 Law of Cosines

The Law of Cosines In triangle ABC,

$$a^2 = b^2 + c^2 - 2bc \cos \alpha \qquad (1)$$
$$b^2 = a^2 + c^2 - 2ac \cos \beta \qquad (2)$$
$$c^2 = a^2 + b^2 - 2ab \cos \gamma \qquad (3)$$

The student is urged to note the *pattern* of the three forms of the law of cosines as an aid in their memorization.

To prove the law of cosines, we deal with the cases shown in Figure 25.

Case 1. The angles of triangle ABC are all acute (Figure 25a). We construct the perpendicular CD to side AB. Applying the Pythagorean theorem to right triangle BDC, we have

$$\begin{aligned}
a^2 &= h^2 + (c - x)^2 \\
&= h^2 + c^2 - 2cx + x^2 \\
&= (h^2 + x^2) + c^2 - 2cx \\
&= b^2 + c^2 - 2cx \qquad (4)
\end{aligned}$$

The last step results from application of the Pythagorean theorem to right triangle ADC. Also,

$$\cos \alpha = \frac{x}{b} \qquad \text{or} \qquad x = b \cos \alpha$$

which we then substitute in Equation (4) to yield

$$a^2 = b^2 + c^2 - 2bc \cos \alpha$$

This establishes the desired result of Equation (1).

Case 2. Triangle ABC has an obtuse angle β (Figure 25b). We construct the perpendicular CD to side AB. The Pythagorean theorem can be applied to right triangle BDC to give

$$a^2 = h^2 + x^2 \qquad (5)$$

Next, we use the trigonometry of the right triangle ADC to obtain

$$\sin \alpha = \frac{h}{b} \qquad \text{or} \qquad h = b \sin \alpha$$

$$\cos \alpha = \frac{c + x}{b} \qquad \text{or} \qquad x = b \cos \alpha - c$$

Substituting for h and x in Equation (5) we have

$$\begin{aligned}
a^2 &= b^2 \sin^2 \alpha + (b \cos \alpha - c)^2 \\
&= b^2 \sin^2 \alpha + b^2 \cos^2 \alpha - 2bc \cos \alpha + c^2 \\
&= b^2(\sin^2 \alpha + \cos^2 \alpha) - 2bc \cos \alpha + c^2 \\
&= b^2 + c^2 - 2bc \cos \alpha \quad (\text{Since } \sin^2 \alpha + \cos^2 \alpha = 1)
\end{aligned}$$

Once again, this is the desired result of Equation (1).

We have thus established the first form of the law of cosines for both cases. A similar argument can be used to establish the other two forms, given in Equations (2) and (3).

Examination of the law of cosines shows that it can be used in the following circumstances.

Applying the Law of Cosines

The law of cosines may be used when

(a) three sides of a triangle are known (SSS), or

(b) two sides of a triangle are known and the measure of the angle formed by those sides is known (SAS).

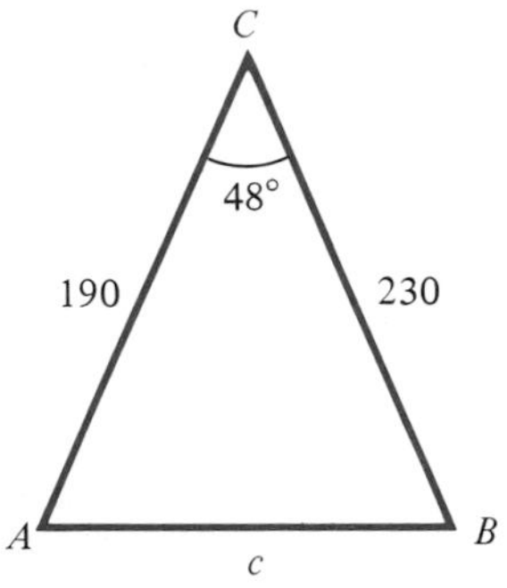

FIGURE 26 Diagram for Example 1

EXAMPLE 1 Applying the Law of Cosines

Highway engineers who are to dig a tunnel through a small mountain wish to determine the length of the tunnel. Points A and B are chosen as the endpoints of the tunnel. Then a point C is selected from which the distances to A and B are found to be 190 feet and 230 feet, respectively. If angle ACB measures $48°$, find the approximate length of the tunnel.

SOLUTION

The known information is displayed in Figure 26. Applying the law of cosines,

$$c^2 = a^2 + b^2 - 2ab\cos\gamma$$
$$c^2 = 230^2 + 190^2 - 2(230)(190)\cos 48°$$
$$c^2 \approx 30{,}518$$
$$c \approx 175 \text{ feet}$$

∎

EXAMPLE 2 The Law of Cosines

Find the approximate measure of the angles of triangle ABC if $a = 150$, $b = 100$, and $c = 75$.

SOLUTION

Substituting in the equation

$$a^2 = b^2 + c^2 - 2bc\cos\alpha$$
$$150^2 = 100^2 + 75^2 - 2(100)(75)\cos\alpha$$
$$22{,}500 = 10{,}000 + 5625 - 15{,}000\cos\alpha$$
$$\cos\alpha = -0.4583$$

Since $\cos\alpha$ is negative, angle α must lie in the second quadrant and is an obtuse angle. Using Table VI in the Tables Appendix, or a calculator,

$$\alpha \approx 180° - 62°40' \approx 117°20'$$

Similarly,

$$b^2 = a^2 + c^2 - 2ac \cos \beta$$
$$100^2 = 150^2 + 75^2 - 2(150)(75) \cos \beta$$
$$10,000 = 22,500 + 5625 - 22,500 \cos \beta$$
$$\cos \beta = 0.8056$$
$$\beta \approx 36°20'$$

Finally, we may easily determine γ since the sum of the angles of a triangle is 180°.

$$\gamma \approx 180° - (117°20' + 36°20') \approx 26°20'$$

The student should verify this result by substituting in the equation

$$c^2 = a^2 + b^2 - 2ab \cos \gamma$$

■

EXERCISE SET 6.4

In Exercises 1–10 use the law of cosines to approximate the required part of triangle ABC.

1. $a = 10, b = 15, c = 21$; find β.

2. $a = 5, b = 12, c = 15$; find γ.

3. $a = 25, c = 30, \beta = 28°30'$; find b.

4. $b = 20, c = 13, \alpha = 19°10'$; find a.

5. $a = 10, b = 12, \gamma = 108°$; find c.

6. $a = 30, c = 40, \beta = 122°$; find b.

7. $b = 6, a = 7, \gamma = 68°$; find α.

8. $a = 6, b = 15, c = 16$; find β.

9. $a = 9, b = 12, c = 15$; find γ.

10. $a = 11, c = 15, \beta = 33°$; find γ.

11. The sides of a parallelogram measure 25 centimeters and 40 centimeters, and the longer diagonal measures 50 centimeters. Find the approximate measure of the smaller angle of the parallelogram.

12. The sides of a parallelogram measure 40 inches and 70 inches, and one of the angles is 108°. Find the approximate length of each diagonal of the parallelogram.

13. A ship leaves port at 9 A.M. and travels due west at a rate of 15 miles per hour. At 11 A.M. the ship changes direction to S 32° W. What is the distance and bearing of the ship from port at 1 P.M.?

14. A ship leaves from port A intending to travel direct to port B, a distance of 25 kilometers. After traveling 12 kilometers the captain finds that his course has been in error by 10°. How far is the ship from port B?

15. Two trains leave Pennsylvania Station in New York City at 2 P.M. and travel in directions that differ by 55°. If the trains travel at constant rates of 50 miles per hour and 80 miles per hour, respectively, what is the distance between them at 2:30 P.M.?

16. Hurricane David has left a telephone pole in a non-vertical position. Workmen place a 30-foot ladder at a point 10 feet from the base of the pole. If the ladder touches the pole at a point 26 feet up the pole, find the angle the pole makes with the ground.

17. Find the approximate perimeter of triangle ABC if $a = 20, b = 30$, and $\gamma = 37°$.

18. A hill makes an angle of 10° with the horizontal. An antenna 50 feet in height is erected at the top of the hill and a guy wire is run to a point 30 feet from the base of the antenna. What is the length of the guy wire?

19. Prove that if ABC is a right triangle, the law of cosines reduces to the Pythagorean theorem.

20. Prove the following in triangle ABC.
   (a) $a^2 + b^2 + c^2 = 2(bc \cos \alpha + ac \cos \beta + ab \cos \gamma)$
   (b) $\dfrac{\cos \alpha}{a} + \dfrac{\cos \beta}{b} + \dfrac{\cos \gamma}{c} = \dfrac{a^2 + b^2 + c^2}{2abc}$

21. Prove that if

$$\frac{\cos \beta}{a} = \frac{\cos \alpha}{b}$$

triangle ABC is either a right triangle or an isosceles triangle.

**6.5
LAW OF SINES**

In the last section we applied the law of cosines to an oblique triangle. That law derives its name from the appearance of the cosine function in its statement.

We will now state and prove the law of sines, which also applies to an oblique triangle. Not surprisingly, the law of sines involves the sine function. Once again, we denote the angles of triangle ABC by α, β, and γ, with opposite sides a, b, and c, respectively.

The Law of Sines	In triangle ABC, $$\frac{a}{\sin \alpha} = \frac{b}{\sin \beta} = \frac{c}{\sin \gamma}$$

The two cases are illustrated in Figure 27.

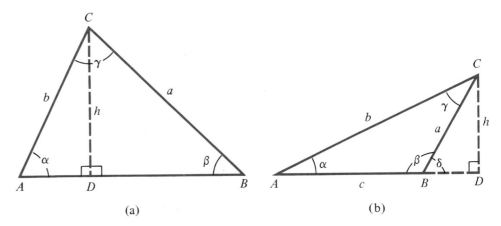

(a)   (b)

FIGURE 27 The Law of Sines

Case 1. The angles of triangle ABC are all acute (Figure 27a). We construct the perpendicular CD to side AB. Then triangles ADC and BDC are both right triangles, and we can apply trigonometry of a right triangle to obtain

$$\sin \alpha = \frac{h}{b} \qquad \text{or} \qquad h = b \sin \alpha$$

$$\sin \beta = \frac{h}{a} \qquad \text{or} \qquad h = a \sin \beta$$

Equating the expressions for h yields

$$b \sin \alpha = a \sin \beta$$

which can be written in the convenient form

$$\frac{a}{\sin \alpha} = \frac{b}{\sin \beta}$$

Case 2. Triangle *ABC* has an obtuse angle β (Figure 27b). We construct the perpendicular *CD* to side *AB*. Applying right triangle trigonometry to triangles *ADC* and *BDC*, and noting that $\delta = 180° - \beta$, we obtain

$$\sin \alpha = \frac{h}{b} \qquad \text{or} \qquad h = b \sin \alpha$$

$$\sin \delta = \sin(180° - \beta) = \frac{h}{a} \qquad \text{or} \qquad h = a \sin(180° - \beta)$$

Equating the expressions for *h* yields

$$b \sin \alpha = a \sin(180° - \beta)$$

Since sine is positive in both the first and second quadrants, the Reference Angle Rule tells us that

$$\sin(180° - \beta) = \sin \beta$$

Substituting, we again obtain

$$b \sin \alpha = a \sin \beta$$

or

$$\frac{a}{\sin \alpha} = \frac{b}{\sin \beta}$$

To complete the proof of the law of sines we need only drop a perpendicular from *A* to *BC* and use a similar argument to show that

$$\frac{b}{\sin \beta} = \frac{c}{\sin \gamma}$$

The law of sines then follows from the transitive property of equality.

The law of sines can be used in the following circumstances.

Applying the Law of Sines	The law of sines may be used when the known parts of a triangle are **(a)** one side and two angles (SAA), or **(b)** two sides and an angle opposite one of these sides (SSA).

Remember that if two angles of a triangle are known, we can immediately determine the third angle. Here is an example.

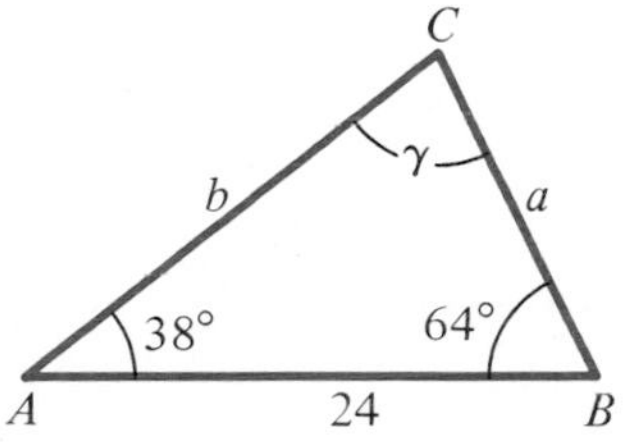

EXAMPLE 1 Using the Law of Sines

In triangle *ABC*, $\alpha = 38°$, $\beta = 64°$, and $c = 24$. Find approximate values for the remaining parts of the triangle.

FIGURE 28 Diagram for Example 1

SOLUTION

(See Figure 28.) Since α and β are known,

$$\gamma = 180° - (\alpha + \beta) = 180° - (38° + 64°) = 78°$$

Applying the law of sines,

$$\frac{a}{\sin \alpha} = \frac{c}{\sin \gamma}$$

$$\frac{a}{\sin 38°} = \frac{24}{\sin 78°}$$

$$a = \frac{24 \sin 38°}{\sin 78°} = \frac{24(0.6157)}{(0.9781)} \approx 15.1$$

Similarly, from

$$\frac{b}{\sin \beta} = \frac{c}{\sin \gamma}$$

we obtain

$$b \approx 22.1$$

∎

When the given parts of a triangle are two sides and an angle opposite one of them, the situation is not straightforward since a *unique* triangle is not always determined. In Figure 29 we have constructed angle α and side b and then used a compass to construct a side of length a with an endpoint at C. In Figure 29a no triangle exists satisfying the given conditions; Figure 29b shows that we may obtain a right triangle; Figure 29c illustrates the possibility that two triangles will satisfy the given conditions; Figure 29d shows that precisely one acute triangle may be possible.

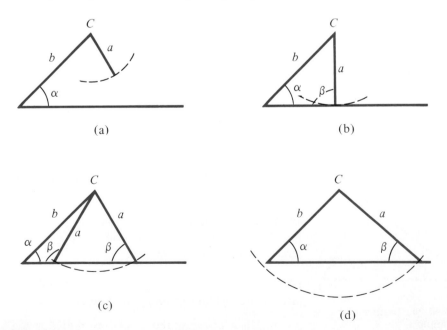

FIGURE 29 Construction Given Two Sides and an Angle

In Exercise 23 you will be asked to prove a number of inequalities that determine which of the four cases applies to a given set of conditions. In practice, we prefer to have you go ahead with the law of sines and let the results lead you to the appropriate answer.

Assume that sides a and b and angle α of triangle ABC are known and that we use the law of sines to determine angle β. These are the results that correspond to the possibilities of Figure 29.

(a) $\sin \beta > 1$. Since $|\sin \theta| \le 1$ for all θ, there is no angle β satisfying the given conditions. This corresponds to the illustration in Figure 29a.

(b) $\sin \beta = 1$. Then $\beta = 90°$ and the given parts determine a unique right triangle (Figure 29b).

(c) $0 < \sin \beta < 1$. There are two possible choices for β, which is why this is called the **ambiguous case.** Since the sine function is positive in quadrants I and II, one choice will be an acute angle and one will be an obtuse angle (Figure 29c).

(d) $0 < \sin \beta < 1$. There are two possible choices for β but the obtuse angle does not form a triangle (Figure 29d). This case is signaled by $\alpha + \beta$ exceeding $180°$.

Here are several illustrations of the law of sines when two sides and an angle opposite one of these sides are known.

EXAMPLE 2 The Law of Sines (No Solution)
In triangle ABC, $\alpha = 60°$, $a = 5$, and $b = 7$. Find angle β.

SOLUTION
Using the law of sines,

$$\frac{a}{\sin \alpha} = \frac{b}{\sin \beta}$$

$$\sin \beta = \frac{b \sin \alpha}{a} = \frac{7 \sin 60°}{5} \approx 1.2$$

Since the sine function has a maximum value of 1, there is no angle β such that $\sin \beta = 1.2$. Hence, there is no triangle with the given parts. This example corresponds to Figure 29a.

■

EXAMPLE 3 The Law of Sines (Ambiguous Case)
In triangle ABC, $a = 5$, $b = 8$, and $\alpha = 22°$. Find the remaining angles of the triangle.

SOLUTION
Using the law of sines,

$$\frac{a}{\sin \alpha} = \frac{b}{\sin \beta}$$

$$\sin \beta = \frac{b \sin \alpha}{a} = \frac{8 \sin 22°}{5} \approx 0.5993$$

Using tables or a calculator, we find that $\beta \approx 36°50'$. Thus, the angles are $\alpha = 22°$, $\beta = 36°50'$, and $\gamma = 121°10'$.

However, the angle $\beta = 180° - 36°50' = 143°10'$ also satisfies the requirement that $\sin \beta = 0.5993$. Therefore, another satisfactory triangle has angles $\alpha = 22°$, $\beta = 143°10'$, and $\gamma = 14°50'$.

This is an example of the ambiguous case, and corresponds to Figure 29c.

EXAMPLE 4 The Law of Sines (Unique Solution)

In triangle ABC, $a = 9$, $b = 6$, and $\alpha = 35°$. Find angles β and γ.

SOLUTION

We again apply the law of sines.

$$\frac{a}{\sin \alpha} = \frac{b}{\sin \beta}$$

$$\sin \beta = \frac{b \sin \alpha}{a} = \frac{6 \sin 35°}{9} \approx 0.3824$$

Using tables or a calculator yields $\beta \approx 22°30'$. A triangle satisfying the given conditions has $\alpha = 35°$, $\beta = 22°30'$, and $\gamma = 122°30'$.

The angle $\beta = 180° - 22°30' = 157°30'$ also satisfies the requirement that $\sin \beta = 0.3824$. But this "solution" must be rejected since $\alpha + \beta > 180°$.

This example corresponds to Figure 29d.

EXERCISE SET 6.5

In Exercises 1–12 use the law of sines to approximate the required part(s) of triangle ABC. Give both solutions if more than one triangle satisfies the given conditions.

1. $\alpha = 25°$, $\beta = 82°$, $a = 12.4$; find b.

2. $\alpha = 74°$, $\gamma = 36°$, $c = 6.8$; find a.

3. $\beta = 23°$, $\gamma = 47°$, $a = 9.3$; find c.

4. $\alpha = 46°$, $\beta = 88°$, $c = 10.5$; find b.

5. $\alpha = 42°20'$, $\gamma = 78°40'$, $b = 20$; find a.

6. $\beta = 16°30'$, $\gamma = 84°40'$, $a = 15$; find c.

7. $\alpha = 65°$, $a = 25$, $b = 30$; find β.

8. $\beta = 32°$, $b = 20$, $c = 14$; find α and γ.

9. $\gamma = 30°$, $a = 12.6$, $c = 6.3$; find b.

10. $\beta = 64°$, $a = 10$, $b = 8$; find c.

11. $\gamma = 45°$, $b = 7$, $c = 6$; find a.

12. $\alpha = 64°$, $a = 11$, $b = 12$; find β and γ.

13. Points A and B are chosen on opposite sides of a rock quarry. A point C is 160 meters from B, and the measures of angles BAC and ABC are found to be $95°$ and $47°$, respectively. Find the width of the quarry.

14. A tunnel is to be dug between points A and B on opposite sides of a hill. A point C is chosen that is 150 meters from A and 180 meters from B. If angle ABC measures $54°$, find the length of the tunnel.

15. A ski lift 750 meters in length rises to the top of a mountain at an angle of inclination of $40°$. A second lift is to be built whose base is in the same horizontal plane as the initial lift. If the angle of elevation of the second lift is $45°$, what is the length of the second lift?

16. A tree leans away from the sun at an angle of $9°$ from the vertical. The tree casts a shadow 20 meters in length when the angle of elevation of the sun is $62°$. Find the height of the tree.

17. A ship is sailing due north at a rate of 22 miles per hour. At 2 P.M. a lighthouse is seen at a bearing of N 15° W. At 4 P.M., the bearing of the same lighthouse is S 65° W. Find the distance of the ship from the lighthouse at 2 P.M.

18. A plane leaves airport A and flies at a bearing of N 32° E. A few moments later, the plane is spotted from airport B at a bearing of N 56° W. If airport B lies 15 miles due east of airport A, find the distance of the plane from airport B at the moment it is spotted.

19. A guy wire attached to the top of a vertical pole has an angle of inclination of 65° with the ground. From a point 10 meters farther from the pole, the angle of elevation of the top of the pole is 45°. Find the height of the pole.

20. At 5 P.M. a sailor on board a ship sailing at a rate of 18 miles per hour spots an island due east of the ship. The ship maintains a bearing of N 26° E. At 6 P.M. the sailor finds the bearing of the island to be S 37° E. Find the distance of the island from the ship at 6 P.M.

21. The short side of a parallelogram and the shorter diagonal measure 80 centimeters and 100 centimeters, respectively. If the angle between the longer side and the shorter diagonal is 43°, find the length of the longer side.

22. An archaeological mound is discovered in a jungle in Central America. To determine the height of the mound, a point A is chosen from which the angle of elevation of the top of the mound is found to be 31°. A second point B is chosen on a line with A and the base of the mound, 30 meters closer to the base of the mound. If the angle of elevation of the top of the mound from point B is 39°, find the height of the mound.

23. In a triangle, sides of length a and b and an angle α are given. Prove the following.
    (a) If $b \sin \alpha > a$, there is no triangle with the given parts.
    (b) If $b \sin \alpha = a$, the parts determine a right triangle.
    (c) If $b \sin \alpha < a < b$, there are two triangles with the given parts.
    (d) If $b \leq a$, there is one acute triangle with the given parts.

TERMS AND SYMBOLS

acute angle **286**	coterminal **289**	negative angle **286**	Reference Angle Rule **295**
ambiguous case (law of sines) **315**	degree, minute, second **286**	oblique triangle **308**	right angle **286**
angle of depression **302**	initial side of an angle **286**	obtuse angle **286**	solving a triangle **300**
angle of elevation **302**		positive angle **286**	standard position of an angle **286**
bearing **304**	law of cosines **309**	quadrantal angle **286**	terminal side of an angle **286**
central angle formula **291**	law of sines **312**	radian **286**	
		reference angle **289**	

KEY IDEAS FOR REVIEW

Topic	Page	Key Idea
Angular measure and conversion	286	An angle may be measured in either degrees or in radians. The two forms of measure are related by the equation $$\pi \text{ radians} = 180°$$
conversion formulas	287	Conversion can be accomplished by the formulas $$1 \text{ radian} = \left(\frac{180}{\pi}\right)°$$ $$1° = \left(\frac{\pi}{180}\right) \text{ radians}$$
Trigonometric function of angles	294	A trigonometric function of an angle is the same as the trigonometric function of the arc on the unit circle that the angle intercepts.
Reference Angle Rule	295	The Reference Angle Rule is analogous to the Reference Number Rule. It enables us to find the value of a trigonometric function of an angle greater than 90° (or $\pi/2$) by reference to the appropriate acute angle.

KEY IDEAS FOR REVIEW

Topic	Page	Key Idea
Right triangle trigonometry	298	Right triangle trigonometry relates a trigonometric function of an angle θ of a right triangle to the ratio of the lengths of two of its sides as follows: $$\sin \theta = \frac{\text{side opposite } \theta}{\text{hypotenuse}}$$ $$\cos \theta = \frac{\text{side adjacent to } \theta}{\text{hypotenuse}}$$ $$\tan \theta = \frac{\text{side opposite to } \theta}{\text{side adjacent to } \theta}$$
Law of Cosines and Law of Sines		The law of cosines and the law of sines are useful in solving problems that involve an oblique triangle. The derivation of these laws is accomplished by using right triangle trigonometry.
law of cosines	309	$$c^2 = a^2 + b^2 - 2ab \cos \gamma$$
law of sines	312	$$\frac{a}{\sin \alpha} = \frac{b}{\sin \beta} = \frac{c}{\sin \gamma}$$

REVIEW EXERCISES

Solutions to exercises whose numbers are in color are in the Solutions section in the back of the book.

6.1 In Exercises 1–4 convert from degree measure to radian measure or from radian measure to degree measure.

1. $-60°$ 2. $\dfrac{3\pi}{2}$ 3. $-\dfrac{5\pi}{12}$ 4. $45°$

In Exercises 5–7 determine if the pair of angles are coterminal.

5. $100°$, $\dfrac{5\pi}{9}$ 6. $\dfrac{4\pi}{3}$, $480°$ 7. $\dfrac{5\pi}{4}$, $-135°$

In Exercises 8–10 find the reference angle of the given angle.

8. $310°$ 9. $-185°$ 10. $405°$

11. If a central angle θ subtends an arc of length 14 centimeters on a circle whose radius is 10 centimeters, find the radian measure of θ.

12. A central angle of $2\pi/3$ radians subtends an arc of length $5\pi/2$ centimeters. Find the radius of the circle.

6.3 In Exercises 13–15 express the required trigonometric function as a ratio of the given parts of the right triangle ABC with $\gamma = 90°$.

13. $a = 5$, $b = 12$; find $\sin \alpha$.

14. $a = 3$, $c = 5$; find $\tan \beta$.

15. $a = 4$, $b = 7$; find $\sec \alpha$.

In Exercises 16–19 the point P lies on the terminal side of the angle θ. Find the value of the required trigonometric function without using tables or a calculator.

16. $P(-\sqrt{3}, 1)$; $\csc \theta$ 17. $P(\sqrt{2}, -\sqrt{2})$; $\cot \theta$

18. $P(-1, -\sqrt{3})$; $\cos \theta$ 19. $P(\sqrt{2}, \sqrt{2})$; $\sin \theta$

In Exercises 20–23 find the required part of triangle ABC with $\gamma = 90°$. Use Table VI in the Tables Appendix, or a calculator.

20. $a = 50$, $b = 60$; find α.

21. $a = 40$, $\beta = 20°$; find b.

22. $a = 20$, $\alpha = 52°$; find c.

23. $b = 15$, $\alpha = 25°$; find c.

24. A ladder 6 meters in length leans against a vertical wall. If the ladder makes an angle of $65°$ with the ground, find the height that the ladder reaches above the ground.

25. Find the angle of elevation of the sun when a tree 25 meters in height casts a horizontal shadow 10 meters in length.

26. A rectangle is 22 centimeters long and 16 centimeters wide. Find the measure of the smaller angle formed by the diagonal with a side.

6.4–6.5 In Exercises 27 and 28 use the law of cosines or the law of sines to approximate the required part of triangle ABC.

27. $a = 12, b = 7, c = 15$; find α.

28. $a = 20, b = 15, \alpha = 55°$; find β.

PROGRESS TEST 6A

In Problems 1–3 convert from degree measure to radian measure or from radian measure to degree measure.

1. $\dfrac{5\pi}{3}$

2. $-200°$

3. $75°$

In Problems 4 and 5 find an angle θ, $0 < \theta < 360°$, that is coterminal with the given angle.

4. $-25°$

5. $\dfrac{17\pi}{4}$

In Problems 6 and 7 find the reference angle of the given angle.

6. $160°$

7. $\dfrac{7\pi}{4}$

8. If a central angle θ subtends an arc of length 12 inches on a circle whose radius is 15 inches, find the radian measure of θ.

In Problems 9 and 10 ABC is a right triangle with $\gamma = 90°$. Express the required trigonometric function as a ratio of the given parts of the triangle.

9. $a = 7, b = 5$; $\tan \alpha$

10. $b = 5, c = 15$; $\sec \alpha$

In Problems 11–13 the point P lies on the terminal side of the angle θ. Find the value of the required trigonometric function without using tables or a calculator.

11. $P(-\sqrt{2}, \sqrt{2})$; $\cot \theta$

12. $P(0, -5)$; $\sin \theta$

13. $P(2, 2\sqrt{3})$; $\sec \theta$

In Problems 14–16 use Table VI in the Tables Appendix, or a calculator, to find the required part of triangle ABC with $\gamma = 90°$.

14. $a = 25, c = 30$; find α.

15. $b = 20, \alpha = 32°$; find c.

16. $a = 15, b = 20$; find β.

17. From the top of a hill 100 meters in height, the angle of depression of the entrance to a castle is $36°$. Find the distance of the castle from the base of the hill.

PROGRESS TEST 6B

In Problems 1–3 convert from degree measure to radian measure or from radian measure to degree measure.

1. $-135°$

2. $\dfrac{3\pi}{4}$

3. $-\dfrac{5\pi}{6}$

In Problems 4 and 5 find an angle θ, $0 \le \theta < 360°$, that is coterminal with the given angle.

4. $430°$

5. $-\dfrac{2\pi}{3}$

In Problems 6 and 7 find the reference angle of the given angle.

6. $\dfrac{9\pi}{16}$

7. $345°$

8. A central angle of $100°$ subtends an arc of length $7\pi/3$ centimeters. Find the radius of the circle.

In Problems 9 and 10 ABC is a right triangle with $\gamma = 90°$. Express the required trigonometric function as a ratio of the given parts of the triangle.

9. $a = 6, b = 8$; $\csc \beta$

10. $a = 7, b = 6$; $\cot \alpha$

In Problems 11–13 the point P lies on the terminal side of the angle θ. Find the value of the required trigonometric function without using tables or a calculator.

11. $P(-3, 0)$; $\csc \theta$

12. $P(2, 2\sqrt{3})$; $\csc \theta$

13. $P(-\sqrt{2}, -\sqrt{2})$; $\tan \theta$

In Problems 14–16 use Table VI in the Tables Appendix, or a calculator, to find the required part of triangle ABC with $\gamma = 90°$.

14. $a = 5, \beta = 61°$; find c.

15. $b = 6, c = 15$; find α.

16. $a = 7, b = 10$; find c.

17. A surveyor finds the angle of elevation of the top of a tree to be $42°$. If the surveyor is 75 feet from the base of the tree, find the height of the tree.

CHAPTER 7

ANALYTIC TRIGONOMETRY

Much of the language and terminology of algebra carries over to trigonometry. For example, we have seen that algebraic expressions involve variables, constants, and algebraic operations. **Trigonometric expressions** involve these same elements but also permit trigonometric functions of variables and constants. They also allow algebraic operations upon these trigonometric functions. Thus,

$$x + \sin x \qquad \sin x + \tan x \qquad \frac{1 - \cos x}{\sec^2 x}$$

are all examples of trigonometric expressions.

The distinction between an identity and an equation also carries over to trigonometry. Thus, a **trigonometric identity** is true for all real values in the domain of the variable, but a **trigonometric equation** is true only for certain values called **solutions.** (Note that the solutions of a trigonometric equation may be expressed as real numbers or as angles.) As usual, the set of all solutions of a trigonometric equation is called the **solution set.**

7.1
TRIGONOMETRIC
IDENTITIES

FUNDAMENTAL IDENTITIES

In Section 5.2 we established the identity

$$\sin^2 t + \cos^2 t = 1 \qquad (1)$$

If $\cos t \neq 0$, we may divide both sides of Equation (1) by $\cos^2 t$ to obtain

$$\frac{\sin^2 t}{\cos^2 t} + \frac{\cos^2 t}{\cos^2 t} = \frac{1}{\cos^2 t}$$

or

$$\tan^2 t + 1 = \sec^2 t \qquad (2)$$

Similarly, if $\sin t \neq 0$, dividing Equation (1) by $\sin^2 t$ yields

$$\frac{\sin^2 t}{\sin^2 t} + \frac{\cos^2 t}{\sin^2 t} = \frac{1}{\sin^2 t}$$

or

$$\cot^2 t + 1 = \csc^2 t \qquad (3)$$

Observe that $\tan t$ and $\cot t$ are undefined for exactly those values of t for which $\cos t$ and $\sin t$ are 0, respectively. It follows that the identities (2) and (3) are true for all values of t for which the trigonometric expressions are defined.

The two identities that we have just established, together with the identities discussed in Sections 5.2 and 5.6, are called the **fundamental identities.** Since we will use these eight identities throughout this chapter, it is essential you know and recognize them in their various forms. They are presented in Table 1 for you to review.

TABLE 1   The Fundamental Identities

Fundamental Identity	Alternate Form(s)
$\tan t = \dfrac{\sin t}{\cos t}$	
$\cot t = \dfrac{\cos t}{\sin t}$	
$\csc t = \dfrac{1}{\sin t}$	$\sin t = \dfrac{1}{\csc t}$
$\sec t = \dfrac{1}{\cos t}$	$\cos t = \dfrac{1}{\sec t}$
$\cot t = \dfrac{1}{\tan t}$	$\tan t = \dfrac{1}{\cot t}$
$\sin^2 t + \cos^2 t = 1$	$\sin^2 t = 1 - \cos^2 t$
	$\cos^2 t = 1 - \sin^2 t$
$\tan^2 t + 1 = \sec^2 t$	$\tan^2 t = \sec^2 t - 1$
$\cot^2 t + 1 = \csc^2 t$	$\cot^2 t = \csc^2 t - 1$

In Section 5.2 we saw that trigonometric identities can be used to simplify a trigonometric expression. Here is another example, in which we use the identities developed in this section.

EXAMPLE 1   Using Trigonometric Identities
Simplify the expression $\sin^2 x + \sin^2 x \tan^2 x$.

SOLUTION
We begin by noting that $\sin^2 x$ appears in both terms, which suggests that we factor.

$$
\begin{aligned}
\sin^2 x + \sin^2 x \tan^2 x &= \sin^2 x(1 + \tan^2 x) \quad &&\text{Factoring} \\
&= \sin^2 x \sec^2 x \quad &&1 + \tan^2 x = \sec^2 x \\
&= \frac{\sin^2 x}{\cos^2 x} \quad &&\sec x = \frac{1}{\cos x} \\
&= \tan^2 x \quad &&\frac{\sin x}{\cos x} = \tan x
\end{aligned}
$$

∎

PROGRESS CHECK

Simplify the expression $\dfrac{\csc \theta}{1 + \cot^2 \theta}$.

ANSWER

$\sin \theta$

TRIGONOMETRIC IDENTITIES

The fundamental identities can be employed to prove or, more properly, to verify various trigonometric identities. The principal reasons for including this topic are (a) to improve your skills in recognizing and using the fundamental identities, and (b) to sharpen your reasoning processes. There are also times in calculus and applied mathematics when simplification of a trigonometric expression may enable us to see a relationship that would otherwise be obscured. Finally, in computer applications it is much more efficient to evaluate a simple trigonometric expression than an involved one.

The preferred method of verifying an identity is to transform one side of the equation into the other. We will use this method whenever practical. Unfortunately, we cannot outline a rigid set of steps that will "work" to transform one side into the other; in fact, there are often many ways to tackle a given identity. We will provide a number of examples that demonstrate some of the techniques that can be employed. If you should make a false start and find yourself trying something that doesn't appear to be working, start again and try another approach. With practice your skills will improve.

EXAMPLE 2   Verifying an Identity
Verify the identity $\cos x \tan x \csc x = 1$.

SOLUTION
It is often helpful to write all of the trigonometric functions in terms of sine and cosine.
The student should supply a reason for each step.

$$\cos x \tan x \csc x = \cos x \frac{\sin x}{\cos x} \frac{1}{\sin x}$$

$$= 1$$

■

 PROGRESS CHECK
Verify the identity $\sin x \sec x = \tan x$.

EXAMPLE 3 Verifying an Identity
Verify the identity $\dfrac{1}{1 - \sin x} + \dfrac{1}{1 + \sin x} = 2\sec^2 x$.

SOLUTION
*Another useful technique is to begin with the more complicated expression and complete
the indicated operations.* We will begin with the left-hand side and will combine the
fractions.

$$\frac{1}{1 - \sin x} + \frac{1}{1 + \sin x} = \frac{1 + \sin x + 1 - \sin x}{(1 - \sin x)(1 + \sin x)}$$

$$= \frac{2}{1 - \sin^2 x} = \frac{2}{\cos^2 x}$$

$$= 2\sec^2 x$$

■

 PROGRESS CHECK
Verify the identity $\cos x + \tan x \sin x = \sec x$.

EXAMPLE 4 Verifying an Identity
Verify the identity $\sin \alpha - \sin^2 \alpha = \dfrac{1 - \sin \alpha}{\csc \alpha}$.

SOLUTION
Factoring will sometimes help to simplify an expression. The student should supply a
reason for each step.

$$\sin \alpha - \sin^2 \alpha = \sin \alpha(1 - \sin \alpha)$$

$$= \frac{1 - \sin \alpha}{\csc \alpha}$$

■

PROGRESS CHECK

Verify the identity $\dfrac{\sin^2 y - 1}{1 - \sin y} = -1 - \sin y$.

EXAMPLE 5   Verifying an Identity

Verify the identity $\dfrac{\cos \theta}{1 - \sin \theta} = \sec \theta + \tan \theta$.

SOLUTION

Multiplying the numerator and denominator of a rational expression by the same quantity is a useful technique. Of course, this quantity should be selected carefully. In this example, multiplying the denominator $1 - \sin \theta$ by $1 + \sin \theta$ will produce $1 - \sin^2 \theta = \cos^2 \theta$. (Similarly, should $\sec x - 1$ appear in a denominator, you might try multiplying by $\sec x + 1$ to obtain $\sec^2 x - 1 = \tan^2 x$.)

The student should supply a reason for each of the following steps.

$$\frac{\cos \theta}{1 - \sin \theta} = \frac{\cos \theta}{1 - \sin \theta} \cdot \frac{1 + \sin \theta}{1 + \sin \theta}$$

$$= \frac{\cos \theta(1 + \sin \theta)}{1 - \sin^2 \theta}$$

$$= \frac{\cos \theta(1 + \sin \theta)}{\cos^2 \theta}$$

$$= \frac{1 + \sin \theta}{\cos \theta}$$

$$= \frac{1}{\cos \theta} + \frac{\sin \theta}{\cos \theta}$$

$$= \sec \theta + \tan \theta$$

∎

PROGRESS CHECK

Verify the identity $\dfrac{1 + \cos t}{\sin t} + \dfrac{\sin t}{1 + \cos t} = 2 \csc t$.

We said earlier that the preferred way of verifying an identity is to transform one side of the equation into the other. At times, both sides may involve complicated expressions and this approach may not be practical. We can then try to transform each side of the equation into the same expression, being careful to use only procedures that are reversible. Here is an example.

EXAMPLE 6   Verifying an Identity

Verify the identity $\dfrac{\cot u - \tan u}{\sin u \cos u} = \csc^2 u - \sec^2 u$.

SOLUTION

Beginning with the left-hand side we have

$$\frac{\cot u - \tan u}{\sin u \cos u} = \frac{\dfrac{\cos u}{\sin u} - \dfrac{\sin u}{\cos u}}{\sin u \cos u}$$

$$= \frac{\cos^2 u - \sin^2 u}{\sin^2 u \cos^2 u}$$

We then transform the right-hand side of the equation by writing all trigonometric functions in terms of sine and cosine.

$$\csc^2 u - \sec^2 u = \frac{1}{\sin^2 u} - \frac{1}{\cos^2 u}$$

$$= \frac{\cos^2 u - \sin^2 u}{\sin^2 u \cos^2 u}$$

We have successfully transformed both sides of the equation into the same expression. Since all the steps are reversible, we have verified the identity.

■

PROGRESS CHECK

Verify the identity $\dfrac{\sin x + \cos x}{\tan^2 x - 1} = \dfrac{\cos^2 x}{\sin x - \cos x}$.

EXERCISE SET 7.1

In Exercises 1–46 verify each of the following identities.

1. $\csc \gamma - \cos \gamma \cot \gamma = \sin \gamma$

2. $\cot x \sec x = \csc x$

3. $\sec v + \tan v = \dfrac{1 + \sin v}{\cos v}$

4. $\cos \theta + \tan \theta \sin \theta = \sec \theta$

5. $\sin \alpha \sec \alpha = \tan \alpha$

6. $\sec \beta - \cos \beta = \sin \beta \tan \beta$

7. $3 - \sec^2 x = 2 - \tan^2 x$

8. $1 - 2\sin^2 t = 2\cos^2 t - 1$

9. $\dfrac{\sec^2 y}{\tan y} = \tan y + \cot y$

10. $\dfrac{\sin x + \cos x}{\cos x} = 1 + \tan x$

11. $\dfrac{\sin u}{\csc u} + \dfrac{\cos u}{\sec u} = 1$

12. $\dfrac{\tan^2 \alpha}{1 + \sec \alpha} = \sec \alpha - 1$

13. $\dfrac{\sec^2 \theta - 1}{\sec^2 \theta} = \sin^2 \theta$

14. $\sin^4 x + 2\sin^2 x \cos^2 x + \cos^4 x = 1$

15. $\cos \gamma + \cos \gamma \tan^2 \gamma = \sec \gamma$

16. $\dfrac{1}{\tan u + \cot u} = \cos u \sin u$

17. $\dfrac{\sec w \sin w}{\tan w + \cot w} = \sin^2 w$

18. $(1 - \cos^2 \beta)(1 + \cot^2 \beta) = 1$

19. $(\sin\alpha + \cos\alpha)^2 + (\sin\alpha - \cos\alpha)^2 = 2$

20. $\dfrac{1 + \tan^2 u}{\csc^2 u} = \tan^2 u$

21. $\sec^2 v + \cos^2 v = \dfrac{\sec^4 v + 1}{\sec^2 v}$

22. $\sin^2\theta - \tan^2\theta = -\tan^2\theta\sin^2\theta$

23. $\dfrac{\sin^2\alpha}{1 + \cos\alpha} = 1 - \cos\alpha$

24. $\cot x \sin^3 x = \cos x - \cos^3 x$

25. $\dfrac{\cos t}{1 + \sin t} = \dfrac{1 - \sin t}{\cos t}$

26. $\dfrac{\sin\beta}{1 + \cos\beta} + \dfrac{1 + \cos\beta}{\sin\beta} = 2\csc\beta$

27. $\csc^2\theta - \dfrac{\cos^2\theta}{\sin^2\theta} = 1$

28. $\dfrac{\cos^2 u}{1 - \sin u} = 1 + \sin u$

29. $\dfrac{\cot y}{1 + \cot^2 y} = \sin y \cos y$

30. $\dfrac{1 + \tan^2 x}{\tan^2 x} = \csc^2 x$

31. $\cos(-t)\csc(-t) = -\cot t$

32. $\sin(-\theta)\sec(-\theta) = -\tan\theta$

33. $\dfrac{\sec x + \csc x}{1 + \tan x} = \csc x$

34. $\dfrac{\sec u}{\sec u - 1} = \dfrac{1}{1 - \cos u}$

35. $\dfrac{1 + \tan x}{1 + \cot x} = \dfrac{\sec x}{\csc x}$

36. $(\tan u + \sec u)^2 = \dfrac{1 + \sin u}{1 - \sin u}$

37. $\dfrac{1 - \sin t}{1 + \sin t} = (\sec t - \tan t)^2$

38. $2\csc^2\theta - \csc^4\theta = 1 - \cot^4\theta$

39. $\dfrac{\sin^2 w}{\cos^4 w + \cos^2 w \sin^2 w} = \tan^2 w$

40. $\dfrac{\sin z + \tan z}{1 + \cos z} = \tan z$

41. $\dfrac{\sec\gamma - \csc\gamma}{\sec\gamma + \csc\gamma} = \dfrac{\tan\gamma - 1}{\tan\gamma + 1}$

42. $\dfrac{\cot x - 1}{1 - \tan x} = \dfrac{\csc x}{\sec x}$

43. $\dfrac{\tan\gamma - \sin\gamma}{\tan\gamma} = \dfrac{\sin^2\gamma}{1 + \cos\gamma}$

44. $\cos^4 u - \sin^4 u = \cos^2 u - \sin^2 u$

45. $\dfrac{\csc x}{1 + \csc x} - \dfrac{\csc x}{1 - \csc x} = 2\sec^2 x$

46. $\sin^3\theta + \cos^3\theta = (1 - \sin\theta\cos\theta)(\sin\theta + \cos\theta)$

In Exercises 47–52 show that each of the following equations is not an identity by finding a value of the variable for which the equation is not true.

47. $\sin x = \sqrt{1 - \cos^2 x}$

48. $\tan x = \sqrt{\sec^2 x - 1}$

49. $(\sin t + \cos t)^2 = \sin^2 t + \cos^2 t$

50. $\sin\theta + \cos\theta = \sec\theta + \csc\theta$

51. $\sqrt{\cos^2 x} = \cos x$

52. $\sqrt{\cot^2 x} = \cot x$

7.2 THE ADDITION FORMULAS

The identities that we verified in the examples and exercises of Section 7.1 were themselves of no special significance; we were primarily interested in having you practice manipulation with the fundamental identities. There are, however, many trigonometric identities that are indeed of importance; these identities are called **trigonometric formulas.** Such formulas are used so frequently that it is probably best for you to memorize them. We will develop these formulas in a logical sequence so that you will be able to derive them yourself should you wish to verify that your memorization is correct.

Our first objective is to develop the **addition formula** for $\cos(s + t)$ where s and t are any real numbers. It happens that it is easier to begin with $\cos(s - t)$, which demonstrates that the mathematician may at times have to take a circuitous route to establish a result!

For convenience, we assume that s, t, and $s - t$ are all positive and less than 2π. We let $P = P(s)$, $Q = P(t)$, $R = P(s - t)$ be the unit circle points determined by s, t, and $s - t$, respectively (see Figure 1). Then $\overparen{ARP} = s$, $\overparen{AQ} = t$, $\overparen{AR} = s - t$, and by the definitions of sine and cosine, the coordinates of the points can be written as

$$P(\cos s, \sin s) \qquad Q(\cos t, \sin t) \qquad R(\cos(s - t), \sin(s - t))$$

Since the arcs $\overparen{QP}$ and $\overparen{AR}$ are both of length $s - t$, the chords QP and AR are also of equal length. By the distance formula, we have

$$\overline{AR} = \overline{QP}$$

$$\sqrt{[\cos(s - t) - 1]^2 + [\sin(s - t)]^2} = \sqrt{(\cos s - \cos t)^2 + (\sin s - \sin t)^2}$$

Squaring both sides and rearranging terms, we have

$$\sin^2(s - t) + \cos^2(s - t) - 2\cos(s - t) + 1$$
$$= \sin^2 s + \cos^2 s + \sin^2 t + \cos^2 t - 2\cos s \cos t - 2\sin s \sin t$$

Since each of the expressions $\sin^2(s - t) + \cos^2(s - t)$, $\sin^2 s + \cos^2 s$, and $\sin^2 t + \cos^2 t$ equals 1, we have

$$2 - 2\cos(s - t) = 2 - 2\cos s \cos t - 2\sin s \sin t$$

Solving for $\cos(s - t)$ yields the formula

$$\cos(s - t) = \cos s \cos t + \sin s \sin t \tag{1}$$

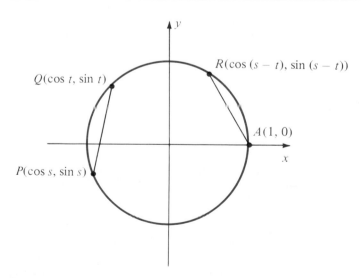

FIGURE 1 $\cos(s - t)$

Now it's easy to obtain the addition formula for $\cos(s + t)$. By writing

$$s + t = s - (-t)$$

we have

$$\cos(s + t) = \cos(s - (-t))$$
$$= \cos s \cos(-t) + \sin s \sin(-t)$$

Since $\cos(-t) = \cos t$ and $\sin(-t) = -\sin t$,

$$\cos(s + t) = \cos s \cos t - \sin s \sin t \qquad (2)$$

EXAMPLE 1 Using the Addition Formulas
Find $\cos 15°$ without the use of tables or a calculator.

SOLUTION
Since $15° = 45° - 30°$, we may use the formula for $\cos(s - t)$ to obtain

$$\cos 15° = \cos(45° - 30°)$$
$$= \cos 45° \cos 30° + \sin 45° \sin 30°$$
$$= \frac{\sqrt{2}}{2} \cdot \frac{\sqrt{3}}{2} + \frac{\sqrt{2}}{2} \cdot \frac{1}{2}$$
$$= \frac{\sqrt{6} + \sqrt{2}}{4}$$

∎

PROGRESS CHECK
Solve Example 1 using the identity $15° = 60° - 45°$.

EXAMPLE 2 Using the Addition Formulas
Find the exact value of $\cos(5\pi/12)$.

SOLUTION
We note that $5\pi/12 = 2\pi/12 + 3\pi/12 = \pi/6 + \pi/4$. Then

$$\cos\left(\frac{5\pi}{12}\right) = \cos\left(\frac{\pi}{6} + \frac{\pi}{4}\right)$$
$$= \cos\frac{\pi}{6}\cos\frac{\pi}{4} - \sin\frac{\pi}{6}\sin\frac{\pi}{4}$$
$$= \frac{\sqrt{3}}{2} \cdot \frac{\sqrt{2}}{2} - \frac{1}{2} \cdot \frac{\sqrt{2}}{2} = \frac{\sqrt{6} - \sqrt{2}}{4}$$

∎

☑ **PROGRESS CHECK**
Solve Example 2 using the identity $5\pi/12 = 9\pi/12 - 4\pi/12$.

Before tackling $\sin(s + t)$, we first establish the following important functional relationships.

$$\cos\left(\frac{\pi}{2} - t\right) = \sin t \qquad (3)$$

$$\sin\left(\frac{\pi}{2} - t\right) = \cos t \qquad (4)$$

$$\tan\left(\frac{\pi}{2} - t\right) = \cot t \qquad (5)$$

Using the difference formula for cosine we have

$$\cos\left(\frac{\pi}{2} - t\right) = \cos\frac{\pi}{2}\cos t + \sin\frac{\pi}{2}\sin t$$
$$= 0 \cdot \cos t + 1 \cdot \sin t$$
$$= \sin t$$

which establishes Equation (3). Replacing t with $\frac{\pi}{2} - t$ in this identity yields

$$\cos\left[\frac{\pi}{2} - \left(\frac{\pi}{2} - t\right)\right] = \sin\left(\frac{\pi}{2} - t\right)$$
$$\cos t = \sin\left(\frac{\pi}{2} - t\right)$$

which establishes Equation (4). The third identity follows from the definition of tangent and from Equations (3) and (4):

$$\tan\left(\frac{\pi}{2} - t\right) = \frac{\sin\left(\dfrac{\pi}{2} - t\right)}{\cos\left(\dfrac{\pi}{2} - t\right)}$$
$$= \frac{\cos t}{\sin t} = \cot t$$

Functions satisfying the properties of the identities (3) and (4) are called **cofunctions.** Thus, sine and cosine are cofunctions. So, too, the tangent and cotangent functions are cofunctions, as are secant and cosecant. This is the origin of the prefix *co* in *co*sine, *co*secant, and *co*tangent.

EXAMPLE 3 Cofunctions
Use trigonometry of the right triangle to show that sine and cosine are cofunctions.

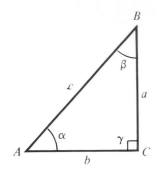

FIGURE 2   Diagram for Example 3

SOLUTION

In right triangle ABC, angle $\gamma = 90°$ (Figure 2). Then $\sin \alpha = a/c = \cos \beta$. But angles α and β are complementary; that is, $\alpha + \beta = 90°$. Thus $\sin \alpha = \cos(90° - \alpha)$ and $\cos \beta = \sin(90° - \beta)$, which establishes that they are cofunctions.   ∎

We are now prepared to prove the following.

$$\sin(s + t) = \sin s \cos t + \cos s \sin t \qquad (6)$$

$$\sin(s - t) = \sin s \cos t - \cos s \sin t \qquad (7)$$

We supply the steps for a proof of Equation (6); the student should supply a reason for each step.

$$\sin(s + t) = \cos\left[\frac{\pi}{2} - (s + t)\right]$$

$$= \cos\left[\left(\frac{\pi}{2} - s\right) - t\right]$$

$$= \cos\left(\frac{\pi}{2} - s\right)\cos t + \sin\left(\frac{\pi}{2} - s\right)\sin t$$

$$= \sin s \cos t + \cos s \sin t$$

The student should now prove Equation (7) by using

$$\sin(s - t) = \sin[s + (-t)]$$

We conclude with the addition formulas for the tangent function.

$$\tan(s + t) = \frac{\tan s + \tan t}{1 - \tan s \tan t} \qquad (8)$$

$$\tan(s - t) = \frac{\tan s - \tan t}{1 + \tan s \tan t} \qquad (9)$$

Again, we supply the steps for a proof of Equation (8) and will let the student supply a reason for each step.

$$\tan(s + t) = \frac{\sin(s + t)}{\cos(s + t)}$$

$$= \frac{\sin s \cos t + \cos s \sin t}{\cos s \cos t - \sin s \sin t}$$

$$= \frac{\left(\dfrac{\sin s}{\cos s} \cdot \dfrac{\cos t}{\cos t}\right) + \left(\dfrac{\cos s}{\cos s} \cdot \dfrac{\sin t}{\cos t}\right)}{\left(\dfrac{\cos s}{\cos s} \cdot \dfrac{\cos t}{\cos t}\right) - \left(\dfrac{\sin s}{\cos s} \cdot \dfrac{\sin t}{\cos t}\right)}$$

$$= \frac{\tan s + \tan t}{1 - \tan s \tan t}$$

COMPUTING SINE AND COSINE

```
10 LET S1 = 0.01745
20 LET C1 = 0.99985
30 PRINT
   "DEGREES",
   "SIN"  "COS"
40 PRINT
   S1, C1
50 LET S2 = S1
60 LET C2 = C1
70 FOR I = 2 TO 90
80   LET S3 = S2
90   LET S2 =
     (S1 * C2) +
     (C1 * S2)
100  LET C2 =
     (C1 * C2) -
     (S1 * S3)
110  PRINT I, S2,
     C2
120 NEXT I
130 END
```

We can make use of the trigonometric formulas to generate a table of sine and cosine values. Suppose we have determined that

$$\sin 1° = 0.01745 \qquad \cos 1° = 0.99985 \tag{1}$$

We can then write

$$\sin(1° + \alpha) = \sin 1° \cos \alpha + \cos 1° \sin \alpha$$
$$\cos(1° + \alpha) = \cos 1° \cos \alpha - \sin 1° \sin \alpha$$

Substituting for $\sin 1°$ and $\cos 1°$ from Equations (1),

$$\sin(1° + \alpha) = 0.01745 \cos \alpha + 0.99985 \sin \alpha \tag{2}$$
$$\cos(1° + \alpha) = 0.99985 \cos \alpha - 0.01745 \sin \alpha \tag{3}$$

Now, if we let $\alpha = 1°$, Equations (2) and (3) can be used to calculate $\sin 2°$ and $\cos 2°$. We can then repeat the process with $\alpha = 2°$ to calculate $\sin 3°$ and $\cos 3°$, and so on. Since this is an iterative procedure well suited for a computer, we are providing a program in BASIC that will calculate sine and cosine values from 2° to 90° in increments of 1°. Although this is a neat illustration of the use of trigonometric formulas, computer subroutines use more sophisticated methods.

The student should now prove Equation (9) by using

$$\tan(s - t) = \tan[s + (-t)]$$

EXAMPLE 4 Applying the Addition Formulas
Show that $\sin(x + 2\pi) = \sin x$.

SOLUTION
Using the addition formula,

$$\sin(x + 2\pi) = \sin x \cos 2\pi + \cos x \sin 2\pi$$
$$= \sin x \cdot 1 + \cos x \cdot 0$$
$$= \sin x$$

■

PROGRESS CHECK
Show that $\tan(x + \pi) = \tan x$.

EXAMPLE 5 Using the Addition Formulas
Given $\sin \alpha = -\frac{4}{5}$, with α an angle in quadrant III, and $\cos \beta = -\frac{5}{13}$, with β an angle in quadrant II, use the addition formula to find $\sin(\alpha + \beta)$ and the quadrant in which $\alpha + \beta$ lies.

SOLUTION

The addition formula

$$\sin(\alpha + \beta) = \sin \alpha \cos \beta + \cos \alpha \sin \beta$$

requires that we know $\sin \alpha$, $\cos \alpha$, $\sin \beta$, and $\cos \beta$. Using the fundamental identity $\sin^2 \alpha + \cos^2 \alpha = 1$, we have

$$\cos^2 \alpha = 1 - \sin^2 \alpha = 1 - \frac{16}{25} = \frac{9}{25}$$

Taking the square root of both sides, we must have $\cos \alpha = -\frac{3}{5}$ since α is in quadrant III. Similarly,

$$\sin^2 \beta = 1 - \cos^2 \beta = 1 - \frac{25}{169} = \frac{144}{169}$$

Taking the square root of both sides, we must have $\sin \beta = \frac{12}{13}$ since β is in quadrant II. Thus,

$$\sin(\alpha + \beta) = \left(-\frac{4}{5}\right)\left(-\frac{5}{13}\right) + \left(-\frac{3}{5}\right)\left(\frac{12}{13}\right)$$

$$= \frac{20}{65} - \frac{36}{65} = -\frac{16}{65}$$

Since $\sin(\alpha + \beta)$ is negative, $\alpha + \beta$ lies in either quadrant III or quadrant IV. However, the sum of an angle that lies in quadrant III and an angle that lies in quadrant II cannot lie in quadrant III. Thus, $\alpha + \beta$ lies in quadrant IV.   ∎

PROGRESS CHECK

Given $\cos \alpha = -\frac{4}{5}$, with α in quadrant III, and $\cos \beta = \frac{3}{5}$, with β in quadrant I, find $\cos(\alpha - \beta)$ and the quadrant in which $\alpha - \beta$ lies.

ANSWER

$-\frac{24}{25}$; quadrant II

EXERCISE SET 7.2

In Exercises 1–6 show that the given equation is not an identity. (*Hint:* For each equation, find a pair of values of s and t for which the equation is not true.)

1. $\cos(s - t) = \cos s - \cos t$

2. $\sin(s + t) = \sin s + \sin t$

3. $\sin(s - t) = \sin s - \sin t$

4. $\cos(s + t) = \cos s + \cos t$

5. $\tan(s + t) = \tan s + \tan t$

6. $\tan(s - t) = \tan s - \tan t$

In Exercises 7–22 use the addition formulas to find exact values.

7. $\cos\left(\dfrac{\pi}{6} + \dfrac{\pi}{4}\right)$

8. $\sin\left(\dfrac{\pi}{6} - \dfrac{\pi}{4}\right)$

9. $\sin\left(\dfrac{\pi}{4} + \dfrac{\pi}{3}\right)$

10. $\cos\left(\dfrac{\pi}{3} - \dfrac{\pi}{4}\right)$

11. $\cos(30° + 180°)$

12. $\tan(60° + 300°)$

13. $\tan(300° - 60°)$

14. $\sin(270° - 45°)$

15. $\sin 11\pi/12$ (*Hint:* $11\pi/12 = \pi/6 + 3\pi/4$)

16. $\tan 7\pi/12$ (*Hint:* $7\pi/12 = \pi/4 + \pi/3$)

17. $\cos 7\pi/12$ (*Hint:* $7\pi/12 = 5\pi/6 - \pi/4$)

18. $\tan 75°$ (*Hint:* $75° = 135° - 60°$)

19. $\sin 7\pi/6$ 20. $\cos 5\pi/6$

21. $\tan 15°$ 22. $\tan 165°$

In Exercises 23–28 write the given expression in terms of cofunctions of complementary angles.

23. $\sin 47°$ 24. $\cos 78°$ 25. $\tan \pi/6$ 26. $\tan 84°$

27. $\cos \pi/3$ 28. $\sin 72°30'$

29. If $\sin t = -\frac{3}{5}$, with $P(t)$ in quadrant III, find $\sin(\pi/2 - t)$.

30. If $\cos t = -\frac{5}{13}$, with $P(t)$ in quadrant II, find $\sin(t - \pi)$.

31. If $\tan \theta = \frac{4}{3}$ and angle θ lies in quadrant III, find $\tan(\theta + \pi/4)$.

32. If $\sec \theta = \frac{5}{3}$ and angle θ lies in quadrant I, find $\sin(\theta + \pi/6)$.

33. If $\cos t = 0.4$, with $P(t)$ in quadrant IV, find $\tan(t + \pi)$.

34. If $\sec \alpha = 1.2$ and angle α lies in quadrant IV, find $\tan(\alpha - \pi)$.

35. If $\sin s = \frac{3}{5}$ and $\cos t = -\frac{12}{13}$, with $P(s)$ in quadrant II and $P(t)$ in quadrant III, find $\sin(s + t)$.

36. If $\sin s = -\frac{4}{5}$ and $\csc t = \frac{13}{5}$, with $P(s)$ in quadrant IV and $P(t)$ in quadrant II, find $\cos(s - t)$.

37. If $\cos \alpha = \frac{5}{13}$ and $\tan \beta = -2$, with angle α in quadrant I and angle β in quadrant II, find $\tan(\alpha + \beta)$

38. If $\sec \alpha = \frac{5}{3}$ and $\cot \beta = \frac{15}{8}$, with angle α in quadrant IV and angle β in quadrant III, find $\tan(\alpha - \beta)$.

In Exercises 39–55 prove each of the following identities by transforming the left-hand side of the equation into the expression on the right-hand side.

39. $\sin 2\alpha = 2 \sin \alpha \cos \alpha$

40. $\cos 2t = \cos^2 t - \sin^2 t$

41. $\tan 2\alpha = \dfrac{2 \tan \alpha}{1 - \tan^2 \alpha}$

42. $\sin(x + y)\sin(x - y) = \sin^2 x - \sin^2 y$

43. $\cos(x - y)\cos(x + y) = \cos^2 x \cos^2 y - \sin^2 x \sin^2 y$

44. $\dfrac{\sin(s + t)}{\sin(s - t)} = \dfrac{\tan s + \tan t}{\tan s - \tan t}$

45. $\csc(t + \pi/2) = \sec t$

46. $\tan(\alpha + 90°) = -\cot \alpha$

47. $\tan(x + \pi/4) = \dfrac{1 + \tan x}{1 - \tan x}$

48. $\csc(t - \pi) = -\csc t$

49. $\cot(s - t) = \dfrac{1 + \tan s \tan t}{\tan s - \tan t}$

50. $\cot(u + v) = \dfrac{\cot u \cot v - 1}{\cot u + \cot v}$

51. $\sin(s + t) + \sin(s - t) - 2 \sin s \cos t$

52. $\cos(s + t) + \cos(s - t) = 2 \cos s \cos t$

53. $\dfrac{\sin(x + h) - \sin x}{h} = \sin x \left(\dfrac{\cos h - 1}{h}\right) + \cos x \left(\dfrac{\sin h}{h}\right)$

54. $\dfrac{\cos(x + h) - \cos x}{h} = \cos x \left(\dfrac{\cos h - 1}{h}\right) - \sin x \left(\dfrac{\sin h}{h}\right)$

55. $\cos(x - y)\cos(x + y) = \cos^2 x + \cos^2 y - 1$

7.3 DOUBLE- AND HALF-ANGLE FORMULAS

Our initial objective in this section is to derive expressions for $\sin 2t$, $\cos 2t$, and $\tan 2t$ in terms of trigonometric functions of t. We will establish the following **double-angle formulas.**

DOUBLE-ANGLE FORMULAS

$$\sin 2t = 2 \sin t \cos t \tag{1}$$

$$\cos 2t = \cos^2 t - \sin^2 t \tag{2}$$

$$\tan 2t = \frac{2 \tan t}{1 - \tan^2 t} \tag{3}$$

Once again, it's best to memorize these formulas. However, the derivations are so straightforward that you can always return to them to verify the results.

To establish Equation (1), we simply rewrite $2t$ as $(t + t)$ and use the addition formula.

$$\sin 2t = \sin(t + t)$$
$$= \sin t \cos t + \cos t \sin t$$
$$= 2 \sin t \cos t$$

We proceed in the same manner to prove Equation (2).

$$\cos 2t = \cos(t + t)$$
$$= \cos t \cos t - \sin t \sin t$$
$$= \cos^2 t - \sin^2 t$$

Using the addition formula for the tangent function yields a proof of Equation (3).

$$\tan 2t = \tan(t + t)$$
$$= \frac{\tan t + \tan t}{1 - \tan t \tan t}$$
$$= \frac{2 \tan t}{1 - \tan^2 t}$$

EXAMPLE 1 Using the Double-Angle Formulas

If $\cos t = -\frac{3}{5}$ and $P(t)$ is in quadrant II, evaluate $\sin 2t$ and $\cos 2t$. In which quadrant does $P(2t)$ lie?

SOLUTION

We first find $\sin t$ by use of the fundamental identity $\sin^2 t + \cos^2 t = 1$. Thus,

$$\sin^2 t + \frac{9}{25} = 1$$

$$\sin^2 t = \frac{16}{25}$$

Since $P(t)$ is in quadrant II, $\sin t$ must be positive. Therefore,

$$\sin t = \frac{4}{5}$$

Applying the double-angle formulas with $\cos t = -\frac{3}{5}$, $\sin t = \frac{4}{5}$, yields

$$\sin 2t = 2 \sin t \cos t = 2\left(\frac{4}{5}\right)\left(-\frac{3}{5}\right) = -\frac{24}{25}$$

$$\cos 2t = \cos^2 t - \sin^2 t = \frac{9}{25} - \frac{16}{25} = -\frac{7}{25}$$

Since $\sin 2t$ and $\cos 2t$ are both negative, we may conclude that $P(2t)$ lies in quadrant III.

PROGRESS CHECK

If $\sin \theta = \frac{5}{13}$ and θ is in quadrant I, evaluate $\sin 2\theta$ and $\tan 2\theta$.

ANSWER

$$\sin 2\theta = \frac{120}{169}; \ \tan 2\theta = \frac{120}{119}$$

EXAMPLE 2   Using the Addition and Double-Angle Formulas

Express $\sin 3t$ in terms of $\sin t$ and $\cos t$.

SOLUTION

We write $3t$ as $(2t + t)$. Then

$$\begin{aligned}
\sin 3t &= \sin(2t + t) \\
&= \sin 2t \cos t + \cos 2t \sin t \\
&= 2 \sin t \cos t \cos t + (\cos^2 t - \sin^2 t) \sin t \\
&= 2 \sin t \cos^2 t + \sin t \cos^2 t - \sin^3 t \\
&= 3 \sin t \cos^2 t - \sin^3 t
\end{aligned}$$

■

PROGRESS CHECK

Express $\cos 3t$ in terms of $\sin t$ and $\cos t$.

ANSWER

$$\cos 3t = 4 \cos^3 t - 3 \cos t$$

If we begin with the formula for $\cos 2t$ and use the fundamental identity $\cos^2 t = 1 - \sin^2 t$, we obtain

$$\begin{aligned}
\cos 2t &= \cos^2 t - \sin^2 t \\
&= (1 - \sin^2 t) - \sin^2 t \\
&= 1 - 2 \sin^2 t
\end{aligned}$$

Similarly, replacing $\sin^2 t$ by $1 - \cos^2 t$ yields

$$\begin{aligned}
\cos 2t &= \cos^2 t - \sin^2 t \\
&= \cos^2 t - (1 - \cos^2 t) \\
&= 2 \cos^2 t - 1
\end{aligned}$$

We then have two additional formulas for $\cos 2t$.

$$\cos 2t = 1 - 2 \sin^2 t \qquad (4)$$

$$\cos 2t = 2 \cos^2 t - 1 \qquad (5)$$

EXAMPLE 3   Double-Angle Formulas in Verifying an Identity

Verify the identity $\dfrac{1 - \cos 2\alpha}{2 \sin \alpha \cos \alpha} = \tan \alpha$.

SOLUTION

Substituting $\cos 2\alpha = 1 - 2 \sin^2 \alpha$ will leave all expressions in terms of α. Then,

$$\frac{1 - \cos 2\alpha}{2 \sin \alpha \cos \alpha} = \frac{1 - (1 - 2 \sin^2 \alpha)}{2 \sin \alpha \cos \alpha}$$

$$= \frac{2 \sin^2 \alpha}{2 \sin \alpha \cos \alpha}$$

$$= \frac{\sin \alpha}{\cos \alpha}$$

$$= \tan \alpha$$

∎

PROGRESS CHECK

Verify the identity $\dfrac{1 + \cos 2\theta}{\sin 2\theta} = \cot \theta$.

HALF-ANGLE FORMULAS   If we begin with the alternative forms for $\cos 2t$ given in Equations (4) and (5), we can obtain the following expressions for $\sin^2 t$ and $\cos^2 t$. The expressions are often used in calculus.

$$\sin^2 t = \frac{1 - \cos 2t}{2} \tag{6}$$

$$\cos^2 t = \frac{1 + \cos 2t}{2} \tag{7}$$

We will use the identities in Equations (6) and (7) to derive formulas for $\sin t/2$, $\cos t/2$, and $\tan t/2$. Substituting $s = 2t$ in Equations (6) and (7) we obtain

$$\sin^2 \frac{s}{2} = \frac{1 - \cos s}{2}$$

$$\cos^2 \frac{s}{2} = \frac{1 + \cos s}{2}$$

Replacing s with t and solving, we have

$$\sin \frac{t}{2} = \pm \sqrt{\frac{1 - \cos t}{2}} \tag{8}$$

$$\cos \frac{t}{2} = \pm \sqrt{\frac{1 + \cos t}{2}} \tag{9}$$

The appropriate sign to use in Equations (8) and (9) depends on the quadrant in which $P(t/2)$ is located. Thus, $\sin t/2$ is positive if $P(t/2)$ lies in quadrant I or II; similarly, we choose the positive root for $\cos t/2$ in Equation (9) if $P(t/2)$ lies in quadrant I or IV.

Using the identity

$$\tan\frac{t}{2} = \frac{\sin\dfrac{t}{2}}{\cos\dfrac{t}{2}}$$

we obtain

$$\tan\frac{t}{2} = \pm\sqrt{\frac{1 - \cos t}{1 + \cos t}} \tag{10}$$

Formulas (8), (9), and (10) are known as the **half-angle formulas.**

EXAMPLE 4   Applying the Half-Angle Formulas

Find the exact values of $\sin 22.5°$ and $\cos 112.5°$.

SOLUTION
Applying the half-angle formulas with $22.5° = 45°/2$ yields

$$\sin 22.5° = \sin\frac{45°}{2}$$

$$= \sqrt{\frac{1 - \cos 45°}{2}}$$

$$= \frac{\sqrt{1 - \sqrt{2}/2}}{2}$$

$$= \frac{\sqrt{2 - \sqrt{2}}}{2}$$

Note that we choose the positive square root since $22.5°$ is in the first quadrant and the sine function is positive in the first quadrant. Similarly,

$$\cos 112.5° = \cos\frac{225°}{2}$$

$$= -\sqrt{\frac{1 + \cos 225°}{2}}$$

$$= -\sqrt{\frac{1 - \cos 45°}{2}}$$

$$= -\sqrt{\frac{1 - \sqrt{2}/2}{2}}$$

$$= -\frac{\sqrt{2 - \sqrt{2}}}{2}$$

The negative square root was selected since $112.5°$ is in the second quadrant and the cosine function is negative in quadrant II.

■

PROGRESS CHECK

Use the half-angle formulas to evaluate $\tan\dfrac{\pi}{8}$.

ANSWER

$\sqrt{2} - 1$

EXAMPLE 5   Applying the Half-Angle Formulas

If $\sin\theta = -\frac{3}{5}$ and θ is in quadrant III, evaluate $\cos\theta/2$.

SOLUTION

We first evaluate $\cos\theta$ by using the identity

$$\cos^2\theta = 1 - \sin^2\theta = 1 - \frac{9}{25} = \frac{16}{25}$$

Since θ is in quadrant III, $\cos\theta$ is negative. Thus, $\cos\theta = -\frac{4}{5}$. We can now employ the half-angle formula

$$\cos\frac{\theta}{2} = \pm\sqrt{\frac{1 + \cos\theta}{2}}$$

$$= \pm\sqrt{\frac{1 - \frac{4}{5}}{2}}$$

$$= \pm\frac{\sqrt{10}}{10}$$

Since $180° < \theta < 270°$, we see that $90° < \theta/2 < 135°$. Thus, $\theta/2$ is in quadrant II and $\cos\theta/2$ is negative. We conclude that $\cos\theta/2 = -\sqrt{10}/10$.

■

PROGRESS CHECK

If $\tan\alpha = \frac{3}{4}$ and α is in quadrant III, evaluate $\tan\alpha/2$.

ANSWER

-3

EXERCISE SET 7.3

In Exercises 1–12 use the given conditions to determine the value of the specified trigonometric function.

1. $\sin u = \frac{3}{5}$ and $P(u)$ is in quadrant II; find $\cos 2u$.

2. $\cos x = -\frac{5}{13}$ and $P(x)$ is in quadrant III; find $\sin 2x$.

3. $\sec\alpha = -2$ and α is in quadrant II; find $\sin 2\alpha$.

4. $\tan\theta = \frac{4}{3}$ and θ is in quadrant I; find $\cos 2\theta$.

5. $\csc t = -\frac{17}{8}$ and $P(t)$ is in quadrant IV; find $\tan 2t$.

6. $\cot \beta = \frac{3}{4}$ and β is in quadrant III; find $\cot 2\beta$.

7. $\sin 2\alpha = -\frac{4}{5}$ and 2α is in quadrant IV; find $\sin 4\alpha$.

8. $\sec 5x = -\frac{13}{12}$ and $P(5x)$ is in quadrant III; find $\tan 10x$.

9. $\cos(\theta/2) = \frac{8}{17}$ and $\theta/2$ is acute; find $\cos \theta$.

10. $\csc(t/4) = -\frac{13}{5}$ and $P(t/4)$ is in quadrant IV; find $\cos(t/2)$.

11. $\sin 42° = 0.67$; find $\cos 84°$.

12. $\cos 77° = 0.22$; find $\cos 154°$.

In Exercises 13–18 use the half-angle formulas to find exact values for each of the following.

13. $\sin 15°$ 14. $\cos 75°$ 15. $\tan \pi/8$ 16. $\sec 5\pi/8$

17. $\csc 165°$ 18. $\cot 7\pi/12$

In Exercises 19–26 use the given conditions to determine the exact value of the specified trigonometric function.

19. $\sin \theta = -\frac{4}{5}$ and θ is in quadrant IV; find $\cos \theta/2$.

20. $\cos \theta = \frac{3}{5}$ and θ is in quadrant I; find $\sin \theta/2$.

21. $\sec t = -3$ and $P(t)$ is in quadrant II; find $\sin t/2$.

22. $\tan x = \frac{4}{3}$ and $P(x)$ is in quadrant III; find $\cos x/2$.

23. $\cot \beta = \frac{3}{4}$ and β is in quadrant III; find $\tan \beta/2$.

24. $\csc \alpha = \frac{13}{5}$ and α is in quadrant II; find $\tan \alpha/2$.

25. $\cos 4x = \frac{1}{3}$ and $P(4x)$ is in quadrant IV; find $\cos 2x$.

26. $\sec 6\alpha = -\frac{13}{12}$ and α is in quadrant III; find $\sin 3\alpha$.

In Exercises 27–46 verify the identity.

27. $\sin 50x = 2 \sin 25x \cos 25x$

28. $(\sin \theta + \cos \theta)^2 = 1 + \sin 2\theta$

29. $\tan 2y = \dfrac{2 \cot y}{\csc^2 y - 2}$

30. $2 \sin^2 2t + \cos 4t = 1$

31. $\sin 4\alpha = 4 \sin \alpha \cos^3 \alpha - 4 \sin^3 \alpha \cos \alpha$

32. $\cos 4\beta = 1 - 8 \sin^2 \beta \cos^2 \beta$

33. $\cos 2u = \dfrac{1 - \tan^2 u}{1 + \tan^2 u}$

34. $\sin 2\theta = \dfrac{2 \tan \theta}{1 + \tan^2 \theta}$

35. $\sin \dfrac{t}{2} \cos \dfrac{t}{2} = \dfrac{\sin t}{2}$

36. $\tan \dfrac{y}{2} = \csc y - \cot y$

37. $\sin \alpha - \cos \alpha \tan \dfrac{\alpha}{2} = \tan \dfrac{\alpha}{2}$

38. $\dfrac{1 - \cos 2\beta}{1 + \cos 2\beta} = \tan^2 \beta$

39. $\cos^4 x - \sin^4 x = \cos 2x$

40. $\dfrac{\sin 2t}{\sin t} - \dfrac{\cos 2t}{\cos t} = \sec t$

41. $\dfrac{2 \tan \alpha}{1 + \tan^2 \alpha} = \sin 2\alpha$

42. $\cos^2 \dfrac{x}{2} = \dfrac{\tan x + \sin x}{2 \tan x}$

43. $\sec 2t = \dfrac{\sec^2 t}{2 - \sec^2 t}$

44. $\cos 2t + \cot 2t = \cot 2t(\sin t + \cos t)^2$

45. $\tan \dfrac{t}{2} = \dfrac{1 - \cos t}{\sin t}$

46. $\tan \dfrac{t}{2} = \dfrac{\sin t}{1 + \cos t}$

In Exercises 47–50 find the requested value.

47. $\sin(2 \arccos \frac{3}{5})$ 48. $\cos(2 \sin^{-1} \frac{3}{5})$ 49. $\tan(2 \arcsin \frac{5}{13})$ 50. $\cos(2 \arctan \frac{12}{5})$

7.4 THE PRODUCT–SUM FORMULAS

The formulas that will be derived in this section are of use in calculus and in other courses in higher mathematics. They are not as important as the formulas that appeared in Sections 7.2 and 7.3 and need not be memorized. Rather, you should be aware of these formulas so that you can look them up when needed.

The following formulas express a product as a sum.

$$\sin s \cos t = \frac{\sin(s + t) + \sin(s - t)}{2} \qquad (1)$$

$$\cos s \sin t = \frac{\sin(s + t) - \sin(s - t)}{2} \qquad (2)$$

$$\cos s \cos t = \frac{\cos(s + t) + \cos(s - t)}{2} \qquad (3)$$

$$\sin s \sin t = \frac{\cos(s - t) - \cos(s + t)}{2} \qquad (4)$$

To prove Equation (1), we begin with the right-hand side of the equation.

$$\frac{\sin(s + t) + \sin(s - t)}{2}$$

$$= \frac{(\sin s \cos t + \cos s \sin t) + (\sin s \cos t - \cos s \sin t)}{2}$$

$$= \frac{2 \sin s \cos t}{2}$$

$$= \sin s \cos t$$

The proofs of Equations (2), (3), and (4) are very similar.

EXAMPLE 1 Applying the Product–Sum Formulas

Express $\sin 4x \cos 3x$ as a sum or a difference.

SOLUTION

Applying Equation (1) we obtain

$$\sin 4x \cos 3x = \frac{\sin(4x + 3x) + \sin(4x - 3x)}{2}$$

$$= \frac{\sin 7x + \sin x}{2}$$

■

PROGRESS CHECK

Express $\sin 5x \sin 2x$ as a sum or as a difference.

ANSWER

$\frac{1}{2}(\cos 3x - \cos 7x)$

EXAMPLE 2   Applying the Product–Sum Formulas

Evaluate the product $\cos \dfrac{5\pi}{8} \cos \dfrac{3\pi}{8}$ by a product–sum formula.

SOLUTION

Using Equation (3) we have

$$
\begin{aligned}
\cos \frac{5\pi}{8} \cos \frac{3\pi}{8} &= \frac{1}{2}\left[\cos\left(\frac{5\pi}{8}+\frac{3\pi}{8}\right) + \cos\left(\frac{5\pi}{8}-\frac{3\pi}{8}\right) \right] \\
&= \frac{1}{2}\left[\cos \pi + \cos \frac{\pi}{4} \right] \\
&= \frac{1}{2}\left[-1 + \frac{\sqrt{2}}{2} \right] = \frac{\sqrt{2}-2}{4}
\end{aligned}
$$

∎

PROGRESS CHECK

Evaluate $\cos \pi/3 \sin \pi/6$ by a product–sum formula.

ANSWER

$\frac{1}{4}$

The following formulas express a sum as a product.

$$
\sin s + \sin t = 2\sin \frac{s+t}{2} \cos \frac{s-t}{2} \tag{5}
$$

$$
\sin s - \sin t = 2\cos \frac{s+t}{2} \sin \frac{s-t}{2} \tag{6}
$$

$$
\cos s + \cos t = 2\cos \frac{s+t}{2} \cos \frac{s-t}{2} \tag{7}
$$

$$
\cos s - \cos t = -2\sin \frac{s+t}{2} \sin \frac{s-t}{2} \tag{8}
$$

To prove the identity in Equation (5), begin with the right-hand side and apply Equation (1). Then

$$
2\sin \frac{s+t}{2} \cos \frac{s-t}{2} = \frac{1}{2}\left[\sin\left(\frac{s+t}{2}+\frac{s-t}{2}\right) + \sin\left(\frac{s+t}{2}-\frac{s-t}{2}\right) \right]
$$

$$
= \sin s + \sin t
$$

This establishes Equation (5).

EXAMPLE 3   Applying the Sum–Product Formulas

Express $\sin 5x - \sin 3x$ as a product.

SOLUTION

Using Equation (6) we have

$$\sin 5x - \sin 3x = 2 \cos \frac{5x + 3x}{2} \sin \frac{5x - 3x}{2}$$

$$= 2 \cos 4x \sin x$$

PROGRESS CHECK

Express $\cos 6x + \cos 2x$ as a product.

ANSWER

$2 \cos 4x \cos 2x$

EXAMPLE 4   Applying the Sum–Product Formulas

Evaluate $\cos 5\pi/12 - \cos \pi/12$ by using a sum–product formula.

SOLUTION

Using Equation (8), we have

$$\cos \frac{5\pi}{12} - \cos \frac{\pi}{12} = -2 \sin \frac{\pi}{4} \sin \frac{\pi}{6}$$

$$= -2 \left(\frac{\sqrt{2}}{2} \right) \frac{1}{2} = -\frac{\sqrt{2}}{2}$$

PROGRESS CHECK

Evaluate $\sin 2\pi/3$ by using a sum–product formula.

ANSWER

$\sqrt{3}/2$

EXERCISE SET 7.4

In Exercises 1–8 express each product as a sum or difference.

1. $2 \sin 5\alpha \cos \alpha$

2. $-3 \cos 6x \sin 2x$

3. $\sin 3x \sin(-2x)$

4. $\cos 7t \cos(-3t)$

5. $-2 \cos 2\theta \cos 5\theta$

6. $\sin \dfrac{5\theta}{2} \sin \dfrac{\theta}{2}$

7. $\cos(\alpha + \beta) \cos(\alpha - \beta)$

8. $-\sin 2u \cos 4u$

In Exercises 9–12 evaluate each product by using a product–sum formula.

9. $\cos \dfrac{7\pi}{8} \sin \dfrac{5\pi}{8}$

10. $\cos \dfrac{\pi}{3} \cos \dfrac{\pi}{6}$

11. $\sin 120° \cos 60°$

12. $\sin \dfrac{13\pi}{12} \sin \dfrac{11\pi}{12}$

In Exercises 13–20 express each sum or difference as a product.

13. $\sin 5x + \sin x$

14. $\cos 8t - \cos 2t$

15. $\cos 2\theta + \cos 6\theta$

16. $\sin 5\alpha - \sin 7\alpha$

17. $\sin(\alpha + \beta) + \sin(\alpha - \beta)$

18. $\cos\dfrac{x}{2} - \cos\dfrac{3x}{2}$

19. $\sin 7x - \sin 3x$

20. $\cos 5\theta + \cos 3\theta$

In Exercises 21–24 evaluate each sum by using a sum–product formula.

21. $\cos 75° + \cos 15°$

22. $\sin\dfrac{5\pi}{12} + \sin\dfrac{\pi}{12}$

23. $\cos\dfrac{3\pi}{4} - \cos\dfrac{\pi}{4}$

24. $\sin\dfrac{13\pi}{12} - \sin\dfrac{5\pi}{12}$

In Exercises 25–34 verify the identities.

25. $\sin 40° + \sin 20° = \cos 10°$

26. $\cos 70° - \cos 10° = -\sin 40°$

27. $\dfrac{\sin 5\theta - \sin 3\theta}{\cos 3\theta - \cos 5\theta} = \cot 4\theta$

28. $\dfrac{\cos 7x - \cos x}{\sin 7x + \sin x} = -\tan 3x$

29. $\dfrac{\sin t - \sin s}{\cos t - \cos s} = -\cot\dfrac{s + t}{2}$

30. $\dfrac{\sin s + \sin t}{\cos s + \cos t} = \tan\dfrac{s + t}{2}$

31. $\dfrac{\sin 50° - \sin 10°}{\cos 50° - \cos 10°} = -\sqrt{3}$

32. $2\sin\left(\theta + \dfrac{\pi}{4}\right)\sin\left(\theta - \dfrac{\pi}{4}\right) = -\cos 2\theta$

33. $\dfrac{\cot x - \tan x}{\cot x + \tan x} = \cos 2x$

34. $\cos 6t \cos 2t + \sin^2 4x = \cos^2 2x$

35. Express $(\sin ax)(\cos bx)$ as a sum.

36. Express $(\cos ax)(\cos bx)$ as a sum.

7.5 TRIGONOMETRIC EQUATIONS

Thus far, this chapter has dealt exclusively with trigonometric identities. We now seek to solve trigonometric equations that are not true for all values of the variable but may be true for some values.

We have seen that algebraic equations may have just one or two solutions. The situation is quite different with trigonometric equations since the periodic nature of the trigonometric functions assures us that if there is a solution, there are an infinite number of solutions. To handle this complication, we simply seek all solutions t such that $0 \le t < 2\pi$. Then for every integer value of n, $t + 2\pi n$ is also a solution. The following example illustrates this convenient means for writing the solution set.

EXAMPLE 1   Solving a Trigonometric Equation

Find all solutions of the equation $\cos t = 0$.

SOLUTION

The only values in the interval $[0, 2\pi)$ for which $\cos t = 0$ are $\pi/2$ and $3\pi/2$. Then every solution is included among those values of t such that

$$t = \dfrac{\pi}{2} + 2\pi n \qquad \text{or} \qquad t = \dfrac{3\pi}{2} + 2\pi n, \quad n \text{ an integer}$$

Since $\dfrac{3\pi}{2} = \dfrac{\pi}{2} + \pi$, the solution set can be written in the more compact form

$$t = \dfrac{\pi}{2} + \pi n, \quad n \text{ an integer}$$

■

Factoring provides the key for solving many trigonometric equations. If we can write the equation in the form $P(x)Q(x) = 0$, we can then find the solutions by setting $P(x) = 0$ and $Q(x) = 0$. Of course, P and Q will themselves generally contain trigonometric functions.

It may also be helpful to think in terms of a substitution of variable. Thus, the equation

$$4 \sin^2 x + 3 \sin x - 1 = 0$$

can be viewed as a quadratic in u

$$4u^2 + 3u - 1 = 0$$

by substituting $u = \sin x$.

EXAMPLE 2   Restricting the Solutions to a Trigonometric Equation

Find all solutions of the equation $2 \cos^2 t - \cos t - 1 = 0$ in the interval $[0, 2\pi)$.

SOLUTION

Factoring the left side of the equation yields

$$(2 \cos t + 1)(\cos t - 1) = 0$$

Setting each factor equal to 0, we have

$$2 \cos t + 1 = 0 \qquad \text{or} \qquad \cos t - 1 = 0$$

so that

$$\cos t = -\frac{1}{2} \qquad \text{or} \qquad \cos t = 1$$

The solutions of $\cos t = -\frac{1}{2}$ in the interval $[0, 2\pi)$ are $t = 2\pi/3$ and $t = 4\pi/3$; the only solution of $\cos t = 1$ in the interval $[0, 2\pi)$ is $t = 0$. The solutions of the original equation in the interval $[0, 2\pi)$ are

$$t = \frac{2\pi}{3}, \qquad t = \frac{4\pi}{3}, \qquad \text{and} \qquad t = 0$$

∎

 PROGRESS CHECK

Find all solutions of the equation $2 \sin^2 t - 3 \sin t + 1 = 0$ in the interval $[0, 2\pi)$.

ANSWER

$$\frac{\pi}{2}, \frac{5\pi}{6}, \frac{\pi}{6}$$

If the solutions of a trigonometric equation are angles, the answer may be given in either radians or degrees.

EXAMPLE 3   Expressing Solutions in Radians and Degrees
Find all solutions of the equation $\tan \theta \cos^2 \theta - \tan \theta = 0$.

SOLUTION
Factoring the left side yields

$$(\tan \theta)(\cos^2 \theta - 1) = 0$$

Setting each factor equal to 0,

$$\tan \theta = 0 \qquad \text{or} \qquad \cos^2 \theta = 1$$

so that

$$\tan \theta = 0, \qquad \cos \theta = 1, \qquad \text{or} \qquad \cos \theta = -1$$

These equations yield the following solutions in the interval $[0, 2\pi)$.

$$\tan \theta = 0: \qquad \theta = 0 \qquad \text{or} \qquad \theta = \pi$$
$$\cos \theta - 1: \qquad \theta = 0$$
$$\cos \theta = -1: \qquad \theta = \pi$$

The solutions of the original equation are

$$\theta = 0 + 2\pi n \qquad \text{and} \qquad \theta = \pi + 2\pi n, \quad n \text{ an integer}$$

which can be expressed more compactly as

$$\theta = \pi n, \quad n \text{ an integer}$$

In degree measure, the solution is

$$\theta = 180°n, \quad n \text{ an integer}$$

$\blacksquare$

EXAMPLE 4   Expressing Restricted Solutions in Radians and Degrees
Find all solutions of the equation $\sin 2\theta - 3 \sin \theta = 0$ in the interval $[0, 2\pi)$.

SOLUTION
Using the identity $\sin 2\theta = 2 \sin \theta \cos \theta$ yields

$$2 \sin \theta \cos \theta - 3 \sin \theta = 0$$
$$\sin \theta(2 \cos \theta - 3) = 0$$
$$\sin \theta = 0 \qquad \text{or} \qquad 2 \cos \theta - 3 = 0$$
$$\sin \theta = 0 \qquad \text{or} \qquad \cos \theta = \frac{3}{2}$$

The equation $\cos \theta = \frac{3}{2}$ has no solutions; the solutions of $\sin \theta = 0$ are $\theta = 0$ and $\theta = \pi$. The solutions of the original equation are

$$\theta = 0 \qquad \text{and} \qquad \theta = \pi$$

or, in degree measure,

$$\theta = 0° \qquad \text{and} \qquad \theta = 180°$$

∎

 PROGRESS CHECK

Find all solutions of the equation $\cos 2\theta + \cos \theta = 0$.

ANSWER

$60° + 360°n, 180°, 300° + 360°n, 360°n$

or

$\dfrac{\pi}{3} + 2\pi n, \pi, \dfrac{5\pi}{3} + 2\pi n, 2\pi n$

Equations involving multiple angles can often be solved by using a substitution of variable. The following example shows that we must proceed with caution when seeking solutions in the interval $[0, 2\pi)$.

EXAMPLE 5 Substitution of Variable in Trigonometric Equations

Find all solutions of the equation $\cos 3x = 0$ in the interval $[0, 2\pi)$.

SOLUTION

We are given

$$\cos 3x = 0, \qquad 0 \le x < 2\pi$$

Substituting $t = 3x$, we obtain

$$\cos t = 0, \qquad 0 \le \frac{t}{3} < 2\pi$$

or

$$\cos t = 0, \qquad 0 \le t < 6\pi$$

Note that we seek solutions of $\cos t = 0$ in the interval $[0, 6\pi)$ rather than $[0, 2\pi)$. The solutions are then

$$t = \frac{\pi}{2} \quad \frac{3\pi}{2} \quad \frac{5\pi}{2} \quad \frac{7\pi}{2} \quad \frac{9\pi}{2} \quad \frac{11\pi}{2}$$

Since $x = t/3$ we obtain

$$x = \frac{\pi}{6} \quad \frac{\pi}{2} \quad \frac{5\pi}{6} \quad \frac{7\pi}{6} \quad \frac{3\pi}{2} \quad \frac{11\pi}{6}$$

∎

It is sometimes possible to treat the trigonometric equation as a quadratic equation after performing a suitable substitution of variable. Here is an example.

EXAMPLE 6   Substitution of Variable
Find all solutions of the equation

$$3 \tan^2 x + \tan x - 1 = 0$$

in the interval $[0, \pi)$.

SOLUTION
The equation doesn't yield to the method of factoring. However, it can be viewed as a quadratic equation in $\tan x$. That is, if we substitute $t = \tan x$ we obtain

$$3t^2 + t - 1 = 0$$

which is a quadratic in t. By the quadratic formula,

$$t = \frac{-1 \pm \sqrt{13}}{6}$$

so that

$$\tan x = \frac{-1 \pm \sqrt{13}}{6}$$

Using Table V in the Tables Appendix, or a calculator, we have

$$x \approx 0.41 \qquad \text{and} \qquad x \approx 2.49$$

as the solutions of the equation.

■

EXERCISE SET 7.5

In Exercises 1–20 find all solutions of the given equation in the interval $[0, 2\pi)$. Express the answers in both radian measure and degree measure.

1. $2 \sin \theta - 1 = 0$
2. $2 \cos \theta + 1 = 0$
3. $\cos \alpha + 1 = 0$
4. $\cot \gamma + 1 = 0$
5. $4 \cos^2 \alpha = 3$
6. $\tan^2 \theta = 3$
7. $3 \tan^2 \alpha = 1$
8. $2 \cos^2 \alpha - 1 = 0$
9. $2 \sin^2 \beta = \sin \beta$
10. $\sin \alpha = \cos \alpha$
11. $2 \cos^2 \theta - 3 \cos \theta + 1 = 0$
12. $2 \sin^2 \theta - \sin \theta - 1 = 0$
13. $\sin 5\theta = 1$
14. $\tan 3\beta = -\sqrt{3}$
15. $2 \sin^2 \alpha - 3 \cos \alpha = 0$
16. $\csc 2\theta = 2$
17. $2 \cos^2 \theta - 1 = \sin \theta$
18. $\cos^2 2\alpha = \frac{1}{4}$
19. $\sin^2 \beta + 3 \cos \beta - 3 = 0$
20. $2 \cos^2 \theta \tan \theta - \tan \theta = 0$

In Exercises 21–38 find all the solutions of the given equation.

21. $3 \tan^2 x - 1 = 0$
22. $2 \sin^2 y - 1 = 0$
23. $3 \cot^2 \theta - 1 = 0$
24. $1 - 4 \cos^2 t = 0$
25. $\sec 2u - 2 = 0$
26. $\tan 3x - 1 = 0$

27. $\sin 4x = 0$

28. $\cos 5t = -1$

29. $4\cos^2 2t - 3 = 0$

30. $\csc^2 2x - 2 = 0$

31. $\sin 2t + 2\cos t = 0$

32. $\sin 2t + 3\cos t = 0$

33. $\cos 2t + \sin t = 0$

34. $2\cos 2t + 2\sin t = 0$

35. $\tan^2 x - \tan x = 0$

36. $\sec^2 x - 3\sec x + 2 = 0$

37. $2\sin^2 x + 3\sin x - 2 = 0$

38. $2\cos^2 x - 5\cos x - 3 = 0$

In Exercises 39–42 find the approximate solutions of the given equations in the interval $[0, 2\pi)$ by using Table V in the Tables Appendix, or a calculator.

39. $5\sin^2 x - \sin x - 2 = 0$

40. $\sec^2 y - 5\sec y + 6 = 0$

41. $3\tan^2 u + 5\tan u + 1 = 0$

42. $\cos^2 t - 2\sin t + 3 = 0$

TERMS AND SYMBOLS

addition formulas **327**	fundamental identities **321**	trigonometric equation **343**	trigonometric formulas **326**
cofunctions **329**	half-angle formulas **336**	trigonometric expression **327**	trigonometric identity **320**
double-angle formulas **333**	product–sum formulas **339**		

KEY IDEAS FOR REVIEW

Topic	Page	Key Idea
Trigonometric identity	320	A trigonometric identity is true for all real values in the domain of the variable.
fundamental identities	321	The fundamental identities are those trigonometric identities that occur so frequently that they must be remembered and recognized.
Verifying an identity	322	The fundamental identities can be used to verify other trigonometric identities. The techniques commonly used to verify identities include the following. ■ Write all of the trigonometric functions in terms of sine and cosine. ■ Factor. ■ Complete the indicated operations, especially when this involves the sum of fractional expressions. ■ Multiply the numerator and denominator of a fractional expression by the same quantity to produce a simpler product such as $1 - \sin^2\theta$, $1 - \cos^2\theta$, or $\sec^2\theta - 1$.
Trigonometric formulas		The most useful of the trigonometric formulas are the following. **Addition Formulas** $$\sin(s + t) = \sin s \cos t + \cos s \sin t$$ $$\cos(s + t) = \cos s \cos t - \sin s \sin t$$ $$\tan(s + t) = \frac{\tan s + \tan t}{1 - \tan s \tan t}$$

Topic	Page	Key Idea
		Double-Angle Formulas $$\sin 2t = 2\sin t \cos t$$ $$\cos 2t = \cos^2 t - \sin^2 t$$ $$\tan 2t = \frac{2\tan t}{1 - \tan^2 t}$$ **Half-Angle Formulas** $$\sin\frac{t}{2} = \pm\sqrt{\frac{1 - \cos t}{2}}$$ $$\cos\frac{t}{2} = \pm\sqrt{\frac{1 + \cos t}{2}}$$ $$\tan\frac{t}{2} = \pm\sqrt{\frac{1 - \cos t}{1 + \cos t}}$$
Trigonometric equations	343	Since the trigonometric functions are periodic, a trigonometric equation has either no solutions or an infinite number of solutions.
restricted answers	344	It is sometimes desirable to restrict the answer(s) to a specific interval.
radian and degree measure	345	Answers may be expressed in either radian or degree measure. When working with a computer, radian measure is assumed.

REVIEW EXERCISES

Solutions to exercises whose numbers are in color are in the Solutions section in the back of the book.

7.1 In Exercises 1–3 verify the given identity.

1. $\sin\sigma\sec\sigma + \tan\sigma = 2\tan\sigma$

2. $\dfrac{\cos^2 x}{1 - \sin x} = 1 + \sin x$

3. $\sin\alpha + \sin\alpha\cot^2\alpha = \csc\alpha$

7.2 In Exercises 4–7 determine the exact value of the given expression by using the addition formulas.

4. $\sin\left(\dfrac{\pi}{6} + \dfrac{\pi}{4}\right)$

5. $\cos(45° + 90°)$

6. $\tan\left(\dfrac{\pi}{3} + \dfrac{\pi}{4}\right)$

7. $\sin\dfrac{7\pi}{12}$

In Exercises 8–11 write the given expression in terms of cofunctions of complementary angles.

8. $\csc 15°$

9. $\cos 23°$

10. $\sin\dfrac{\pi}{8}$

11. $\tan\dfrac{2\pi}{7}$

12. If $\cos\sigma = -\frac{12}{13}$ and $0 \le \sigma \le 180°$, find $\sin(\pi - \sigma)$.

13. If $\sec\sigma = \frac{10}{8}$ and σ lies in quadrant IV, find $\csc(\sigma + \pi/3)$.

14. If $\sin t = -\frac{3}{5}$ and $P(t)$ is in quadrant III, find $\tan(t + \pi)$.

15. If $\cos\alpha = -\frac{12}{13}$ and $\tan\beta = -\frac{5}{2}$, with angles α and β in quadrant II, find $\tan(\alpha + \beta)$.

16. If $\sin x = \frac{3}{5}$ and $\csc y = \frac{13}{12}$, with $P(x)$ in quadrant II and $P(y)$ in quadrant I, find $\cos(x - y)$.

7.3 **17.** If $\csc u = -\frac{5}{4}$ and $P(u)$ is in quadrant IV, find $\cos 2u$.

18. If $\tan\sigma = -\frac{3}{4}$ and $0 \le \sigma \le 180°$, find $\sin 2\sigma$.

19. If $\sin 2t = \frac{3}{5}$ and $P(2t)$ is in quadrant I, find $\sin 4t$.

20. If $\sin\sigma = 0.5$ and $\pi/2 \le \sigma \le \pi$, find $\sin 2\sigma$.

21. If $\cos(\sigma/2) = \frac{12}{13}$ and σ is acute, find $\sin\sigma$.

22. If $\sin\alpha = -\frac{3}{5}$ and α is in quadrant III, find $\cos(\alpha/2)$.

23. If $\cot t = -\frac{4}{3}$ and $P(t)$ is in quadrant IV, find $\tan(t/2)$.

24. If $\cos 4x = \frac{2}{3}$ and $P(4x)$ is in quadrant IV, find $\cos 2x$.

25. Find the exact value of $\cos 15°$ by using a half-angle formula.

26. Find the exact value of $\sin \pi/8$ by using a half-angle formula.

27. Find the exact value of $\tan 112.5°$ by using a half-angle formula.

In Exercises 28–30 verify the given identity.

28. $\cos 30x = 1 - 2\sin^2 15x$

29. $\frac{1}{2}\sin 2y = \frac{\sin y}{\sec y}$

30. $\tan\frac{\alpha}{2} = \frac{(1 - \cos \alpha)}{\sin \alpha}$

7.4 **31.** Express $\sin\frac{3\alpha}{2}\sin\frac{\alpha}{2}$ as a sum or difference.

32. Express $\cos 3x - \cos x$ as a product.

33. Evaluate $\sin 75° \sin 15°$ by using a product–sum formula.

34. Evaluate $\cos\frac{3\pi}{4} + \cos\frac{\pi}{4}$ by using a sum–product formula.

7.5 In Exercises 35–37 find all solutions of the given equation in the interval $[0, 2\pi)$. Express the answers in radian measure.

35. $2\cos^2 \alpha - 1 = 0$

36. $2\sin \sigma \cos \sigma = 0$

37. $\sin 2t - \sin t = 0$

In Exercises 38–40 find all solutions of the given equation. Express the answers in degree measure.

38. $\cos^2 \alpha - 2\cos \alpha = 0$

39. $\tan 3x + 1 = 0$

40. $4\sin^2 2t = 3$

PROGRESS TEST 7A

1. Verify the identity $4 - \tan^2 x = 5 - \sec^2 x$.

In Problems 2 and 3 determine exact values of the given expressions by using the addition formulas.

2. $\cos(270° + 30°)$ 3. $\tan\left(\frac{\pi}{4} - \frac{\pi}{3}\right)$

4. Write $\sin 47°$ in terms of its cofunction.

5. If $\cos \theta = \frac{4}{5}$ and θ lies in quadrant IV, find $\sin(\theta - \pi)$.

6. If $\sin x = -\frac{5}{13}$ and $\tan y = \frac{8}{3}$ with angles x and y in quadrant III, find $\tan(x - y)$.

7. If $\sin v = -\frac{12}{13}$ and $P(v)$ is in quadrant IV, find $\cos 2v$.

8. If $\cos 2\alpha = -\frac{4}{5}$ and 2α is in quadrant II, find $\cos 4\alpha$.

9. If $\csc \alpha = -2$ and α is in quadrant III, find $\cos(\alpha/2)$.

10. Find the exact value of $\tan 15°$ by using a half-angle formula.

11. Verify the identity $\sin\frac{x}{4} = 2\sin\frac{x}{8}\cos\frac{x}{8}$.

12. Express $\sin 2x + \sin 3x$ as a product.

13. Express $\sin 150° - \sin 30°$ by using a sum–product formula.

14. Find all solutions of the equation $4\sin^2 \alpha = 3$ in the interval $[0, 2\pi)$. Express the answers in radian measure.

15. Find all solutions of the equation $\sin^2 \theta - \cos^2 \theta = 0$ and express the answers in degree measure.

PROGRESS TEST 7B

1. Verify the identity $\dfrac{\tan u + \cot u}{\sec u \sin u} = \csc^2 u$.

In Problems 2 and 3 determine exact values of the given expression by using the addition formulas.

2. $\csc(180° - 30°)$ 3. $\sin\frac{7\pi}{12}$

4. Write $\tan 71°$ in terms of its cofunction.

5. If $\sin t = -\frac{5}{13}$ and $P(t)$ lies in quadrant III, find $\sec(t + \pi/4)$.

6. If $\cos \alpha = -0.6$ and $\csc \beta = \frac{5}{4}$ with angles α and β in quadrant II, find $\sin(\alpha - \beta)$.

7. If $\sec \theta = \frac{5}{4}$ and $0 \le \theta \le 180°$, find $\sin 2\theta$.

8. If $\sin \theta/2 = \frac{3}{5}$ and θ is acute, find $\sin 2\theta$.

9. If $\sin 6x = -\frac{12}{13}$ and $P(6x)$ is in quadrant IV, find $\cos 3x$.

10. Find the exact value of $\sin \pi/8$ by using a half-angle formula.

11. Verify the identity $\sec 2t = \dfrac{1 + \tan^2 t}{1 - \tan^2 t}$.

12. Express $\sin \pi/4 \cos \pi/3$ as a sum or difference.

13. Express $\cos 75° \cos 15°$ by using a product–sum formula.

14. Find all solutions of the equation $\tan^2 x + \tan x = 0$ in the interval $[0, 2\pi)$. Express the answers in radian measure.

15. Find all solutions of the equation
$$2 \sin^2 \alpha - \sin \alpha - 1 = 0$$
and express the answers in degree measure.

UNIT 2 CUMULATIVE PROGRESS TEST
CHAPTERS 5–7

1. Write each of the following trigonometric functions in terms of an angle t where t is in the first quadrant. Your answers may use only the forms $\sin t$, $-\sin t$, $\cos t$, $-\cos t$, $\tan t$, and $-\tan t$.
   (a) $\sin -\dfrac{9\pi}{5}$ (b) $\cos \dfrac{8\pi}{3}$ (c) $\tan -\dfrac{11\pi}{7}$

2. Evaluate without using a calculator:
   (a) $\sec -\dfrac{37\pi}{3}$ (b) $\cot \dfrac{23\pi}{6}$ (c) $\sin -\dfrac{29\pi}{6}$

3. Sketch the graph of $f(x) = 3 \sin\left(2x + \dfrac{\pi}{2}\right)$

4. Find the values of t in the interval $[0, 2\pi]$ that satisfy the given equation.
   (a) $\cos t = -\dfrac{\sqrt{2}}{2}$ (b) $\csc t = 2$ (c) $\tan t = -\sqrt{3}$

5. Evaluate
   (a) $\tan\left(\cos^{-1}\dfrac{3}{5}\right)$ (b) $\sec\left(\sin^{-1}\dfrac{5}{13}\right)$
   (c) $\cos^{-1}\left(\cot -\dfrac{\pi}{4}\right)$

6. Find t in the interval $[0, 2\pi]$ if
   (a) $\tan t = \sqrt{3}$ and $\cos t < 0$
   (b) $\sec t = 2$ and $\tan t < 0$
   (c) $\cos t = -\sqrt{3}$ and $\tan t > 0$

7. Prove:
   (a) $\sec t \cot t = \csc t$ (b) $\dfrac{1}{\sec t - \tan t} = \sec t + \tan t$

8. Determine the amplitude, period, and phase shift:
   (a) $f(x) = -\sin\left(\dfrac{x}{2} + \dfrac{\pi}{2}\right)$ (b) $f(x) = 3 \cos\left(2x - \dfrac{\pi}{2}\right)$

9. Solve and express the exact solution in terms of inverse trigonometric functions.
   (a) $3 \sin^2 x = 2$ (b) $2 \cos^2 x - \cos x - 3 = 0$

10. Use a calculator to find all solutions in the interval $[0, 2\pi]$:
$$\tan^2 x - 4 \tan x + 2 = 0$$

11. Find the reference angle for the given angle.
    (a) $265°$ (b) $-7\pi/3$ (c) $-120°$

12. Convert from radians to degrees or degrees to radians.
    (a) $5°$ (b) $-\pi/6$ (c) 2

13. A right triangle has legs of length 3 and 5. Let θ be the angle between the leg of length 5 and the hypotenuse. Evaluate:
    (a) $\sec \theta$ (b) $\sin\left(\dfrac{\pi}{2} - \theta\right)$ (c) $\cot \theta$

14. To bring the seat of a ferris wheel to a point at which it can be serviced, the ferris wheel must go through a central angle of $30°$. The distance covered is 20 feet. What is the diameter of the ferris wheel?

15. Find the angle of elevation of the sun when a building 50′ in height casts a horizontal shadow 30′ in length. (Use a calculator to obtain an answer accurate to the nearest degree.)

16. The sides of a rectangle are 6 and 9. What is the degree measure of the angle between the shorter side and the diagonal?

17. Use the Law of Cosines or Law of Sines to approximate the required part of $\triangle ABC$.
    (a) $a = 5, b = 12, c = 8$; find γ.
    (b) $a = 15, b = 18, \alpha = 50°$; find γ.

18. Verify the identity: $\dfrac{\tan^2 t - 1}{\tan^2 t + 1} = 1 - 2 \cos^2 t$

19. Write the following in terms of trigonometric functions of θ alone:
    (a) $\cos\left(\dfrac{\pi}{6} - \theta\right)$ (b) $\tan\left(\theta - \dfrac{\pi}{4}\right)$

20. Find exact values without the use of a calculator.
    (a) $\cos\left(-\dfrac{\pi}{8}\right)$ (b) $\tan 255°$

21. Write in terms of the cofunction:
    (a) $\cot 62°$ (b) $\sin \dfrac{\pi}{12}$

22. Given $\sin t = -\dfrac{3}{5}$ and $\cos t > 0$, find
    (a) $\sin \dfrac{t}{2}$ (b) $\tan 2t$

23. Express as a product: $\cos 40° + \cos 20°$

24. Express as a sum: $\sin 2t \cos 3t$

25. Find all solutions of the equation $\sin \theta - \cos 2\theta = 0$ in the interval $[0, 2\pi]$

Ezra Stoller © ESTO

TOPICS IN ALGEBRA

From the intrinsic evidence of his creation, the Great Architect of the Universe now begins to appear as a pure mathematician.

Sir James Hopwood Jeans

God made integers, all else is the work of man.

Leopold Kronecker

The four chapters that comprise this unit provide a fascinating look at the diversity of topics in algebra. In Chapter 8 we deal with the roots of polynomial equations and provide insight into the Fundamental Theorem of Algebra. How very different is the material of Chapter 9 where we study analytic geometry, sometimes called the marriage of algebra and geometry! Its creator, René Descartes, sought to unify all of science through the medium of mathematics.

In Chapter 10 we deal with *systems* of equations and inequalities and are led into the study of matrices and determinants. Our final chapter deals with sequences and series. This brings us full circle, since we will be dealing once again with functions, albeit specifically functions whose domain is the set of natural numbers.

CHAPTER 8

THE FUNDAMENTAL THEOREM OF ALGEBRA

Does a polynomial equation always have a root? The answer to this question depends on the number system under consideration. For example, the equation

$$x^2 + x + 1 = 0$$

has no real roots but does have two complex roots, which we can find by using the quadratic formula. On the other hand, the seemingly simple equation

$$x^4 - 1 = 0$$

has two real roots, ± 1, and two complex roots, $\pm i$.

Questions about the roots of polynomial equations attracted the attention of mathematicians for several hundred years. The examples we have just looked at tell us that our search must involve the set of complex numbers. We will therefore begin this chapter by delving more deeply into this number system and its properties. We will then be in a better position to extract some fundamental relationships between the complex number system and the roots of polynomial equations.

8.1
COMPLEX NUMBERS
AND THEIR PROPERTIES

We introduced the complex number system in Section 1.3 and then used this number system in Section 2.6 to provide solutions to quadratic equations. Recall that $z = a + bi$ is said to be a complex number where a and b are real numbers and the imaginary unit $i = \sqrt{-1}$ has the property that $i^2 = -1$. We say that $a + bi$ is the **algebraic form** of z. We defined fundamental operations with complex numbers in the following way.

Equality:	$a + bi = c + di$ if $a = c$ and $b = d$
Addition:	$(a + bi) + (c + di) = (a + c) + (b + d)i$
Multiplication:	$(a + bi)(c + di) = (ac - bd) + (ad + bc)i$

With this background, we can now explore further properties of the complex number system.

The complex number $a - bi$ is called the **complex conjugate** (or simply the **conjugate**) of the complex number $a + bi$. For example, $3 - 2i$ is the conjugate of $3 + 2i$, $4i$ is the conjugate of $-4i$, and 2 is the conjugate of 2. Forming the product $(a + bi)(a - bi)$ we have

$$(a + bi)(a - bi) = a^2 - abi + abi - b^2i^2$$
$$= a^2 + b^2 \quad \text{Since } i^2 = -1$$

Because a and b are real numbers, $a^2 + b^2$ is also a real number. We can summarize this result as follows.

The product of a complex number and its conjugate is a real number.

$$(a + bi)(a - bi) = a^2 + b^2$$

We can now demonstrate that the quotient of two complex numbers is also a complex number. The quotient

$$\frac{q + ri}{s + ti}$$

can be written in the form $a + bi$ by multiplying both numerator and denominator by $s - ti$, the conjugate of the denominator. We then have

$$\frac{q + ri}{s + ti} = \frac{q + ri}{s + ti} \cdot \frac{s - ti}{s - ti} = \frac{(qs + rt) + (rs - qt)i}{s^2 + t^2}$$
$$= \frac{qs + rt}{s^2 + t^2} + \frac{(rs - qt)}{s^2 + t^2}i$$

which is a complex number of the form $a + bi$. Of course, the reciprocal of the complex number $s + ti$ is the quotient $1/(s + ti)$, which can also be written as a complex number by using the same technique. In summary,

---

- The quotient of two complex numbers is a complex number.
- The reciprocal of a nonzero complex number is a complex number.

EXAMPLE 1   Division of Complex Numbers

(a) Write the quotient $\dfrac{-2 + 3i}{3 - 2i}$ in the form $a + bi$.

(b) Write the reciprocal of $2 - 5i$ in the form $a + bi$.

SOLUTION

(a) Multiplying numerator and denominator by the conjugate $3 + 2i$ of the denominator, we have

$$\frac{-2 + 3i}{3 - 2i} = \frac{-2 + 3i}{3 - 2i} \cdot \frac{3 + 2i}{3 + 2i} = \frac{-6 - 4i + 9i + 6i^2}{3^2 + 2^2} = \frac{-6 + 5i + 6(-1)}{9 + 4}$$

$$= \frac{-12 + 5i}{13} = -\frac{12}{13} + \frac{5}{13}i$$

(b) The reciprocal is $1/(2 - 5i)$. Multiplying both numerator and denominator by the conjugate $2 + 5i$, we have

$$\frac{1}{2 - 5i} \cdot \frac{2 + 5i}{2 + 5i} = \frac{2 + 5i}{2^2 + 5^2} = \frac{2 + 5i}{29} = \frac{2}{29} + \frac{5}{29}i$$

Verify that $(2 - 5i)\left(\dfrac{2}{29} + \dfrac{5}{29}i\right) = 1$.

■

☑ **PROGRESS CHECK**

Write the following in the form $a + bi$.

(a) $\dfrac{4 - 2i}{5 + 2i}$ (b) $\dfrac{1}{2 - 3i}$ (c) $\dfrac{-3i}{3 + 5i}$

ANSWERS

(a) $\dfrac{16}{29} - \dfrac{18}{29}i$ (b) $\dfrac{2}{13} + \dfrac{3}{13}i$ (c) $-\dfrac{15}{34} - \dfrac{9}{34}i$

If we let $z = a + bi$, it is customary to write the conjugate $a - bi$ as $\bar{z}$. We will have need to use the following properties of complex numbers and their conjugates.

Properties of Complex Numbers	If z and w are complex numbers, then

(1) $\bar{z} = \bar{w}$ if and only if $z = w$

(2) $\bar{z} = z$ if and only if z is a real number

(3) $\overline{z + w} = \bar{z} + \bar{w}$

(4) $\overline{z \cdot w} = \bar{z} \cdot \bar{w}$

(5) $\overline{z^n} = \bar{z}^n$, n a positive integer

To prove Properties (1)–(5), let $z = a + bi$ and $w = c + di$. Properties (1) and (2) follow directly from the definition of equality of complex numbers. To prove Property (3), we note that $z + w = (a + c) + (b + d)i$. Then, by the definition of a complex conjugate,

$$\overline{z + w} = (a + c) - (b + d)i$$
$$= (a - bi) + (c - di)$$
$$= \bar{z} + \bar{w}$$

Properties (4) and (5) can be proved in a similar manner, although a rigorous proof of Property (5) requires the use of mathematical induction, a method we will discuss in a later chapter.

EXAMPLE 2 Properties of Complex Conjugates

If $z = 1 + 2i$ and $w = 3 - i$, verify that

(a) $\overline{z + w} = \bar{z} + \bar{w}$ **(b)** $\overline{z \cdot w} = \bar{z} \cdot \bar{w}$ **(c)** $\overline{z^2} = \bar{z}^2$

SOLUTION

(a) $z + w = (1 + 2i) + (3 - i)$ $\bar{z} = 1 - 2i$

$\qquad\qquad = 4 + i$ $\bar{w} = 3 + i$

$\overline{z + w} = 4 - i$ $\bar{z} + \bar{w} = (1 - 2i) + (3 + i)$

$\qquad\qquad\qquad\qquad\qquad\qquad\qquad = 4 - i$

Thus $\overline{z + w} = \bar{z} + \bar{w}$.

(b) $z \cdot w = (1 + 2i)(3 - i) = 5 + 5i$ $\bar{z} = 1 - 2i$

$\overline{z \cdot w} = 5 - 5i$ $\bar{w} = 3 + i$

$\qquad\qquad\qquad\qquad\qquad\qquad\qquad \bar{z} \cdot \bar{w} = (1 - 2i)(3 + i)$

$\qquad\qquad\qquad\qquad\qquad\qquad\qquad\qquad = 5 - 5i$

Thus, $\overline{z \cdot w} = \bar{z} \cdot \bar{w}$.

(c) $z^2 = (1 + 2i)(1 + 2i)$ $\bar{z}^2 = (1 - 2i)(1 - 2i)$

$\qquad = -3 + 4i$ $= -3 - 4i$

$\overline{z^2} = -3 - 4i$

Thus, $\overline{z^2} = \bar{z}^2$.

◼

✓ PROGRESS CHECK

If $z = 2 + 3i$ and $w = \frac{1}{2} - 2i$, verify that

(a) $\overline{z + w} = \bar{z} + \bar{w}$ **(b)** $\overline{z \cdot w} = \bar{z} \cdot \bar{w}$ **(c)** $\overline{z^2} = \bar{z}^2$ **(d)** $\overline{w^3} = \bar{w}^3$

THE COMPLEX PLANE

We associate the complex number $a + bi$ with the point in the plane whose coordinates are (a, b). Figure 1 illustrates the geometric representation of several complex numbers. Conversely, every point (a, b) in the plane represents a complex number, $a + bi$. When a rectangular coordinate system is used to represent complex numbers, it is called a **complex plane** and the x- and y-axes are called the **real axis** and the **imaginary axis,** respectively.

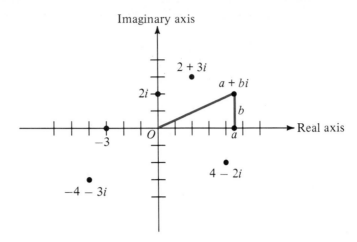

FIGURE 1 **The Complex Plane**

ABSOLUTE VALUE

We can extend the concept of absolute value to complex numbers in a natural manner. Since $|x|$ represents the distance on a real number line from the origin to a point that corresponds to x, it would be consistent to define the **absolute value $|a + bi|$** as the distance from the origin to the point corresponding to $a + bi$. Applying the distance formula (see Figure 1) we are led to the following definition.

> The absolute value of a complex number $a + bi$ is denoted by $|a + bi|$ and is defined by
>
> $$|a + bi| = \sqrt{a^2 + b^2}$$

EXAMPLE 3 Absolute Value of a Complex Number

Find the absolute value of each of the following complex numbers.

(a) $2 - 3i$ (b) $4i$ (c) -2

SOLUTION

Applying the definition of absolute value,

(a) $|2 - 3i| = \sqrt{4 + 9} = \sqrt{13}$ (b) $|4i| = \sqrt{0 + 16} = 4$

(c) $|-2| = \sqrt{4 + 0} = 2$

∎

EXERCISE SET 8.1

In Exercises 1–6 multiply by the conjugate and simplify.

1. $2 - i$ 2. $3 + i$ 3. $3 + 4i$ 4. $2 - 3i$

5. $-4 - 2i$ 6. $5 + 2i$

In Exercises 7–15 perform the indicated operations and write the answer in the form $a + bi$.

7. $\dfrac{2 + 5i}{1 - 3i}$

8. $\dfrac{1 + 3i}{2 - 5i}$

9. $\dfrac{3 - 4i}{3 + 4i}$

10. $\dfrac{4 - 3i}{4 + 3i}$

11. $\dfrac{3 - 2i}{2 - i}$

12. $\dfrac{2 - 3i}{3 - i}$

13. $\dfrac{2 + 5i}{3i}$

14. $\dfrac{5 - 2i}{-3i}$

15. $\dfrac{4i}{2 + i}$

In Exercises 16–21 find the reciprocal and write the answer in the form $a + bi$.

16. $3 + 2i$

17. $4 + 3i$

18. $\frac{1}{2} - i$

19. $1 - \frac{1}{3}i$

20. $-7i$

21. $-5i$

22. Prove that the multiplicative inverse of the complex number $a + bi$ (a and b not both 0) is

$$\frac{a}{a^2 + b^2} - \frac{b}{a^2 + b^2}i$$

23. If z and w are complex numbers, prove that

$$\overline{z \cdot w} = \bar{z} \cdot \bar{w}$$

24. If z is a complex number, verify that $\overline{z^2} = \bar{z}^2$ and $\overline{z^3} = \bar{z}^3$.

In Exercises 25–30 find the absolute value of the given complex number.

25. $3 - 2i$

26. $-7 + 6i$

27. $1 + i$

28. $\dfrac{1}{2} + \dfrac{1}{2}i$

29. $-6 - 2i$

30. $3 - i$

8.2
TRIGONOMETRY AND COMPLEX NUMBERS

The representation of a complex number as a point in a coordinate plane can be used to link complex numbers with trigonometry of the right triangle. In Figure 2, $a + bi$ is any nonzero complex number, and we consider the line segment OP to be the terminal side of an angle θ in standard position. Using trigonometry of the right triangle, we see that

$$a = r \cos \theta \qquad \text{and} \qquad b = r \sin \theta$$

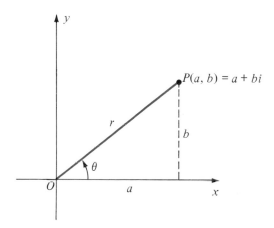

FIGURE 2 Polar Form

We may then write

$$a + bi = (r \cos \theta) + (r \sin \theta)i$$

or

$$a + bi = r(\cos \theta + i \sin \theta) \qquad (1)$$

where $r = \overline{OP} = |a + bi| = \sqrt{a^2 + b^2}$. If $a + bi = 0$, then $r = 0$, and θ may assume any value.

Equation (1) is known as the **trigonometric form** or **polar form** of a complex number. Since we have an infinite number of choices for the angle θ, the polar form of a complex number is not unique. We call r the **modulus** and θ the **argument** of the complex number $r(\cos \theta + i \sin \theta)$. If θ is in the interval $(-\pi, \pi]$, then θ is called the **principal argument.**

EXAMPLE 1   Converting to Trigonometric Form
Write the complex number $-2 + 2i$ in trigonometric form.

SOLUTION
The geometric representation is shown in Figure 3. The modulus of $-2 + 2i$ is

$$r = |-2 + 2i| = \sqrt{4 + 4} = 2\sqrt{2}$$

The principal argument θ is an angle in the second quadrant such that

$$\tan \theta = \frac{2}{-2} = -1$$

Thus, $\theta = 135°$, and using the trigonometric form of a complex number of Equation (1), we have

$$-2 + 2i = 2\sqrt{2}(\cos 135° + i \sin 135°)$$

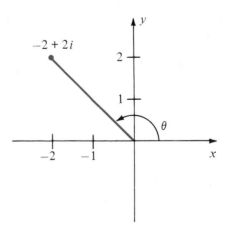

FIGURE 3   Trigonometric Form of $-2 + 2i$

PROGRESS CHECK

Write the complex number $1 - \sqrt{3}i$ in trigonometric form.

ANSWER

$2(\cos 300° + i \sin 300°)$

EXAMPLE 2   Converting from Trigonometric Form

Write the complex number $2\sqrt{3}(\cos 150° + i \sin 150°)$ in the form $a + bi$.

SOLUTION

We need only substitute $\cos 150° = -\dfrac{\sqrt{3}}{2}$ and $\sin 150° = \dfrac{1}{2}$. Thus,

$$2\sqrt{3}(\cos 150° + i \sin 150°) = 2\sqrt{3}\left(-\frac{\sqrt{3}}{2} + \frac{1}{2}i\right)$$

$$= -3 + \sqrt{3}i$$

PROGRESS CHECK

Write the complex number $\sqrt{2}\left(\cos\dfrac{\pi}{4} + i \sin\dfrac{\pi}{4}\right)$ in the form $a + bi$.

ANSWER

$1 + i$

MULTIPLICATION AND DIVISION

Why have we introduced the trigonometric form of a complex number? Because multiplication and division of complex numbers is very simple when this form is used. If $r_1(\cos\theta_1 + i\sin\theta_1)$ and $r_2(\cos\theta_2 + i\sin\theta_2)$ are any two complex numbers, the rules for their multiplication and division are

$$r_1(\cos\theta_1 + i\sin\theta_1) \cdot r_2(\cos\theta_2 + i\sin\theta_2)$$
$$= r_1 r_2[\cos(\theta_1 + \theta_2) + i\sin(\theta_1 + \theta_2)] \tag{2}$$

$$\frac{r_1(\cos\theta_1 + i\sin\theta_1)}{r_2(\cos\theta_2 + i\sin\theta_2)} = \frac{r_1}{r_2}[\cos(\theta_1 - \theta_2) + i\sin(\theta_1 - \theta_2)] \tag{3}$$

Note that the rule for multiplication requires the multiplication of the moduli and addition of the arguments. To prove this we see that

$$r_1(\cos\theta_1 + i\sin\theta_1) \cdot r_2(\cos\theta_2 + i\sin\theta_2)$$
$$= r_1 r_2[(\cos\theta_1\cos\theta_2 - \sin\theta_1\sin\theta_2) + i(\sin\theta_1\cos\theta_2 + \cos\theta_1\sin\theta_2)]$$
$$= r_1 r_2[\cos(\theta_1 + \theta_2) + i\sin(\theta_1 + \theta_2)]$$

where the last step results from the addition formulas.

The rule for division requires the division of moduli and the subtraction of the arguments. The proof is left as an exercise.

EXAMPLE 3 Product of Complex Numbers

Find the product of the complex numbers $1 + i$ and $-2i$ (a) by writing the numbers in trigonometric form and (b) by multiplying the numbers algebraically.

SOLUTION

(a) The trigonometric forms of these complex numbers are

$$1 + i = \sqrt{2}(\cos 45° + i \sin 45°)$$

and

$$-2i = 2(\cos 270° + i \sin 270°)$$

Multiplying, we have

$$\sqrt{2}(\cos 45° + i \sin 45°) \cdot 2(\cos 270° + i \sin 270°)$$
$$= 2\sqrt{2}(\cos 315° + i \sin 315°)$$
$$= 2\sqrt{2}\left(\frac{\sqrt{2}}{2} - i\frac{\sqrt{2}}{2}\right)$$
$$= 2 - 2i$$

(b) Multiplying algebraically,

$$(1 + i)(-2i) = -2i - 2i^2 = -2i + 2 = 2 - 2i$$

∎

PROGRESS CHECK

Express the complex numbers $1 + \sqrt{3}i$ and $1 - \sqrt{3}i$ in trigonometric form and find their product.

ANSWER

2(cos 60° + i sin 60°); 2(cos 300° + i sin 300°); 4

DE MOIVRE'S THEOREM

Since exponentiation is repeated multiplication, we are led to anticipate a simple result when a complex number in trigonometric form is raised to a power. The theorem that states this result is credited to Abraham De Moivre, a French mathematician. In this theorem $r(\cos \theta + i \sin \theta)$ is a complex number and n is a natural number.

De Moivre's Theorem	$[r(\cos \theta + i \sin \theta)]^n = r^n(\cos n\theta + i \sin n\theta)$

We can verify the theorem for some values of n. Thus, by Equation (2),

$$[r(\cos \theta + i \sin \theta)]^2 = r(\cos \theta + i \sin \theta) \cdot r(\cos \theta + i \sin \theta)$$
$$= r^2[\cos(\theta + \theta) + i \sin(\theta + \theta)]$$
$$= r^2(\cos 2\theta + i \sin 2\theta)$$

which is precisely what we obtain by using De Moivre's theorem. If we multiply again by $r(\cos\theta + i\sin\theta)$ and again apply Equation (2), we have

$$[r(\cos\theta + i\sin\theta)]^3 = r^2(\cos 2\theta + i\sin 2\theta) \cdot r(\cos\theta + i\sin\theta)$$
$$= r^3(\cos 3\theta + i\sin 3\theta)$$

Thus, De Moivre's theorem seems "reasonable." A rigorous *proof* requires the application of the method of mathematical induction, which will be discussed in a later chapter.

EXAMPLE 4   Applying De Moivre's Theorem
Evaluate $(1 - i)^{10}$.

SOLUTION
Writing $1 - i$ in trigonometric form we have

$$1 - i = \sqrt{2}(\cos 315° + i\sin 315°)$$

and

$$(1 - i)^{10} = [\sqrt{2}(\cos 315° + i\sin 315°)]^{10}$$

Applying De Moivre's theorem,

$$(1 - i)^{10} = (\sqrt{2})^{10}[\cos 3150° + i\sin 3150°]$$
$$= 32[\cos 270° + i\sin 270°]$$
$$= 32[0 + i(-1)] = -32i$$

■

PROGRESS CHECK
Evaluate $(\sqrt{3} + i)^6$.

ANSWER
-64

---

nth ROOT OF A COMPLEX NUMBER

Recall that a real number a is said to be an nth root of the real number b if $a^n = b$ for a positive integer n. In an analogous manner, we say that the complex number u is an **nth root** of the nonzero complex number z if $u^n = z$. If we express u and z in trigonometric form as

$$u = s(\cos\phi + i\sin\phi) \qquad z = r(\cos\theta + i\sin\theta) \tag{4}$$

we can then apply De Moivre's theorem to obtain

$$u^n = s^n(\cos n\phi + i\sin n\phi) = r(\cos\theta + i\sin\theta) \tag{5}$$

Since the two complex numbers u^n and z are equal, they are represented by the same point in the complex plane. Hence, the moduli must be equal, since the modulus is the distance of the point from the origin. Therefore, $s^n = r$ or

$$s = \sqrt[n]{r}$$

Since $z \neq 0$, we know that $r \neq 0$. We may therefore divide Equation (5) by r to obtain

$$\cos n\phi + i \sin n\phi = \cos \theta + i \sin \theta$$

By the definition of equality of complex numbers, we must have

$$\cos n\phi = \cos \theta \qquad \sin n\phi = \sin \theta$$

Since both sine and cosine are periodic functions with period 2π, we conclude that

$$n\phi = \theta + 2\pi k$$

or

$$\phi = \frac{\theta + 2\pi k}{n}$$

where k is an integer. Substituting for s and for ϕ in the trigonometric form of u given in Equation (4) yields

---

The nth Roots of a Complex Number

The n distinct roots of $r(\cos \theta + i \sin \theta)$ are given by

$$\sqrt[n]{r}\left[\cos\left(\frac{\theta + 2\pi k}{n}\right) + i \sin\left(\frac{\theta + 2\pi k}{n}\right)\right]$$

where $k = 0, 1, 2, \ldots, n - 1$.

Note that when k exceeds $n - 1$, we repeat a previous root. For example, when $k = n$, the angle is

$$\frac{\theta + 2\pi n}{n} = \frac{\theta}{n} + 2\pi = \frac{\theta}{n}$$

which is the same result that is obtained when $k = 0$.

EXAMPLE 5   The Cube Roots of a Complex Number
Find the cube roots of $-8i$.

SOLUTION
For $z = a + bi = 0 - 8i$, we have $a = 0$, $b = -8$, and thus

$$r = \sqrt{a^2 + b^2} = \sqrt{0^2 + (-8)^2} = 8$$

From Figure 4, we see that $\theta = 270°$ is an acceptable value for θ. Then, in trigonometric form,

$$-8i = 8(\cos 270° + i \sin 270°)$$

With $r = 8$, $\theta = 270°$, and $n = 3$, the cube roots are

$$\sqrt[3]{8}\left[\cos\left(\frac{270° + 360°k}{3}\right) + i \sin\left(\frac{270° + 360°k}{3}\right)\right]$$

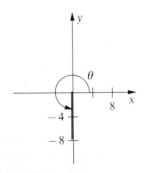

FIGURE 4   Trigonometric Form of $-8i$

for $k = 0, 1, 2$. Substituting for each value of k we have

$$2(\cos 90° + i \sin 90°) = 2i$$
$$2(\cos 210° + i \sin 210°) = -\sqrt{3} - i$$
$$2(\cos 330° + i \sin 330°) = \sqrt{3} - i$$

When $z = 1$, we call the n distinct nth roots the **nth roots of unity.**

EXAMPLE 6   The Fourth Roots of Unity
Find the four fourth roots of unity.

SOLUTION
In trigonometric form,

$$1 = 1(\cos 0° + i \sin 0°)$$

so that $r = 1$, $\theta = 0°$, and $n = 4$. The fourth roots are then given by

$$\sqrt[4]{1}\left[\cos\left(\frac{0° + 360°k}{4}\right) + i \sin\left(\frac{0° + 360°k}{4}\right)\right]$$

for $k = 0, 1, 2, 3$. Substituting these values for k yields

$$\cos 0° + i \sin 0° = 1$$
$$\cos 90° + i \sin 90° = i$$
$$\cos 180° + i \sin 180° = -1$$
$$\cos 270° + i \sin 270° = -i$$

It is easy to verify that each of these answers is indeed a fourth root of unity.

PROGRESS CHECK
Find the two square roots of $\dfrac{\sqrt{3}}{2} - \dfrac{1}{2}i$. Express the answers in trigonometric form.

ANSWER
$\cos 165° + i \sin 165°, \cos 345° + i \sin 345°$

EXERCISE SET 8.2
In Exercises 1–8 express the given complex number in trigonometric form.

1. $3 - 3i$
2. $2 + 2i$
3. $\sqrt{3} - i$
4. $-2 - 2\sqrt{3}i$
5. $-1 + i$
6. $-2i$
7. -4
8. $3i$

In Exercises 9–14 convert the given complex number from trigonometric form to the algebraic form $a + bi$.

9. $4(\cos 180° + i \sin 180°)$

10. $\dfrac{1}{2}\left(\cos\dfrac{\pi}{2} + i \sin\dfrac{\pi}{2}\right)$

11. $\sqrt{2}(\cos 135° + i \sin 135°)$ 12. $2(\cos 120° + i \sin 120°)$

13. $5\left(\cos\dfrac{3\pi}{2} + i \sin\dfrac{3\pi}{2}\right)$ 14. $4(\cos 240° + i \sin 240°)$

In Exercises 15–18 find the product of the given complex numbers. Express the answers in trigonometric form.

15. $2(\cos 150° + i \sin 150°) \cdot 3(\cos 210° + i \sin 210°)$ 16. $3(\cos 120° + i \sin 120°) \cdot 3(\cos 150° + i \sin 150°)$

17. $2(\cos 10° + i \sin 10°) \cdot (\cos 320° + i \sin 320°)$ 18. $3(\cos 230° + i \sin 230°) \cdot 4(\cos 250° + i \sin 250°)$

In Exercises 19–24 express the given complex numbers in trigonometric form, compute the product, and write the answer in the form $a + bi$.

19. $1 - i, 2i$ 20. $-\sqrt{3} + i, -2$ 21. $-2 + 2\sqrt{3}i, 3 + 3i$ 22. $1 - \sqrt{3}i, 1 + \sqrt{3}i$

23. $5, -2 - 2i$ 24. $-4i, -3i$

In Exercises 25–30 use De Moivre's theorem to express the given number in the form $a + bi$.

25. $(-2 + 2i)^6$ 26. $(\sqrt{3} - i)^{10}$ 27. $(1 - i)^9$ 28. $(-1 + \sqrt{3}i)^{10}$

29. $(-1 - i)^7$ 30. $(-\sqrt{2} + \sqrt{2}i)^6$

In Exercises 31–34 find the indicated roots of the given complex number. Express the answer in the indicated form.

31. The fourth roots of -16; algebraic form $a + bi$.   32. The square roots of -25; trigonometric form.

33. The square roots of $1 - \sqrt{3}i$; trigonometric form.   34. The four fourth roots of unity; algebraic form.

In Exercises 35–38 determine all roots of the given equation.

35. $x^3 + 8 = 0$ 36. $x^3 + 125 = 0$ 37. $x^4 - 16 = 0$ 38. $x^4 + 16 = 0$

39. Prove $\dfrac{r_1(\cos\theta_1 + i \sin\theta_1)}{r_2(\cos\theta_2 + i \sin\theta_2)} = \dfrac{r_1}{r_2}[\cos(\theta_1 - \theta_2) + i \sin(\theta_1 - \theta_2)]$.

40. Prove that the sum of the nth roots of unity is 0 for every positive integer n.

8.3
THE FUNDAMENTAL THEOREM OF ALGEBRA

We began this chapter with the question, Does a polynomial equation always have a root? The answer was supplied by Carl Friedrich Gauss in his doctoral dissertation in 1799. Unfortunately, the proof of this theorem is beyond the scope of this book.

The Fundamental Theorem of Algebra—Part I	Every polynomial $P(x)$ of degree $n \geq 1$ has at least one root among the complex numbers.

Note that the root guaranteed by this theorem may be a real number since the real numbers are a subset of the complex number system.

Gauss, who is considered by many to have been the greatest mathematician of all time, supplied the proof at age 22. The importance of the theorem is reflected in its title. We now see why it was necessary to create the complex numbers and that we need not create any other number system beyond the complex numbers in order to solve polynomial equations.

How many roots does a polynomial of degree n have? The next theorem will bring us closer to an answer.

Linear Factor Theorem	A polynomial $P(x)$ of degree $n \geq 1$ can be written as the product of n linear factors. $$P(x) = a(x - r_1)(x - r_2)\ldots(x - r_n)$$

Note that a is the leading coefficient of $P(x)$ and that $r_1, r_2, \ldots, r_n$ are, in general, complex numbers.

To prove this theorem, we first note that the Fundamental Theorem of Algebra guarantees us the existence of a root r_1. By the Factor Theorem, $x - r_1$ is a factor and, consequently,

$$P(x) = (x - r_1)Q_1(x) \tag{1}$$

where $Q_1(x)$ is a polynomial of degree $n - 1$. If $n - 1 \geq 1$, then $Q_1(x)$ must have a root r_2. Thus

$$Q_1(x) = (x - r_2)Q_2(x) \tag{2}$$

where $Q_2(x)$ is of degree $n - 2$. Substituting in Equation (1) for $Q_1(x)$ we have

$$P(x) = (x - r_1)(x - r_2)Q_2(x) \tag{3}$$

This process is repeated n times until $Q_n(x) = a$ is of degree 0. Hence,

$$P(x) = a(x - r_1)(x - r_2)\ldots(x - r_n) \tag{4}$$

Since a is the leading coefficient of the polynomial on the right side of Equation (4), it must also be the leading coefficient of $P(x)$.

EXAMPLE 1   Finding a Polynomial with Given Roots
Find the polynomial $P(x)$ of degree 3 that has the roots -2, i, and $-i$, and satisfies $P(1) = -3$.

SOLUTION
Since -2, i, and $-i$ are roots of $P(x)$, we may write

$$P(x) = a(x + 2)(x - i)(x + i)$$

To find the constant a, we use the condition $P(1) = -3$.

$$P(1) = -3 = a(1 + 2)(1 - i)(1 + i) = 6a$$

$$a = -\frac{1}{2}$$

so that

$$P(x) = -\frac{1}{2}(x + 2)(x - i)(x + i)$$

∎

MULTIPLICITY OF A ROOT   Recall that the roots of a polynomial need not be distinct from each other. The polynomial

$$P(x) = x^2 - 2x + 1$$

can be written in the factored form

$$P(x) = (x - 1)(x - 1)$$

which shows that the roots of $P(x)$ are 1 and 1. Since a root is associated with a factor and a factor may be repeated, we may have repeated roots. If the factor $x - r$ appears k times, we say that r is a **root of multiplicity k.**

It is now easy to establish the following, which may be thought of as an alternative form of the Fundamental Theorem of Algebra.

The Fundamental Theorem of Algebra—Part II   If $P(x)$ is a polynomial of degree $n \geq 1$, then $P(x)$ has precisely n roots among the complex numbers when a root of multiplicity k is counted k times.

We may prove this theorem as follows. If we write $P(x)$ in the form of Equation (4), we see that $r_1, r_2, \ldots, r_n$ are roots of the equation $P(x) = 0$ and hence there exist n roots. If there is an additional root r that is distinct from the roots $r_1, r_2, \ldots, r_n$, then $r - r_1$, $r - r_2, \ldots, r - r_n$ are all different from 0. Substituting r for x in Equation (4) yields

$$P(r) = a(r - r_1)(r - r_2)\ldots(r - r_n) \tag{5}$$

which cannot equal 0, since the product of nonzero numbers cannot equal 0. Thus, $r_1, r_2, \ldots, r_n$ are roots of $P(x)$ and there are no other roots. We conclude that $P(x)$ has precisely n roots.

EXAMPLE 2   Polynomials in Factored Form

Find all roots of the polynomial

$$P(x) = \left(x - \frac{1}{2}\right)^3 (x + i)(x - 5)^4$$

SOLUTION

The distinct roots are $\frac{1}{2}$, $-i$, and 5. Further, $\frac{1}{2}$ is a root of multiplicity 3; $-i$ is a root of multiplicity 1; 5 is a root of multiplicity 4.

∎

EXAMPLE 3   Applying the Fundamental Theorem

If -1 is a root of multiplicity 2 of $P(x) = x^4 + 4x^3 + 2x^2 - 4x - 3$, find the remaining roots and write $P(x)$ as a product of linear factors.

SOLUTION

Since -1 is a double root of $P(x)$, then $(x + 1)^2$ is a factor of $P(x)$. Therefore,

$$P(x) = (x + 1)^2 Q(x)$$

or

$$P(x) = (x^2 + 2x + 1)Q(x)$$

Using polynomial division, we can divide both sides of the last equation by $x^2 + 2x + 1$ to obtain

$$Q(x) = \frac{x^4 + 4x^3 + 2x^2 - 4x - 3}{x^2 + 2x + 1}$$

$$- x^2 + 2x - 3$$

$$= (x - 1)(x + 3)$$

The roots of the depressed equation $Q(x) = 0$ are 1 and -3, and these are the remaining roots of $P(x)$. By the Linear Factor Theorem,

$$P(x) = (x + 1)^2(x - 1)(x + 3)$$

■

PROGRESS CHECK

If -2 is a root of multiplicity 2 of $P(x) = x^4 + 4x^3 + 5x^2 + 4x + 4$, write $P(x)$ as a product of linear factors.

ANSWER

$P(x) = (x + 2)(x + 2)(x + i)(x - i)$

We know from the quadratic formula that if a quadratic equation with real coefficients has a complex root $a + bi$, then the conjugate $a - bi$ is the other root. The following theorem extends this result to a polynomial of degree n with real coefficients.

Conjugate Roots Theorem	If $P(x)$ is a polynomial of degree $n \geq 1$ with real coefficients, and if $a + bi, b \neq 0$, is a root of $P(x)$, then the complex conjugate $a - bi$ is also a root of $P(x)$.

PROOF OF CONJUGATE ROOTS THEOREM (Optional)

To prove the Conjugate Roots Theorem, we let $z = a + bi$ and make use of the properties of complex conjugates developed earlier in this section. We may write

$$P(x) = a_n x^n + a_{n-1}x^{n-1} + \cdots + a_1 x + a_0 \qquad (6)$$

and, since z is a root of $P(x)$,

$$a_n z^n + a_{n-1}z^{n-1} + \cdots + a_n z + a_0 = 0 \qquad (7)$$

But if $z = w$, then $\bar{z} = \bar{w}$. Applying this property of complex numbers to both sides of Equation (7), we have

$$\overline{a_n z^n + a_{n-1}z^{n-1} + \cdots + a_1 z + a_0} = \bar{0} = 0 \qquad (8)$$

We also know that $\overline{z + w} = \bar{z} + \bar{w}$. Applying this property to the left side of Equation (8) we see that

$$\overline{a_n z^n} + \overline{a_{n-1}z^{n-1}} + \cdots + \overline{a_1 z} + \overline{a_0} = 0 \qquad (9)$$

Further, $\overline{z \cdot w} = \bar{z} \cdot \bar{w}$, so we may rewrite Equation (9) as

$$\overline{a_n}\overline{z^n} + \overline{a_{n-1}}\overline{z^{n-1}} + \cdots + \overline{a_1}\bar{z} + \overline{a_0} = 0 \tag{10}$$

Since $a_0, a_1, \ldots, a_n$ are all real numbers, we know that $\overline{a_0} = a_0$, $\overline{a_1} = a_1, \ldots, \overline{a_n} = a_n$. Finally, we use the property $\overline{z^n} = \bar{z}^n$ to rewrite Equation (10) as

$$a_n\bar{z}^n + a_{n-1}\bar{z}^{n-1} + \cdots + a_1\bar{z} + a_0 = 0$$

which establishes that $\bar{z}$ is a root of $P(x)$.

EXAMPLE 4   Applying the Conjugate Roots Theorem
Find a polynomial $P(x)$ with real coefficients that is of degree 3 and whose roots include -2 and $1 - i$.

SOLUTION
Since $1 - i$ is a root, it follows from the Conjugate Roots Theorem that $1 + i$ is also a root of $P(x)$. By the Factor Theorem, $(x + 2)$, $[x - (1 - i)]$, and $[x - (1 + i)]$ are factors of $P(x)$. Therefore,

$$\begin{aligned} P(x) &= (x + 2)[x - (1 - i)][x - (1 + i)] \\ &= (x + 2)(x^2 - 2x + 2) \\ &= x^3 - 2x + 4 \end{aligned}$$

■

PROGRESS CHECK
Find a polynomial $P(x)$ with real coefficients that is of degree 4 and whose roots include i and $-3 + i$.

ANSWER
$P(x) = x^4 + 6x^3 + 11x^2 + 6x + 10$

The following is a corollary of the Conjugate Roots Theorem.

> A polynomial $P(x)$ of degree $n \geq 1$ with real coefficients can be written as a product of linear and quadratic factors with real coefficients so that the quadratic factors have no real roots.

By the Linear Factor Theorem, we may write

$$P(x) = a(x - r_1)(x - r_2) \cdots (x - r_n)$$

where $r_1, r_2, \ldots, r_n$ are the n roots of $P(x)$. Of course, some of these roots may be complex numbers. A complex root $a + bi$, $b \neq 0$, may be paired with its conjugate $a - bi$ to provide the quadratic factor

$$[x - (a + bi)][x - (a - bi)] = x^2 - 2ax + a^2 + b^2 \tag{11}$$

A "NATURAL" MATHEMATICIAN

Srinivasa Ramanujan was born in India in 1887. At age 15 he borrowed a copy of Carr's *Synopsis of Pure Mathematics* from a library. Thus began what is possibly the strangest story in the history of mathematics.

Through the study of this one-volume compendium of theorems, which was already 40 years out-of-date, Ramanujan jumped to the forefront of the mathematicians of his time. With no formal training and little idea of what constituted a mathematical proof, this "natural" mathematician presented to G. H. Hardy of Cambridge a list of 120 theorems, many of remarkable difficulty and insight.

Hardy brought Ramanujan to England in 1913 and acted as his tutor, friend, and coauthor until Ramanujan's return to India, where he died in 1920 at age 33. When visiting Ramanujan in an English hospital, Hardy remarked that the taxicab he had ridden in was No. 1729, a rather dull number.

"No," replied Ramanujan, "it is a very interesting number. It is the smallest number that can be written as the sum of two cubes in two different ways."

Ramanujan was asserting that

$$1729 = a^3 + b^3 \qquad \text{and} \qquad 1729 = c^3 + d^3$$

where a, b, c and d are distinct positive integers, and that 1729 is the smallest number that can be written in this manner. By generating a list of cubes less than 1729, you can easily find a, b, c and d. You might want to devise a computer program to verify Ramanujan's assertion that 1729 is the smallest such number.

which has real coefficients. Thus, a *quadratic* factor with real coefficients results from each pair of complex conjugate roots; a *linear* factor with real coefficients results from each real root. Further, the discriminant of the quadratic factor in Equation (11) is $-4b^2$ and is therefore always negative, which shows that the quadratic factor has no real roots.

POLYNOMIALS WITH COMPLEX COEFFICIENTS

Although the definition of a polynomial given in Section 3.1 permits the coefficients to be complex numbers, we have limited our examples to polynomials with real coefficients because beginning calculus courses adhere to this restriction. To round out our work, we point out that both the Linear Factor Theorem and the Fundamental Theorem of Algebra hold for polynomials with complex coefficients.

On the other hand, the Conjugate Roots Theorem may not hold if the polynomial $P(x)$ has complex coefficients. To see this, consider the polynomial

$$P(x) = x - (2 + i)$$

which has a complex coefficient and has the root $2 + i$. Note that the complex conjugate $2 - i$ is *not* a root of $P(x)$ and that, therefore, the Conjugate Roots Theorem fails to apply to $P(x)$.

EXAMPLE 5   Polynomials with Complex Coefficients

Find a polynomial $P(x)$ of degree 2 that has the roots -1 and $1 - i$.

SOLUTION

Since -1 is a root of $P(x)$, $x + 1$ is a factor. Similarly, $[x - (1 - i)]$ is also a factor of $P(x)$. We can then write

$$P(x) = (x + 1)[x - (1 - i)]$$
$$= x^2 + ix - 1 + i$$

which is a polynomial of degree 2 (with complex coefficients) that has the desired roots.

∎

EXERCISE SET 8.3

In Exercises 1–6 find a polynomial $P(x)$ of lowest degree that has the indicated roots.

1. $2, -4, 4$
2. $5, -5, 1, -1$
3. $-1, -2, -3$
4. $-3, \sqrt{2}, -\sqrt{2}$
5. $4, 1 \pm \sqrt{3}$
6. $1, 2, 2 \pm \sqrt{2}$

In Exercises 7–10 find the polynomial $P(x)$ of lowest degree that has the indicated roots and satisfies the given condition.

7. $\frac{1}{2}, \frac{1}{2}, -2; P(2) = 3$
8. $3, 3, -2, 2; P(4) = 12$
9. $\sqrt{2}, -\sqrt{2}, 4; P(-1) = 5$
10. $\frac{1}{2}, -2, 5; P(0) = 5$

In Exercises 11–18 find the roots of the given equation.

11. $(x - 3)(x + 1)(x - 2) = 0$
12. $(x - 3)(x^2 - 3x - 4) = 0$
13. $(x + 2)(x^2 - 16) = 0$
14. $(x^2 - x)(x^2 - 2x + 5) = 0$
15. $(x^2 + 3x + 2)(2x^2 + x) = 0$
16. $(x^2 + x + 4)(x - 3)^2 = 0$
17. $(x - 5)^3(x + 5)^2 = 0$
18. $(x + 1)^2(x + 3)^4(x - 2) = 0$

In Exercises 19–22 find a polynomial that has the indicated roots and no others.

19. -2 of multiplicity 3
20. 1 of multiplicity 2, -4 of multiplicity 1
21. $\frac{1}{2}$ of multiplicity 2, -1 of multiplicity 2
22. -1 of multiplicity 2, 0 and 2 each of multiplicity 1

In Exercises 23–28 find a polynomial that has the indicated roots and no others.

23. $1 + 3i, -2$
24. $1, -1, 2 - i$
25. $1 + i, 2 - i$
26. $-2, 3, 1 + 2i$
27. -2 is a root of multiplicity 2, $3 - 2i$
28. 3 is a triple root, $-i$

In Exercises 29–34 use the given root(s) to help in writing the given equation as a product of linear and quadratic factors with real coefficients.

29. $x^3 - 7x^2 + 16x - 10 = 0; 3 - i$
30. $x^3 + x^2 - 7x + 65 = 0; 2 + 3i$
31. $x^4 + 4x^3 + 13x^2 + 18x + 20 = 0; -1 - 2i$
32. $x^4 + 3x^3 - 5x^2 - 29x - 30 = 0; -2 + i$
33. $x^5 + 3x^4 - 12x^3 - 42x^2 + 32x + 120 = 0; -3 - i, -2$
34. $x^5 - 8x^4 + 29x^3 - 54x^2 + 48x - 16 = 0; 2 + 2i, 2$

35. Write a polynomial $P(x)$ with complex coefficients that has the root $a + bi$, $b \neq 0$, and does not have $a - bi$ as a root.

36. Prove that a polynomial equation of degree 4 with real coefficients has 4 real roots, 2 real roots, or no real roots.

37. Prove that a polynomial equation of odd degree with real coefficients has at least one real root.

8.4
DESCARTES'S RULE OF SIGNS

In this section we will restrict our investigation to polynomials with real coefficients. Our objective is to obtain some information concerning the number of positive real roots and the number of negative real roots of such polynomials.

If the terms of a polynomial with real coefficients are written in descending order, then a **variation in sign** occurs whenever two successive terms have opposite signs. In determining the number of variations in sign, we ignore terms with zero coefficients. The polynomial

$$4x^5 - 3x^4 - 2x^2 + 1$$

has two variations in sign. The French mathematician René Descartes (1596–1650), who provided us with the foundations of analytic geometry, also gave us a theorem that relates the nature of the real roots of polynomials to the variations in sign. The proof of Descartes's theorem is outlined in Exercises 19–24.

Descartes's Rule of Signs	If $P(x)$ is a polynomial with real coefficients, then

(i) the number of positive roots either is equal to the number of variations in sign of $P(x)$ or is less than the number of variations in sign by an even number, and

(ii) the number of negative roots either is equal to the number of variations in sign of $P(-x)$ or is less than the number of variations in sign by an even number.

If it is determined that a polynomial of degree n has r real roots, then the remaining $n - r$ roots must be complex numbers.

To apply Descartes's Rule of Signs to the polynomial

$$P(x) = 3x^5 + 2x^4 - x^3 + 2x - 3$$

we first note that there are 3 variations in sign as indicated. Thus, either there are 3 positive roots or there is 1 positive root. Next, we form $P(-x)$,

$$P(-x) = 3(-x)^5 + 2(-x)^4 - (-x)^3 + 2(-x) - 3$$
$$= -3x^5 + 2x^4 + x^3 - 2x - 3$$

which can be obtained by negating the coefficients of the odd-power terms. We see that $P(-x)$ has two variations in sign and conclude that $P(x)$ has either 2 negative roots or no negative roots.

EXAMPLE 1 Applying Descartes's Rule of Signs
Use Descartes's Rule of Signs to analyze the roots of the equation

$$2x^5 + 7x^4 + 3x^2 - 2 = 0$$

SOLUTION
Since

$$P(x) = 2x^5 + 7x^4 + 3x^2 - 2$$

has 1 variation in sign, there is precisely 1 positive root. The polynomial $P(-x)$ is

formed—

$$P(-x) = -2x^5 + 7x^4 + 3x^2 - 2$$

and is seen to have 2 variations in sign, so that $P(-x)$ has either 2 negative roots or no negative roots. Since $P(x)$ has 5 roots, the possibilities are

1 positive root, 2 negative roots, 2 complex roots

1 positive root, 0 negative roots, 4 complex roots

■

 PROGRESS CHECK

Use Descartes's Rule of Signs to analyze the nature of the roots of the equation

$$x^6 + 5x^4 - 4x^2 - 3 = 0$$

ANSWER

1 positive root, 1 negative root, 4 complex roots

EXAMPLE 2   Using Descartes's Rule of Signs and the Rational Root Theorem

Write the polynomial

$$P(x) = 3x^4 + 2x^3 + 2x^2 + 2x - 1$$

as a product of linear and quadratic factors with real coefficients such that the quadratic factors have no real roots.

SOLUTION

We first use the Rational Root Theorem to list the possible rational roots.

possible numerators: ± 1 (factors of 1)

possible denominators: $\pm 1, \pm 3$ (factors of 3)

possible rational roots: $\pm 1, \pm \frac{1}{3}$

Next, we note that $P(x)$ has real coefficients and that Descartes's Rule of Signs may therefore be employed. Since $P(x)$ has 1 variation in sign, there is precisely 1 positive real root. If this real root is a rational number, it must be either $+1$ or $+\frac{1}{3}$. Trying $+1$, we quickly see that $P(1) = 8$ and that $+1$ is not a root. Using synthetic division,

$$
\begin{array}{r|rrrrr}
\frac{1}{3} & 3 & 2 & 2 & 2 & -1 \\
 & & 1 & 1 & 1 & 1 \\
\hline
 & 3 & 3 & 3 & 3 & 0 \\
\end{array}
$$

coefficients of depressed equation

we see that $\frac{1}{3}$ is a root and the depressed equation is

$$Q_1(x) = 3x^3 + 3x^2 + 3x + 3 = 0$$

which has the same roots as

$$Q_2(x) = x^3 + x^2 + x + 1 = 0$$

FERMAT'S LAST THEOREM

If you were asked to find natural numbers a, b, and c that satisfy the equation

$$a^2 + b^2 = c^2$$

you would have no trouble coming up with "triplets" such as $(3, 4, 5)$ and $(5, 12, 13)$. In fact, there are an infinite number of solutions, since any multiple of $(3, 4, 5)$ such as $(6, 8, 10)$ is also a solution.

Generalizing the above problem, suppose we seek natural numbers a, b, and c that satisfy the equation

a^n

$+$

b^n

$=$

c^n

$$a^n + b^n = c^n$$

for integer values of $n > 2$. Pierre Fermat, a great French mathematician of the seventeenth century, stated that there are no natural numbers a, b, and c that satisfy this equation for any integer $n > 2$. This seductively simple conjecture is known as Fermat's Last Theorem. Fermat wrote in his notebook that he had a proof but it was too long to include in the margin.

A proof of this theorem or a counterexample has eluded mathematicians for 300 years! In 1983, the German mathematician Gerd Faltings proved that the equation $x^n + y^n = 1$ has only a finite number of rational solutions, which may well be a step in establishing the theorem. In 1988, *The New York Times* published a report that a Japanese number theoretician had submitted a proof of the theorem for review by other mathematicians. The "proof" had errors, leaving Fermat's Last Theorem as an ongoing challenge.

Since any root of $Q_2(x)$ is also a root of $P(x)$ and since we have removed the only positive root, we know that $Q_2(x)$ cannot have any positive roots. (Verify that $Q_2(x)$ has no variations in sign!) However, forming

$$Q_2(-x) = -x^3 + x^2 - x + 1$$

we see that $Q_2(x)$ has at least 1 negative root. By the Rational Root Theorem, the only possible rational roots of $Q_2(x)$ are ± 1. Using synthetic division,

$$
\begin{array}{r|rrrr}
-1 & 1 & 1 & 1 & 1 \\
   &   & -1 & 0 & -1 \\
\hline
   & 1 & 0 & 1 & 0
\end{array}
$$

coefficients of depressed equation

we verify that -1 is indeed a root. Finally, we note that the depressed equation $x^2 + 1 = 0$ has no real roots, since the discriminant is negative. Thus,

$$P(x) = 3x^4 + 2x^3 + 2x^2 + 2x - 1 = 3\left(x - \frac{1}{3}\right)(x + 1)(x^2 + 1)$$

∎

EXERCISE SET 8.4

In Exercises 1–12 use Descartes's Rule of Signs to analyze the nature of the roots of the given equation. List all possibilities.

1. $3x^4 - 2x^3 + 6x^2 + 5x - 2 = 0$

2. $2x^6 + 5x^5 + x^3 - 6 = 0$

3. $x^6 + 2x^4 + 4x^2 + 1 = 0$

4. $3x^3 - 2x + 2 = 0$

5. $x^5 - 4x^3 + 7x - 4 = 0$

6. $2x^3 - 5x^2 + 8x - 2 = 0$

7. $5x^3 + 2x^2 + 7x - 1 = 0$

8. $x^5 + 6x^4 - x^3 - 2x - 3 = 0$

9. $x^4 - 2x^3 + 5x^2 + 2 = 0$

10. $3x^4 - 2x^3 - 1 = 0$

11. $x^8 + 7x^3 + 3x - 5 = 0$

12. $x^7 + 3x^5 - x^3 - x + 2 = 0$

In Exercises 13–18 use Descartes's Rule of Signs, the Rational Root Theorem, and the depressed equation to find all roots of the given equation.

13. $x^4 - 6x^3 + 10x^2 - 6x + 9 = 0$

14. $2x^4 - 3x^3 + 5x^2 - 6x + 2 = 0$

15. $x^4 - 6x^2 + 8 = 0$

16. $x^4 - 4x^3 + 7x^2 - 6x + 2 = 0$

17. $4x^4 + 4x^3 - 3x^2 - 4x - 1 = 0$

18. $x^5 + x^4 - 7x^3 - 11x^2 - 8x - 12 = 0$

19. Prove that if $P(x)$ is a polynomial with real coefficients and r is a positive root of $P(x)$, then the depressed equation

$$Q(x) = \frac{P(x)}{(x - r)}$$

has at least one fewer variation in sign than $P(x)$. (*Hint:* Assume the leading coefficient of $P(x)$ to be positive and use synthetic division to obtain $Q(x)$. Note that the coefficients of $Q(x)$ remain positive at least until there is a variation in sign in $P(x)$.)

20. Prove that if $P(x)$ is a polynomial with real coefficients, then the number of positive roots is not greater than the number of variations in sign in $P(x)$. (*Hint:* Let $r_1, r_2, \ldots, r_k$ be the positive roots of $P(x)$, and let

$$P(x) = (x - r_1)(x - r_2) \cdots (x - r_k)Q(x)$$

Use the result of Exercise 19 to show that $Q(x)$ has at least k fewer variations in sign than does $P(x)$.)

21. Prove that if $r_1, r_2, \ldots, r_k$ are positive numbers, then

$$P(x) = (x - r_1)(x - r_2) \cdots (x - r_k)$$

has alternating signs. (*Hint:* Use the result of Exercise 20.)

22. Prove that the number of variations in sign of a polynomial with real coefficients is even if the first and last coefficients have the same sign and is odd if they are of opposite sign.

23. Prove that if the number of positive roots of the polynomial $P(x)$ with real coefficients is less than the number of variations in sign, then it is less by an even number. (*Hint:* Write $P(x)$ as a product of linear factors corresponding to the positive and negative roots, and of quadratic factors corresponding to complex roots. Apply the results of Exercises 21 and 22.)

24. Prove that the positive roots of $P(-x)$ correspond to the negative roots of $P(x)$; that is, if $a > 0$ is a root of $P(-x)$, then $-a$ is a root of $P(x)$.

TERMS AND SYMBOLS

absolute value of a
   complex number **358**
algebraic form **355**
argument **360**
complex conjugate **355**
complex plane **357**

De Moivre's
   theorem **362**
imaginary axis **357**
modulus **360**
nth roots of a complex
   number **363**

nth roots of unity **365**
polar form **360**
principal argument **360**
real axis **357**
root of multiplicity
   k **368**

trigonometric form **360**
variation in sign **373**
$\bar{z}$ **356**

KEY IDEAS FOR REVIEW

Topic	Page	Key Idea
Trigonometric form of $a + bi$	360	The complex number $a + bi$ can be associated with the point $P(a, b)$. The trigonometric or polar form of the complex number $a + bi$ is given by $$a + bi = r(\cos \theta + i \sin \theta)$$ where r is the length of the line segment OP and θ is the measure of the angle in standard position whose terminal side is OP.
De Moivre's theorem	362	The trigonometric form of a complex number is useful since multiplication and division of complex numbers take on simple forms. In particular, exponentiation of complex numbers is handled by De Moivre's theorem, which states $$[r(\cos \theta + i \sin \theta)]^n = r^n(\cos n\theta + i \sin n\theta)$$
nth roots of a complex number	363	The complex number u is an nth root of the complex number z if $u^n = z$. De Moivre's theorem can be used to find a formula for determining u.
Linear Factor Theorem	367	A polynomial $P(x)$ of degree $n \geq 1$ can be written as the product of n linear factors.
Fundamental Theorem of Algebra	368	If $P(x)$ is a polynomial of degree $n \geq 1$, then $P(x)$ has precisely n roots among the complex numbers.
Conjugate Roots Theorem	369	If $a + bi$, $b \neq 0$, is a root of the polynomial $P(x)$ with real coefficients, then $a - bi$ is also a root of $P(x)$.
Descartes's Rule of Signs	373	Relates the maximum number of positive roots and the maximum number of negative roots of the polynomial equation $P(x) = 0$ to the number of variations in sign in $P(x)$ and $P(-x)$, respectively.

REVIEW EXERCISES

Solutions to exercises whose numbers are in color are in the Solutions section in the back of the book.

8.1   In Exercises 1–3 write the given quotient in the form $a + bi$.

1. $\dfrac{3 - 2i}{4 + 3i}$

2. $\dfrac{2 + i}{-5i}$

3. $\dfrac{-5}{1 + i}$

In Exercises 4–6 write the reciprocal of the given complex number in the form $a + bi$.

4. $1 + 3i$

5. $-4i$

6. $2 - 5i$

In Exercises 7–9 determine the absolute value of the given complex number.

7. $2 - i$

8. $-3 + 2i$

9. $-4 - 5i$

8.2   In Exercises 10–13 convert from trigonometric to algebraic form and vice versa.

10. $-3 + 3i$

11. $2(\cos 90° + i \sin 90°)$

12. $\sqrt{2}(\cos 315° + i \sin 315°)$

13. -2

In Exercises 14–16 find the indicated product or quotient. Express the answer in trigonometric form.

14. $4(\cos 22° + i \sin 22°) \cdot 6(\cos 15° + i \sin 15°)$

15. $\dfrac{5(\cos 71° + i \sin 71°)}{3(\cos 50° + i \sin 50°)}$

16. $2(\cos 210° + i \sin 210°) \cdot (\cos 240° + i \sin 240°)$

In Exercises 17 and 18 use De Moivre's theorem to express the given number in the form $a + bi$.

17. $(3 - 3i)^5$

18. $[2(\cos 90° + i \sin 90°)]^3$

19. Express the two square roots of -9 in trigonometric form.

20. Determine all roots of the equation $x^3 - 1 = 0$.

8.3 In Exercises 21–23 find a polynomial of lowest degree that has the indicated roots.

21. $-3, -2, -1$ 22. $3, \pm\sqrt{-3}$

23. $-2, \pm\sqrt{3}, 1$

In Exercises 24–26 find a polynomial that has the indicated roots and no others.

24. $\frac{1}{2}$ of multiplicity 2, -1 of multiplicity 2

25. $i, -i$, each of multiplicity 2

26. -1 of multiplicity 3, 3 of multiplicity 1

8.4 In Exercises 27–30 use Descartes's Rule of Signs to determine the maximum number of positive and negative real roots of the given equation.

27. $x^4 - 2x - 1 = 0$

28. $x^5 - x^4 + 3x^3 - 4x^2 + x - 5 = 0$

29. $x^3 - 5 = 0$

30. $3x^4 - 2x^2 + 1 = 0$

In Exercises 31 and 32 find all roots of the given equation.

31. $6x^3 + 15x^2 - x - 10 = 0$

32. $2x^4 - 3x^3 - 10x^2 + 19x - 6 = 0$

PROGRESS TEST 8A

1. Write the quotient $\dfrac{1 - i}{3 + 2i}$ in the form $a + bi$.

2. Write the reciprocal of $-2 + i$ in the form $a + bi$.

3. Determine the absolute value of $3 - 4i$.

In Problems 4 and 5 find the indicated product or quotient. Express the answer in trigonometric form.

4. $\dfrac{1}{2}(\cos 14° + i \sin 14°) \cdot 10(\cos 72° + i \sin 72°)$

5. $\dfrac{3(\cos 85° + i \sin 85°)}{6(\cos 8° + i \sin 8°)}$

6. Use De Moivre's theorem to express

$$\left[\frac{1}{5}(\cos 120° + i \sin 120°)\right]^4$$

in the form $a + bi$.

7. Express the three cube roots of -27 in trigonometric form.

In Problems 8 and 9 find a polynomial of lowest degree that has the indicated roots.

8. $-2, 1, 3$ 9. $-1, 1, 3 \pm \sqrt{2}$

In Problems 10 and 11 find the roots of the given equation.

10. $(x^2 + 1)(x - 2) = 0$

11. $(x + 1)^2(x - 3x - 2) = 0$

In Problems 12–14 find a polynomial that has the indicated roots and no others.

12. -3 of multiplicity 2; 1 of multiplicity 3

13. $-\frac{1}{4}$ of multiplicity 2; $i, -i$, and 1

14. $i, 1 + i$

15. If $2 + i$ is a root of $x^3 - 6x^2 + 13x - 10 = 0$, write the equation as a product of linear and quadratic factors with real coefficients.

In Problems 16 and 17 determine the maximum number of roots, of the type indicated, of the given equation.

16. $2x^5 - 3x^4 + 1 = 0$; positive real roots

17. $3x^4 + 2x^3 - 2x^2 - 1 = 0$; negative real roots

18. Find all roots of the equation

$$3x^4 + 7x^3 - 3x^2 + 7x - 6 = 0$$

PROGRESS TEST 8B

1. Write the quotient $\dfrac{-1}{2 + 2i}$ in the form $a + bi$.

2. Write the reciprocal of $3 - 4i$ in the form $a + bi$.

3. Determine the absolute of $-2 + 2i$.

In Problems 4 and 5 find the indicated product or quotient. Express the answer in trigonometric form.

4. $(\cos 125° + i \sin 125°) \cdot 5(\cos 125° + i \sin 125°)$

5. $\dfrac{\frac{1}{2}(\cos 67° + i \sin 67°)}{\frac{1}{4}(\cos 12° + i \sin 12°)}$

6. Use De Moivre's theorem to express $(-2i)^6$ in the form $a + bi$.

7. Determine all roots of the equation $x^3 + 1 = 0$.

In Problems 8 and 9 find a polynomial of lowest degree that has the indicated roots.

8. $-\frac{1}{2}, 1, 1, -1$

9. $2, 1 \pm \sqrt{3}$

In Problems 10 and 11 find the roots of the given equation.

10. $(x^2 - 3x + 2)(x - 2)^2$

11. $(x^2 + 3x - 1)(x - 2)(x + 3)^2$

In Problems 12–14 find a polynomial that has the indicated roots and no others.

12. $\frac{1}{2}$ of multiplicity 3; -2 of multiplicity 1

13. -3 of multiplicity 2; $1 + i, 1 - i$

14. $3 \pm \sqrt{-1}$; -1 of multiplicity 2

15. If $1 - i$ is a root of $2x^4 - x^3 - 4x^2 + 10x - 4 = 0$, write the equation as a product of linear and quadratic factors with real coefficients.

In Problems 16 and 17 determine the maximum number of roots, of the type indicated, of the given equation.

16. $3x^4 + 3x - 1 = 0$; positive real roots

17. $2x^4 + x^3 - 3x^2 + 2x + 1 = 0$; negative real roots

18. Find all roots of the equation $2x^4 - x^3 - 4x^2 + 2x = 0$.

CHAPTER 9

ANALYTIC GEOMETRY: THE CONIC SECTIONS

In 1637 the great French philosopher and scientist René Descartes developed an idea that the nineteenth-century British philosopher John Stuart Mill described as "the greatest single step ever made in the progress of the exact sciences." Descartes combined the techniques of algebra with those of geometry and created a new field of study called **analytic geometry.** Analytic geometry enables us to apply algebraic methods and equations to the solution of problems in geometry and, conversely, to obtain geometric representations of algebraic equations.

We will first develop a formula for the coordinates of the midpoint of a line segment. We will then use the distance and midpoint formulas as tools to illustrate the usefulness of analytic geometry by proving a number of general theorems from plane geometry.

The power of the methods of analytic geometry is also very well demonstrated, as we shall see in this chapter, in a study of the conic sections. We will find in the course of that study that (a) a geometric definition can be converted into an algebraic equation and (b) an algebraic equation can be classified by the type of graph it represents.

9.1
ANALYTIC GEOMETRY

We have previously seen that the length d of the line segment joining points $P_1(x_1, y_1)$ and $P_2(x_2, y_2)$ is given by

$$d = \sqrt{(x_2 - x_1)^2 + (y_2 - y_1)^2}$$

It is also possible to obtain a formula for the coordinates (x, y) of the midpoint P of the line segment whose endpoints are P_1 and P_2 (see Figure 1). Passing lines through P and P_2 parallel to the y-axis and a line through P_1 parallel to the x-axis results in the similar right triangles $P_1 A P$ and $P_1 B P_2$. Using the fact that corresponding sides of similar triangles are in proportion, we can write

$$\frac{\overline{P_1 P_2}}{\overline{P_2 B}} = \frac{\overline{P_1 P}}{\overline{PA}}$$

Since P is the midpoint of $P_1 P_2$, the length of $P_1 P$ is $d/2$, so

$$\frac{d}{y_2 - y_1} = \frac{\dfrac{d}{2}}{y - y_1}$$

Solving for y we have

$$y = \frac{y_1 + y_2}{2}$$

Similarly,

$$\frac{\overline{P_1 P_2}}{\overline{P_1 B}} = \frac{\overline{P_1 P}}{\overline{P_1 A}} \qquad \text{or} \qquad \frac{d}{x_2 - x_1} = \frac{\dfrac{d}{2}}{x - x_1}$$

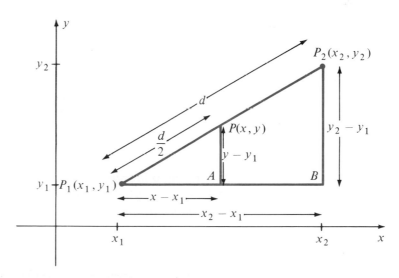

FIGURE 1 The Midpoint Formula

We solve for x to obtain

$$x = \frac{x_1 + x_2}{2}$$

We have established the following formula.

The Midpoint Formula	If $P(x, y)$ is the midpoint of the line segment whose endpoints are $P_1(x_1, y_1)$ and $P_2(x_2, y_2)$, then $$x = \frac{x_1 + x_2}{2} \qquad y = \frac{y_1 + y_2}{2}$$

EXAMPLE 1   The Midpoint Formula

Find the midpoint of the line segment whose endpoints are $P_1(3, 4)$ and $P_2(-2, -6)$.

SOLUTION

If $P(x, y)$ is the midpoint, then

$$x = \frac{x_1 + x_2}{2} = \frac{3 + (-2)}{2} = \frac{1}{2}$$

$$y = \frac{y_1 + y_2}{2} = \frac{4 + (-6)}{2} = -1$$

Thus, the midpoint is $(\frac{1}{2}, -1)$.

■

PROGRESS CHECK

Find the midpoint of the line segment whose endpoints are the following.
(a) $(0, -4), (-2, -2)$ **(b)** $(-10, 4), (7, -5)$

ANSWERS

(ɐ) $(-1, -3)$ (q) $(-\frac{3}{2}, -\frac{1}{2})$

The formulas for distance, midpoint of a line segment, and slope of a line are sufficient to allow us to demonstrate the beauty and power of analytic geometry. With these tools, we can prove theorems from plane geometry by placing the figures on a rectangular coordinate system.

EXAMPLE 2   Proving a Theorem Using Analytic Geometry

Prove that the line joining the midpoints of two sides of a triangle is parallel to the third side and has length equal to one-half the third side.

SOLUTION

We place the triangle OAB in a convenient location, namely, with one vertex at the origin and one side on the positive x-axis (Figure 2). If Q and R are the midpoints of OB

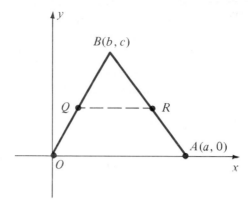

FIGURE 2 Diagram for Example 2

and AB, then, by the midpoint formula, the coordinates of Q and R are

$$Q\left(\frac{b}{2}, \frac{c}{2}\right) \qquad R\left(\frac{a+b}{2}, \frac{c}{2}\right)$$

We see that the line joining Q and R has slope 0, since the difference of the y-coordinates is

$$\frac{c}{2} - \frac{c}{2} = 0$$

But side OA also has slope 0, which proves that QR is parallel to OA.

Applying the distance formula to QR, we have

$$\overline{QR} = \sqrt{(x_2 - x_1)^2 + (y_2 - y_1)^2}$$
$$= \sqrt{\left(\frac{a+b}{2} - \frac{b}{2}\right)^2 + \left(\frac{c}{2} - \frac{c}{2}\right)^2}$$
$$= \sqrt{\left(\frac{a}{b}\right)^2} = \frac{a}{2}$$

Since $\overline{OA}$ has length a, we have shown that $\overline{QR}$ is one-half of $\overline{OA}$.

∎

PROGRESS CHECK

Prove that the midpoint of the hypotenuse of a right triangle is equidistant from all three vertices. (*Hint:* Place the triangle so that two legs coincide with the positive x- and y-axes. Find the coordinates of the midpoint of the hypotenuse by the midpoint formula. Finally, compute the distance from the midpoint to each vertex by the distance formula.)

EXERCISE SET 9.1

In Exercises 1–12 find the midpoint of the line segment whose endpoints are given.

1. $(2, 6), (3, 4)$
2. $(1, 1), (-2, 5)$
3. $(2, 0), (0, 5)$
4. $(-3, 0), (-5, 2)$
5. $(-2, 1), (-5, -3)$
6. $(2, 3), (-1, 3)$
7. $(0, -4), (0, 3)$
8. $(1, -3), (3, 2)$
9. $(-1, 3), (-1, 6)$
10. $(3, 2), (0, 0)$
11. $(1, -1), (-1, 1)$
12. $(2, 4), (2, -4)$

13. Prove that the medians from the equal angles of an isosceles triangle are of equal length. (*Hint:* Place the triangle so that its vertices are at the points $A(-a, 0)$, $B(a, 0)$, and $C(0, b)$.)

14. Show that the midpoints of the sides of a rectangle are the vertices of a rhombus (a quadrilateral with four equal sides). (*Hint:* Place the rectangle so that its vertices are at the points $(0, 0), (a, 0), (0, b)$, and (a, b).)

15. Prove that a triangle with two equal medians is isosceles.

16. Show that the sum of the squares of the lengths of the medians of a triangle equals three-fourths the sum of the squares of the lengths of the sides. (*Hint:* Place the triangle so that its vertices are at the points $(-a, 0), (b, 0)$, and $(0, c)$.)

17. Prove that the diagonals of a rectangle are equal in length. (*Hint:* Place the rectangle so that its vertices are at the points $(0, 0), (a, 0), (0, b)$, and (a, b).)

9.2
THE CIRCLE

The conic sections provide us with an outstanding opportunity to illustrate the double-edged power of analytic geometry. We will see that a geometric figure defined as a set of points can often be described analytically by an algebraic equation; conversely, we can start with an algebraic equation and use graphing procedures to study the properties of the curve.

First, let's see how the term conic section originates. If we pass a plane through a cone at various angles, as shown in Figure 3, the intersections are called **conic sections.** (In exceptional cases, the intersection of a plane and a cone is a point, a line, or a pair of lines.) In the coming sections we will show that each of these geometric configurations can be defined in purely analytic terms, that is, via algebraic expressions.

Let's begin with the geometric definition of a circle.

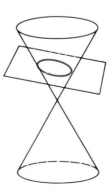
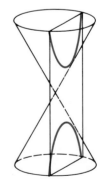

Circle   Parabola   Ellipse   Hyperbola

FIGURE 3   The Conic Sections

Definition of the Circle | A **circle** is the set of all points in a plane that are at a given distance from a fixed point. The fixed point is called the **center** of the circle and the given distance is called the **radius.**

Using the methods of analytic geometry, we place the center at a point (h, k) as in Figure 4. If $P(x, y)$ is a point on the circle, then the distance from P to the center (h, k) must be equal to the radius r. By the distance formula,

$$\sqrt{(x - h)^2 + (y - k)^2} = r$$

or

$$(x - h)^2 + (y - k)^2 = r^2$$

Since $P(x, y)$ is any point on the circle, we say that

Equation of the Circle |
$$(x - h)^2 + (y - k)^2 = r^2$$

is the **standard form of the equation of the circle** with center (h, k) and radius r.

EXAMPLE 1   Finding the Equation of a Circle
Write the equation of the circle with center at $(2, -5)$ and radius 3.

SOLUTION
Substituting $h = 2$, $k = -5$, and $r = 3$ in the equation

$$(x - h)^2 + (y - k)^2 = r^2$$

yields

$$(x - 2)^2 + (y + 5)^2 = 9$$ ■

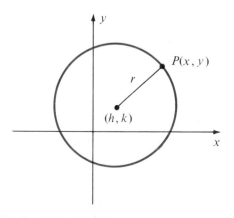

FIGURE 4   Deriving the Equation of the Circle

EXAMPLE 2   Finding the Center and Radius of a Circle

Find the center and radius of the circle whose equation is

$$(x + 1)^2 + (y - 3)^2 = 4$$

SOLUTION

Since the standard form is

$$(x - h)^2 + (y - k)^2 = r^2$$

we must have

$$x - h = x + 1 \qquad y - k = y - 3 \qquad r^2 = 4$$

Solving, we find that

$$h = -1 \qquad k = 3 \qquad r = 2$$

The center is at $(-1, 3)$ and the radius is 2.

∎

 PROGRESS CHECK

Find the center and radius of the circle whose equation is

$$\left(x - \frac{1}{2}\right)^2 + (y + 5)^2 = 15$$

ANSWER

center $(\frac{1}{2}, -5)$, radius $\sqrt{15}$

When we are given the equation of a circle in the **general form**

$$Ax^2 + Ay^2 + Dx + Ey + F = 0, \qquad A \neq 0$$

in which the coefficients of x^2 and y^2 are the same, we may rewrite the equation in standard form. The process involves completing the square in each variable. (If necessary, review Section 2.6.)

EXAMPLE 3   Standard Form of the Equation of a Circle

Write the equation of the circle $2x^2 + 2y^2 - 12x + 16y - 31 = 0$ in standard form.

SOLUTION

Grouping the terms in x and y and factoring produces

$$2(x^2 - 6x) + 2(y^2 + 8y) = 31$$

Completing the square in both x and y, we have

$$2(x^2 - 6x + 9) + 2(y^2 + 8y + 16) = 31 + 18 + 32$$
$$2(x - 3)^2 + 2(y + 4)^2 = 81$$

Note that the quantities 18 and 32 were added to the right-hand side because each factor is multiplied by 2. The last equation can be written as

$$(x - 3)^2 + (y + 4)^2 = \frac{81}{2}$$

This is the standard form of the equation of the circle with center at $(3, -4)$ and radius $9\sqrt{2}/2$.

∎

PROGRESS CHECK

Write the equation of the circle $4x^2 + 4y^2 - 8x + 4y = 103$ in standard form, and determine the center and radius.

ANSWER

$(x - 1)^2 + \left(y + \frac{1}{2}\right)^2 = 27$, center $\left(1, -\frac{1}{2}\right)$, radius $\sqrt{27}$

EXAMPLE 4   Standard Form of the Equation of a Circle
Write the equation $3x^2 + 3y^2 - 6x + 15 = 0$ in standard form.

SOLUTION
Regrouping, we have

$$3(x^2 - 2x) + 3y^2 = -15$$

We then complete the square in x and y:

$$3(x^2 - 2x + 1) + 3y^2 = -15 + 3$$
$$3(x - 1)^2 + 3y^2 = -12$$
$$(x - 1)^2 + y^2 = -4$$

Since $r^2 = -4$ is an impossible situation, the graph of the equation is not a circle. Note that the left-hand side of the equation in standard form is a sum of squares and is therefore nonnegative, while the right-hand side is negative. Thus, there are no real values of x and y that satisfy the equation. This is an example of an equation that does not have a graph!

∎

PROGRESS CHECK

Write the equation $x^2 + y^2 - 12y + 36 = 0$ in standard form, and analyze its graph.

ANSWER

The standard form is $x^2 + (y - 6)^2 = 0$. The equation is that of a "circle" with center at $(0, 6)$ and radius of 0. The "circle" is actually the point $(0, 6)$.

EXERCISE SET 9.2

In Exercises 1–8 find an equation of the circle with center at (h, k) and radius r.

1. $(h, k) = (2, 3), r = 2$

2. $(h, k) = (-3, 0), r = 3$

3. $(h, k) = (-2, -3), r = \sqrt{5}$

4. $(h, k) = (2, -4), r = 4$

5. $(h, k) = (0, 0), r = 3$

6. $(h, k) = (0, -3), r = 2$

7. $(h, k) = (-1, 4), r = 2\sqrt{2}$

8. $(h, k) = (2, 2), r = 2$

In Exercises 9–16 find the center and radius of the circle with the given equation.

9. $(x - 2)^2 + (y - 3)^2 = 16$

10. $(x + 2)^2 + y^2 = 9$

11. $(x - 2)^2 + (y + 2)^2 = 4$

12. $\left(x + \dfrac{1}{2}\right)^2 + (y - 2)^2 = 8$

13. $(x + 4)^2 + \left(y + \dfrac{3}{2}\right)^2 = 18$

14. $x^2 + (y - 2)^2 = 4$

15. $\left(x - \dfrac{1}{3}\right)^2 + y^2 = -\dfrac{1}{9}$

16. $(x - 1)^2 + \left(y - \dfrac{1}{2}\right)^2 = 3$

In Exercises 17–24 write the equation of each given circle in standard form, and determine the center and radius if they exist.

17. $x^2 + y^2 + 4x - 8y + 4 = 0$

18. $x^2 + y^2 - 2x + 6y - 15 = 0$

19. $2x^2 + 2y^2 - 6x - 10y + 6 = 0$

20. $2x^2 + 2y^2 + 8x - 12y - 8 = 0$

21. $2x^2 + 2y^2 - 4x - 5 = 0$

22. $4x^2 + 4y^2 - 2y + 7 = 0$

23. $3x^2 + 3y^2 - 12x + 18y + 15 = 0$

24. $4x^2 + 4y^2 + 4x + 4y - 4 = 0$

In Exercises 25–36 write the given equation in standard form, and determine if the graph of the equation is a circle, a point, or neither.

25. $x^2 + y^2 - 6x + 8y + 7 = 0$

26. $x^2 + y^2 + 4x + 6y + 5 = 0$

27. $x^2 + y^2 + 3x - 5y + 7 = 0$

28. $x^2 + y^2 - 4x - 6y - 13 = 0$

29. $2x^2 + 2y^2 - 12x - 4 = 0$

30. $2x^2 + 2y^2 + 4x - 4y + 25 = 0$

31. $2x^2 + 2y^2 - 6x - 4y - 2 = 0$

32. $2x^2 + 2y^2 - 10y + 6 = 0$

33. $3x^2 + 3y^2 + 12x - 4y - 20 = 0$

34. $x^2 + y^2 + x + y = 0$

35. $4x^2 + 4y^2 + 12x - 20y + 38 = 0$

36. $4x^2 + 4y^2 - 12x - 36 = 0$

9.3
THE PARABOLA

We begin our study of the parabola with the geometric definition.

Definition of the Parabola	A **parabola** is the set of all points that are equidistant from a given point and a given line. The given point is called the **focus** and the given line is called the **directrix** of the parabola.

In Figure 5 all points P on the parabola are equidistant from the focus F and the directrix L, that is, $\overline{PF} = \overline{PQ}$. The line through the focus that is perpendicular to the directrix is called the **axis of the parabola** (or simply the **axis**), and the parabola is seen to be symmetric with respect to the axis. The point V (Figure 5), where the parabola intersects its axis, is called the **vertex** of the parabola. The vertex, then, is the point from

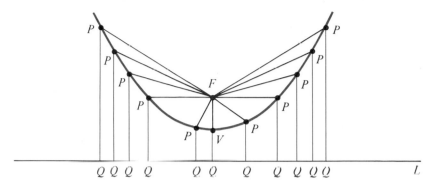

FIGURE 5 Focus and Directrix of a Parabola

which the parabola opens. Note that the vertex is the point on the parabola that is closest to the directrix.

We can apply the methods of analytic geometry to find an equation of the parabola. We choose the y-axis as the axis of the parabola and the origin as the vertex (Figure 6). Since the vertex is on the parabola, it is equidistant from the focus and the directrix. Thus, if the coordinates of the focus F are $(0, p)$, then the equation of the directrix is $y = -p$. We then let $P(x, y)$ be any point on the parabola, and we equate the distance from P to the focus F and the distance from P to the directrix L. Using the distance formula,

$$\overline{PF} = \overline{PQ}$$
$$\sqrt{(x - 0)^2 + (y - p)^2} = \sqrt{(x - x)^2 + (y + p)^2}$$

Squaring both sides,

$$x^2 + y^2 - 2py + p^2 = y^2 + 2py + p^2$$
$$x^2 = 4py$$

We have obtained an important form of the equation of a parabola.

Standard Form of the Equation of a Parabola	$$x^2 - 4py$$ is the standard form of the equation of a parabola whose vertex is at the origin, whose focus is at $(0, p)$, and whose axis is vertical.

Conversely, it can be shown that the graph of the equation $x^2 = 4py$ is a parabola. Note that substituting $-x$ for x leaves the equation unchanged, verifying symmetry with respect to the y-axis. If $p > 0$, the parabola opens upward as shown in Figure 6a, while if $p < 0$, the parabola opens downward, as shown in Figure 6b.

EXAMPLE 1 Finding the Focus and Directrix of a Parabola
Determine the focus and directrix of the parabola $x^2 = 8y$, and sketch its graph.

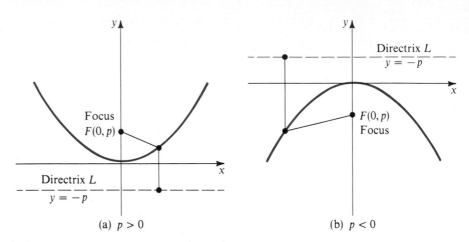

FIGURE 6   Deriving the Equation of the Parabola

SOLUTION

The equation of the parabola is of the form

$$x^2 = 4py = 8y$$

so $p = 2$. The equation of the directrix is $y = -p = -2$, and the focus is at $(0, p) = (0, 2)$. Since $p > 0$, the parabola opens upward. The graph of the parabola is shown in Figure 7.

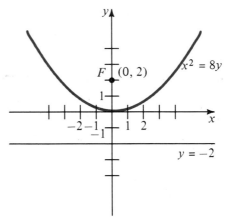

FIGURE 7   Diagram for Example 1

  PROGRESS CHECK

Determine the focus and directrix of the parabola $x^2 = -3y$.

ANSWER

focus at $(0, -\frac{3}{4})$, directrix $y = \frac{3}{4}$

DEVICES WITH A PARABOLIC SHAPE

The properties of the parabola are used in the design of some important devices. For example, by rotating a parabola about its axis, we obtain a **parabolic reflector,** a shape used in the headlight of an automobile. In the accompanying figure, the light source (the bulb) is placed at the focus of the parabola. The headlight is coated with a reflecting material, and the rays of light bounce back in lines that are parallel to the axis of the parabola. This permits a headlight to disperse light in front of the auto where it is needed.

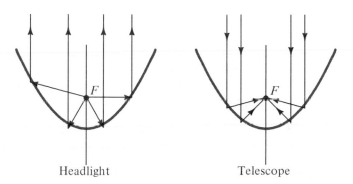

Headlight   Telescope

A reflecting telescope reverses the use of these same properties. Here, the rays of light from a distant star, which are nearly parallel to the axis of the parabola, are reflected by the mirror to the focus (see accompanying figure). The eyepiece is placed at the focus, where the rays of light are gathered.

EXAMPLE 2   Finding the Equation of a Parabola
Find the equation of the parabola with vertex at $(0, 0)$ and focus at $(0, -\frac{3}{2})$.

SOLUTION
Since the focus is at $(0, p)$ we have $p = -\frac{3}{2}$. The equation of the parabola is

$$x^2 = 4py = 4(-\tfrac{3}{2}y) = -6y$$

∎

PROGRESS CHECK
Find the equation of the parabola with vertex at $(0, 0)$ and focus at $(0, 3)$.

ANSWER
$x^2 = 12y$

If we place the parabola as shown in Figure 8, we can proceed as above to obtain the following result.

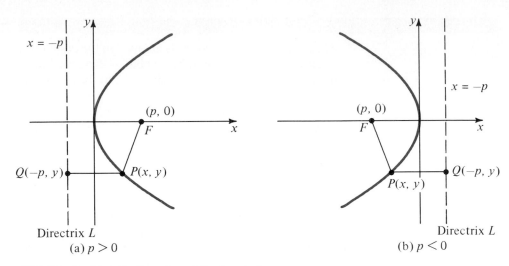

(a) $p > 0$ (b) $p < 0$

FIGURE 8 Deriving the Equation of the Parabola

Standard Form of the Equation of a Parabola	$$y^2 = 4px$$
	is the standard form of the equation of a parabola whose vertex is at the origin, whose focus is at $(p, 0)$, and whose axis is horizontal.

Note that substituting $-y$ for y leaves this equation unchanged, verifying symmetry with respect to the x-axis. If $p > 0$, the parabola opens to the right, as shown in Figure 8a, while if $p < 0$, the parabola opens to the left, as shown in Figure 8b.

EXAMPLE 3 Finding the Equation of a Parabola

Find the equation of the parabola with vertex at $(0, 0)$ and directrix $x = \frac{1}{2}$.

SOLUTION

The directrix is $x = -p$, so $p = -\frac{1}{2}$. The equation of the parabola is then

$$y^2 = 4px = -2x$$

∎

EXAMPLE 4 Finding the Equation of a Parabola

Find the equation of the parabola that has the x-axis as its axis, has vertex at $(0, 0)$, and passes through the point $(-2, 3)$.

SOLUTION

Since the axis of the parabola is the x-axis, the equation of the parabola is $y^2 = 4px$. The parabola passes through the point $(-2, 3)$, so the coordinates of this point must satisfy the equation of the parabola. Thus,

$$y^2 = 4px$$
$$(3)^2 = 4p(-2)$$
$$4p = -\frac{9}{2}$$

and the equation of the parabola is

$$y^2 = 4px = -\frac{9}{2}x$$

■

 PROGRESS CHECK

Find the equation of the parabola that has the y-axis as its axis, has vertex at $(0, 0)$, and passes through the point $(1, -2)$.

ANSWER

$x^2 = -\dfrac{1}{2}y$

EXERCISE SET 9.3

In Exercises 1–8 determine the focus and directrix of the given parabola, and sketch the graph.

1. $x^2 = 4y$
2. $x^2 = -4y$
3. $y^2 = 2x$
4. $y^2 = -\dfrac{3}{2}x$

5. $x^2 + 5y = 0$
6. $2y^2 - 3x = 0$
7. $y^2 - 12x = 0$
8. $x^2 - 9y = 0$

In Exercises 9–20 determine the equation of the parabola that has its vertex at the origin and that satisfies the given conditions.

9. Focus at $(1, 0)$

10. Focus at $(0, -3)$

11. Directrix $x = -\dfrac{3}{2}$

12. Directrix $y = \dfrac{5}{2}$

13. Axis is the x-axis, and parabola passes through the point $(2, 1)$.

14. Axis is the y-axis, and parabola passes through the point $(4, -2)$.

15. Axis is the x-axis, and $p = -\dfrac{5}{4}$.

16. Axis is the y-axis, and $p = 2$.

17. Focus at $(-1, 0)$ and directrix $x = 1$.

18. Focus at $\left(0, -\dfrac{5}{2}\right)$ and directrix $y = \dfrac{5}{2}$.

19. Axis is the x-axis, and parabola passes through the point $(4, 2)$.

20. Axis is the y-axis, and parabola passes through the point $(2, 4)$.

In Exercises 21–24 determine whether the given parabola opens upward, downward, to the left, or to the right.

21. $4x^2 + y = 0$
22. $4x^2 - y = 0$
23. $2x + y^2 = 0$
24. $2x - 5y^2 = 0$

9.4
THE ELLIPSE

The geometric definition of an ellipse is as follows.

Definition of the Ellipse   An **ellipse** is the set of all points the sum of whose distances from two fixed points is a constant. The fixed points are called the **foci** of the ellipse.

An ellipse may be constructed in the following way. Place a thumbtack at each of the foci F_1 and F_2, and attach one end of a string to each of the thumbtacks. Hold a pencil tight against the string, as shown in Figure 9, and move the pencil. The point P will describe an ellipse, since the sum of the distances from P to the foci is always a constant, namely, the length of the string.

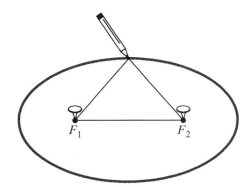

FIGURE 9   The Foci of an Ellipse

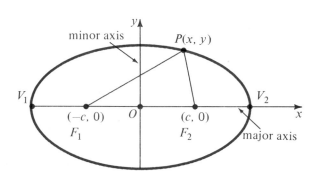

FIGURE 10   Deriving the Equation of the Ellipse

We will use Figure 10 to obtain an equation for an ellipse. The line segment joining the foci F_1 and F_2 is called the **major axis** of the ellipse, and the midpoint of the major axis is called the **center** of the ellipse. The line segment through the center and perpendicular to the major axis is called the **minor axis.** The points at which the ellipse intersects the major axis (V_1 and V_2 in Figure 10) are called the **vertices** of the ellipse.

It is easiest to derive the equation of an ellipse when the center is at the origin and the major axis is one of the coordinate axes. We first consider the case when the major axis is the x-axis. If the focus F_2 is at $(c, 0)$, then the other focus F_1 is at $(-c, 0)$, as in Figure 10. Let $P(x, y)$ be a point on the ellipse, and let the constant sum of the distances from P to the foci be denoted by $2a$. Then we have

$$\overline{F_1P} + \overline{F_2P} = 2a$$

Using the distance formula, we may rewrite this as

$$\sqrt{(x - c)^2 + (y - 0)^2} + \sqrt{(x + c)^2 + (y - 0)^2} = 2a$$

or

$$\sqrt{(x - c)^2 + y^2} = 2a - \sqrt{(x + c)^2 + y^2}$$

The domed roof in the accompanying figure has the shape of an ellipse that has been rotated about its major axis. It can be shown, using basic laws of physics, that a sound uttered at one focus will be reflected to the other focus, where it will be clearly heard. This property of such rooms is known as the "whispering gallery effect."

Famous whispering galleries include the dome of St. Paul's Cathedral, London; St. John Lateran, Rome; the Salle des Cariatides in the Louvre, Paris; and the original House of Representatives (now the National Statuary Hall in the United States Capitol), Washington, D.C.

Squaring both sides, we obtain

$$(x - c)^2 + y^2 = 4a^2 - 4a\sqrt{(x + c)^2 + y^2} + (x + c)^2 + y^2$$
$$x^2 - 2cx + c^2 + y^2 = 4a^2 - 4a\sqrt{(x + c)^2 + y^2} + x^2 + 2cx + c^2 + y^2$$

Simplifying, we have

$$a\sqrt{(x + c)^2 + y^2} = a^2 + cx$$

Squaring both sides, we now have

$$a^2[(x + c)^2 + y^2] = (a^2 + cx)^2$$
$$a^2[x^2 + 2cx + c^2 + y^2] = a^4 + 2a^2cx + c^2x^2$$
$$a^2x^2 + 2a^2cx + a^2c^2 + a^2y^2 = a^4 + 2a^2cx + c^2x^2$$
$$(a^2 - c^2)x^2 + a^2y^2 = a^2(a^2 - c^2) \tag{1}$$

Since the vertex V_1 lies on the ellipse and since the sum of the distances from any point on the ellipse to the foci is $2a$, we must have

$$\overline{F_1V_1} + \overline{F_2V_1} = 2a$$

From Figure 10,

$$\overline{F_1F_2} = 2c$$

and

$$\overline{F_1V_1} + \overline{F_2V_1} > \overline{F_1F_2}$$

Substituting the values from the above equations yields

$$2a > 2c$$

so $a > c$ and, therefore, $a^2 > c^2$. Dividing both sides of Equation (1) by $a^2(a^2 - c^2)$, we obtain

$$\frac{x^2}{a^2} + \frac{a^2y^2}{a^2 - c^2} = 1$$

Letting

$$b^2 = a^2 - c^2, \qquad b > 0 \tag{2}$$

we obtain the **standard form** of the equation of an ellipse with center at $(0,0)$ and foci at $(-c, 0)$ and $(c, 0)$.

Standard Form of the Equation of an Ellipse	$$\frac{x^2}{a^2} + \frac{y^2}{b^2} = 1, \qquad b \leq a$$	(3)

Conversely, we can show that the graph of an equation of the form given in Equation (3) is an ellipse with center at $(0,0)$ and foci at $(-c, 0)$ and $(c, 0)$.

The vertices of the ellipse are the x-intercepts of the ellipse, and these values may be obtained by letting $y = 0$ in Equation (3). We have $x^2/a^2 = 1$, so $x = \pm a$. Thus, V_1 is the point $(-a, 0)$ and V_2 is the point $(a, 0)$; the length of the major axis is $2a$ (Figure 11). When $x = 0$, we obtain $y = \pm b$, so the ellipse intersects the y-axis at the points $(0, -b)$ and $(0, b)$. Thus, the length of the minor axis is $2b$. Observe also that since $c \geq 0$, $b^2 \leq a^2$, so $b \leq a$. Thus, the major axis is longer than the minor axis. Figure 11 shows the ellipse given by Equation (3), along with the relationships between a, b, and c.

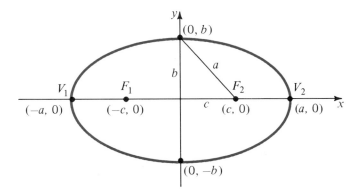

FIGURE 11   Deriving the Equation of the Ellipse

Since Equation (3) remains unchanged when we replace x with $-x$, or y with $-y$, we conclude that the ellipse is symmetric with respect to both the x-axis and the y-axis and, therefore, with respect to the origin.

EXAMPLE 1   Using the Equation of an Ellipse
Discuss and sketch the graph of the equation

$$9x^2 + 25y^2 = 225$$

SOLUTION
We obtain the standard form by dividing both sides by 225, obtaining

$$\frac{x^2}{25} + \frac{y^2}{9} = 1$$

which is in the form of Equation (3). This is the equation of an ellipse whose major axis is the x-axis. We have $a^2 = 25$ and $b^2 = 9$, so $a = 5$, and $b = 3$. Since $c^2 = a^2 - b^2$, we have

$$c^2 = 25 - 9 = 16$$

so $c = 4$. Thus, the foci are at $(-4, 0)$ and $(4, 0)$. The vertices are at $(-5, 0)$ and $(5, 0)$. The ellipse is sketched in Figure 12.

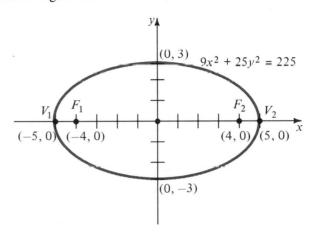

FIGURE 12 Diagram for Example 1

■

EXAMPLE 2 Finding the Equation of an Ellipse

Find the equation of an ellipse with center at $(0, 0)$, a focus at $(-3, 0)$, and a vertex at $(6, 0)$.

SOLUTION

Since a focus is at $(-3, 0)$ and the center is at the origin, $c = 3$. Since a vertex is at $(6, 0)$, $a = 6$. We obtain b^2 by using Equation (2):

$$b^2 = a^2 - c^2 = 6^2 - 3^2 = 27$$

Then the equation of the ellipse is

$$\frac{x^2}{36} + \frac{y^2}{27} = 1$$

■

When the major axis of an ellipse is located on the y-axis and the center of the ellipse is at the origin, the foci F_1 and F_2 are at $(0, -c)$ and $(0, c)$ as shown in Figure 13. If the constant sum of the distances is again denoted by $2a$, a similar derivation shows that the standard form of the equation of such an ellipse is as follows.

Standard Form of the Equation of an Ellipse	$\dfrac{x^2}{b^2} + \dfrac{y^2}{a^2} = 1, \qquad b \le a$	(4)

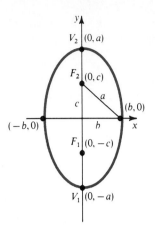

FIGURE 13   Deriving the Equation of the Ellipse

The length of the major axis is $2a$, and the length of the minor axis is $2b$. Thus, *the major axis is always the longer axis.*

EXAMPLE 3   Using the Equation of an Ellipse
Discuss and sketch the graph of the equation

$$9x^2 + 4y^2 = 36$$

SOLUTION
We obtain the standard form by dividing both sides by 36, obtaining

$$\frac{x^2}{4} + \frac{y^2}{9} = 1$$

which is of the form in Equation (4). Hence, the graph of this equation is an ellipse. We have $a^2 = 9$ and $b^2 = 4$, so $a = 3$ and $b = 2$. From Equation (2) we have

$$c^2 = a^2 - b^2 = 9 - 4 = 5$$
$$c = \sqrt{5}$$

The major axis of this ellipse lies along the y-axis; the foci are at $(0, -\sqrt{5})$ and $(0, \sqrt{5})$; the vertices are at $(0, -3)$ and $(0, 3)$. The ellipse is sketched in Figure 14.

∎

PROGRESS CHECK
Discuss and sketch the graph of the equation $9x^2 + y^2 = 9$, indicating the foci and vertices.

ANSWER
See Figure 15.

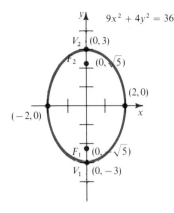

FIGURE 14 Diagram for Example 3

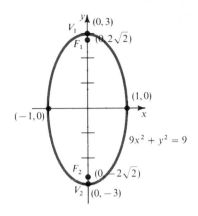

FIGURE 15 Diagram for Progress Check

EXAMPLE 4 Working with the Equation of an Ellipse

Is the major axis of the ellipse

$$25x^2 + 4y^2 = 100$$

along the x-axis or along the y-axis?

SOLUTION

We first obtain the standard form of the equation by dividing both sides by 100:

$$\frac{x^2}{4} + \frac{y^2}{25} = 1$$

Since we must have $a \geq b$, the equation is of the form

$$\frac{x^2}{b^2} + \frac{y^2}{a^2} = 1$$

and the major axis is along the y-axis.　　　　　　　　　　　　　　　■

If $a = b$, then the equation of the ellipse becomes

$$\frac{x^2}{a^2} + \frac{y^2}{a^2} = 1$$

or

$$x^2 + y^2 = a^2$$

which we identify as a circle with center at the origin and radius a. The circle is seen to be a special case of an ellipse in which the major and minor axes are equal.

EXERCISE SET 9.4

In Exercises1–10 sketch and discuss the graph of the given equation, giving the coordinates of the foci and vertices.

1. $25x^2 + 16y^2 - 400 = 0$

2. $x^2 + 4y^2 - 16 = 0$

3. $9x^2 + y^2 - 36 = 0$

4. $3x^2 + y^2 - 81 = 0$

5. $25x^2 + 64y^2 - 1600 = 0$

6. $9x^2 + y^2 - 36 = 0$

7. $4x^2 + y^2 - 9 = 0$

8. $5x^2 + 12y^2 - 45 = 0$

9. $2x^2 + 10y^2 - 40 = 0$

10. $25x^2 + 169y^2 - 4225 = 0$

In Exercises 11–24 find an equation of the ellipse satisfying the given conditions.

11. Center at $(0,0)$, focus at $(0,3)$, and vertex at $(0,7)$

12. Center at $(0,0)$, focus at $(2,0)$, and vertex at $(-4,0)$

13. Foci at $(0, \pm\sqrt{5})$ and vertices at $(0, \pm 3)$

14. Foci at $(0, \pm\sqrt{7})$ and vertices at $(0, \pm 4)$

15. Foci at $(\pm\sqrt{15}, 0)$ and vertex at $(4,0)$

16. Foci at $(\pm 2\sqrt{6}, 0)$ and vertex at $(-5,0)$

17. Focus at $(0, 2\sqrt{2})$ and vertices at $(0, \pm 3)$

18. Focus at $(0, -2\sqrt{15})$ and vertices at $(0, \pm 8)$

19. Foci at $(\pm 4, 0)$ and vertices at $(\pm 5, 0)$

20. Foci at $(\pm 5, 0)$ and vertices at $(\pm 6, 0)$

21. Foci at $(\pm 4\sqrt{5}, 0)$, length of major axis is 18

22. Foci at $(\pm 6, 0)$, length of minor axis is $\sqrt{13}$

23. Vertices at $(\pm 7, 0)$ and passing through the point $\left(1, \frac{6}{7}\sqrt{3}\right)$

24. Vertices at $(0, \pm 1)$ and passing through the point $\left(\frac{1}{4}, \frac{\sqrt{3}}{2}\right)$

In Exercises 25–28 determine whether the major axis of the given ellipse lies along the x-axis or along the y-axis.

25. $7x^2 + 4y^2 - 81 = 0$

26. $16x^2 + y^2 - 2 = 0$

27. $x^2 + 4y^2 - 6 = 0$

28. $4x^2 + 5y^2 - 1 = 0$

29. The **eccentricity** of an ellipse is defined as the ratio c/a. Describe the general shape of the ellipse when its eccentricity is (a) almost 1, (b) almost 0.

9.5
THE HYPERBOLA

The hyperbola is the remaining conic section that we will consider in this chapter.

Definition of the Hyperbola	A **hyperbola** is the set of all points the difference of whose distances from two fixed points is a positive constant. The two fixed points are called the **foci** of the hyperbola.

The line through the foci is called the **transverse axis,** the midpoint of the line segment between the foci is called the **center,** and the line through the center and perpendicular to the transverse axis is called the **conjugate axis.** The two separate parts of the hyperbola are called its **branches** (Figure 16).

It is easiest to derive the equation of a hyperbola when the center is at the origin and the transverse axis is one of the coordinate axes. If the foci lie on the x-axis and one focus F_2 is at $(c, 0)$, $c > 0$, then the other focus F_1 is at $(-c, 0)$. (See Figures 16a and 16b). Let $P(x, y)$ be a point on the hyperbola, and let the constant difference of the distances from P to the foci be denoted by $2a$. If P is on the right branch, we have

$$\overline{PF_1} - \overline{PF_2} = 2a \tag{1a}$$

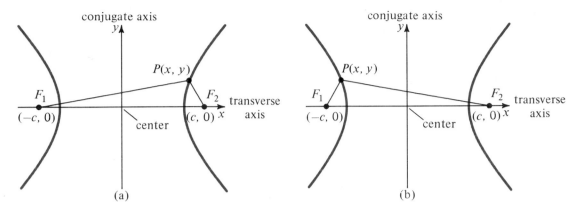

conjugate axis

conjugate axis

FIGURE 16 Deriving the Equation of the Hyperbola

whereas if P is on the left branch, we have

$$\overline{PF_2} - \overline{PF_1} = 2a \qquad (1b)$$

Both of these equations can be expressed by the single equation

$$|\overline{PF_1} - \overline{PF_2}| = 2a$$

or

$$\left|\sqrt{(x + c)^2 + y^2} - \sqrt{(x - c)^2 + y^2}\right| = 2a \qquad (2)$$

Squaring both sides of Equation (2), we have

$$(x + c)^2 + y^2 - 2\sqrt{[(x + c)^2 + y^2][(x - c)^2 + y^2]} + (x - c)^2 + y^2 = 4a^2$$

We next expand this equation, cancel terms, isolate the radical on the left, and square both sides, obtaining

$$(c^2 - a^2)x^2 - a^2y^2 = a^2(c^2 - a^2)$$

We now focus on triangle F_1PF_2 in Figure 16a. Since the sum of the lengths of two sides of a triangle is always greater than the length of the third side, we have

$$\overline{PF_1} < \overline{PF_2} + \overline{F_1F_2}$$
$$\overline{PF_1} - \overline{PF_2} < \overline{F_1F_2}$$
$$2a < 2c \quad \text{by Equation (1a)}$$

so $a < c$. Similarly, if P is on the left branch (Figure 16b), we again find that $a < c$. Thus, $a^2 < c^2$. Dividing both sides of Equation (3) by $a^2(c^2 - a^2)$, we obtain

$$\frac{x^2}{a^2} - \frac{y^2}{c^2 - a^2} = 1$$

Letting

$$b^2 = c^2 - a^2 > 0 \qquad (4)$$

we obtain the standard form of the equation of a hyperbola with center at $(0, 0)$ and foci at $(-c, 0)$ and $(c, 0)$:

Standard Form of the Equation of a Hyperbola	$$\dfrac{x^2}{a^2} - \dfrac{y^2}{b^2} = 1 \qquad\qquad (5)$$

Conversely, it can be shown that the graph of an equation of the form given in Equation (5) is a hyperbola with center at $(0, 0)$ and foci at $(-c, 0)$ and $(c, 0)$.

The **vertices** of the hyperbola are the x-intercepts of the hyperbola, and these values may be obtained by letting $y = 0$ in Equation (5). We have

$$\frac{x^2}{a^2} = 1 \qquad \text{or} \qquad x = \pm a$$

Thus, V_1 is the point $(-a, 0)$ and V_2 is the point $(a, 0)$ (Figure 17a). When $x = 0$, we obtain the equation $-\dfrac{y^2}{b^2} = 1$, which has no solutions, so there are no y-intercepts.

Since Equation (5) remains unchanged when we replace x with $-x$ and y with $-y$, we conclude that the hyperbola is symmetric with respect to both the x-axis and the y-axis and, therefore, with respect to the origin.

To see why there are no points on the hyperbola when $|x| < a$, we proceed as follows. Solve Equation (5) for y, obtaining

$$y = \pm \frac{b}{a}\sqrt{x^2 - a^2}$$

The radical is defined only if $x^2 - a^2 \geq 0$. That is, there is no value of y when $x^2 < a^2$ or when $|x| < a$. Thus, there are no points on the hyperbola for $-a < x < a$. Since $x^2 - a^2 \geq 0$ when $x \geq a$ or $x \leq -a$, we obtain the two branches shown in Figure 17a.

When the foci of a hyperbola are located on the y-axis and the center of the hyperbola is at the origin, the foci are at $(0, -c)$ and $(0, c)$ as shown in Figure 17b. If the

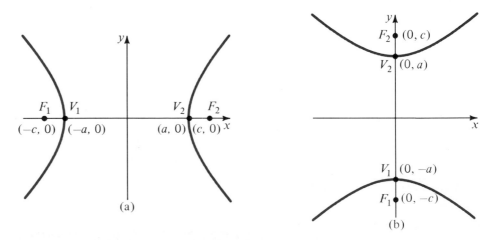

(a)

(b)

FIGURE 17   The Vertices and Foci of the Hyperbola

constant difference of the distances is again denoted by $2a$, a similar derivation shows that the standard form of the equation of such a hyperbola is as follows:

Standard Form of the Equation of a Hyperbola	$$\frac{y^2}{a^2} - \frac{x^2}{b^2} = 1$$	(6)

EXAMPLE 1 Working with the Equation of a Hyperbola

Discuss and sketch the graph of the equation $4x^2 - 9y^2 = 36$.

SOLUTION

We obtain the standard form by dividing both sides by 36, which yields

$$\frac{x^2}{9} - \frac{y^2}{4} = 1$$

which is of the form in Equation (5). This is the equation of a hyperbola whose foci lie on the x-axis. We have $a^2 = 9$ and $b^2 = 4$, so $a = 3$, and the vertices are at $(-3, 0)$ and $(3, 0)$. From Equation (4) we obtain

$$c^2 = a^2 + b^2 = 9 + 4 = 13$$

Then $c = \pm\sqrt{13}$, and the foci are at $F_1(-\sqrt{13}, 0)$ and $F_2(\sqrt{13}, 0)$. The hyperbola is sketched in Figure 18a. (Don't worry about the "width" of the hyperbola—we'll provide guidance in the next topic material.)

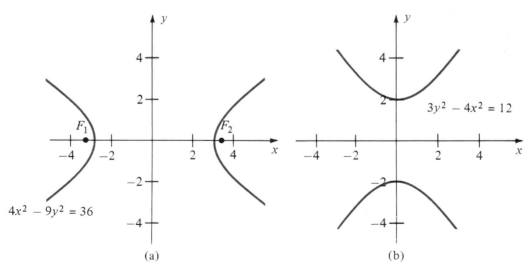

(a)   (b)

FIGURE 18 Diagrams for Examples 1 and 2

EXAMPLE 2 Working with the Equation of a Hyperbola

Discuss and sketch the graph of the equation $3y^2 - 4x^2 = 12$.

SOLUTION

We obtain the standard form by dividing both sides by 12, which yields

$$\frac{y^2}{4} - \frac{x^2}{3} = 1$$

which is of the form in Equation (6). This is the equation of a hyperbola whose foci lie on the y-axis. We have $a^2 = 4$ and $b^2 = 3$, so $a = 2$, and the vertices are at $(0, -2)$ and $(0, 2)$. From Equation (4) we obtain

$$c^2 = a^2 + b^2 = 4 + 3 = 7$$

so $c = \pm\sqrt{7}$. The foci are at $(0, -\sqrt{7})$ and $(0, \sqrt{7})$. The hyperbola is sketched in Figure 18b.

$\blacksquare$

PROGRESS CHECK

Discuss the graph of the equation $16x^2 - y^2 = 16$.

ANSWER

Hyperbola: $a = 1$, $b = 4$, vertices at $(-1, 0)$ and $(1, 0)$, foci at $(-\sqrt{17}, 0)$ and $(\sqrt{17}, 0)$.

EXAMPLE 3   Finding the Equation of a Hyperbola

Find the equation of a hyperbola with center at $(0, 0)$, a focus at $(0, 4)$, and a vertex at $(0, 3)$.

SOLUTION

Since a focus is at $(0, 4)$, the foci lie on the y-axis, and $c = 4$. Since a vertex is at $(0, 3)$, $a = 3$. We obtain b^2 by using Equation (4):

$$b^2 = c^2 - a^2 = 16 - 9 = 7$$

Then the equation of the hyperbola is

$$\frac{y^2}{9} - \frac{x^2}{7} = 1$$

$\blacksquare$

ASYMPTOTES

There is a feature of the hyperbola that distinguishes it from the parabola and the ellipse. By examining the corresponding values of y for values of x that are very far away from the origin, we obtain certain lines that enable us to sketch a given hyperbola readily. Solving Equation (5) for y, we obtain

$$y = \pm\frac{b}{a}\sqrt{x^2 - a^2}$$

$$y = \pm\frac{b}{a}\sqrt{x^2\left(1 - \frac{a^2}{x^2}\right)}$$

$$y = \pm\frac{b}{a}x\sqrt{1 - \frac{a^2}{x^2}} \tag{7}$$

For large values of $|x|$, the term a^2/x^2 in Equation (7) is very small, and the graph of Equation (5) approaches the asymptotes

$$y = \pm \frac{b}{a} x$$

A similar discussion shows that the asymptotes of the hyperbola given by Equation (6) are

$$y = \pm \frac{a}{b} x$$

The asymptotes are shown in Figure 19.

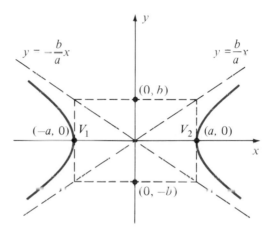

FIGURE 19   The Asymptotes of the Hyperbola

An easy way to obtain the asymptotes of a hyperbola whose equation is in standard form is to replace the 1 on the right side with 0 and then to solve for y in terms of x.

To summarize,

Asymptotes of the Hyperbola		
$\dfrac{x^2}{a^2} - \dfrac{y^2}{b^2} = 1$	has asymptotes	$y = \pm \dfrac{b}{a} x$
$\dfrac{y^2}{a^2} - \dfrac{x^2}{b^2} = 1$	has asymptotes	$y = \pm \dfrac{a}{b} x$

The asymptotes of a hyperbola can be readily constructed. We consider the hyperbola given by Equation (5). Plot the points (a, b), $(-a, b)$, $(-a, -b)$, and $(a, -b)$, which form a rectangle. The lines determined by the diagonals of this rectangle have slopes b/a and $-b/a$ (Figure 19) and are thus the asymptotes of the hyperbola. The hyperbola is then sketched by using the asymptotes as an aid.

EXAMPLE 4   Finding the Asymptotes of a Hyperbola

Find the asymptotes of the hyperbola

$$\frac{y^2}{9} - \frac{x^2}{4} = 1$$

SOLUTION

Setting the right-hand side equal to 0 and solving for y, we find that

$$y = \pm \frac{3}{2} x$$

so the asymptotes are

$$y = \frac{3}{2} x \qquad \text{and} \qquad y = -\frac{3}{2} x$$

∎

EXAMPLE 5   Analyzing the Equation of a Hyperbola

Discuss and sketch the graph of the equation $25x^2 - 4y^2 = 100$, including the asymptotes of the hyperbola.

SOLUTION

We first write the standard form of the equation by dividing both sides by 100, obtaining

$$\frac{x^2}{4} - \frac{y^2}{25} = 1 \tag{8}$$

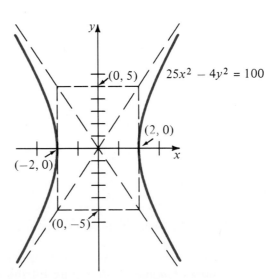

FIGURE 20   Diagram for Example 5

Then $a^2 = 4$ and $b^2 = 25$, so $a = 2$ and $b = 5$. The asymptotes are obtained by replacing the 1 on the right side of (8) with 0 and solving for y, obtaining

$$y = \pm \frac{5}{2}x$$

To construct the asymptotes, we first plot the four points $(2, 5), (-2, 5), (-2, -5)$, and $(2, -5)$. The diagonals of the rectangle determined by these four points determine the asymptotes of the hyperbola, which can then be sketched by using the asymptotes as aids. The graph is shown in Figure 20.

∎

 PROGRESS CHECK

Discuss and sketch the graph of the equation $4x^2 - 9y^2 = 144$, including the asymptotes of the hyperbola.

ANSWER

See Figure 21.

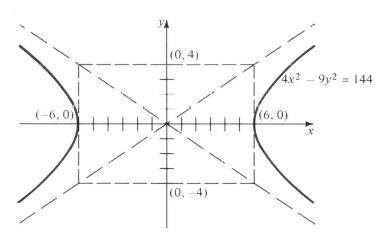

FIGURE 21   Diagram for Progress Check

EXERCISE SET 9.5

In Exercises 1–8 sketch the graph of the given equation, giving the coordinates of the foci and vertices and including the asymptotes.

1. $x^2 - 9y^2 - 9 = 0$
2. $9x^2 - y^2 - 36 = 0$
3. $9y^2 - 4x^2 - 36 = 0$
4. $9x^2 - 16y^2 - 144 = 0$
5. $x^2 - y^2 - 9 = 0$
6. $y^2 - 4x^2 - 100 = 0$
7. $-x^2 + 4y^2 - 100 = 0$
8. $x^2 - 2y^2 + 18 = 0$

In Exercises 9–12 determine whether the foci of the given hyperbola lie on the x-axis or on the y-axis.

9. $2x^2 - 3y^2 - 5 = 0$
10. $3x^2 - 3y^2 + 4 = 0$
11. $y^2 - 4x^2 - 20 = 0$
12. $4y^2 - 9x^2 + 36 = 0$

In Exercises 13–28 find the equation of the hyperbola satisfying the given conditions.

13. Center at $(0, 0)$, focus at $(0, 5)$, and vertex at $(0, 4)$.
14. Center at $(0, 0)$, focus at $(0, 7)$, and vertex at $(0, 5)$.

15. Foci at $(\pm 3, 0)$ and vertex at $(2, 0)$.

16. Foci at $(\pm 4, 0)$ and vertex at $(-1, 0)$.

17. Focus at $(0, 9)$ and vertices at $(0, \pm 1)$.

18. Focus at $(0, 4)$ and vertices at $(0, \pm 1)$.

19. Vertices at $(0, \pm 3)$ and asymptote $y = x$.

20. Vertices at $(\pm 2, 0)$ and asymptote $y = -2x$.

21. Center at $(0, 0)$, vertex at $(0, -2)$, and asymptote $y = \frac{1}{2}x$.

22. Center at $(0, 0)$, focus at $(2, 0)$, and asymptote $y = -5x$.

23. Center at $(0, 0)$, focus at $(0, 3)$, and asymptote $y = 3x$.

24. Center at $(0, 0)$, focus at $(4, 0)$, and asymptote $y = x/5$.

25. Vertices at $(0, \pm 4)$ and passing through the point $(5, 5)$.

26. Vertices at $(\pm 2, 0)$ and passing through the point $(3, 1)$.

27. Focus at $(0, 3)$ and asymptotes $y = \pm x$.

28. Focus at $(6, 0)$ and asymptotes $y = \pm 2x$.

29. The **eccentricity** of a hyperbola is defined as the quantity $e = c/a$. Since $c > a, e > 1$. Describe the general shape of a hyperbola for (a) e near 1 and (b) very large e.

9.6
TRANSLATION OF AXES

In Section 2.4, horizontal and vertical shifts were used to aid in sketching the graph of a function. A technique known as translation of axes extends this idea to aid in analyzing and sketching graphs in general.

In Figure 22, the x and y coordinate axes are displayed; they intersect, as usual, at the origin O. Also displayed are an additional set of coordinate axes, x' and y', which are parallel to the x-axis and the y-axis, respectively, and which intersect at the point O'. We may think of the x'- and y'-axes as the result of shifting the x- and y-axes parallel to themselves until they intersect at the point O'. This process is called **translation of axes,** and we say that the x- and y-axes have been **translated.**

A point P in the plane has coordinates (x, y) with respect to the xy coordinate system and coordinates (x', y') with respect to the $x'y'$ coordinate system. There is a straightforward relationship between these pairs of coordinates. Let the origin O' of the $x'y'$ coordinate system have coordinates (h, k) with respect to the xy coordinate system. From Figure 22 we arrive at these useful formulas.

Translation of Axes Formulas			
$x = x' + h$	and	$y = y' + k$	(1)
$x' = x - h$	and	$y' = y - k$	(2)

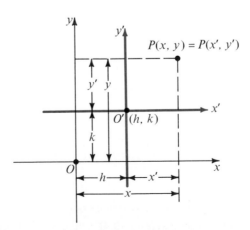

FIGURE 22   Deriving the Formulas for Translation of Axes

EXAMPLE 1   Using the Formulas for Translation of Axes
The origin O' of the $x'y'$ coordinate system is at $(-2, 4)$.
(a) Express x' and y' in terms of x and y, respectively.
(b) Find the $x'y'$ coordinates of the point P whose xy coordinates are $(4, -6)$.

SOLUTION
(a) Substituting $h = -2$ and $k = 4$ in Equation (2), we obtain the translation formulas

$$x' = x + 2 \qquad y' = y - 4$$

(b) Substituting $x = 4$ and $y = -6$ yields

$$x' = 4 + 2 = 6 \qquad y' = -6 - 4 = -10$$

The $x'y'$ coordinates of P are $(6, -10)$.

■

PROGRESS CHECK
The original O' of the $x'y'$ coordinate system is at $(-1, -2)$.
(a) Express x and y in terms of x' and y', respectively.
(b) Find the xy coordinates of the point P whose $x'y'$ coordinates are $(3, -3)$.

ANSWERS
(a) $x = x' - 1, y = y' - 2$ (b) $(2, -5)$

We can use the translation formulas to transform equations given in the xy coordinate system to equations in the $x'y'$ coordinate system, and vice versa. For example, in the $x'y'$ system, the equation of the circle with center at O' and radius r is

$$x'^2 + y'^2 = r^2$$

Substituting the translation formulas

$$x' = x - h \qquad \text{and} \qquad y' = y - k$$

we find that the equation of this circle in xy coordinates is

$$(x - h)^2 + (y - k)^2 = r^2$$

which is precisely the standard form of the equation of the circle with center at (h, k) and radius r, as discussed in Section 9.2.

EXAMPLE 2   Analyzing and Sketching a Conic Section
Discuss and sketch the graph of the equation

$$x^2 - 4x + y^2 + 2y + 1 = 0$$

SOLUTION
We group the terms in x and in y

$$(x^2 - 4x \quad) + (y^2 + 2y \quad) = -1$$

in preparation for completing the square in each variable:

$$(x^2 - 4x + 4) + (y^2 + 2y + 1) = -1 + 4 + 1 = 4$$

(Note that the equation is balanced by adding $4 + 1$ to the right-hand side.) We then have

$$(x - 2)^2 + (y + 1)^2 = 4$$

which is the equation of a circle with center at $(2, -1)$ and radius 2. In terms of the $x'y'$ coordinate system with origin O' at $(2, -1)$, the equation becomes

$$x'^2 + y'^2 = 4$$

We see that the equation and analysis are simplified by translating the axes to the point $(h, k) = (2, -1)$ as shown in Figure 23.

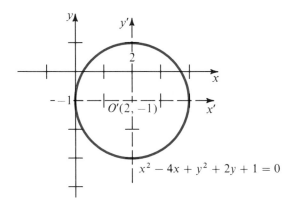

FIGURE 23   Diagram for Example 2

■

The technique of translation of axes can be applied to each of the conic sections. The results can be summarized as follows.

Standard Forms of the Conic Sections— Center or Vertex at (h, k)		
Circle:	$(x - h)^2 + (y - k)^2 = r^2$ Center: (h, k) Radius: r	
Parabola:	$(x - h)^2 = 4p(y - k)$ Vertex: (h, k) Axis: $x = h$ Directrix: $y = k - p$	$(y - k)^2 = 4p(x - h)$ Vertex: (h, k) Axis: $y = k$ Directrix: $x = h - p$
Ellipse:	$\dfrac{(x - h)^2}{a^2} + \dfrac{(y - k)^2}{b^2} = 1$ Center: (h, k)	
Hyperbola:	$\dfrac{(x - h)^2}{a^2} - \dfrac{(y - k)^2}{b^2} = 1$ $\dfrac{(y - k)^2}{a^2} - \dfrac{(x - h)^2}{b^2} = 1$ Center: (h, k)	

If we write the equation of a conic in standard form, we can perform a translation of axes to the origin $O'(h, k)$ and then analyze and sketch the graph in the simplified form relative to the $x'y'$ coordinate system.

EXAMPLE 3   Analyzing and Sketching a Conic Section

Sketch the graph of the equation

$$x^2 + 4x - 4y^2 + 24y - 48 = 0$$

SOLUTION

We rewrite the equation as

$$(x^2 + 4x \quad\ ) - 4(y^2 - 6y \quad\ ) = 48$$

and complete the square in both x and y.

$$(x^2 + 4x + 4) - 4(y^2 - 6y + 9) = 48 + 4 - 36$$
$$(x + 2)^2 - 4(y - 3)^2 = 16$$

Letting

$$x' = x + 2 \qquad \text{and} \qquad y' = y - 3$$

we rewrite the last equation as

$$x'^2 - 4y'^2 = 16$$

and after we divide both sides by 16,

$$\frac{x'^2}{16} - \frac{y'^2}{4} = 1$$

On the $x'y'$ coordinate system this is seen to be the equation of a hyperbola with center at $O'(-2, 3)$, $a = 4$, $b = 2$. Using the intercepts and asymptotes, the graph is sketched in Figure 24.

∎

WARNING

To complete the square in the equation

$$2(x^2 + 2x \quad\ ) - 3(y^2 - 4y \quad\ ) = 16$$

we must add 1 to the terms in x and 4 to the terms in y:

$$2(x^2 + 2x + 1) - 3(y^2 - 4y + 4) = 16 + 2 - 12$$

Note that adding 1 in the first parenthesis results in adding 2 to the left-hand side and is balanced by adding 2 to the right-hand side. Similarly, adding 4 in the second parenthesis results in adding -12 to the left-hand side, and this is also balanced on the right-hand side.

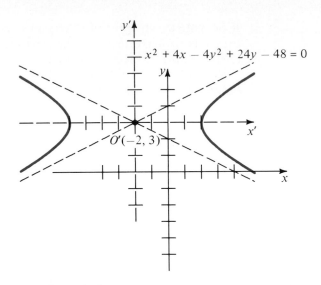

$$x^2 + 4x - 4y^2 + 24y - 48 = 0$$

$O'(-2, 3)$

FIGURE 24   **Diagram for Example 3**

PROGRESS CHECK

(a) Show that the graph of the equation

$$4x^2 + 16x + y^2 + 2y + 13 = 0$$

is an ellipse.

(b) Show that the graph of the equation

$$4y^2 - 24y - 25x^2 - 50x - 89 = 0$$

is a hyperbola.

EXAMPLE 4   **Analyzing and Sketching a Conic Section**

Sketch the graph of the equation

$$x^2 - 4x - 4y - 4 = 0$$

SOLUTION

We rewrite the equation as

$$(x^2 - 4x) = 4y + 4$$

and complete the square in x.

$$(x^2 - 4x + 4) = 4y + 4 + 4$$
$$(x - 2)^2 = 4y + 8 = 4(y + 2)$$

Letting

$$x' = x - 2 \qquad \text{and} \qquad y' = y + 2$$

we rewrite the last equation as

$$x'^2 = 4y'$$

On the $x'y'$ coordinate system this is seen to be the equation of a parabola with vertex at $O'(2, -2)$, $4p = 4$, so $p = 1$. The graph is sketched in Figure 25.

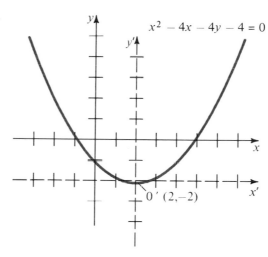

FIGURE 25 Diagram for Example 4

PROGRESS CHECK

Show that the graph of the equation

$$y^2 + 4y - 6x + 22 = 0$$

is a parabola.

It can be shown that the graph of the equation

$$Ax^2 + Cy^2 + Dx + Ey + F = 0$$

is a conic section—or a degenerate form of a conic section, such as a point, a line; a pair of lines, or no graph.

EXAMPLE 5 Identifying the Conic Section

Identify the graph of the equation

$$x^2 - 4x + y^2 - 2y + 9 = 0$$

SOLUTION

We rewrite the equation as

$$(x^2 - 4x) + (y^2 - 2y) = -9$$

and complete the square in both x and y.

$$(x^2 - 4x + 4) + (y^2 - 2y + 1) = -9 + 4 + 1$$
$$(x - 2)^2 + (y + 1)^2 = -4$$

Since the sum of two squares is nonnegative, this equation has no graph.

■

EXERCISE SET 9.6

In Exercises 1–4 the origin O' of the $x'y'$ coordinate system is at $(-1, 4)$. Find the $x'y'$ coordinates of the point whose xy coordinates are given.

1. $(0, 0)$ 2. $(-2, 1)$ 3. $(4, 3)$ 4. $(-6, -2)$

In Exercises 5–8 the origin O' of the $x'y'$ coordinate system is at $(-3, 4)$. Find the xy coordinates of the point whose $x'y'$ coordinates are given.

5. $(0, 0)$ 6. $(-2, 1)$ 7. $(4, 3)$ 8. $(-6, -2)$

In Exercises 9–18 sketch the graph of the given equation.

9. $36x^2 - 100y^2 + 216x + 99 = 0$

10. $x^2 - 4y^2 + 10x - 16y + 25 = 0$

11. $x^2 + 4x - y + 5 = 0$

12. $2x^2 - 12x + y + 21 = 0$

13. $y^2 + 2x + 15 = 0$

14. $16x^2 + 4y^2 + 12y - 7 = 0$

15. $x^2 + 4y^2 + 10x - 8y + 13 = 0$

16. $9x^2 + 25y^2 - 36x + 50y - 164 = 0$

17. $x^2 + 9y^2 - 54y + 72 = 0$

18. $x^2 - y^2 + 4x + 8y - 11 = 0$

In Exercises 19–26 identify the conic section whose equation is given.

19. $y^2 - 8x + 6y + 17 = 0$

20. $4x^2 + 4y^2 - 12x + 16y - 11 = 0$

21. $4x^2 + y^2 + 24x - 4y + 24 = 0$

22. $4x^2 - y^2 - 40x - 4y + 80 = 0$

23. $x^2 - y^2 + 6x + 4y - 4 = 0$

24. $2x^2 + y^2 - 4x + 4y - 12 = 0$

25. $25x^2 - 16y^2 + 210x + 96y + 656 = 0$

26. $x^2 - 10x + 8y + 1 = 0$

9.7
ROTATION OF AXES

In Section 9.6 we observed that the equation of a conic whose axis (or axes) is (or are) parallel to one of the coordinate axes can always be written in the form

$$Ax^2 + Cy^2 + Dx + Ey + F = 0 \tag{1}$$

where A and C are not both zero. Conversely, the graph of such an equation is a conic section or a degenerate conic section (a point, a line, a pair of lines, or no graph). The particular conic section can be obtained by completing the square and translating axes.

If the axis (or axes) of a conic section is (or are) not parallel to the coordinate axes, it can be shown that the equation of the conic section will then be of the form

$$Ax^2 + Bxy + Cy^2 + Dx + Ey + F = 0, \qquad B \neq 0 \tag{2}$$

Notice that Equation (2) differs from Equation (1) in the presence of the "cross-product" term Bxy.

An equation of the form given in Equation (2), where A, B, and C are not all zero, is called the **general second-degree equation** in two variables. We shall find that the graph of the general second-degree equation is the same as that of Equation (1). To sketch the graph of Equation (2) when $B \neq 0$, we perform a transformation to eliminate the xy

term, and we will show that this transformation is equivalent to a rotation of the coordinate axes.

In Figure 26a we show the original coordinate system, and in Figure 26b we obtain the new $x'y'$ coordinate system by rotating the x- and y-axes counterclockwise about the origin through an angle θ. Let P be a point in the plane. Then P has coordinates (x, y) in the original coordinate system and coordinates (x', y') in the rotated system. We can derive the relationship between the xy and $x'y'$ coordinate systems with the aid of Figure 27. Let r be the distance from the origin to P, and let α be the angle between the positive x'-axis and the line OP. Then

$$x' = r \cos \alpha \qquad y' = r \sin \alpha \qquad (3)$$

and

$$x = r \cos(\theta + \alpha) \qquad y = r \sin(\theta + \alpha) \qquad (4)$$

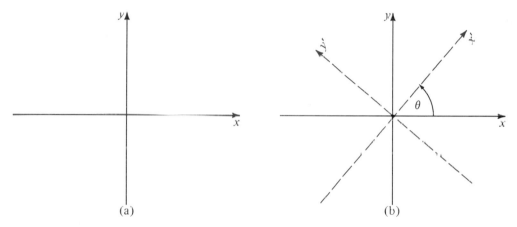

(a) (b)

FIGURE 26   **Rotation of Axes**

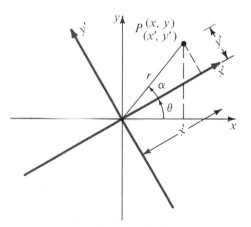

FIGURE 27   **Deriving the Formulas for Rotation of Axes**

Using the formulas from Section 7.2, we can write the Equations in (4) as

$$x = r[\cos\theta\cos\alpha - \sin\theta\sin\alpha]$$
$$y = r[\sin\theta\cos\alpha + \cos\theta\sin\alpha]$$

and, by substituting the expressions in (3), we obtain these formulas.

Rotation of Axes Formulas	$x = x'\cos\theta - y'\sin\theta$ $y = x'\sin\theta + y'\cos\theta$	(5)

EXAMPLE 1 Using the Formulas for Rotation of Axes

Suppose the coordinate axes are rotated counterclockwise through an angle $\theta = 45°$ to obtain the $x'y'$ coordinate system. Determine the equation of the curve

$$x^2 + xy + y^2 = 6$$

in the $x'y'$ system.

SOLUTION

Letting $\theta = 45°$ in Equation (5). we obtain the expressions

$$x = x'\cos 45° - y'\sin 45° = \frac{\sqrt{2}}{2}x' - \frac{\sqrt{2}}{2}y'$$

$$y = x'\sin 45° + y'\cos 45° = \frac{\sqrt{2}}{2}x' + \frac{\sqrt{2}}{2}y'$$

Substituting these expressions for x and y in the given equation we have

$$\left(\frac{\sqrt{2}}{2}x' - \frac{\sqrt{2}}{2}y'\right)^2 + \left(\frac{\sqrt{2}}{2}x' - \frac{\sqrt{2}}{2}y'\right)\left(\frac{\sqrt{2}}{2}x' + \frac{\sqrt{2}}{2}y'\right) + \left(\frac{\sqrt{2}}{2}x' + \frac{\sqrt{2}}{2}y'\right)^2 = 6$$

$$\frac{1}{2}x'^2 - x'y' + \frac{1}{2}y'^2 + \frac{1}{2}x'^2 - \frac{1}{2}y'^2 + \frac{1}{2}x'^2 + x'y' + \frac{1}{2}y'^2 = 6$$

$$\frac{3}{2}x'^2 + \frac{1}{2}y'^2 = 6$$

$$\frac{x'^2}{4} + \frac{y'^2}{12} = 1$$

which is the equation of an ellipse centered at the origin of the $x'y'$ system (Figure 28). ∎

Observe that in Example 1 we obtained an equation of the curve in the $x'y'$ system that had no cross-product term. This was accomplished by rotating the x and y coordinate axes in the counterclockwise direction through an angle θ until they were parallel to the axis or axes of the given curve. Our objective, then, is to determine the appropriate angle θ such that the substitutions

$$x = x'\cos\theta - y'\sin\theta$$
$$y = x'\sin\theta + y'\cos\theta$$

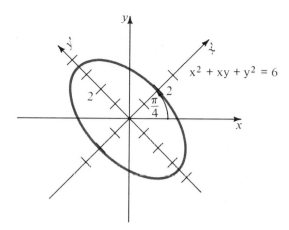

FIGURE 28   Diagram for Example 1

in the general second-degree equation

$$Ax^2 + Bxy + Cy^2 + Dx + Ey + F = 0, \qquad B \neq 0$$

will "drop out" the xy cross product. Substituting, we have

$$A(x'\cos\theta - y'\sin\theta)^2 + B(x'\cos\theta - y'\sin\theta)(x'\sin\theta + y'\cos\theta)$$
$$+ C(x'\sin\theta + y'\cos\theta)^2 + D(x'\cos\theta - y'\sin\theta)$$
$$+ E(x'\sin\theta + y'\cos\theta) + F = 0$$

or

$$A'x'^2 + B'x'y' + C'y'^2 + D'x' + E'y' + F' = 0$$

where (verify)

$$A' = A\cos^2\theta + B\cos\theta\sin\theta + C\sin^2\theta$$
$$B' = 2(C - A)\sin\theta\cos\theta + B(\cos^2\theta - \sin^2\theta)$$
$$C' = A\sin^2\theta - B\sin\theta\cos\theta + C\cos^2\theta$$
$$D' = D\cos\theta + E\sin\theta$$
$$E' = -D\sin\theta + E\cos\theta$$
$$F' = F$$

To eliminate the $x'y'$ term, we must have $B' = 0$ or

$$2(C - A)\sin\theta\cos\theta + B(\cos^2\theta - \sin^2\theta) = 0 \qquad (6)$$

Using the double-angle formulas (see Section 7.3), we can rewrite Equation (6) as

$$(C - A)\sin 2\theta + B\cos 2\theta = 0$$

which yields

$$\cot 2\theta = \frac{A - C}{B}, \qquad B \neq 0 \qquad (7)$$

Thus, by rotating the coordinate axes counterclockwise through the angle θ satisfying Equation (7), we obtain an equation in x' and y' that has no cross-product term. It is easy to see that we can always choose θ so that $0 < \theta < \pi/2$, and we will always seek an angle θ that satisfies this condition.

EXAMPLE 2   Using the Rotation Formulas to Graph a Conic Section

Discuss and sketch the graph of the equation $xy = 1$.

SOLUTION

We describe and illustrate the steps of the procedure.

<table>
<tr><td colspan="2" align="center">**Rotation of Axes**</td></tr>
<tr>
<td>*Step 1.* Determine the coefficients A, B, and C.</td>
<td>*Step 1.* $A = 0$ $B = 1$ $C = 0$</td>
</tr>
<tr>
<td>*Step 2.* Determine θ from the equation

$$\cot 2\theta = \frac{A - C}{B}$$</td>
<td>*Step 2.*

$$\cot 2\theta = \frac{A - C}{B} = 0$$

Then

$$2\theta = \frac{\pi}{2}$$

$$\theta = \frac{\pi}{4} \text{ or } 45°$$</td>
</tr>
<tr>
<td>*Step 3.* Form

$$x = x' \cos\theta - y' \sin\theta$$
$$y = x' \sin\theta + y' \cos\theta$$</td>
<td>*Step 3.*

$$x = x' \cos 45° - y' \sin 45°$$
$$= \frac{\sqrt{2}}{2} x' - \frac{\sqrt{2}}{2} y'$$
$$y = x' \sin 45° + y' \cos 45°$$
$$= \frac{\sqrt{2}}{2} x' + \frac{\sqrt{2}}{2} y'$$</td>
</tr>
<tr>
<td>*Step 4.* Substitute the expressions of Step 3 for x and y in the original equation.</td>
<td>*Step 4.*

$$xy = 1$$
$$\left(\frac{\sqrt{2}}{2} x' - \frac{\sqrt{2}}{2} y' \right)\left(\frac{\sqrt{2}}{2} x' + \frac{\sqrt{2}}{2} y' \right) = 1$$
$$\frac{x'^2}{2} - \frac{y'^2}{2} = 1$$</td>
</tr>
<tr>
<td>*Step 5.* Analyze the equation in terms of the $x'y'$ coordinate system. Sketch the graph.</td>
<td>*Step 5.* The equation is that of a hyperbola whose foci and intercepts are on the x'-axis. See Figure 29.</td>
</tr>
</table>

■

PROGRESS CHECK

Discuss and sketch the graph of the equation

$$31x^2 + 10\sqrt{3}xy + 21y^2 = 144$$

ANSWER

$$\frac{x'^2}{4} + \frac{y'^2}{9} = 1 \quad \text{(Figure 30)}$$

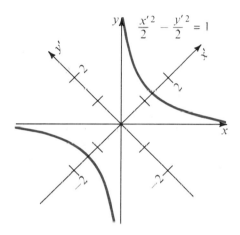

$$\frac{x'^2}{2} - \frac{y'^2}{2} = 1$$

FIGURE 29   Diagram for Example 2

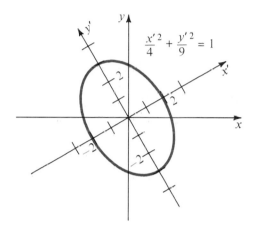

$$\frac{x'^2}{4} + \frac{y'^2}{9} = 1$$

FIGURE 30   Diagram for Progress Check

In the examples of rotation of axes considered so far, we first determined $\cot 2\theta$, then the angle θ, and finally $\sin \theta$ and $\cos \theta$. This is easy to do when the angle 2θ is one of the familiar angles. When it is not readily clear how to obtain the angle 2θ from its cotangent, we find $\cos 2\theta$ and then use the half-angle formulas (see Section 7.3):

$$\sin \theta = \sqrt{\frac{1 - \cos 2\theta}{2}} \qquad \cos \theta = \sqrt{\frac{1 + \cos 2\theta}{2}}$$

Note that θ is always a first-quadrant angle and we can therefore take positive values for $\sin \theta$ and $\cos \theta$.

EXAMPLE 3   Using the Rotation Formulas to Graph a Conic Section

Discuss and sketch the graph of the equation

$$97x^2 + 192xy + 153y^2 - 80x + 60y - 125 = 0$$

SOLUTION

We have $A = 97$, $B = 192$, and $C = 153$. Then

$$\cot 2\theta = \frac{A - C}{B} = \frac{97 - 153}{192} = -\frac{56}{192} = -\frac{7}{24}$$

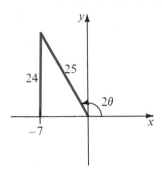

FIGURE 31 Diagram for Example 3

Using the triangle of Figure 31, $\cos 2\theta = -7/25$, so

$$\cos \theta = \sqrt{\frac{1 + \cos 2\theta}{2}} = \sqrt{\frac{1 - 7/25}{2}} = \frac{3}{5}$$

$$\sin \theta = \sqrt{\frac{1 - \cos 2\theta}{2}} = \sqrt{\frac{1 + 7/25}{2}} = \frac{4}{5}$$

Then the equations of rotation are

$$x = x'\cos\theta - y'\sin\theta = \frac{3}{5}x' - \frac{4}{5}y'$$

$$y = x'\sin\theta + y'\cos\theta = \frac{4}{5}x' + \frac{3}{5}y'$$

Substituting these expressions in the original equation, we have

$$97\left(\frac{3}{5}x' - \frac{4}{5}y'\right)^2 + 192\left(\frac{3}{5}x' - \frac{4}{5}y'\right)\left(\frac{4}{5}x' + \frac{3}{5}y'\right)$$

$$+ 153\left(\frac{4}{5}x' + \frac{3}{5}y'\right)^2 - 80\left(\frac{3}{5}x' - \frac{4}{5}y'\right) + 60\left(\frac{4}{5}x' + \frac{3}{5}y'\right) - 125 = 0$$

Expanding and simplifying, we obtain

$$9x'^2 + y'^2 + 4y' = 5$$

(Verify that this is correct.) Completing the square in y',

$$9x'^2 + (y' + 2)^2 = 9$$

or

$$x'^2 + \frac{(y' + 2)^2}{9} = 1$$

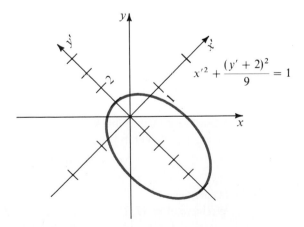

FIGURE 32 Diagram for Example 3

The graph is an ellipse whose center in the $x'y'$ system is at $(0, -2)$; $a = 3, b = 1$, and the major axis is along the y'-axis (Figure 32).

■

PROGRESS CHECK

Perform a rotation of axes so that the equation

$$x^2 + 2xy + y^2 + 4\sqrt{2}x - 4\sqrt{2}y = 0$$

does not have a cross-product term in the $x'y'$ coordinate system.

ANSWER

$x'^2 = 4y'$

EXERCISE SET 9.7

In Exercises 1–10 the coordinate axes are rotated counterclockwise through an angle θ to obtain the $x'y'$ coordinate system. Determine the equation of the given curve in the $x'y'$ system.

1. $\theta = 45°$; $5x^2 - 6xy + 5y^2 - 36 = 0$

2. $\theta = 30°$; $x^2 + y^2 = 25$

3. $\theta = 30°$; $x - \sqrt{3}y - 4 = 0$

4. $\theta - 45°$; $x^2 + 4xy + y^2 + \sqrt{2}x - 3\sqrt{2}y + 6 = 0$

5. $\theta = \cot^{-1}\left(\dfrac{4}{3}\right)$; $135x^2 + 65y^2 + 240xy - 270x - 390y - 225 = 0$

6. $\theta = \cot^{-1}\left(\dfrac{7}{24}\right)$; $1201x^2 - 336xy + 674y^2 - 625$

7. $\theta = \tan^{-1}\left(\dfrac{24}{7}\right)$; $66x + 137y - 0$

8. $\theta = 30°$; $x^2 + 2\sqrt{3}xy - y^2 + 6x = 0$

9. $\theta = 30°$; $-4x^2 + 2\sqrt{3}xy - 6y^2 - 6y = 0$

10. $\theta = 90°$; $x^2 - y^2 + 6x - 5y + 1 = 0$

In Exercises 11–20 perform a rotation of axes to obtain an equation without an xy term. Sketch the graph of the given equation.

11. $9x^2 + 4xy + 6y^2 - 5 = 0$

12. $154x^2 + 240xy + 54y^2 + 117 = 0$

13. $9x^2 + y^2 + 6xy - 10\sqrt{10}x + 10\sqrt{10}y + 90 = 0$

14. $4x^2 + 8xy + 4y^2 + 25\sqrt{2}x + 23\sqrt{2}y + 38 = 0$

15. $52x^2 - 72xy + 73y^2 - 120x - 90y + 125 = 0$

16. $5x^2 - 6xy + 5y^2 - 30\sqrt{2}x + 18\sqrt{2}y + 82 = 0$

17. $3x^2 + 4xy - \dfrac{38}{\sqrt{5}}x - \dfrac{4}{\sqrt{5}}y + 11 = 0$

18. $5x^2 + 12xy - 12\sqrt{13}x = 36$

19. $73x^2 + 72xy + 52y^2 + 510x + 320y + 825 = 0$

20. $8x^2 - 16xy + 8y^2 + 33\sqrt{2}x - 31\sqrt{2}y + 70 = 0$

21. Show that $B^2 - 4AC = B'^2 - 4A'C'$ for any angle θ. (*Hint:* Substitute Equations (5) in Equation (2).)

22. Show that $A + C = A' + C'$ for any angle θ. (*Hint:* Substitute Equations (5) in Equation (2).)

9.8
POLAR COORDINATES

Thus far we have represented a point P in the plane by the ordered pair (x, y), where x and y represent the distances of P from the y-axis and the x-axis, respectively. In this section we discuss the polar coordinate system, another useful way of representing points in the plane.

We start with a point O, called the **origin** or **pole**, draw a fixed ray, called the **polar axis**, with an endpoint at O, and select a unit of length for measuring distance. For any

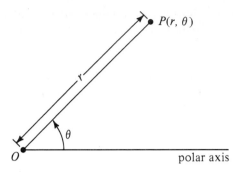

FIGURE 33   The Pole and Polar Axis

point P in the plane other than O, we draw a ray from O to P (Figure 33). If $r = \overline{OP}$, the length of the segment OP, and θ is the angle between the polar axis and the ray OP, then r and θ are called the **polar coordinates** of P, and P is denoted by $P(r, \theta)$, or simply by (r, θ). As usual, the angle θ is positive if it is measured in a counterclockwise direction and negative if measured in a clockwise direction. The angle θ may be expressed in degrees or in radians.

EXAMPLE 1   Plotting Polar Coordinates
Plot the points with the given polar coordinates.

(a) $P(3, 45°)$ **(b)** $P\left(3, \dfrac{\pi}{6}\right)$ **(c)** $P(2, -120°)$ **(d)** $P\left(2, -\dfrac{\pi}{6}\right)$

SOLUTION
See Figure 34.

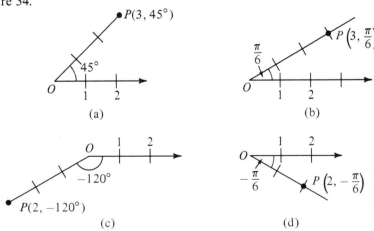

FIGURE 34   Diagram for Example 1

    There is an important difference between the polar coordinates and the rectangular coordinates of a point P. The rectangular coordinates of P are unique. However, P does not have unique polar coordinates. In fact, P has infinitely many polar coordinates since $(r, \theta + 2\pi n)$ designates the same point for all integer values of n.

PROGRESS CHECK

Give three other polar coordinate representations for the point $(3, -30°)$.

ANSWER

Possibilities are $(3, 330°), (3, -390°), (3, 690°)$.

When P is the origin, we find that $r = 0$, so any angle θ can be assigned to the ray OP. We therefore say that the pole has polar coordinates $(0, \theta)$, where θ is any angle.

It will also be convenient to consider polar coordinates where r is negative. We shall consider the point $(-r, \theta)$ to be the point $(r, \theta + 180°)$ where r is a positive real number.

EXAMPLE 2   Working with Polar Coordinates

Find polar coordinates for the point $P\left(-2, \dfrac{\pi}{4}\right)$ so that r is nonnegative.

SOLUTION

By definition, we have

$$P\left(-2, \frac{\pi}{4}\right) = P\left(2, \frac{\pi}{4} + \pi\right) = P\left(2, \frac{5}{4}\pi\right)$$

Moreover, the given point can also be designated as $P\left(2, \dfrac{5}{4}\pi + 2\pi n\right)$ for all integer values of n.

∎

PROGRESS CHECK

Which of the following polar coordinates represent the point $(-3, -45°)$?
(a) $(3, 45°)$ **(b)** $(3, 135°)$ **(c)** $(-3, 315°)$ **(d)** $(3, 225°)$

ANSWER

(b) and (c)

RECTANGULAR AND POLAR COORDINATES

It is often convenient to be able to convert between polar coordinates and rectangular coordinates. To obtain the relationship between the two coordinate systems, we let the polar axis coincide with the positive x-axis and the pole with the origin of the rectangular system. From Figure 35 we obtain the following relations.

Conversion Equations		
	$x = r \cos \theta \qquad y = r \sin \theta$	(1)
	$\tan \theta = \dfrac{y}{x} \qquad r = \sqrt{x^2 + y^2}$	(2)

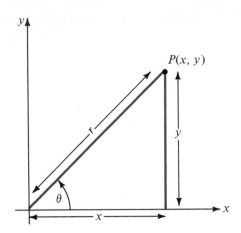

FIGURE 35 **Polar and Rectangular Coordinates**

EXAMPLE 3 Finding the Rectangular Coordinates

Find the rectangular coordinates of the point P whose polar coordinates are $(3, 225°)$.

SOLUTION

We let $r = 3$ and $\theta = 225°$ in Equation (1), obtaining

$$x = 3\cos 225° = 3\left(-\frac{\sqrt{2}}{2}\right) = -\frac{3}{2}\sqrt{2}.$$

$$y = 3\sin 225° = 3\left(-\frac{\sqrt{2}}{2}\right) = -\frac{3}{2}\sqrt{2}$$

Thus, the rectangular coordinates of P are $\left(-\frac{3}{2}\sqrt{2}, -\frac{3}{2}\sqrt{2}\right)$.

∎

Converting from rectangular to polar coordinates is a bit more involved, since a point has many polar coordinate representations.

EXAMPLE 4 Finding the Polar Coordinates

Find polar coordinates of the point P whose rectangular coordinates are $(\sqrt{3}, -1)$.

SOLUTION

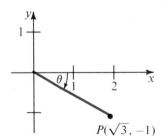

FIGURE 36 **Diagram for Example 4**

The given point is shown in Figure 36. Using the equations in (2) with $x = \sqrt{3}$ and $y = -1$, we have

$$r = \sqrt{x^2 + y^2} = \sqrt{4} = 2$$

$$\tan \theta = \frac{y}{x} = \frac{-1}{\sqrt{3}} = -\frac{\sqrt{3}}{3}$$

Since P lies in the fourth quadrant, θ may be taken as $-30°$ or as $330°$. Thus, two possible polar coordinate representations for P are

$$(2, -30°) \qquad \text{and} \qquad (2, 330°)$$

∎

PROGRESS CHECK

(**a**) Find the rectangular coordinates of the point P whose polar coordinates are $\left(2, \dfrac{\pi}{3}\right)$.

(**b**) Find polar coordinates of the point P whose rectangular coordinates are $(-1, -1)$.

ANSWERS

(a) $(1, \sqrt{3})$ (b) $(\sqrt{2}, 5\pi/4)$ (not unique)

POLAR EQUATIONS

An equation in the variables r and θ is called a **polar equation**. A **solution** of a polar equation is an ordered pair (a, b) that satisfies the polar equation when $r = a$ and $\theta = b$ are substituted in the equation. Just as the graph of an equation in x and y is obtained by plotting all the solutions of the equation on a rectangular coordinate system, the **graph of a polar equation** is obtained by plotting all the solutions of the equation on a **polar coordinate system.** When sketching the graph of a polar equation, it is convenient to superimpose the rectangular system on the polar coordinate system so that the origin coincides with the pole and the positive x-axis coincides with the polar axis.

EXAMPLE 5   Sketching a Polar Equation

Sketch the graph of the polar equation $r = 4\cos\theta$.

SOLUTION

We form the following table of values.

θ	0	$\dfrac{\pi}{6}$	$\dfrac{\pi}{4}$	$\dfrac{\pi}{3}$	$\dfrac{\pi}{2}$	$\dfrac{2}{3}\pi$	$\dfrac{3}{4}\pi$	$\dfrac{5}{6}\pi$	π
r	4	$2\sqrt{3}$	$2\sqrt{2}$	2	0	-2	$-2\sqrt{2}$	$-2\sqrt{3}$	-4

Plotting the points in this table, we seem to obtain a circle of radius 2, with center at $(2, 0)$, shown in Figure 37. As θ takes values from π to 2π, we obtain the same circle

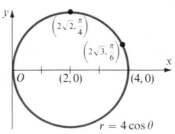

FIGURE 37   Diagram for Example 5

again. In Example 9 we shall show that the graph of this polar equation is indeed a circle of radius 2, with center at $(2, 0)$ in the xy coordinate system.

■

EXAMPLE 6   Sketching a Polar Equation

Sketch the graph of the polar equation $r = 1 + \sin \theta$.

SOLUTION

We form the following table of values.

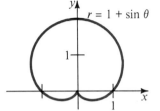

θ	0	$\dfrac{\pi}{6}$	$\dfrac{\pi}{3}$	$\dfrac{\pi}{2}$	$\dfrac{2}{3}\pi$	$\dfrac{5}{6}\pi$	π	$\dfrac{7}{6}\pi$	$\dfrac{4}{3}\pi$	$\dfrac{3}{2}\pi$	$\dfrac{5}{3}\pi$	$\dfrac{11}{6}\pi$	2π
r	1	$\dfrac{3}{2}$	$1 + \dfrac{\sqrt{3}}{2}$	2	$\dfrac{3}{2}$	$1 + \dfrac{\sqrt{3}}{2}$	1	$\dfrac{1}{2}$	$1 - \dfrac{\sqrt{3}}{2}$	0	$1 - \dfrac{\sqrt{3}}{2}$	$\dfrac{1}{2}$	1

FIGURE 38   Diagram for Example 6

Plotting these points, we obtain the heart-shaped curve shown in Figure 38, which is called a **cardioid.**

■

EXAMPLE 7   Sketching a Polar Equation

Sketch the graph of the polar equation $r = 2 \sin 2\theta$.

SOLUTION

Instead of using a table of values, we proceed as follows. When $\theta = 0$, $r = 0$. As θ increases from 0 to $\pi/4$, r increases from 0 to 2, since $\sin 2\theta$ increases from 0 to 1. As θ increases from $\pi/4$ to $\pi/2$, $\sin 2\theta$ decreases from 1 to 0, so r decreases from 2 to 0. Plotting these points, we obtain the loop shown in Figure 39a. Similarly, as θ increases from $\frac{1}{2}\pi$ to $\frac{3}{4}\pi$, r decreases from 0 to -2, and as θ increases from $\frac{3}{4}\pi$ to π, we find that r increases from -2 to 0. These points yield the second loop, shown in Figure 39b. As θ increases from π to 2π, we obtain two additional loops. The final figure, shown in Figure 39c, is called a **four-leaved rose.**

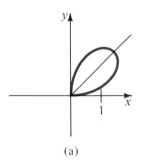

(a)

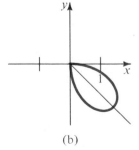

(b)

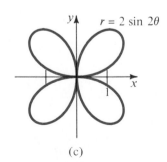

(c)

FIGURE 39   Diagram for Example 7

■

Many interesting curves are obtained as the graphs of polar equations, and some of these are given as exercises.

PROGRESS CHECK

Sketch the graph of the given polar equation.

(a) $r = 2 \sin \theta$ (b) $r = 1 + 2 \cos \theta$ (c) $r = \cos 3\theta$

ANSWER

See Figure 40.

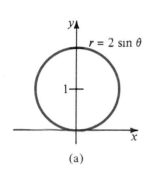

(a)

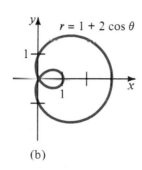

(b)

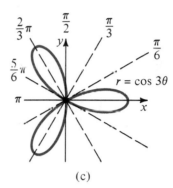

(c)

FIGURE 40 Diagram for Progress Check

WARNING

When an equation is given in the variables r and θ, we assume that the graph is to be sketched using polar coordinates. Don't use rectangular coordinates to sketch the equation

$$r = 2 \cos 2\theta$$

The conversion formulas enable us to transform polar equations to equations in x and y, and vice versa.

EXAMPLE 8 Sketching a Polar Equation

Transform the polar equation $r = 4 \cos \theta$ to an equation in x and y.

SOLUTION

This equation was plotted in Example 5. Multiplying both sides of the given equation by r, we have

$$r^2 = 4r \cos \theta$$

or

$$x^2 + y^2 = 4x$$

Completing the square in the last equation, we have

$$(x - 2)^2 + y^2 = 4$$

whose graph is a circle of radius 2 with center at $(2, 0)$. This result verifies our solution of Example 5.

∎

✓ PROGRESS CHECK

Transform the rectangular equation $x = 5$ to a polar equation.

ANSWER

$r \cos \theta = 5$

EXERCISE SET 9.8

1. Plot the points with the given polar coordinates.
   (a) $(4, 30°)$ (b) $(-2, 60°)$
   (c) $\left(5, \dfrac{\pi}{4}\right)$ (d) $\left(-3, -\dfrac{\pi}{2}\right)$

2. Plot the points with the given polar coordinates.
   (a) $(2, 225°)$ (b) $(-4, -150°)$
   (c) $\left(5, \dfrac{2}{3}\pi\right)$ (d) $\left(-4, -\dfrac{3}{4}\pi\right)$

3. For each point given in polar coordinates, give two other polar coordinate representations.
   (a) $(6, 135°)$ (b) $(-2, 120°)$
   (c) $\left(4, \dfrac{5}{6}\pi\right)$ (d) $\left(-4, -\dfrac{7}{4}\pi\right)$

4. For each point given in polar coordinates, give two other polar coordinate representations.
   (a) $(4, 315°)$ (b) $(-3, -150°)$
   (c) $\left(1, \dfrac{11}{6}\pi\right)$ (d) $\left(-1, -\dfrac{3}{2}\pi\right)$

5. For each point given in polar coordinates, give a polar coordinate representation with $r \geq 0$.
   (a) $(-2, 30°)$ (b) $(-4, -60°)$
   (c) $\left(-3, \dfrac{2}{3}\pi\right)$ (d) $\left(-1, -\dfrac{7}{6}\pi\right)$

6. For each point given in polar coordinates, give a polar coordinate representation with $r \geq 0$.
   (a) $(-2, 60°)$ (b) $(-3, 45°)$
   (c) $\left(-5, \dfrac{7}{6}\pi\right)$ (d) $\left(-4, -\dfrac{2}{3}\pi\right)$

7. Which of the following polar coordinates represent the point $(2, -30°)$?
   (a) $(-2, 150°)$ (b) $(-2, 330°)$
   (c) $(2, 330°)$ (d) $(2, 510°)$

8. Which of the following polar coordinates represent the point $\left(4, \dfrac{2}{3}\pi\right)$?
   (a) $\left(4, \dfrac{5}{3}\pi\right)$ (b) $\left(-4, \dfrac{5}{3}\pi\right)$
   (c) $\left(4, -\dfrac{4}{3}\pi\right)$ (d) $\left(4, -\dfrac{2}{3}\pi\right)$

9. Find the rectangular coordinates of the points with the given polar coordinates.
   (a) $(5, 330°)$ (b) $(2, 270°)$
   (c) $\left(4, \dfrac{\pi}{6}\right)$ (d) $\left(-3, -\dfrac{2}{3}\pi\right)$

10. Find the rectangular coordinates of the points with the given polar coordinates.
    (a) $(1, 315°)$ (b) $(3, 150°)$
    (c) $\left(2, \dfrac{3}{4}\pi\right)$ (d) $\left(-4, -\dfrac{4}{3}\pi\right)$

11. Find polar coordinates of the points with the given rectangular coordinates.
    (a) $(-2, 2)$ (b) $(1, -\sqrt{3})$
    (c) $(\sqrt{3}, 1)$ (d) $(-4, -4)$

12. Find polar coordinates of the points with the given rectangular coordinates.
    (a) $(-1, -\sqrt{3})$ (b) $(-\sqrt{3}, 1)$
    (c) $(3, -3)$ (d) $(3, 3)$

In Exercises 13–26 sketch the graph of the given polar equation.

13. $\theta = 45°$ 14. $\theta = \dfrac{2}{3}\pi$ 15. $r = 3$ 16. $r - 1 - \sin\theta$

17. $r = 2\sin\theta$ 18. $r = 2$ 19. $r = -3\cos\theta$ 20. $r = 2 + 4\sin\theta$

21. $r = 3\sin 5\theta$ 22. $r = 3\cos 4\theta$ 23. $r^2 = 1 + \sin 2\theta$ 24. $r^2 = 4\sin 2\theta$

25. $r = \theta, \theta \geq 0$ 26. $r\theta = 2$

In Exercises 27–32, transform the given polar equation to an equation in x and y.

27. $r = 4$ 28. $\theta = \dfrac{\pi}{4}$ 29. $r + 2\sin\theta = 0$ 30. $r = 3\sec\theta$

31. $r\cos\theta - 2$ 32. $r = 2\tan\theta$

In Exercises 33–38 transform the given rectangular equation to a polar equation.

33. $x^2 + y^2 = 25$ 34. $y = 3$ 35. $x = -5$ 36. $y^2 = 2x$

37. $y = 3x$ 38. $x^2 - y^2 = 9$

9.9 PARAMETRIC EQUATIONS

There are situations in which it is convenient to describe movement along a curve as a function of time. We may then consider each of the coordinates of a point $P(x, y)$ to be a function of time t, that is,

$$x = f(t) \qquad y = g(t)$$

These are called **parametric equations,** and the variable t is called a **parameter.** The equations may hold for all real values of t or, quite commonly, for all real values of t in an interval $[a, b]$. The following example illustrates how the graph of a curve described in parametric form can be sketched

EXAMPLE 1 Sketching from Parametric Equations
Sketch the curve whose parametric equations are

$$x = 2t \qquad y = t^2$$

for $1 \leq t \leq 4$.

SOLUTION
For each value of t in the interval $[1, 4]$ we obtain corresponding values of x and y. Some of these values are shown in Table 1. Plotting the points (x, y) and connecting

TABLE 1 Coordinates Expressed in Parametric Form

t	1	$\dfrac{3}{2}$	2	$\dfrac{5}{2}$	3	$\dfrac{7}{2}$	4
$x = 2t$	2	3	4	5	6	7	8
$y = t^2$	1	$\dfrac{9}{4}$	4	$\dfrac{25}{4}$	9	$\dfrac{49}{4}$	16
(x, y)	$(2, 1)$	$\left(3, \dfrac{9}{4}\right)$	$(4, 4)$	$\left(5, \dfrac{25}{4}\right)$	$(6, 9)$	$\left(7, \dfrac{49}{4}\right)$	$(8, 16)$

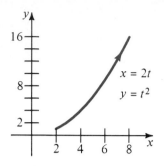

FIGURE 41   **Diagram for Example 1**

them by a smooth curve, we obtain the curve shown in Figure 41. The arrow on the curve indicates the direction of motion along the curve as the parameter t increases from 1 to 4.

To see if we can further identify the curve in Example 1, we eliminate the parameter t. Thus,

$$y = t^2 = \left(\frac{x}{2}\right)^2 = \frac{x^2}{4}$$

which is the familiar rectangular equation of a parabola.

■

Observe that a curve frequently has many different parametric equations. For example, the curve given in Example 1 can also be described by the parametric equations

$$x = 3t \qquad y = \frac{9}{4}t^2, \qquad -2 \le t \le 2$$

One of the main advantages of describing a curve by means of parametric equations with time t as the parameter is that as we move along the curve, we know *when* we arrive at each point on the curve.

EXAMPLE 2   Sketching from Parametric Equations

Discuss and sketch the curve whose parametric equations are

$$x = r \cos t \qquad y = r \sin t, \qquad 0 \le t \le 2\pi$$

where r is a positive constant.

SOLUTION

We show that the curve is the circle of radius r centered at the origin. We have

$$\begin{aligned} x^2 + y^2 &= r^2 \cos^2 t + r^2 \sin^2 t \\ &= r^2(\cos^2 t + \sin^2 t) \\ &= r^2 \end{aligned}$$

Thus, if $P(x, y)$ is a point on the curve, then $x^2 + y^2 = r^2$, so P lies on the circle of radius r centered at the origin. Observe that as t increases from 0 to 2π, the point P starts at $(r, 0)$ and traverses the circle in a counterclockwise direction, ending at $(r, 0)$, as shown in Figure 42a.

■

Consider now the curve whose parametric equations are

$$x = r \cos t \qquad y = r \sin t, \qquad 0 \le t \le 4\pi$$

In this case, the parametric equations describe the circle of radius r centered at the

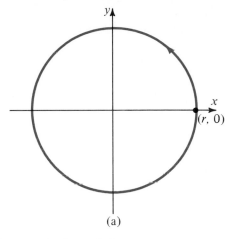

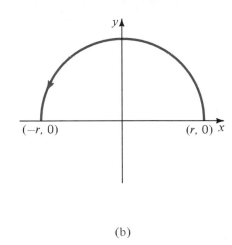

FIGURE 42   Diagram for Example 2

origin, but the circle is now traversed twice in a counterclockwise direction. Similarly, the parametric equations

$$x = r\cos t \qquad y = r\sin t, \qquad 0 \le t \le \pi$$

describe the upper half of the circle of radius r centered at the origin (Figure 42b).

PROGRESS CHECK
Discuss the curve whose parametric equations are

$$x = 2 + 3t \qquad y = 1 - 3t, \qquad t \ge 0$$

Eliminate the parameter t.

ANSWER
The curve consists of all points on the line $y = 3 - x$ for which $x \ge 2$.

EXAMPLE 3   Eliminating the Parameter
Discuss the curve whose parametric equations are

$$x = 4\cos t \qquad y = 3\sin t, \qquad 0 < t < 2\pi$$

SOLUTION
We eliminate the parameter t by first writing

$$\frac{x}{4} = \cos t \qquad \frac{y}{3} = \sin t$$

and then using the familiar trigonometric identity

$$\sin^2 t + \cos^2 t = 1$$

or

$$\left(\frac{x}{4}\right)^2 + \left(\frac{y}{3}\right)^2 = 1$$

$$\frac{x^2}{16} + \frac{y^2}{9} = 1$$

whose graph is an ellipse. The point P that traverses the curve starts when $t = 0$, so P starts at $(4, 0)$ and moves in a counterclockwise direction. The y-intercept $(0, 3)$ is reached when $t = \pi/2$.

∎

EXERCISE SET 9.9

In Exercises 1–12 sketch the curve with the given parametric equations, and find a rectangular equation for the curve by eliminating the parameter.

1. $x = 4t - 1$, $y = 2t$

2. $x = \sin t$, $y = \cos t$

3. $x = t^2$, $y = t^3$

4. $x = 3 \sin t + 1$, $y = 2 \cos t + 2$

5. $x = \dfrac{t}{2}$, $y = t^3$

6. $x = 2t$, $y = 3t$

7. $x = \dfrac{3t}{1 + t^3}$, $y = \dfrac{3t^2}{1 + t^3}$

8. $x = 3\sqrt{t - 3}$, $y = 2\sqrt{4 - t}$

9. $x = 4 \sec t$, $y = 3 \tan t$

10. $x = 2 \cos t$, $y = 2 \cos\left(\dfrac{1}{2}t\right)$

11. $x = 2 \sin t$, $y = 3 \cos t$

12. $x = \sin 3t$, $y = \cos 3t$

In Exercises 13–18 eliminate the parameter t to show that the given parametric equations describe the stated curve.

13. $x = \sec t$, $y = \tan t$; hyperbola

14. $x = 5 \sin t - 5$, $y = 4 \cos t + 4$; ellipse

15. $x = t$, $y = \dfrac{2}{t}$; hyperbola

16. $x = 3t^2 - 1$, $y = t^2 + 1$; line

17. $x = 2t + 1$, $y = -t^2$; parabola

18. $x = \sqrt{\dfrac{t - 1}{t + 1}}$, $y = \dfrac{1}{\sqrt{t + 1}}$; ellipse

TERMS AND SYMBOLS

analytic geometry **380**
asymptotes of a
    hyperbola **404**
axes of an ellipse **394**
axes of a hyperbola **400**
axis of a parabola **388**
center of a circle **235**
center of an ellipse **394**
conic sections **384**
conjugate axis **400**

directrix **388**
ellipse **394**
equation of a circle **385**
equation of a hyperbola
    402
equation of a parabola
    389
equation of an ellipse **395**
foci of a hyperbola **400**
foci of an ellipse **394**

focus of a parabola **388**
general second-degree
    equation **414**
hyperbola **400**
midpoint formula **382**
parabola **388**
parametric equation **429**
polar axis **421**
polar coordinates **422**
polar equation **425**

pole **421**
radius of a circle **385**
rotation of axes **414**
translation of axes **408**
transverse axis **400**
vertex of a parabola **388**
vertices of an ellipse **394**

KEY IDEAS FOR REVIEW

Topic	Page	Key Idea
The Midpoint Formula	382	The midpoint of the line segment joining the points $P_1(x_1, y_1)$ and $P_2(x_2, y_2)$ has coordinates $$\left(\frac{x_1 + x_2}{2}, \frac{y_1 + y_2}{2} \right)$$
Analytic geometry	380	Analytic geometry applies algebraic techniques to the study of geometry. Theorems from plane geometry can be proved using the methods of analytic geometry. In general, place the given geometric figure in a convenient position relative to the origin and axes. The distance formula, the midpoint formula, and the computation of slope are the basic tools to apply in proving a theorem.
Conic sections	384	The conic sections represent the possible intersections of a plane and a cone. The conic sections are the circle, parabola, ellipse, and hyperbola. (In special cases, these may reduce to a point, a line, two lines, or no graph.) Each conic section has a geometric definition that can be used to derive a second-degree equation in two variables whose graph corresponds to the conic.
equation (axis parallel to a coordinate axis)	414	The equation of a conic with an axis parallel to one of the coordinate axes can always be written as $$Ax^2 + Cy^2 + Dx + Ey + F = 0$$ where A and C are not both zero. Conversely, the graph of such an equation is a conic section or a degenerate conic section (a point, a line, a pair of lines, or no graph). The particular conic section can be found by completing the square and translating axes.
equation (no axis parallel to a coordinate axis)	414	If no axis of a conic section is parallel to a coordinate axis, its equation is of the form $$Ax^2 + Bxy + Cy^2 + Dx + Ey + F = 0, \qquad B \neq 0$$ Conversely, the graph of such an equation, where A, B, and C are not all zero, is a conic section, a line, a pair of lines, or no graph. The particular conic section can be obtained by first rotating axes to eliminate the xy term and then translating the resulting axes.
Polar coordinates	421	Polar coordinates specify a point P in the plane in terms of an angle θ and the distance r from the origin.
Parametric equations	429	Parametric equations provide an alternative way to describe a curve in the plane.

REVIEW EXERCISES

Solutions to exercises whose numbers are in color are in the Solutions section in the back of the book.

9.1 In Exercises 1 and 2 find the midpoint of the line segment whose endpoints are given.

   1. $(-5, 4), (3, -6)$ **2.** $(-2, 0), (-3, 5)$

   3. Find the coordinates of the point P_2 if $(2, 2)$ are the coordinates of the midpoint of the line segment joining $P_1(-6, -3)$ and P_2.

4. Use the distance formula to show that $P_1(-1, 2)$, $P_2(4, 3)$, $P_3(1, -1)$, and $P_4(-4, -2)$ are the coordinates of a parallelogram.

5. Show that the points $A(-8, 4)$, $B(5, 3)$, and $C(2, -2)$ are the vertices of a right triangle.

6. Find an equation of the perpendicular bisector of the line segment joining the points $A(-4, -3)$ and $B(1, 3)$. (The perpendicular bisector passes through the midpoint of AB and is perpendicular to AB.)

9.2　**7.** Write an equation of the circle whose center is at $(-5, 2)$ and whose radius is 4.

8. Write an equation of the circle whose center is at $(-3, 3)$ and whose radius is 2.

In Exercises 9 and 10 determine the center and radius of the circle with the given equation.

9. $(x - 2)^2 + (y + 3)^2 = 9$

10. $x^2 + y^2 + 4x - 6y = -10$

9.3　In Exercises 11 and 12 determine the focus and directrix of the given parabola, and sketch the graph.

11. $x^2 = -\dfrac{2}{3}y$　　　　**12.** $3y^2 + 2x = 0$

In Exercises 13 and 14 determine the equation of the parabola with its vertex at the origin that satisfies the given conditions.

13. directrix $y = \dfrac{7}{4}$

14. axis the y-axis; parabola passing through the point $\left(1, \dfrac{5}{2}\right)$

9.4　In Exercises 15 and 16 sketch and discuss the graph of the given ellipse, giving the coordinates of the foci and vertices.

15. $81x^2 + 4y^2 = 9$

16. $4x^2 + y^2 + 8x - 2y = 4$

In Exercises 17 and 18 find an equation of the ellipse satisfying the given conditions.

17. foci at $(\pm 3, 0)$, vertices at $(\pm 9, 0)$

18. foci at $(0, \pm 2)$, length of minor axis $= 2$

9.5　In Exercises 19 and 20 sketch and discuss the graph of the given hyperbola, giving the coordinates of the foci and vertices and including the asymptotes.

19. $4x^2 - y^2 - 2y = 4$　　　**20.** $7y^2 - x^2 = 4$

In Exercises 21 and 22 find an equation of the hyperbola with center at the origin that satisfies the given conditions.

21. foci at $(0, \pm 7)$, vertices at $(0, \pm 2)$

22. vertex at $(2, 0)$; hyperbola passing through the point $(3, 1)$

In Exercises 23–34 identify the conic section whose equation is given.

9.6　**23.** $2x^2 - 4x + y^2 = 0$

24. $4y + x^2 - 2x = 1$

25. $x^2 - 2x + y^2 - 4y = -6$

26. $-x^2 - 4x + y^2 + 4 = 0$

27. $y^2 + 2x - 4 = x^2 - 2x$

28. $x^2 - 4x + 2y = 6 + y^2$

9.7　**29.** $xy + 2x - 2y = 6$

30. $x^2 + y^2 + xy + x - y = 3$

31. $2x^2 - y^2 + 4xy - 2x + 3y = 6$

32. $x^2 + 4xy + 4y^2 - 3x = 6$

33. $13x^2 + 10xy + 13y^2 = 72$

34. $13x^2 - 32xy + 37y^2 - 4\sqrt{5}\,x - 2\sqrt{5}\,y = 40$

9.8　In Exercises 35 and 36 sketch the graph of the given polar equation.

35. $r = \sec\theta\tan\theta$　　　　**36.** $r = |\theta|$

37. Transform the equation $r = 3\csc\theta$ to an equation in x and y.

38. Transform the equation $x^2 + 9y^2 = 9$ to a polar equation.

9.9　**39.** Sketch the curve with parametric equations

$$x = t^2 - 2t + 1$$
$$y = (t - 1)^4 - 1$$

40. Show that the curve with parametric equations

$$x = \frac{\sqrt{t - 1}}{\sqrt{t + 2}} \qquad y = \frac{1}{\sqrt{t + 2}}$$

is an ellipse by eliminating the parameter t.

PROGRESS TEST 9A

1. Find the midpoint of the line segment whose endpoints are $(2, 4)$ and $(-2, 4)$.

2. Find the coordinates of the point P if $(-3, 3)$ are the coordinates of the midpoint of the line segment joining P and $Q\,(-5, 4)$.

3. By using the slope of a line, show that the points $A(-3, -1)$, $B(-5, 4)$, $C(2, 6)$, and $D(4, 1)$ determine a parallelogram.

4. Write an equation of the circle of radius 6 whose center is at $(2, -3)$.

5. Determine the center and radius of the circle $x^2 + y^2 - 2x + 4y = 1$.

6. Determine the vertex and axis of the parabola $x^2 + 6x + 2y + 7 = 0$, and sketch the graph.

In Problems 7–9 write the given equation in standard form and determine the intercepts.

7. $x^2 + 4y^2 = 4$ 8. $4y^2 - 9x^2 = 36$

9. $4x^2 - 4y^2 = 1$

10. Use the intercepts and asymptotes of the hyperbola $9x^2 - y^2 = 9$ to sketch its graph.

In Problems 11–15 identify the conic section.

11. $x^2 + 9y^2 - 4x + 6y + 4 = 0$

12. $x^2 - 3x + 4y = 2$

13. $3x^2 + 6xy + 3y^2 - 4x + 5y = 12$

PROGRESS TEST 9B

1. Find the midpoint of the line segment whose endpoints are $(-5, -3)$ and $(4, 1)$.

2. Find the coordinates of the point A if $(-2, -\frac{1}{2})$ are the coordinates of the midpoint of the line segment joining A and $B(3, -2)$.

3. Show that the diagonals of the quadrilateral whose vertices are $P(-3, 1)$, $Q(-1, 4)$, $R(5, 0)$, and $S(3, -3)$ are equal.

4. Write an equation of the circle of radius 5 whose center is at $(-2, -5)$.

5. Determine the center and radius of the circle $4x^2 + 4y^2 - 4x - 8y = 35$.

6. Determine the vertex and axis of the parabola $9x^2 + 18y - 6x + 7 = 0$, and sketch the graph.

In Problems 7–9 write the given equation in standard form and determine the intercepts.

7. $5x^2 + 9y^2 = 25$ 8. $7x^2 + 6y^2 = 21$

9. $y^2 + 3x^2 = 9$

10. Use the intercepts and asymptotes of the hyperbola $4y^2 - x^2 = 1$ to sketch its graph.

14. $x^2 - 2xy + y^2 = 4$

15. $x^2 + xy + y^2 = 1$

16. Sketch the graph of the polar equation $r = 4\csc\theta$.

17. Transform the equation $r = \sec^2\theta/2$ to an equation in x and y.

18. Transform the equation $x^2 + xy + y^2 = 1$ to a polar equation.

19. Sketch the curve with the parametric equations
$$x = \sqrt{t}, \qquad y = t - 1$$

20. Identify the curve with the parametric equations
$$x = 4\sin t + 1, \qquad y = 3\cos t - 1$$
by eliminating the parameter t.

In Problems 11–15 identify the conic section.

11. $5y^2 - 4x^2 - 6x + 2 = 0$

12. $x^2 + y^2 + 4x - 6y = -13$

13. $-x^2 - 2xy + y^2 = -1$

14. $2x^2 - 5xy + 2y^2 = 0$

15. $-x^2 + 4\sqrt{3}xy + 3y^2 = 7$

16. Sketch the graph of the polar equation $r = \csc^2\theta/2$.

17. Transform the equation $r = -\sin\theta + 4\cos\theta$ to an equation in x and y.

18. Transform the equation $x^2 - xy - y^2 = 1$ to a polar equation.

19. Sketch the curve with the parametric equations
$$x = \sin t, \qquad y = \cos 2t$$

20. Identify the curve with the parametric equations
$$x = t^2 - 2t, \qquad y = t + 1$$
by eliminating the parameter t.

CHAPTER 10

SYSTEMS OF EQUATIONS AND INEQUALITIES

Many problems in business and engineering require the solution of systems of equations and inequalities. In fact, systems of linear equations and inequalities occur with such frequency that mathematicians and computer scientists have devoted considerable energy to devising methods for their solution. With the aid of large-scale computers it is possible to solve systems involving thousands of equations or inequalities, a task that previous generations would not have dared tackle.

We begin with the study of the methods of substitution and elimination, methods that are applicable to all types of systems. We then introduce graphical methods for solving systems of linear inequalities and apply this technique to linear programming problems, a type of optimization problem.

The method of Gaussian elimination is used to introduce the material on matrices and determinants, which serves as an introduction to linear algebra. We will show that matrices and determinants provide neat schemes for automating the computational procedures for solving systems of linear equations.

10.1 SYSTEMS OF EQUATIONS

A pile of 9 coins consists of nickels and quarters. If the total value of the coins is $1.25, how many of each type of coin are there?

The natural way to approach this problem is to let

$$x = \text{the number of nickels}$$

and

$$y = \text{the number of quarters}$$

that is, to use two variables. The requirements can then be expressed as

$$x + \quad y = \quad 9$$
$$5x + 25y = 125$$

This is an example of a **system of equations,** and we seek values of x and y that satisfy *both* equations. An ordered pair (a, b) such that $x = a$, $y = b$ satisfies both equations is called a **solution** of the system. Thus,

$$x = 5 \qquad y = 4$$

is a solution because substituting in the equations of the system gives

$$5 + \quad 4 = \quad 9$$
$$5(5) + 25(4) = 125$$

SOLVING BY SUBSTITUTION

If we can use one of the equations of a system to express one variable in terms of the other variable, then we can *substitute* this expression in the other equation.

EXAMPLE 1 Solving by Substitution
Solve the system of equations.

$$x^2 + y^2 = 25$$
$$x + y = -1$$

SOLUTION
From the second equation we have

$$y = -1 - x$$

Substituting for y in the first equation,

$$x^2 + (-1 - x)^2 = 25$$
$$x^2 + 1 + 2x + x^2 = 25$$
$$2x^2 + 2x - 24 = 0$$
$$x^2 + x - 12 = 0$$
$$(x + 4)(x - 3) = 0$$

which yields $x = -4$ and $x = 3$. Substituting these values for x in the second equation, we obtain the corresponding values of y.

$$x = -4: \quad -4 + y = -1 \qquad x = 3: \quad 3 + y = -1$$
$$y = 3 \qquad\qquad\qquad\qquad y = -4$$

Thus, $x = -4$, $y = 3$ and $x = 3$, $y = -4$ are solutions of the system of equations.

Note that the equations represent a circle and a line; the algebraic solution tells us that they intersect in two points. See Figure 1.

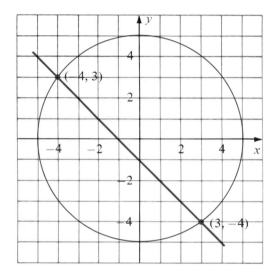

FIGURE 1 Diagram for Example 1

 PROGRESS CHECK
Solve the system of equations.

(a) $x^2 + 3y^2 = 12$ **(b)** $x^2 + y^2 = 34$
      $x + 3y = 6$ $x - y = 2$

ANSWERS
(b) $x = -3$, $y = -5$; $x = 5$, $y = 3$
(a) $x = 3$, $y = 1$; $x = 0$, $y = 2$

It is possible for a system of equations to have no solutions. Surprisingly, a system of equations may even have an infinite number of solutions. The following terminology is used to distinguish these situations.

**Consistent and
Inconsistent Systems**

- A **consistent** system of equations has one or more solution.

- An **inconsistent** system of equations has no solutions.

EXAMPLE 2 Consistent and Inconsistent Systems
Solve the system of equations.

Solve the system of equations.

(a) $x^2 - 2x - y + 3 = 0$ (b) $x + 4y = 10$
$ x + y - 1 = 0$ $ -2x - 8y = -20$

SOLUTION
(a) Solving the second equation for y, we have

$$y = 1 - x$$

and substituting in the first equation yields

$$x^2 - 2x - (1 - x) + 3 = 0$$
$$x^2 - x + 2 = 0$$

Since the discriminant of this quadratic equation is negative, the equation has no real roots. But any solution of the system of equations must satisfy this quadratic equation. We can therefore conclude that the system is inconsistent. The graphs of the equations are a parabola and a line that do not intersect (see Figure 2).

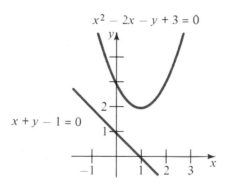

FIGURE 2 Diagram for Example 2a

(b) Solving the first equation for x, we have

$$x = 10 - 4y$$

and substituting in the second equation gives

$$-2(10 - 4y) - 8y = -20$$
$$-20 + 8y - 8y = -20$$
$$-20 = -20$$

The substitution procedure has resulted in an identity, indicating that any solution of the first equation will also satisfy the second equation. Since there are an infinite number of ordered pairs $x = a$, $y = b$ satisfying the first equation, the system is consistent and has an infinite number of solutions.

∎

PROGRESS CHECK
Solve by substitution.

(a) $\begin{aligned} 3x - y &= 7 \\ -9x + 3y &= -22 \end{aligned}$ (b) $\begin{aligned} -5x + 2y &= -4 \\ \frac{5}{2}x - y &= 2 \end{aligned}$

ANSWERS

(a) no solution   (b) any point on the line $-5x + 2y = -4$

WARNING
The expression for x or y obtained from an equation *must not be substituted in the same equation*. From the first equation of the system

$$x + 2y = -1$$
$$3x^2 + y = 2$$

we obtain

$$x = -1 - 2y$$

Substituting (*incorrectly*) in the same equation would result in

$$(-1 - 2y) + 2y = -1$$
$$-1 = -1$$

The substitution $x = -1 - 2y$ must be made in the *second* equation.

SOLVING BY ELIMINATION   The method of elimination seeks to combine the equations of a system in such a way as to *eliminate* one of the variables.

EXAMPLE 3   Solving by Elimination
Solve the system.

$$4x^2 + 9y^2 = 36$$
$$-9x^2 + 18y^2 = 4$$

SOLUTION
We can employ the method of elimination to obtain an equation that has just one variable. If we multiply the first equation by -2 and add the result to the second

equation, we have

$$-8x^2 - 18y^2 = -72$$
$$\underline{-9x^2 + 18y^2 = 4}$$
$$-17x^2 = -68$$
$$x^2 = 4$$
$$x = \pm 2$$

We can now substitute $+2$ and -2 for x in either of the original equations. Using the first equation,

$$4x^2 + 9y^2 = 36 \qquad\qquad 4x^2 + 9y^2 = 36$$
$$4(2)^2 + 9y^2 = 36 \qquad\qquad 4(-2)^2 + 9y^2 = 36$$
$$y = \pm\frac{2}{3}\sqrt{5} \qquad\qquad\qquad y = \pm\frac{2}{3}\sqrt{5}$$

We then have four solutions: $x = 2$, $y = 2\sqrt{5}/3$; $x = 2$, $y = -2\sqrt{5}/3$; $x = -2$, $y = 2\sqrt{5}/3$; $x = -2$, $y = -2\sqrt{5}/3$. The graphs of the equations are an ellipse and a hyperbola that intersect in four points as shown in Figure 3.

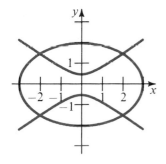

FIGURE 3 Diagram for Example 3

■

☑ **PROGRESS CHECK**
Find the real solutions of the system.

$$x^2 - 4x + y^2 - 4y = 1$$
$$x^2 - 4x + y = -5$$

ANSWER
x = 2, y = − − 1 (The parabola is tangent to the circle.)

SYSTEMS OF LINEAR EQUATIONS

If the equations of a system are of the first degree in x and y, we call this a **system of linear equations** or simply a **linear system.** The methods of substitution and elimination can be used to solve a linear system.

When we graph two linear equations on the same set of coordinate axes, there are three possibilities.

(**a**) The two lines intersect in a point (Figure 4a). The system is consistent and has a unique solution, the point of intersection.

(**b**) The two lines are parallel (Figure 4b). Since the lines do not intersect, the linear system has no solution and is inconsistent.

(**c**) The equations are different forms of the same line (Figure 4c). The system is consistent and has an infinite number of solutions, namely, any point on the line.

EXAMPLE 4 Applying Linear Systems
If 3 sulfa pills and 4 penicillin pills cost 69 cents, while 5 sulfa pills and 2 penicillin pills cost 73 cents, what is the cost of each type of pill?

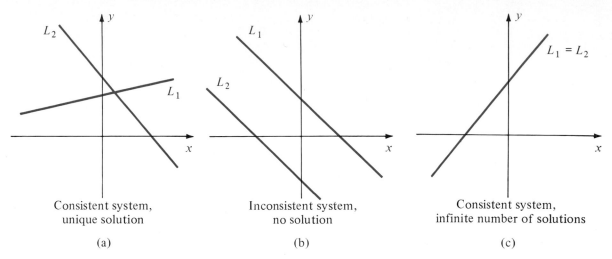

FIGURE 4 **Possible Graphs of a Pair of Linear Equations**

SOLUTION
Using two variables, we let

$$x = \text{the cost of each sulfa pill}$$
$$y = \text{the cost of each penicillin pill}$$

Then

$$3x + 4y = 69$$
$$5x + 2y = 73$$

We multiply the second equation by -2 and add to eliminate y:

$$
\begin{array}{rcr}
3x + 4y = & & 69 \\
-10x - 4y = & & -146 \\
\hline
7x \quad\quad = & & -77 \\
x = & & 11
\end{array}
$$

Substituting in the first equation, we have

$$3(11) + 4y = 69$$
$$4y = 36$$
$$y = 9$$

Thus, each sulfa pill costs 11 cents and each penicillin pill costs 9 cents.

∎

EXERCISE SET 10.1

In Exercises 1–10 solve the system of equations by the method of substitution.

1. $x + y = 1$
   $x - y = 3$

2. $x + 2y = 8$
   $3x - 4y = 4$

3. $x^2 + y^2 = 13$
   $2x - y = 4$

4. $x^2 + 4y^2 = 32$
   $x + 2y = 0$

5. $y^2 - x = 0$
   $y - 4x = -3$

6. $xy = -4$
   $4x - y - 8$

7. $x^2 - 2x + y^2 = 3$
   $2x + y = 4$

8. $4x^2 + y^2 = 4$
   $x - y = 3$

9. $xy = 1$
   $x - y + 1 = 0$

10. $\frac{1}{2}x - \frac{3}{2}y = 4$
    $\frac{3}{2}x + y = 1$

In Exercises 11–20 solve the system of equations by the method of elimination.

11. $x + 2y = 1$
    $5x + 2y - 13$

12. $x - 4y = -7$
    $2x + 3y = 8$

13. $25y^2 - 16x^2 = 400$
    $9y^2 - 4x^2 - 36$

14. $x^2 - y^2 = 3$
    $x^2 + y^2 = 5$

15. $4x^2 + 9y^2 = 72$
    $4x - 3y^2 = 0$

16. $x^2 + y^2 + 2y = 9$
    $y - 2x = 4$

17. $3x - y = 4$
    $6x - 2y = -8$

18. $2x + 3y = -2$
    $-3x - 5y = 4$

19. $2y^2 - x^2 = -1$
    $4y^2 + x^2 = 25$

20. $x^2 + 4y^2 = 25$
    $4x^2 + y^2 = 25$

In Exercises 21–32 determine whether the system is consistent (C) or inconsistent (I). If the system is consistent, find all solutions.

21. $2x + 2y = 6$
    $3x + 3y = 6$

22. $2x + y - 2$
    $3x - y = 8$

23. $y^2 - 8x^2 = 9$
    $y^2 + 3x^2 = -31$

24. $4y^2 + 3x^2 = 24$
    $3y^2 - 2x^2 = 35$

25. $3x + 3y = 9$
    $2x + 2y = -6$

26. $x - 4y = -7$
    $2x - 8y = -4$

27. $3x - y = 18$
    $\frac{3}{2}x - \frac{1}{2}y - 9$

28. $2x + y = 6$
    $x + \frac{1}{2}y = 3$

29. $x^2 - 3xy - 2y^2 - 2 = 0$
    $x - y - 2 = 0$

30. $3x^2 + 8y^2 = 20$
    $x^2 + 4y^2 = 10$

31. $2x - 2y = 4$
    $x - y = 8$

32. $2x - 3y = 8$
    $4x - 6y = 16$

In Exercises 33–42 use a pair of equations to solve the given problem.

33. A pile of 34 coins worth $4.10 consists of nickels and quarters. Find the number of each type of coin.

34. Car A can travel 20 kilometers per hour faster than car B. If car A travels 240 kilometers in the same time that car B travels 200 kilometers, what is the speed of each car?

35. How many pounds of nuts worth $2.10 per pound and how many pounds of raisins worth $0.90 per pound must be mixed to obtain a mixture of two pounds that is worth $1.62 per pound?

36. A part of $8000 was invested at an annual interest of 7% and the remainder at 8%. If the total interest received at the end of one year is $590, how much was invested at each rate?

37. The owner of a service station sold 1325 gallons of gasoline and collected 200 ration tickets. If Type-A ration tickets are used to purchase 10 gallons of gasoline and Type-B are used to purchase 1 gallon of gasoline, how many of each type of ration ticket did the station collect?

38. A bank is paying 12% annual interest on one-year certificates, and treasury notes are paying 10% annual interest. An investor received $620 interest at the end of one year by investing a total of $6000. How much was invested at each rate?

39. The sum of the squares of the sides of a rectangle is 100 square meters. If the area of the rectangle is 48 square meters, find the length of each side of the rectangle.

40. Find the dimensions of a rectangle with an area of 30 square feet and a perimeter of 22 feet.

41. Find two numbers such that their product is 20 and their sum is 9.

42. Find two numbers such that the sum of their squares is 65 and their sum is 11.

**10.2
SYSTEMS OF LINEAR
INEQUALITIES**

**GRAPHING LINEAR
INEQUALITIES**

When we draw the graph of a linear equation, say

$$y = 2x - 1$$

we can readily see that the graph of the line divides the plane into two regions, which we call **half-planes** (see Figure 5). If, in the equation $y = 2x - 1$, we replace the equals sign with any of the symbols $<, >, \leq,$ or $\geq$, we have a **linear inequality in two variables.** By the **graph of a linear inequality** such as

$$y < 2x - 1$$

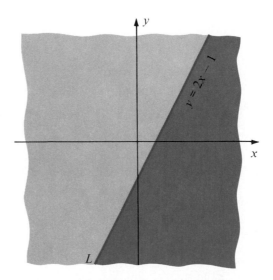

FIGURE 5   Half-planes Formed by the Graph of a Linear Equation

we mean the set of all points whose coordinates satisfy the inequality. Thus, the point $(4, 2)$ lies on the graph of $y < 2x - 1$, since

$$2 < (2)(4) - 1 = 7$$

shows that $x = 4$, $y = 2$ satisfies the inequality. However, the point $(1, 5)$ does *not* lie on the graph of $y < 2x - 1$ since

$$5 < (2)(1) - 1 = 1$$

is not true. Since the coordinates of every point on the line L in Figure 5 satisfy the *equation* $y = 2x - 1$, we readily see that the coordinates of those points in the half-plane below the line must satisfy the *inequality* $y < 2x - 1$. Similarly, the coordinates of those points in the half-plane above the line must satisfy the *inequality* $y > 2x - 1$. This suggests that the graph of a linear inequality in two variables is a half-plane, and it leads to a straightforward method for graphing linear inequalities.

EXAMPLE 1   Graphing a Linear Inequality
Sketch the graph of the inequality $x + y \geq 1$.

SOLUTION

Graphing Linear Inequalities	
Step 1. Replace the inequality sign with an equals sign and plot the line. (a) If the inequality is $\leq$ or $\geq$, plot a solid line (points on the line will satisfy the inequality). (b) If the inequality is $<$ or $>$, plot a dashed line (points on the line will not satisfy the inequality).	*Step 1.* $x + y = 1$ 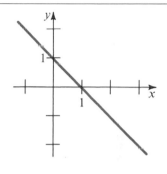
Step 2. Choose any point that is not on the line as a test point. If the origin is not on the line, it is the most convenient choice.	*Step 2.* Choose $(0, 0)$ as a test point.
Step 3. Substitute the coordinates of the test point into the inequality. (a) If the test point satisfies the inequality, then the coordinates of every point in the half-plane that contains the test point will satisfy the inequality.	*Step 3.* Substituting $(0, 0)$ in $$x + y \geq 1$$ gives $$0 + 0 \geq 1 \quad (?)$$ $$0 \geq 1$$ which is false.
(b) If the test point does not satisfy the inequality, then the half-plane on the other side of the line contains all the points satisfying the inequality.	Since $(0, 0)$ is in the half-plane below the line and does not satisfy the inequality, all the points above the line will satisfy the inequality. See Figure 6. 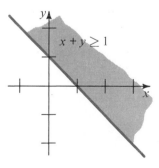 **FIGURE 6 Diagram for Example 1**

■

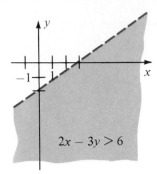

FIGURE 7 Diagram for Example 2

EXAMPLE 2 Graphing a Linear Inequality

Sketch the graph of the inequality $2x - 3y > 6$.

SOLUTION

We first graph the line $2x - 3y = 6$. We draw a dashed or broken line to indicate that $2x - 3y = 6$ is not part of the graph (see Figure 7). Since $(0, 0)$ is not on the line, we can use it as a test point.

$$2x - 3y > 6$$
$$2(0) - 3(0) > 6 \quad (?)$$
$$0 - 0 > 6 \quad (?)$$
$$0 > 6$$

is false. Since $(0, 0)$ is in the half-plane above the line, the graph consists of the half-plane below the line.

∎

PROGRESS CHECK

Graph the inequalities.
(a) $y \leq 2x + 1$ (b) $y + 3x > -2$ (c) $y \geq -x + 1$

ANSWERS

(a)

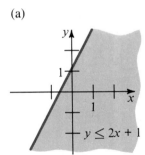

(b)

(c)

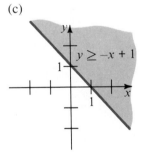

EXAMPLE 3 Graphing a Linear Inequality

Graph the inequalities.
(a) $y > x$ (b) $2x \geq 5$

SOLUTION

(a) Since the origin lies on the line $y = x$, we choose another test point, say $(0, 1)$ above the line. Since $(0, 1)$ does satisfy the inequality, the graph of the inequality is the half-plane above the line. See Figure 8a.

(b) The graph of $2x = 5$ is a vertical line, and the graph of $2x \geq 5$ is the half-plane to the right of the line and also the line itself. See Figure 8b.

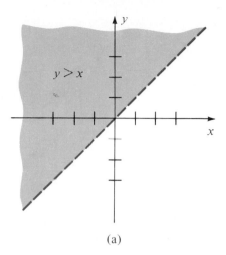
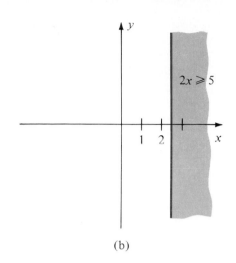

(a)

(b)

FIGURE 8　Diagram for Example 3

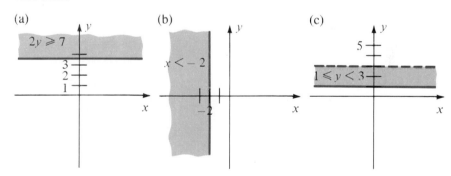

✓ **PROGRESS CHECK**

Graph the inequalities.

(a) $2y \geq 7$ **(b)** $x < -2$ **(c)** $1 \leq y < 3$

ANSWERS

(a)

(b)

(c)

SYSTEMS OF LINEAR INEQUALITIES

We may also consider **systems of linear inequalities** in two variables, x and y. Examples of such systems are

$$
\begin{array}{ll}
2x - 3y > 6 & 2x - 5y \leq 12 \\
x + 2y < 2 & 2x + y \leq 18 \\
& x \geq 0 \\
& y \geq 0
\end{array}
$$

The **solution of a system of linear inequalities** consists of all ordered pairs (a, b) such that the substitution $x = a, y - b$ satisfies *all* the inequalities. Thus, the ordered pair $(2, 1)$ is a solution of the system

$$2x - 3y \leq 2$$
$$x + y \leq 6$$

since the substitution $x = 2$, $y = 1$ satisfies both inequalities.

$$(2)(2) - (3)(1) = 1 \leq 2$$
$$2 + 1 = 3 \leq 6$$

We can graph the solution set of a system of linear inequalities by graphing the solution set of each inequality and marking that portion of the graph that satisfies *all* the inequalities.

EXAMPLE 4   Graphing a System of Linear Inequalities

Graph the solution set of the system.

$$2x - 3y \leq 2$$
$$x + y \leq 6$$

SOLUTION

In Figure 9 we have graphed the solution set of each of the inequalities. The cross-

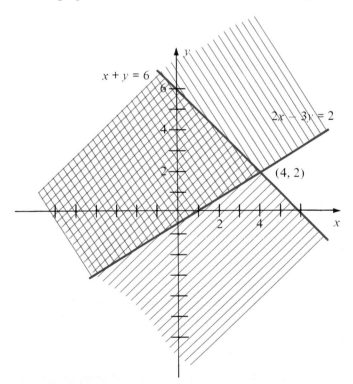

FIGURE 9   Diagram for Example 4

hatched region indicates those points that satisfy both inequalities and is therefore the solution set of the system of inequalities.

■

EXAMPLE 5 Graphing a System of Linear Inequalities

Graph the solution set of the system.

$$x + y < 2$$
$$2x + 3y \geq 9$$
$$x \geq 1$$

SOLUTION

See Figure 10. Since there are no points satisfying *all* the inequalities, we conclude that the system is inconsistent and has no solutions.

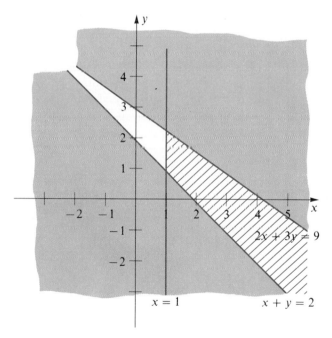

FIGURE 10 Diagram for Example 5

■

☑ PROGRESS CHECK

Graph the solution set of the given system.

(a) $x + y \geq 3$ (b) $2x + y \leq 4$

     $x + 2y < 8$ $x + y \leq 3$

                        $x \geq 0$

                        $y \geq 0$

ANSWERS

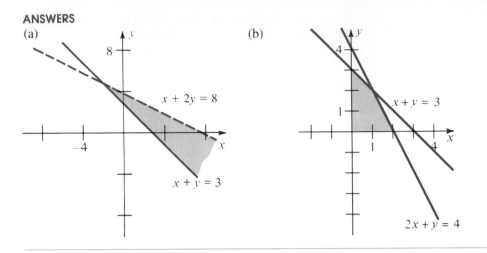

(a)

$x + 2y = 8$

$x + y = 3$

(b)

$x + y = 3$

$2x + y = 4$

EXAMPLE 6 Applying Systems of Linear Inequalities

A dietitian at a university is planning a menu for a meal to consist of two primary foods, A and B, whose nutritional contents are shown in the table. The dietitian insists that the meal provide at most 12 units of fat, at least 2 units of carbohydrate, and at least 1 unit of protein. If x and y represent the number of grams of food types A and B, respectively, write a system of linear inequalities expressing the restrictions. Graph the solution set.

	Nutritional Content in Units per Gram		
	Fat	Carbohydrate	Protein
A	2	2	0
B	3	1	1

SOLUTION

The number of units of fat contained in the meal is $2x + 3y$, so x and y must satisfy the inequality

$$2x + 3y \leq 12 \quad \text{fat requirement}$$

Similarly, the requirements for carbohydrate and protein result in the inequalities

$$2x + y \geq 2 \quad \text{carbohydrate requirement}$$
$$y \geq 1 \quad \text{protein requirement}$$

Of course, we must also have $x \geq 0$, since negative quantities of food type A would make no sense. The system of linear inequalities is then

$$2x + 3y \leq 12$$
$$2x + y \geq 2$$
$$x \geq 0$$
$$y \geq 1$$

and the graph is shown in Figure 11.

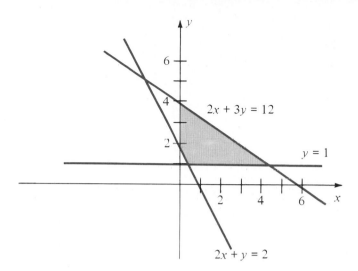

FIGURE 11 **Diagram for Example 6**

EXERCISE SET 10.2

In Exercises 1–18 graph the solution set of the given inequality.

1. $y \le x + 2$
2. $y \ge x + 3$
3. $y > x - 4$
4. $y < x - 5$
5. $y \le 4 - x$
6. $y \ge 2 - x$
7. $y > x$
8. $y \le 2x$
9. $3x - 5y > 15$
10. $2y - 3x < 12$
11. $x \le 4$
12. $3x > -2$
13. $y > -3$
14. $5y \le 25$
15. $x > 0$
16. $y < 0$
17. $-2 \le x \le 3$
18. $-6 < y < -2$

19. A steel producer makes two types of steel, regular and special. A ton of regular steel requires 2 hours in the open-hearth furnace and a ton of special steel requires 5 hours. Let x and y denote the number of tons of regular and special steel, respectively, made per day. If the open-hearth furnace is available at most 15 hours per day, write an inequality that must be satisfied by x and y. Graph this inequality.

20. A patient is placed on a diet that restricts caloric intake to 1500 calories per day. The patient plans to eat x ounces of cheese, y slices of bread, and z apples on the first day of the diet. If cheese contains 100 calories per ounce, bread 110 calories per slice, and apples 80 calories each, write an inequality that must be satisfied by x, y, and z.

In Exercises 21–32 graph the solution set of the system of linear inequalities.

21. $2x - y \le 3$
    $2x + 3y \ge -3$

22. $x - y \le 4$
    $2x + y \ge 6$

23. $3x - y \ge -7$
    $3x + y \le -2$

24. $3x - 2y > 1$
    $2x + 3y \le 18$

25. $3x - 2y \ge -4$
    $2x - y \le 5$
    $y \ge 1$

26. $2x - y \ge -3$
    $x + y \le 5$
    $y \ge 1$

27. $2x - y \le 5$
    $x + 2y \ge 1$
    $x \ge 0$
    $y \ge 0$

28. $-x + 3y \le 2$
    $4x + 3y \le 18$
    $x \ge 0$
    $y \ge 0$

29. $3x + y \le 6$
    $x - 2y \le -1$
    $x \ge 2$

30. $x - y \ge -2$
    $x + y \ge -5$
    $y \ge 0$

31. $3x - 2y \leq -6$
    $8x + 3y \leq 24$
    $5x + 4y \geq 20$
    $x \geq 0$
    $y \geq 0$

32. $2x + 3y \geq 18$
    $x + 3y \geq 12$
    $4x + 3y \geq 24$
    $x \geq 0$
    $y \geq 0$

33. A farmer has 10 quarts of milk and 15 quarts of cream, which he will use to make ice cream and yogurt. Each quart of ice cream requires 0.4 quart of milk and 0.2 quart of cream, and each quart of yogurt requires 0.2 quart of milk and 0.4 quart of cream. Graph the set of points representing the possible production of ice cream and of yogurt.

34. A coffee packer uses Jamaican and Colombian coffee to prepare a mild blend and a strong blend. Each pound of mild blend contains $\frac{1}{2}$ pound of Jamaican coffee and $\frac{1}{2}$ pound of Colombian coffee, and each pound of the strong blend requires $\frac{1}{4}$ pound of Jamaican coffee and $\frac{3}{4}$ pound of Colombian coffee. The packer has available 100 pounds of Jamaican coffee and 125 pounds of Colombian coffee. Graph the set of points representing the possible production of the two blends.

35. A trust fund of $100,000 that has been established to provide university scholarships must adhere to certain restrictions.
    (a) No more than half of the fund may be invested in common stocks.
    (b) No more than $35,000 may be invested in preferred stocks.

(c) No more than $60,000 may be invested in all types of stocks.

(d) The amount invested in common stock cannot be more than twice the amount invested in preferred stocks.

Graph the solution set representing the possible investments in common and preferred stocks.

36. An institution serves a luncheon consisting of two dishes, A and B, whose nutritional content in grams per unit served is given in the accompanying table.

	Fat	Carbohydrate	Protein
A	1	1	2
B	2	1	6

The luncheon is to provide no more than 10 grams of fat, no more than 7 grams of carbohydrate, and at least 6 grams of protein. Graph the solution set of possible quantities of dishes A and B.

10.3
SYSTEMS OF LINEAR EQUATIONS IN THREE UNKNOWNS

The method of substitution and the method of elimination can both be applied to systems of linear equations in three unknowns and, more generally, to systems of linear equations in any number of unknowns. There is yet another method, ideally suited for computers, which we will now apply to solving linear systems in three unknowns.

GAUSSIAN ELIMINATION AND TRIANGULAR FORM

In solving equations, we found it convenient to transform an equation into an equivalent equation having the same solution set. Similarly, we can attempt to transform a system of equations into another system, called an **equivalent system,** that has the same solution set. In particular, the objective of **Gaussian elimination** is to transform a linear system into an equivalent system in triangular form such as

$$3x - y + 3z = -11$$
$$2y + z = 2$$
$$2z = -4$$

A linear system is in **triangular form** when the only nonzero coefficient of x appears in the first equation, the only nonzero coefficients of y appear in the first and second equations, and so on.

Note that when a linear system is in triangular form, the last equation immediately yields the value of an unknown. In our example, we see that

$$2z = -4$$
$$z = -2$$

Substituting $z = -2$ in the second equation yields

$$2y + (-2) = 2$$
$$y = 2$$

Finally, substituting $z = -2$ and $y = 2$ in the first equation yields

$$3x - (2) + 3(-2) = -11$$
$$3x = -3$$
$$x = -1$$

This process of **back-substitution** thus allows us to solve a linear system quickly when it is in triangular form.

The challenge, then, is to find a means of transforming a linear system into triangular form. We now offer (without proof) a list of operations that transform a system of linear equations into an equivalent system.

(1) Interchange any two equations.

(2) Multiply an equation by a nonzero constant.

(3) Replace an equation by the sum of itself plus a constant times another equation. Using these operations we can now demonstrate the method of Gaussian elimination.

EXAMPLE 1 Solving a Linear System by Gaussian Elimination

Solve the linear system

$$2y - z = -5$$
$$x - 2y + 2z = 9$$
$$2x - 3y + 3z = 14$$

SOLUTION

Gaussian Elimination	
Step 1. (a) If necessary, interchange equations to obtain a nonzero coefficient for x in the first equation.	*Step 1.* (a) Interchanging the first two equations yields $$x - 2y + 2z = 9$$ $$2y - z = -5$$ $$2x - 3y + 3z = 14$$

(b) Replace the second equation with the sum of itself plus an appropriate multiple of the first equation, which will result in a zero coefficient for x.
(c) Replace the third equation with the sum of itself plus an appropriate multiple of the first equation, which will result in a zero coefficient for x.

(b) The coefficient of x in the second equation is already 0.

(c) Replace the third equation with the sum of itself plus -2 times the first equation.

$$x - 2y + 2z = 9$$
$$2y - z = -5$$
$$y - z = -4$$

Step 2. Apply the procedures of Step 1 to the second and third equations.

Step 2. Replace the third equation with the sum of itself and $-\frac{1}{2}$ times the second equation.

$$x - 2y + 2z = 9$$
$$2y - z = -5$$
$$-\frac{1}{2}z = -\frac{3}{2}$$

Step 3. The system is now in triangular form. The solution is obtained by back-substitution.

Step 3. From the third equation,

$$-\frac{1}{2}z = -\frac{3}{2}$$
$$z = 3$$

Substituting this value of z in the second equation, we have

$$2y - (3) = -5$$
$$y = -1$$

Substituting for y and for z in the first equation, we have

$$x - 2(-1) + 2(3) = 9$$
$$x + 8 = 9$$
$$x = 1$$

The solution is $x = 1, y = -1, z = 3$.

■

✓ **PROGRESS CHECK**
Solve by Gaussian elimination.

(a) $2x - 4y + 2z = 1$
$3x + y + 3z = 5$
$x - y - 2z = -8$

(b) $-2x + 3y - 12z = -17$
$3x - y - 15z = 11$
$-x + 5y + 3z = -9$

ANSWERS

(a) $x = -\frac{3}{2}, y = \frac{1}{2}, z = 3$ (b) $x = 5, y = -1, z = \frac{1}{3}$

**CONSISTENT AND
INCONSISTENT SYSTEMS**

The graph of a linear equation in three unknowns is a plane in three-dimensional space. A system of three linear equations in three unknowns corresponds to three planes (Figure 12). If the planes intersect in a point P (Figure 12a), the coordinates of the point P are a solution of the system and can be found by Gaussian elimination. The cases of no solution and of an infinite number of solutions are signaled as follows.

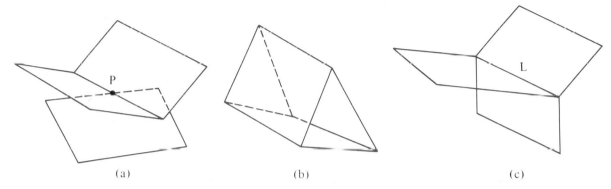

FIGURE 12 (a) Unique Solution (b) No Solution (c) Infinite Number of Solutions

Consistent and Inconsistent Systems	■ If Gaussian elimination results in an equation of the form

- If Gaussian elimination results in an equation of the form

$$0x + 0y + 0z = c, \qquad c \neq 0$$

then the system is inconsistent (Figure 12b).

- If Gaussian elimination results in no equation of the type above but results in an equation of the form

$$0x + 0y + 0z = 0$$

then the system is consistent and has an infinite number of solutions (Figure 12c).

- Otherwise, the system is consistent and has a unique solution.

EXAMPLE 2 Consistent System with an Infinite Number of Solutions
Solve the linear system.

$$\begin{aligned}
x - 2y + 2z &= -4 \\
x + y - 7z &= 8 \\
-x - 4y + 16z &= -20
\end{aligned}$$

SOLUTION
Replacing the second equation by itself minus the first equation, and replacing the third equation by itself plus the first equation, we have

$$\begin{aligned}
x - 2y + 2z &= -4 \\
3y - 9z &= 12 \\
-6y + 18z &= -24
\end{aligned}$$

Replacing the third equation of the last system by itself plus 2 times the second equation results in the system

$$
\begin{aligned}
x - 2y + 2z &= -4 \\
3y - 9z &= 12 \\
0x + 0y + 0z &= 0
\end{aligned}
$$

in which the last equation indicates that the system is consistent and has an infinite number of solutions. If we solve the second equation of the last system for y, we have

$$y = 3z + 4$$

Then, solving the first equation for x, we have

$$
\begin{aligned}
x &= 2y - 2z - 4 \\
&= 2(3z + 4) - 2z - 4 \quad \text{Substituting for } y \\
&= 4z + 4
\end{aligned}
$$

The equations

$$
\begin{aligned}
x &= 4z + 4 \\
y &= 3z + 4
\end{aligned}
$$

yield a solution of the original system for every real value of z. For example, if $z = 0$, then $x = 4$, $y = 4$, $z = 0$ satisfies the original system; if $z = -2$, then $x = -4$, $y = -2$, $z = -2$ is another solution.

■

 PROGRESS CHECK

(a) Verify that the linear system

$$
\begin{aligned}
x - 2y + z &= 3 \\
2x + y - 2z &= -1 \\
-x - 8y + 7z &= 5
\end{aligned}
$$

is consistent.

(b) Verify that the linear system

$$
\begin{aligned}
2x + y + 2z &= 1 \\
x - 4y + 7z &= -4 \\
x - y + 3z &= -1
\end{aligned}
$$

has an infinite number of solutions.

EXERCISE SET 10.3

In Exercises 1–18 solve by Gaussian elimination. Indicate if the system is inconsistent or has an infinite number of solutions.

1. $\begin{aligned} x + 2y + 3z &= -6 \\ 2x - 3y - 4z &= 15 \\ 3x + 4y + 5z &= -8 \end{aligned}$

2. $\begin{aligned} 2x + 3y + 4z &= -12 \\ x - 2y + z &= -5 \\ 3x + y + 2z &= 1 \end{aligned}$

3. $\begin{aligned} x + y + z &= 1 \\ x + y - 2z &= 3 \\ 2x + y + z &= 2 \end{aligned}$

4. $\begin{aligned} 2x - y + z &= 3 \\ x - 3y + z &= 4 \\ -5x - 2z &= -5 \end{aligned}$

5. $\begin{aligned} x + y + z &= 2 \\ x - y + 2z &= 3 \\ 3x + 5y + 2z &= 6 \end{aligned}$

6. $\begin{aligned} x + y + z &= 0 \\ x + y &= 3 \\ y + z &= 1 \end{aligned}$

7. $\begin{aligned} x + 2y + z &= 7 \\ x + 2y + 3z &= 11 \\ 2x + y + 4z &= 12 \end{aligned}$

8. $\begin{aligned} 4x + 2y - z &= 5 \\ 3x + 3y + 6z &= 1 \\ 5x + y - 8z &= 8 \end{aligned}$

9. $x + y + z = 2$
   $x + 2y + z = 3$
   $x + y - z = 2$

10. $x + y - z = 2$
    $x + 2y + z = 3$
    $x + y + 4z = 3$

11. $2x + y + 3z = 8$
    $-x + y + z = 10$
    $x + y + z = 12$

12. $2x - 3z = 4$
    $x + 4y - 5z = -6$
    $3x + 4y - z = -2$

13. $x + 3y + 7z = 1$
    $3x - y - 5z = 9$
    $2x + y + z = 4$

14. $2x - y + z = 2$
    $3x + y + 2z = 3$
    $x + y - z = -1$

15. $x - 2y + 3z = -2$
    $x - 5y + 9z = 4$
    $2x - y = 6$

16. $x + 2y - 2z = 8$
    $5y - z = 6$
    $-2x + y + 3z = -2$

17. $x - 2y + z = -5$
    $2x + z = -10$
    $y - z = 15$

18. $2y - 3z - 4$
    $x + 2z = -2$
    $x - 8y + 14z = -18$

19. A special low-calorie diet consists of dishes A, B, and C. Each unit of A has 2 grams of fat, 1 gram of carbohydrate, and 3 grams of protein. Each unit of B has 1 gram of fat, 2 grams of carbohydrate, and 1 gram of protein. Each unit of C has 1 gram of fat, 2 grams of carbohydrate, and 3 grams of protein. The diet must provide exactly 10 grams of fat, 14 grams of carbohydrate, and 18 grams of protein. How much of each dish should be used?

20. A furniture manufacturer makes chairs, coffee tables, and dining room tables. Each chair requires 2 minutes of sanding, 2 minutes of staining, and 4 minutes of varnishing. Each coffee table requires 5 minutes of sanding, 4 minutes of staining, and 3 minutes of varnishing. Each dining room table requires 5 minutes of sanding, 4 minutes of staining, and 6 minutes of varnishing. The sanding benches, staining benches, and varnishing benches are available 6, 5, and 6 hours per day, respectively. How many of each type of furniture can be made if all facilities are used to capacity?

21. A manufacturer produces 12″, 16″, and 19″ television sets that require assembly, testing, and packing. The 12″ sets each require 45 minutes to assemble, 30 minutes to test, and 10 minutes to package. The 16″ sets each require 1 hour to assemble, 45 minutes to test, and 15 minutes to package. The 19″ sets each require $1\frac{1}{2}$ hours to assemble, 1 hour to test, and 15 minutes to package. If the assembly line operates for $17\frac{3}{4}$ hours per day, the test facility is used for $12\frac{1}{2}$ hours per day, and the packing equipment is used for $3\frac{3}{4}$ hours per day, how many of each type of set can be produced?

10.4 MATRICES AND LINEAR SYSTEMS

We have already studied several methods for solving a linear system such as

$$2x + 3y = -7$$
$$3x - y = 17$$

This system can be displayed by a **matrix**, which is simply a rectangular array of mn real numbers arranged in m horizontal rows and n vertical columns. The numbers are called the **entries** or **elements** of the matrix and are enclosed within brackets. Thus,

$$A = \begin{bmatrix} 2 & 3 & -7 \\ 3 & -1 & 17 \end{bmatrix} \leftarrow \text{rows}$$

columns

is a matrix consisting of two rows and three columns, whose entries are obtained from the two given equations. In general, a matrix of m rows and n columns is said to be of **dimension m by n**, written $m \times n$. The matrix A is seen to be of dimension 2×3. If the numbers of rows and columns of a matrix are both equal to n, the matrix is called a **square matrix** of **order** n.

EXAMPLE 1 **Dimension of a Matrix**

(a)
$$A = \begin{bmatrix} -1 & 4 \\ 0.1 & -2 \end{bmatrix}$$

is a 2×2 matrix. Since matrix A has two rows and two columns, it is a square matrix of order 2.

(b)
$$B = \begin{bmatrix} 4 & -5 \\ -2 & 1 \\ 3 & 0 \end{bmatrix}$$

has three rows and two columns and is a 3×2 matrix.

(c)
$$C = \begin{bmatrix} -8 & 6 & 1 \end{bmatrix}$$

is a 1×3 matrix and is called a **row matrix** since it has precisely one row.

(d)
$$D = \begin{bmatrix} 2 \\ -4 \end{bmatrix}$$

is a 2×1 matrix is called a **column matrix** since it has precisely one column.   ∎

There is a convenient way of denoting a general $m \times n$ matrix, using "double subscripts."

$$A = \begin{bmatrix} a_{11} & a_{12} & \cdots & a_{1j} & \cdots & a_{1n} \\ a_{21} & a_{22} & \cdots & a_{2j} & \cdots & a_{2n} \\ \vdots & \vdots & & \vdots & & \vdots \\ a_{i1} & a_{i2} & \cdots & a_{ij} & \cdots & a_{in} \\ \vdots & \vdots & & \vdots & & \vdots \\ a_{m1} & a_{m2} & \cdots & a_{mj} & \cdots & a_{mn} \end{bmatrix} \begin{matrix} \leftarrow \text{first row} \\ \leftarrow \text{second row} \\ \\ \leftarrow i\text{th row} \\ \\ \leftarrow m\text{th row} \end{matrix}$$

$$\begin{matrix} \uparrow & \uparrow & \uparrow & \uparrow \\ \text{first} & \text{second} & j\text{th} & n\text{th} \\ \text{column} & \text{column} & \text{column} & \text{column} \end{matrix}$$

Thus, a_{ij} is the entry in the ith row and jth column of the matrix A. It is customary to write $A = [a_{ij}]$ to indicate that a_{ij} is the entry in row i and column j of matrix A.

EXAMPLE 2 **Matrix Dimension and Element Notation**

Let

$$A = \begin{bmatrix} 3 & -2 & 4 & 5 \\ 9 & 1 & 2 & 0 \\ -3 & 2 & -4 & 8 \end{bmatrix}$$

Matrix A is of dimension 3×4. The element a_{12} is found in the first row and second column and is seen to be -2. Similarly, we see that $a_{31} = -3$, $a_{33} = -4$, and $a_{34} = 8$.

■

PROGRESS CHECK

Let

$$B = \begin{bmatrix} 4 & 8 & 1 \\ 2 & -5 & 3 \\ -8 & 6 & -4 \\ 0 & 1 & -1 \end{bmatrix}$$

Find **(a)** b_{11} **(b)** b_{23} **(c)** b_{31} **(d)** b_{42}

ANSWERS

(a) 4   (b) 3   (c) −8   (d) 1

If we begin with the system of linear equations

$$2x + 3y = -7$$
$$3x - y = 17$$

the matrix

$$\begin{bmatrix} 2 & 3 \\ 3 & -1 \end{bmatrix}$$

in which the first column is formed from the coefficients of x and the second column is formed from the coefficients of y, is called the **coefficient matrix.** The matrix

$$\begin{bmatrix} 2 & 3 & \vdots & -7 \\ 3 & -1 & \vdots & 17 \end{bmatrix}$$

which includes the column consisting of the right-hand sides of the equations separated by a dashed line, is called the **augmented matrix.**

EXAMPLE 3   Linear Systems and the Augmented Matrix
Write a system of linear equations that corresponds to the augmented matrix.

$$\begin{bmatrix} -5 & 2 & -1 & \vdots & 15 \\ 0 & -2 & 1 & \vdots & -7 \\ \frac{1}{2} & 1 & -1 & \vdots & 3 \end{bmatrix}$$

SOLUTION
We attach the unknown x to the first column, the unknown y to the second column, and

the unknown z to the third column. The resulting system is

$$-5x + 2y - z = 15$$
$$-2y + z = -7$$
$$\tfrac{1}{2}x + y - z = 3$$

■

Now that we have seen how a matrix can be used to represent a system of linear equations, we next proceed to show how routine operations on that matrix can yield the solution of the system. These "matrix methods" are simply a clever streamlining of the methods already studied in this chapter.

In Section 3 of this chapter we used three elementary operations to transform a system of linear equations into triangular form. When applying the same procedures to a matrix, we speak of rows, columns, and elements instead of equations, variables, and coefficients. The three elementary operations that yield an equivalent system now become the **elementary row operations.**

Elementary Row Operations

The following elementary row operations transform an augmented matrix into an equivalent system.

(1) Interchange any two rows.

(2) Multiply each element of any row by a constant $k \neq 0$.

(3) Replace each element of a given row by the sum of itself plus k times the corresponding element of any other row.

The method of Gaussian elimination introduced in Section 3 of this chapter can now be restated in terms of matrices. By use of elementary row operations we seek to transform an augmented matrix into a matrix for which $a_{ij} = 0$ when $i > j$. The resulting matrix will have the following appearance for a system of three linear equations in three unknowns.

$$\begin{bmatrix} * & * & * & \vdots & * \\ 0 & * & * & \vdots & * \\ 0 & 0 & * & \vdots & * \end{bmatrix}$$

Since this matrix represents a linear system in triangular form, back-substitution will provide a solution of the original system. We will illustrate the process with an example.

EXAMPLE 4 Elementary Row Operations and Gaussian Elimination
Solve the system.

$$x - y + 4z = 4$$
$$2x + 2y - z = 2$$
$$3x - 2y + 3z = -3$$

SOLUTION
We describe and illustrate the steps of the procedure.

Gaussian Elimination

Step 1. Form the augmented matrix.	*Step 1.* The augmented matrix is $$\begin{bmatrix} 1 & -1 & 4 & \vdots & 4 \\ 2 & 2 & -1 & \vdots & 2 \\ 3 & -2 & 3 & \vdots & -3 \end{bmatrix}$$
Step 2. If necessary, interchange rows to make sure that a_{11}, the first element of the first row, is nonzero. We call a_{11} the **pivot element** and row 1 the **pivot row.**	*Step 2.* We see that $a_{11} - 1 \neq 0$. The pivot element is a_{11} and is shown in color.
Step 3. Arrange to have 0 as the first element of every row below row 1. This is done by replacing row 2, row 3, and so on by the sum of itself and an appropriate multiple of row 1.	*Step 3.* To make $a_{21} = 0$, replace row 2 by the sum of itself and (-2) times row 1; to make $a_{31} = 0$, replace row 3 by the sum of itself and (-3) times row 1. $$\begin{bmatrix} 1 & -1 & 4 & \vdots & 4 \\ 0 & 4 & -9 & \vdots & -6 \\ 0 & 1 & -9 & \vdots & -15 \end{bmatrix}$$
Step 4. Repeat the process defined by Steps 2 and 3, allowing row 2, row 3, and so on to play the role of the first row. Thus row 2, row 3, and so on serve as the pivot rows.	*Step 4.* Since $a_{22} = 4 \neq 0$, it will serve as the next pivot element and is shown in color. To make $a_{32} = 0$, replace row 3 by the sum of itself and $(-\frac{1}{4})$ times row 2. $$\begin{bmatrix} 1 & -1 & 4 & \vdots & 4 \\ 0 & 4 & -9 & \vdots & 6 \\ 0 & 0 & -\frac{27}{4} & \vdots & -\frac{27}{2} \end{bmatrix}$$
Step 5. The corresponding linear system is in triangular form. Solve by back-substitution.	*Step 5.* The third row of the final matrix yields $$-\frac{27}{4}z = -\frac{27}{2}$$ $$z = 2$$ Substituting $z = 2$, we obtain from the second row of the final matrix $$4y - 9z = -6$$ $$4y - 9(2) = -6$$ $$y = 3$$ Substituting $y = 3$, $z = 2$, we obtain from the first row of the final matrix $$x - y + 4z = 4$$ $$x - 3 + 4(2) = 4$$ $$x = -1$$ The solution is $x - -1$, $y = 3$, $z = 2$.

☑ **PROGRESS CHECK**

Solve the linear system by matrix methods.

$$2x + 4y - z = 0$$
$$x - 2y - 2z = 2$$
$$-5x - 8y + 3z = -2$$

ANSWER

$x = 6, \quad y = -2, \quad z = 4$

Note that we have described the process of Gaussian elimination in a manner that will apply to any augmented matrix that is $n \times (n + 1)$; that is, Gaussian elimination may be used on any system of n linear equations in n unknowns that has a unique solution.

It is also permissible to perform elementary row operations in clever ways to simplify the arithmetic. For instance, you may wish to interchange rows, or to multiply a row by a constant to obtain a pivot element equal to 1. We will illustrate these ideas with an example.

EXAMPLE 5 Elementary Row Operations and Gaussian Elimination

Solve by matrix methods.

$$2y + 3z = 4$$
$$4x + y + 8z + 15w = -14$$
$$x - y + 2z = 9$$
$$-x - 2y - 3z - 6w = 10$$

SOLUTION

We begin with the augmented matrix and perform a sequence of elementary row operations. The pivot element is shown in color.

$$\begin{bmatrix} 0 & 2 & 3 & 0 & | & 4 \\ 4 & 1 & 8 & 15 & | & -14 \\ 1 & -1 & 2 & 0 & | & 9 \\ -1 & -2 & -3 & -6 & | & 10 \end{bmatrix}$$
Augmented matrix.
Note that $a_{11} = 0$.

$$\begin{bmatrix} 1 & -1 & 2 & 0 & | & 9 \\ 4 & 1 & 8 & 15 & | & -14 \\ 0 & 2 & 3 & 0 & | & 4 \\ -1 & -2 & -3 & -6 & | & 10 \end{bmatrix}$$
Interchanged rows 1 and 3 so that $a_{11} = 1$.

$$\begin{bmatrix} 1 & -1 & 2 & 0 & | & 9 \\ 0 & 5 & 0 & 15 & | & -50 \\ 0 & 2 & 3 & 0 & | & 4 \\ 0 & -3 & -1 & -6 & | & 19 \end{bmatrix}$$
To make $a_{21} = 0$, replaced row 2 by the sum of itself and (-4) times row 1.
To make $a_{41} = 0$, replaced row 4 by the sum of itself and row 1.

$$\begin{bmatrix} 1 & -1 & 2 & 0 & | & 9 \\ 0 & 1 & 0 & 3 & | & -10 \\ 0 & 2 & 3 & 0 & | & 4 \\ 0 & -3 & -1 & -6 & | & 19 \end{bmatrix}$$ Multiplied row 2 by $\frac{1}{5}$ so that $a_{22} = 1$.

$$\begin{bmatrix} 1 & -1 & 2 & 0 & | & 9 \\ 0 & 1 & 0 & 3 & | & -10 \\ 0 & 0 & 3 & -6 & | & 24 \\ 0 & 0 & -1 & 3 & | & -11 \end{bmatrix}$$ To make $a_{32} = 0$, replaced row 3 by the sum of itself and (-2) times row 2.
To make $a_{42} = 0$, replaced row 4 by the sum of itself and 3 times row 2.

$$\begin{bmatrix} 1 & -1 & 2 & 0 & | & 9 \\ 0 & 1 & 0 & 3 & | & -10 \\ 0 & 0 & -1 & 3 & | & -11 \\ 0 & 0 & 3 & -6 & | & 24 \end{bmatrix}$$ Interchanged rows 3 and 4 so that the next pivot will be $a_{33} = -1$.

$$\begin{bmatrix} 1 & -1 & 2 & 0 & | & 9 \\ 0 & 1 & 0 & 3 & | & -10 \\ 0 & 0 & -1 & 3 & | & -11 \\ 0 & 0 & 0 & 3 & | & 9 \end{bmatrix}$$ To make $a_{43} = 0$, replaced row 4 by the sum of itself and 3 times row 3.

The last row of the matrix indicates that

$$3w = -9$$
$$w = -3$$

The remaining variables are found by back-substitution.

Third row of final matrix	Second row of final matrix	First row of final matrix
$-z + 3w = -11$	$y + 3w = -10$	$x - y + 2z = 9$
$-z + 3(-3) = -11$	$y + 3(-3) = -10$	$x - (-1) + 2(2) = 9$
$z = 2$	$y = -1$	$x = 4$

The solution is $x = 4$, $y = -1$, $z = 2$, $w = -3$.

■

There is an important variant of Gaussian elimination known as **Gauss–Jordan elimination.** The objective is to transform a linear system into a form that yields a solution without back-substitution. For a 3×3 system that has a unique solution, the final matrix and equivalent linear system will look like this.

$$\begin{bmatrix} 1 & 0 & 0 & | & c_1 \\ 0 & 1 & 0 & | & c_2 \\ 0 & 0 & 1 & | & c_3 \end{bmatrix} \qquad \begin{matrix} x + 0y + 0z = c_1 \\ 0x + y + 0z = c_2 \\ 0x + 0y + z = c_3 \end{matrix}$$

The solution is then seen to be $x = c_1$, $y = c_2$, and $z = c_3$.

The execution of the Gauss–Jordan method is essentially the same as that of Gaussian elimination except that

(a) the pivot elements are always required to be equal to 1, and

(b) all elements in a column, other than the pivot element, are forced to be 0.

These objectives are accomplished by the use of elementary row operations, as illustrated in the following example.

EXAMPLE 6 Gauss–Jordan Elimination

Solve the linear system by the Gauss–Jordan method.

$$x - 3y + 2z = 12$$
$$2x + y - 4z = -1$$
$$x + 3y - 2z = -8$$

SOLUTION

We begin with the augmented matrix. At each stage, the pivot element is shown in color and is used to force all elements in that column (other than the pivot element itself) to be zero.

$$\begin{bmatrix} 1 & -3 & 2 & \vdots & 12 \\ 2 & 1 & -4 & \vdots & -1 \\ 1 & 3 & -2 & \vdots & -8 \end{bmatrix}$$
Pivot element is a_{11}.

$$\begin{bmatrix} 1 & -3 & 2 & \vdots & 12 \\ 0 & 7 & -8 & \vdots & -25 \\ 0 & 6 & -4 & \vdots & -20 \end{bmatrix}$$
To make $a_{21} = 0$, replaced row 2 by the sum of itself and -2 times row 1.
To make $a_{31} = 0$, replaced row 3 by the sum of itself and -1 times row 1.

$$\begin{bmatrix} 1 & -3 & 2 & \vdots & 12 \\ 0 & 1 & -4 & \vdots & -5 \\ 0 & 6 & -4 & \vdots & -20 \end{bmatrix}$$
Replaced row 2 by the sum of itself and -1 times row 3 to yield the next pivot, $a_{22} = 1$.

$$\begin{bmatrix} 1 & 0 & -10 & \vdots & -3 \\ 0 & 1 & -4 & \vdots & -5 \\ 0 & 0 & 20 & \vdots & 10 \end{bmatrix}$$
To make $a_{12} = 0$, replaced row 1 by the sum of itself and 3 times row 2.
To make $a_{32} = 0$, replaced row 3 by the sum of itself and -6 times row 2.

$$\begin{bmatrix} 1 & 0 & -10 & \vdots & -3 \\ 0 & 1 & -4 & \vdots & -5 \\ 0 & 0 & 1 & \vdots & \frac{1}{2} \end{bmatrix}$$
Multiplied row 3 by $\frac{1}{20}$ so that $a_{33} = 1$.

$$\begin{bmatrix} 1 & 0 & 0 & \vdots & 2 \\ 0 & 1 & 0 & \vdots & -3 \\ 0 & 0 & 1 & \vdots & \frac{1}{2} \end{bmatrix}$$
To make $a_{13} = 0$, replaced row 1 by the sum of itself and 10 times row 3.
To make $a_{23} = 0$, replaced row 2 by the sum of itself and 4 times row 3.

We can see the solution directly from the final matrix: $x = 2$, $y = -3$, and $z = \frac{1}{2}$.

EXERCISE SET 10.4

In Exercises 1–6 state the dimension of each matrix.

1. $\begin{bmatrix} 3 & -1 \\ 2 & 4 \end{bmatrix}$

2. $\begin{bmatrix} 1 & 2 & 3 & -1 \end{bmatrix}$

3. $\begin{bmatrix} 4 & 2 & 3 \\ 5 & -1 & 4 \\ 2 & 3 & 6 \\ -8 & -1 & 2 \end{bmatrix}$

4. $\begin{bmatrix} -1 \\ 3 \\ 2 \end{bmatrix}$

5. $\begin{bmatrix} 4 & 2 & 1 \\ 3 & 1 & 5 \\ -4 & -2 & 3 \end{bmatrix}$

6. $\begin{bmatrix} 3 & -1 & 2 & 6 \\ 2 & 8 & 4 & 1 \end{bmatrix}$

7. Given

$$A = \begin{bmatrix} 3 & -4 & -2 & 5 \\ 8 & 7 & 6 & 2 \\ 1 & 0 & 9 & -3 \end{bmatrix}$$

find   (a) a_{12} (b) a_{22} (c) a_{23} (d) a_{34}

8. Given

$$B = \begin{bmatrix} -5 & 6 & 8 \\ 4 & 1 & 3 \\ 0 & 2 & -6 \\ -3 & 9 & 7 \end{bmatrix}$$

find   (a) b_{13} (b) b_{21} (c) b_{33} (d) b_{42}

In Exercises 9–12 write the coefficient matrix and the augmented matrix for each given linear system.

9. $3x - 2y = 12$
   $5x + y = 8$

10. $3x - 4y = 15$
    $4x - 3y = 12$

11. $\frac{1}{2}x + y + z = 4$
    $2x - y - 4z = 6$
    $4x + 2y - 3z = 8$

12. $2x + 3y - 4z = 10$
    $-3x + y = 12$
    $5x - 2y + z = -8$

In Exercises 13–16 write the linear system whose augmented matrix is given.

13. $\left[\begin{array}{cc|c} \frac{3}{2} & 6 & -1 \\ 4 & 5 & 3 \end{array}\right]$

14. $\left[\begin{array}{cc|c} 4 & 0 & 2 \\ -7 & 8 & 3 \end{array}\right]$

15. $\left[\begin{array}{ccc|c} 1 & 1 & 3 & -4 \\ -3 & 4 & 0 & 8 \\ 2 & 0 & 7 & 6 \end{array}\right]$

16. $\left[\begin{array}{ccc|c} 4 & 8 & 3 & 12 \\ 1 & -5 & 3 & -14 \\ 0 & 2 & 7 & 18 \end{array}\right]$

In Exercises 17–20 the augmented matrix corresponding to a linear system has been transformed to the given matrix by elementary row operations. Find a solution of the original linear system.

17. $\left[\begin{array}{ccc|c} 1 & 2 & 0 & 3 \\ 0 & 1 & -2 & 4 \\ 0 & 0 & 1 & 2 \end{array}\right]$

18. $\left[\begin{array}{ccc|c} 1 & 0 & 2 & -1 \\ 0 & 1 & 3 & 2 \\ 0 & 0 & 1 & 5 \end{array}\right]$

19. $\left[\begin{array}{ccc|c} 1 & -2 & 1 & 3 \\ 0 & 1 & 3 & 2 \\ 0 & 0 & 1 & -4 \end{array}\right]$

20. $\left[\begin{array}{ccc|c} 1 & -4 & 2 & -4 \\ 0 & 1 & 3 & -2 \\ 0 & 0 & 1 & 5 \end{array}\right]$

In Exercises 21–30 solve the given linear system by applying Gaussian elimination to the augmented matrix.

21. $x - 2y = -4$
    $2x + 3y = 13$

22. $2x + y = -1$
    $3x - y = -7$

23. $\begin{aligned} x + y + z &= 4 \\ 2x - y + 2z &= 11 \\ x + 2y + 2z &= 6 \end{aligned}$

24. $\begin{aligned} x - y + z &= -5 \\ 3x + y + 2z &= -5 \\ 2x - y - z &= -2 \end{aligned}$

25. $\begin{aligned} 2x + y - z &= 9 \\ x - 2y + 2z &= -3 \\ 3x + 3y + 4z &= 11 \end{aligned}$

26. $\begin{aligned} 2x + y - z &= -2 \\ -2x - 2y + 3z &= 2 \\ 3x + y - z &= -4 \end{aligned}$

27. $\begin{aligned} -x - y + 2z &= 9 \\ x + 2y - 2z &= -7 \\ 2x - y + z &= -9 \end{aligned}$

28. $\begin{aligned} 4x + y - z &= -1 \\ x - y + 2z &= 3 \\ -x + 2y - z &= 0 \end{aligned}$

29. $\begin{aligned} x + y - z + 2w &= 0 \\ 2x + y \quad\quad - w &= -2 \\ 3x \quad\quad + 2z \quad\quad &= -3 \\ -x + 2y \quad\quad + 3w &= 1 \end{aligned}$

30. $\begin{aligned} 2x + y \quad\quad - 3w &= -7 \\ 3x \quad\quad + 2z + w &= 0 \\ -x + 2y \quad\quad + 3w &= 10 \\ -2x - 3y + 2z - w &= 7 \end{aligned}$

31–40. Solve the linear systems of Exercises 21–30 by Gauss–Jordan elimination applied to the augmented matrix.

10.5 MATRIX OPERATIONS AND APPLICATIONS (Optional)

After defining a new type of mathematical entity, it is useful to define operations using this entity. It is common practice to begin with a definition of equality.

Equality of Matrices	Two matrices are equal if they are of the same dimension and their corresponding entries are equal.

EXAMPLE 1 Matrix Equality
Solve for all unknowns.

$$\begin{bmatrix} -2 & 2x & 9 \\ y - 1 & 3 & -4s \end{bmatrix} = \begin{bmatrix} z & 6 & 9 \\ -4 & r & 7 \end{bmatrix}$$

SOLUTION
Equating corresponding elements, we must have

$$\begin{aligned} -2 &= z & \text{or} & \quad z = -2 \\ 2x &= 6 & \text{or} & \quad x = 3 \\ y - 1 &= -4 & \text{or} & \quad y = -3 \\ 3 &= r & \text{or} & \quad r = 3 \\ -4s &= 7 & \text{or} & \quad s = -\tfrac{7}{4} \end{aligned}$$

Matrix addition can be performed only when the matrices are of the same dimension.

Matrix Addition	The sum of two $m \times n$ matrices A and B is the $m \times n$ matrix obtained by adding the corresponding elements of A and B.

EXAMPLE 2 Matrix Addition

Given the following matrices,

$$A = [2 \quad -3 \quad 4] \qquad B = [5 \quad 3 \quad 2]$$

$$C = \begin{bmatrix} 1 & 6 & -1 \\ -2 & 4 & 5 \end{bmatrix} \qquad D = \begin{bmatrix} 16 & 2 & 9 \\ 4 & -7 & -1 \end{bmatrix}$$

find (if possible): **(a)** $A + B$ **(b)** $A + D$ **(c)** $C + D$

SOLUTION

(a) Since A and B are both 1×3 matrices, they can be added, giving

$$A + B = [2 + 5 \quad -3 + 3 \quad 4 + 2] = [7 \quad 0 \quad 6]$$

(b) Matrices A and D are not of the same dimension and cannot be added.

(c) C and D are both 2×3 matrices. Thus,

$$C + D = \begin{bmatrix} 1 + 16 & 6 + 2 & -1 + 9 \\ -2 + 4 & 4 + (-7) & 5 + (-1) \end{bmatrix} = \begin{bmatrix} 17 & 8 & 8 \\ 2 & -3 & 4 \end{bmatrix}$$

■

Matrices are a natural way of writing the information displayed in a table. For example, Table 1 displays the current inventory of the Quality TV Company at its

TABLE 1 Inventory of Television Sets

TV Sets	Boston	Miami	Chicago
17″	140	84	25
19″	62	17	48

various outlets. The same data is displayed by the matrix.

$$S = \begin{bmatrix} 140 & 84 & 25 \\ 62 & 17 & 48 \end{bmatrix}$$

where we understand the columns to represent the cities and the rows to represent the sizes of the television sets. If the matrix

$$M = \begin{bmatrix} 30 & 46 & 15 \\ 50 & 25 & 60 \end{bmatrix}$$

specifies the number of sets of each size received at each outlet the following month, then the matrix

$$T = S + M = \begin{bmatrix} 170 & 130 & 40 \\ 112 & 42 & 108 \end{bmatrix}$$

gives the revised inventory.

Suppose the salespeople at each outlet are told that half of the revised inventory is to be placed on sale. To determine the number of sets of each size to be placed on sale, we need to multiply each element of the matrix T by 0.5. When working with matrices, we call a real number such as 0.5 a **scalar** and define **scalar multiplication** as follows.

Scalar Multiplication	To multiply a matrix A by a scalar c, multiply each entry of A by c.

EXAMPLE 3 Scalar Multiplication

The matrix Q

$$Q = \begin{matrix} \text{Regular} & \text{Unleaded} & \text{Premium} \\ \begin{bmatrix} 130 & 250 & 60 \\ 110 & 180 & 40 \end{bmatrix} & & \begin{matrix} \text{City A} \\ \text{City B} \end{matrix} \end{matrix}$$

shows the quantity (in thousands of gallons) of the principal types of gasolines stored by a refiner at two different locations. It is decided to increase the quantity of each type of gasoline stored at each site by 10%. Use scalar multiplication to determine the desired inventory levels.

SOLUTION

To increase each entry of matrix Q by 10%, we compute the scalar product $1.1Q$.

$$1.1Q = 1.1\begin{bmatrix} 130 & 250 & 60 \\ 110 & 180 & 40 \end{bmatrix}$$

$$= \begin{bmatrix} 1.1(130) & 1.1(250) & 1.1(60) \\ 1.1(110) & 1.1(180) & 1.1(40) \end{bmatrix} = \begin{bmatrix} 143 & 275 & 66 \\ 121 & 198 & 44 \end{bmatrix}$$

We denote $A + (-1)B$ by $A - B$ and refer to this as the **difference** of A and B.

Matrix Subtraction	The difference of two $m \times n$ matrices A and B is the $m \times n$ matrix obtained by subtracting each entry of B from the corresponding entry of A.

EXAMPLE 4 Matrix Subtraction

Using the matrices C and D of Example 2, find $C - D$.

SOLUTION

By definition,

$$C - D = \begin{bmatrix} 1 - 16 & 6 - 2 & -1 - 9 \\ -2 - 4 & 4 - (-7) & 5 - (-1) \end{bmatrix} = \begin{bmatrix} -15 & 4 & -10 \\ -6 & 11 & 6 \end{bmatrix}$$

MATRIX MULTIPLICATION   We will use the Quality TV Company again, this time to help us arrive at a definition of matrix multiplication. Suppose

$$S = \begin{array}{c} \\ \end{array} \begin{array}{ccc} \text{Boston} & \text{Miami} & \text{Chicago} \end{array}$$

$$S = \begin{bmatrix} 60 & 85 & 70 \\ 40 & 100 & 20 \end{bmatrix} \begin{array}{c} 17'' \\ 19'' \end{array}$$

is a matrix representing the supply of television sets at the end of the year. Further, suppose the cost of each 17″ set is $80 and the cost of each 19″ set is $125. To find the total cost of the inventory at each outlet, we need to multiply the number of 17″ sets by $80, the number of 19″ set by $125, and sum the two products. If we let

$$C = [80 \quad 125]$$

be the cost matrix, we seek to define the product

$$[80 \quad 125] \begin{bmatrix} 60 & 85 & 70 \\ 40 & 100 & 20 \end{bmatrix}$$

so that the result will be a matrix displaying the total cost at each outlet. To find the total cost at the Boston outlet, we need to calculate

$$(80)(60) + (125)(40) = 9800$$

$$[80 \quad 125] \begin{bmatrix} 60 & 85 & 70 \\ 40 & 100 & 20 \end{bmatrix}$$

At the Miami outlet, the total cost is

$$(80)(85) + (125)(100) = 19,300$$

$$[80 \quad 125] \begin{bmatrix} 60 & 85 & 70 \\ 40 & 100 & 20 \end{bmatrix}$$

At the Chicago outlet, the total cost is

$$(80)(70) + (125)(20) = 8100$$

$$[80 \quad 125] \begin{bmatrix} 60 & 85 & 70 \\ 40 & 100 & 20 \end{bmatrix}$$

The total cost at each outlet can then be displayed by the 1×3 matrix

$$[9800 \quad 19,300 \quad 8100]$$

which is the product of C and S. Thus,

$$CS = [80 \quad 125] \begin{bmatrix} 60 & 85 & 70 \\ 40 & 100 & 20 \end{bmatrix}$$
$$= [(80)(60) + (125)(40) \quad (80)(85) + (125)(100) \quad (80)(70) + (125)(201)]$$
$$= [9800 \quad 19,300 \quad 8100]$$

Our example illustrates the process for multiplying a matrix by a row matrix. If the matrix C had more than one row, we would repeat the process using each row of C. Here is an example.

EXAMPLE 5 Matrix Multiplication
Find the product AB if

$$A = \begin{bmatrix} 2 & 1 \\ 3 & 5 \end{bmatrix} \qquad B = \begin{bmatrix} 4 & -6 & -2 & 4 \\ 2 & 0 & 1 & -5 \end{bmatrix}$$

SOLUTION

$$AB = \begin{bmatrix} (2)(4) + (1)(2) & (2)(-6) + (1)(0) & (2)(-2) + (1)(1) & (2)(4) + (1)(-5) \\ (3)(4) + (5)(2) & (3)(-6) + (5)(0) & (3)(-2) + (5)(1) & (3)(4) + (5)(-5) \end{bmatrix}$$

$$= \begin{bmatrix} 10 & -12 & -3 & 3 \\ 22 & -18 & -1 & 13 \end{bmatrix}$$

■

PROGRESS CHECK
Find the product AB if

$$A = \begin{bmatrix} -2 & -1 & 2 \\ 4 & 3 & 1 \end{bmatrix} \qquad B = \begin{bmatrix} 5 & -4 \\ 3 & 1 \\ -1 & 0 \end{bmatrix}$$

ANSWER

$$AB = \begin{bmatrix} -15 & 7 \\ 28 & -13 \end{bmatrix}$$

It is important to note that the product AB of an $m \times n$ matrix A and an $n \times r$ matrix B exists only when the number of columns of A equals the number of rows of B (see Figure 13). The product AB will then be of dimension $m \times r$.

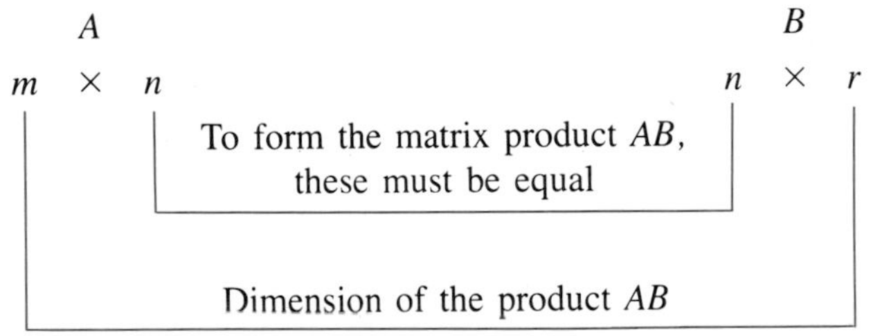

$$
\begin{array}{ccc}
A & & B \\
m \times n & & n \times r
\end{array}
$$

To form the matrix product AB, these must be equal

Dimension of the product AB

FIGURE 13 **Dimension of the Product Matrix**

EXAMPLE 6 Matrix Multiplication
Given the matrices

$$A = \begin{bmatrix} 1 & -1 \\ 2 & 3 \end{bmatrix} \qquad B = \begin{bmatrix} 5 & -3 \\ -2 & 2 \end{bmatrix} \qquad C = \begin{bmatrix} 3 & -1 & -2 \\ 1 & 0 & 4 \end{bmatrix} \qquad D = \begin{bmatrix} 1 \\ 2 \\ 3 \end{bmatrix}$$

(a) Show that $AB \neq BA$.

(b) Determine the dimension of AC.

SOLUTION

(a) $AB = \begin{bmatrix} (1)(5) + (-1)(-2) & (1)(-3) + (-1)(2) \\ (2)(5) + (3)(-2) & (2)(-3) + (3)(2) \end{bmatrix} = \begin{bmatrix} 7 & -5 \\ 4 & 0 \end{bmatrix}$

$BA = \begin{bmatrix} (5)(1) + (-3)(2) & (5)(-1) + (-3)(3) \\ (-2)(1) + (2)(2) & (-2)(-1) + (2)(3) \end{bmatrix} = \begin{bmatrix} -1 & -14 \\ 2 & 8 \end{bmatrix}$

Since the corresponding elements of AB and BA are not equal, $AB \neq BA$.

(b) The product of a 2×2 matrix and a 2×3 matrix is a 2×3 matrix. ∎

PROGRESS CHECK

If possible, find the dimension of CD and of CB, using the matrices of Example 6.

ANSWER

pəuɥəp ʇou ‛1 × 2

We saw in Example 6 that $AB \neq BA$, that is, the commutative law does not hold for matrix multiplication. However, the associative law $A(BC) = (AB)C$ does hold when the dimensions of A, B, and C permit us to find the necessary products.

PROGRESS CHECK

Verify that $A(BC) = (AB)C$ for the matrices A, B, and C of Example 6.

MATRICES AND LINEAR SYSTEMS

Matrix multiplication provides a convenient shorthand means of writing a linear system. For example, the linear system

$$\begin{aligned} 2x - y - 2z &= 3 \\ 3x + 2y + z &= -1 \\ x + y - 3z &= 14 \end{aligned}$$

can be expressed as

$$AX = B$$

where

$$A = \begin{bmatrix} 2 & -1 & -2 \\ 3 & 2 & 1 \\ 1 & 1 & -3 \end{bmatrix} \quad X = \begin{bmatrix} x \\ y \\ z \end{bmatrix} \quad B = \begin{bmatrix} 3 \\ -1 \\ 14 \end{bmatrix}$$

To verify this, simply form the matrix product AX and then apply the definition of matrix equality to the matrix equation $AX = B$.

EXAMPLE 7   Matrices and Linear Systems

Write out the linear system $AX = B$ if

$$A = \begin{bmatrix} -2 & 3 \\ 1 & 4 \end{bmatrix} \qquad X = \begin{bmatrix} x \\ y \end{bmatrix} \qquad B = \begin{bmatrix} 16 \\ -3 \end{bmatrix}$$

SOLUTION

Equating corresponding elements of the matrix equation $AX = B$ yields

$$-2x + 3y = 16$$
$$x + 4y = -3$$

■

EXERCISE SET 10.5

1. For what values of $a, b, c,$ and d are the matrices A and B equal?

$$A = \begin{bmatrix} a & b \\ 6 & -2 \end{bmatrix} \qquad B = \begin{bmatrix} 3 & -4 \\ c & d \end{bmatrix}$$

2. For what values of $a, b, c,$ and d are the matrices A and B equal?

$$A = \begin{bmatrix} a+b & 2c \\ a & c-d \end{bmatrix} \qquad B = \begin{bmatrix} -1 & 6 \\ 5 & 10 \end{bmatrix}$$

In Exercises 3–18 the following matrices are given.

$$A = \begin{bmatrix} 2 & 3 & 1 \\ -3 & 4 & 1 \end{bmatrix} \qquad B = \begin{bmatrix} 2 & -1 \\ 3 & 2 \\ 4 & 1 \end{bmatrix} \qquad C = \begin{bmatrix} 1 & 2 & 3 \\ 4 & -1 & 2 \\ 3 & 2 & 5 \end{bmatrix} \qquad D = \begin{bmatrix} -3 & 2 \\ 4 & 1 \end{bmatrix}$$

$$E = \begin{bmatrix} 1 & -3 & 2 \\ 3 & 2 & 4 \\ 1 & 1 & 2 \end{bmatrix} \qquad F = \begin{bmatrix} 1 & 3 \\ -2 & 4 \end{bmatrix} \qquad G = \begin{bmatrix} -2 & 4 & 2 \\ 1 & 0 & 3 \end{bmatrix}$$

In Exercises 3–18, if possible, compute the indicated matrix.

3. $C + E$
4. $C - E$
5. $2A + 3G$
6. $3G - 4A$
7. $A + F$
8. $2B - D$
9. AB
10. BA
11. $CB + D$
12. $EB - FA$
13. $DF + AB$
14. $AC + 2DG$
15. $DA + EB$
16. $FG + B$
17. $2GE - 3A$
18. $AB + FG$

19. If $A = \begin{bmatrix} -2 & 3 \\ 2 & -3 \end{bmatrix}$, $B = \begin{bmatrix} -1 & 3 \\ 2 & 0 \end{bmatrix}$, $C = \begin{bmatrix} -4 & -3 \\ 0 & -4 \end{bmatrix}$, show that $AB = AC$.

20. If $A = \begin{bmatrix} 1 & 2 \\ 3 & 2 \end{bmatrix}$ and $B = \begin{bmatrix} 2 & -1 \\ -3 & 4 \end{bmatrix}$, show that $AB \neq BA$.

21. If $A = \begin{bmatrix} -2 & 3 \\ 2 & -3 \end{bmatrix}$ and $B = \begin{bmatrix} 3 & 6 \\ 2 & 4 \end{bmatrix}$, show that $AB = \begin{bmatrix} 0 & 0 \\ 0 & 0 \end{bmatrix}$.

22. If $A = \begin{bmatrix} 0 & 1 \\ 1 & 0 \end{bmatrix}$, show that $A \cdot A = \begin{bmatrix} 1 & 0 \\ 0 & 1 \end{bmatrix}$.

23. If $I = \begin{bmatrix} 1 & 0 & 0 \\ 0 & 1 & 0 \\ 0 & 0 & 1 \end{bmatrix}$ and $A = \begin{bmatrix} a_{11} & a_{12} & a_{13} \\ a_{21} & a_{22} & a_{23} \\ a_{31} & a_{32} & a_{33} \end{bmatrix}$, show that $AI = A$ and $IA = A$.

24. Pesticides are sprayed on plants to eliminate harmful insects. However, some of the pesticide is absorbed by the plant, and the pesticide is then absorbed by herbivores (plant-eating animals such as cows) when they eat the plants that have been sprayed. Suppose that we have three pesticides and four plants and that the amounts of pesticide absorbed by the different plants are given by the matrix

Plant 1 Plant 2 Plant 3 Plant 4

$$A = \begin{bmatrix} 3 & 2 & 4 & 3 \\ 6 & 5 & 2 & 4 \\ 4 & 3 & 1 & 5 \end{bmatrix} \begin{matrix} \text{Pesticide 1} \\ \text{Pesticide 2} \\ \text{Pesticide 3} \end{matrix}$$

where a_{ij} denotes the amount of presticide i in milligrams that has been absorbed by plant j. Thus, plant 4 has absorbed 5 mg of pesticide 3. Now suppose that we have three herbivores and that the numbers of plants eaten by

these animals are given by the matrix

Herbivore 1 Herbivore 2 Herbivore 3

$$B = \begin{bmatrix} 18 & 30 & 20 \\ 12 & 15 & 10 \\ 16 & 12 & 8 \\ 6 & 4 & 12 \end{bmatrix} \begin{matrix} \text{Plant 1} \\ \text{Plant 2} \\ \text{Plant 3} \\ \text{Plant 4} \end{matrix}$$

How much of pesticide 2 has been absorbed by herbivore 3?

25. What does entry $(2,3)$ in the matrix product AB of Exercise 24 represent?

In Exercises 26–29 indicate the matrices A, X, and B so that the matrix equation $AX = B$ is equivalent to the given linear system.

26. $7x - 2y = 6$
$-2x + 3y = -2$

27. $3x + 4y = -3$
$3x - y = 5$

28. $5x + 2y - 3z = 4$
$2x - \frac{1}{2}y + z = 10$
$x + y - 5z = -3$

29. $3x - y + 4z = 5$
$2x + 2y + \frac{3}{4}z = -1$
$x - \frac{1}{4}y + z = \frac{1}{2}$

In Exercises 30–33 write out the linear system that it represented by the matrix equation $AX = B$.

30. $A = \begin{bmatrix} 2 & -1 \\ -3 & 4 \end{bmatrix}$ $X = \begin{bmatrix} x \\ y \end{bmatrix}$ $B = \begin{bmatrix} -2 \\ 10 \end{bmatrix}$

31. $A = \begin{bmatrix} 1 & -5 \\ 4 & 3 \end{bmatrix}$ $X = \begin{bmatrix} x_1 \\ x_2 \end{bmatrix}$ $B = \begin{bmatrix} 0 \\ 2 \end{bmatrix}$

32. $A = \begin{bmatrix} 1 & 7 & -2 \\ 3 & 6 & 1 \\ -4 & 2 & 0 \end{bmatrix}$ $X = \begin{bmatrix} x \\ y \\ z \end{bmatrix}$ $B = \begin{bmatrix} 3 \\ -3 \\ 2 \end{bmatrix}$

33. $A = \begin{bmatrix} 4 & 5 & -2 \\ 0 & 3 & -1 \\ 0 & 0 & 2 \end{bmatrix}$ $X = \begin{bmatrix} x_1 \\ x_2 \\ x_3 \end{bmatrix}$ $B = \begin{bmatrix} 2 \\ -5 \\ 4 \end{bmatrix}$

34. The $m \times n$ matrix all of whose elements are zero is called the **zero matrix** and is denoted by O. Show that $A + O = A$ for every $m \times n$ matrix A.

35. The square matrix of order n such that $a_{ii} = 1$ and $a_{ij} = 0$ when $i \neq j$ is called the **identity matrix** of order n and is denoted by I_n. (Note: The definition indicates that the diagonal elements are all equal to 1 and all elements off

the diagonal are 0.) Show that $AI_n = I_nA$ for every square matrix A of order n.

36. The matrix B, each of whose entries is the negative of the corresponding entry of matrix A, is called the **additive inverse** of the matrix A. Show that $A + B = O$ where O is the zero matrix (see Exercise 34).

10.6 DETERMINANTS AND CRAMER'S RULE

In this section we will define a determinant and will develop manipulative skills for evaluating determinants. We will then show that determinants have important applications and can be used to solve linear systems.

Associated with every square matrix A is a number called the **determinant** of A, denoted by $|A|$. If A is the 2×2 matrix

$$A = \begin{bmatrix} a_{11} & a_{12} \\ a_{21} & a_{22} \end{bmatrix}$$

then $|A|$ is said to be a **determinant of second order** and is defined by the rule

$$|A| = \begin{vmatrix} a_{11} & a_{12} \\ a_{21} & a_{22} \end{vmatrix} = a_{11}a_{22} - a_{21}a_{12}$$

EXAMPLE 1 Determinant of Second Order
Compute the real number represented by

$$\begin{vmatrix} 4 & -5 \\ 3 & -1 \end{vmatrix}$$

SOLUTION
We apply the rule for a determinant of second order.

$$\begin{vmatrix} 4 & -5 \\ 3 & -1 \end{vmatrix} = (4)(-1) - (3)(-5) = 11$$

∎

PROGRESS CHECK
Compute the real number represented by

(a) $\begin{vmatrix} -6 & 2 \\ -1 & -2 \end{vmatrix}$ (b) $\begin{vmatrix} \frac{1}{2} & \frac{1}{4} \\ -4 & -2 \end{vmatrix}$

ANSWERS
(a) 14   (b) 0

To simplify matters, when we want to compute the determinant of a matrix we will say "evaluate the determinant." This is not technically correct, however, since a determinant *is* a real number.

MINORS AND COFACTORS

The rule for evaluating a determinant of order 3 is

$$\begin{vmatrix} a_{11} & a_{12} & a_{13} \\ a_{21} & a_{22} & a_{23} \\ a_{31} & a_{32} & a_{33} \end{vmatrix} = a_{11}a_{22}a_{33} - a_{11}a_{32}a_{23} - a_{12}a_{21}a_{33} \\ + a_{12}a_{31}a_{23} + a_{13}a_{21}a_{32} - a_{13}a_{31}a_{22}$$

The situation becomes even more cumbersome for determinants of higher order! Fortunately, we don't have to memorize this rule; instead, we shall see that it is possible to evaluate a determinant of order 3 by reducing the problem to that of evaluating three determinants of order 2.

The **minor of an element** a_{ij} is the determinant of the matrix remaining after deleting the row and column in which the element a_{ij} appears. Given the matrix

$$\begin{bmatrix} 4 & 0 & -2 \\ 1 & -6 & 7 \\ -3 & 2 & 5 \end{bmatrix}$$

the minor of the element in row 2, column 3, is

$$\begin{vmatrix} 4 & 0 & -2 \\ 1 & -6 & 7 \\ -3 & 2 & 5 \end{vmatrix} = \begin{vmatrix} 4 & 0 \\ -3 & 2 \end{vmatrix} = 8 - 0 = 8$$

The **cofactor** of the element a_{ij} is the minor of the element a_{ij} multiplied by $(-1)^{i+j}$. Since $(-1)^{i+j}$ is $+1$ if $i+j$ is even and -1 if $i+j$ is odd, we see that the cofactor is the minor with a sign attached. The cofactor attaches the sign to the minor according to this pattern.

$$
\begin{array}{ccccc}
+ & - & + & - & \cdots \\
- & + & - & + & \cdots \\
+ & - & + & - & \cdots \\
- & + & - & + & \cdots
\end{array}
$$

EXAMPLE 2 Determining Cofactors

Find the cofactor of each element in the first row of the matrix.

$$
\begin{bmatrix}
-2 & 0 & 12 \\
-4 & 5 & 3 \\
7 & 8 & -6
\end{bmatrix}
$$

SOLUTION

The cofactors are

$$
(-1)^{1+1}\begin{vmatrix} -2 & 0 & 12 \\ -4 & 5 & 3 \\ 7 & 8 & -6 \end{vmatrix} = \begin{vmatrix} 5 & 3 \\ 8 & -6 \end{vmatrix} = 30 - 24 = -54
$$

$$
(-1)^{1+2}\begin{vmatrix} -2 & 0 & 12 \\ -4 & 5 & 3 \\ 7 & 8 & -6 \end{vmatrix} = -\begin{vmatrix} -4 & 3 \\ 7 & -6 \end{vmatrix} = -(24 - 21) = -3
$$

$$
(-1)^{1+3}\begin{vmatrix} -2 & 0 & 12 \\ -4 & 5 & 3 \\ 7 & 8 & -6 \end{vmatrix} = \begin{vmatrix} -4 & 5 \\ 7 & 8 \end{vmatrix} = -32 - 35 = -67
$$

■

PROGRESS CHECK

Find the cofactor of each entry in the second column of the matrix.

$$
\begin{bmatrix}
16 & -9 & 3 \\
-5 & 2 & 0 \\
-3 & 4 & -1
\end{bmatrix}
$$

ANSWER

Cofactor of -9 is -5; cofactor of 2 is -7; cofactor of 4 is -15.

The cofactor is the key to the process of evaluating determinants of order 3 or higher.

| **Expansion by Cofactors** | To evaluate a determinant, form the sum of the products obtained by multiplying each entry of any row or any column by its cofactor. This process is called **expansion by cofactors.** |

Let's illustrate the process with an example.

EXAMPLE 3   Expansion by Cofactors

Evaluate the determinant by cofactors.

$$\begin{vmatrix} -2 & 7 & 2 \\ 6 & -6 & 0 \\ 4 & 10 & -3 \end{vmatrix}$$

SOLUTION

Expansion by Cofactors	
Step 1. Choose a row or column about which to expand. In general, a row or column containing zeros will simplify the work.	*Step 1.* We will expand about column 3.
Step 2. Expand about the cofactors of the chosen row or column by multiplying each entry of the row or column by its cofactor.	*Step 2.* The expansion about column 3 is $$(2)(-1)^{1+3}\begin{vmatrix} 6 & -6 \\ 4 & 10 \end{vmatrix}$$ $$+(0)(-1)^{2+3}\begin{vmatrix} -2 & 7 \\ 4 & 10 \end{vmatrix}$$ $$+(-3)(-1)^{3+3}\begin{vmatrix} -2 & 7 \\ 6 & -6 \end{vmatrix}$$
Step 3. Evaluate the cofactors and form the sum indicated in Step 2.	*Step 3.* Using the rule for evaluating a determinant of order 2, we have $$(2)(1)[(6)(10)-(4)(-6)]+0$$ $$+(-3)(1)[(-2)(-6)-(6)(7)]$$ $$=2(60+24)-3(12-42)$$ $$=258$$

■

Note that expansion by cofactors of *any row or any column* will produce the same result. This important property of determinants can be used to simplify the arithmetic. The best choice of a row or column about which to expand is the one that has the most zero entries. The reason for this is that if an entry is zero, the entry times its cofactor will be zero, so we don't have to evaluate that cofactor.

PROGRESS CHECK

Evaluate the determinant of Example 3 by expanding about the second row.

ANSWER

258

EXAMPLE 4   Expansion by Cofactors

Verify the rule for evaluating a determinant of order 3.

$$\begin{vmatrix} a_{11} & a_{12} & a_{13} \\ a_{21} & a_{22} & a_{23} \\ a_{31} & a_{32} & a_{33} \end{vmatrix} = a_{11}a_{22}a_{33} - a_{11}a_{32}a_{23} - a_{12}a_{21}a_{33} \\ + a_{12}a_{31}a_{23} + a_{13}a_{21}a_{32} - a_{13}a_{31}a_{22}$$

SOLUTION

Expanding about the first row, we have

$$\begin{vmatrix} a_{11} & a_{12} & a_{13} \\ a_{21} & a_{22} & a_{23} \\ a_{31} & a_{32} & a_{33} \end{vmatrix} = a_{11}\begin{vmatrix} a_{22} & a_{23} \\ a_{32} & a_{33} \end{vmatrix} - a_{12}\begin{vmatrix} a_{21} & a_{23} \\ a_{31} & a_{33} \end{vmatrix} + a_{13}\begin{vmatrix} a_{21} & a_{22} \\ a_{31} & a_{32} \end{vmatrix}$$

$$= a_{11}(a_{22}a_{33} - a_{32}a_{23}) - a_{12}(a_{21}a_{33} - a_{31}a_{23}) + a_{13}(a_{21}a_{32} - a_{31}a_{22})$$

$$= a_{11}a_{22}a_{33} - a_{11}a_{32}a_{23} - a_{12}a_{21}a_{33} + a_{12}a_{31}a_{23} + a_{13}a_{21}a_{32} - a_{13}a_{31}a_{22}$$

∎

PROGRESS CHECK

Show that the determinant is equal to zero.

$$\begin{vmatrix} a & b & c \\ a & b & c \\ d & e & f \end{vmatrix}$$

The process of expanding by cofactors works for determinants of any order. If we apply the method to a determinant of order 4, we will produce determinants of order 3; applying the method again will result in determinants of order 2.

EXAMPLE 5   Expansion by Cofactors

Evaluate the determinant.

$$\begin{vmatrix} -3 & 5 & 0 & -1 \\ 1 & 2 & 3 & -3 \\ 0 & 4 & -6 & 0 \\ 0 & -2 & 1 & 2 \end{vmatrix}$$

SOLUTION

Expanding about the cofactors of the first column, we have

$$\begin{vmatrix} -3 & 5 & 0 & -1 \\ 1 & 2 & 3 & -3 \\ 0 & 4 & -6 & 0 \\ 0 & -2 & 1 & 2 \end{vmatrix} = -3 \begin{vmatrix} 2 & 3 & -3 \\ 4 & -6 & 0 \\ -2 & 1 & 2 \end{vmatrix} - 1 \begin{vmatrix} 5 & 0 & -1 \\ 4 & -6 & 0 \\ -2 & 1 & 2 \end{vmatrix}$$

Each determinant of order 3 can then be evaluated.

$$-3 \begin{vmatrix} 2 & 3 & -3 \\ 4 & -6 & 0 \\ -2 & 1 & 2 \end{vmatrix} = (-3)(-24) \qquad -1 \begin{vmatrix} 5 & 0 & -1 \\ 4 & -6 & 0 \\ -2 & 1 & 2 \end{vmatrix} = (-1)(-52)$$

$$= 72 \qquad\qquad\qquad\qquad = 52$$

The original determinant has the value $72 + 52 = 124$.

◼

 PROGRESS CHECK

Evaluate.

$$\begin{vmatrix} 0 & -1 & 0 & 2 \\ 3 & 0 & 4 & 0 \\ 0 & 5 & 0 & -3 \\ 1 & 0 & 1 & 0 \end{vmatrix}$$

ANSWER

L

CRAMER'S RULE

Determinants provide a convenient way of expressing formulas in many areas of mathematics, particularly in geometry. One of the best-known uses of determinants is in solving systems of linear equations, a procedure known as **Cramer's rule.**

In an earlier section we solved systems of linear equations by the method of elimination. Let's apply this method to the general system of two equations in two unknowns.

$$a_{11}x + a_{12}y = c_1 \qquad\qquad (1)$$

$$a_{21}x + a_{22}y = c_2 \qquad\qquad (2)$$

If we multiply Equation (1) by a_{22} and Equation (2) by $-a_{12}$ and add, we will eliminate y.

$$a_{11}a_{22}x + a_{12}a_{22}y = \quad c_1 a_{22}$$

$$\underline{-a_{21}a_{12}x - a_{12}a_{22}y = -c_2 a_{12}}$$

$$a_{11}a_{22}x - a_{21}a_{12}x = c_1 a_{22} - c_2 a_{12}$$

Thus,

$$x(a_{11}a_{22} - a_{21}a_{12}) = c_1 a_{22} - c_2 a_{12}$$

or

$$x = \frac{c_1 a_{22} - c_2 a_{12}}{a_{11} a_{22} - a_{21} a_{12}}$$

Similarly, multiplying Equation (i) by a_{21} and Equation (2) by $-a_{11}$ and adding, we can eliminate x and solve for y.

$$y = \frac{c_2 a_{11} - c_1 a_{21}}{a_{11} a_{22} - a_{21} a_{12}}$$

The denominators in the expressions for x and y are identical and can be written as the determinant of the matrix

$$A = \begin{bmatrix} a_{11} & a_{12} \\ a_{21} & a_{22} \end{bmatrix}$$

If we apply this same idea to the numerators, we have

$$x = \frac{\begin{vmatrix} c_1 & a_{12} \\ c_2 & a_{22} \end{vmatrix}}{|A|} \qquad y = \frac{\begin{vmatrix} a_{11} & c_1 \\ a_{21} & c_2 \end{vmatrix}}{|A|} \qquad |A| \neq 0$$

What we have arrived at is Cramer's rule, which is a means of expressing the solution of a system of linear equations in determinant form.

The following example outlines the steps for using Cramer's rule.

EXAMPLE 6 Cramer's Rule

Solve by Cramer's rule.

$$3x - y = 9$$
$$x + 2y = -4$$

SOLUTION

Cramer's Rule	
Step 1. Compute $|A|$, the determinant of the coefficient matrix A. If $|A| = 0$, Cramer's rule cannot be used. Use Gaussian elimination or Gauss–Jordan elimination.	*Step 1.* $$|A| = \begin{vmatrix} 3 & -1 \\ 1 & 2 \end{vmatrix} = 7$$
Step 2. The value of x is the quotient whose numerator is the determinant of the matrix obtained from A by replacing the column of coefficients of x with the column of right-hand sides of the equations.	*Step 2.* $$x = \frac{\begin{vmatrix} 9 & -1 \\ -4 & 2 \end{vmatrix}}{|A|}$$ $$= \frac{18 - 4}{7} = \frac{14}{7} = 2$$

Step 3. The value of y is the quotient whose numerator is the determinant of the matrix obtained from A by replacing the column of coefficients of y with the column of right-hand sides of the equations.

Step 3.

$$y = \frac{\begin{vmatrix} 3 & 9 \\ 1 & -4 \end{vmatrix}}{|A|}$$

$$= \frac{-12 - 9}{7} = \frac{-21}{7} = -3$$

Thus, $x = 2$, $y = 3$

 PROGRESS CHECK
Solve by Cramer's rule.

$$2x + 3y = -4$$
$$3x + 4y = -7$$

ANSWER

$z = \text{ʎ 'ϛ}- = x$

The steps outlined in Example 6 can be applied to solve any system of linear equations in which the number of equations is the same as the number of variables and in which $|A| \neq 0$. Here is an example with three equations and three unknowns.

EXAMPLE 7 Cramer's Rule
Solve by Cramer's rule.

$$\begin{align} 3x \quad\quad + 2z &= -2 \\ 2x - \ y \quad\quad &= \ \ 0 \\ 2y + 6z &= -1 \end{align}$$

SOLUTION
We compute the determinant of the matrix of coefficients by expanding by cofactors.

$$|A| = \begin{vmatrix} 3 & 0 & 2 \\ 2 & -1 & 0 \\ 0 & 2 & 6 \end{vmatrix} = -10$$

Then

$$x = \frac{|A_1|}{|A|} \qquad y = \frac{|A_2|}{|A|} \qquad z = \frac{|A_3|}{|A|}$$

where A_1 is obtained from A by replacing its first column by the column of right-hand sides, A_2 is obtained from A by replacing the second column of A with the column of right-hand sides, and A_3 is obtained from A by replacing its third column with the

column of right-hand sides. Thus

$$x = \frac{\begin{vmatrix} -2 & 0 & 2 \\ 0 & -1 & 0 \\ -1 & 2 & 6 \end{vmatrix}}{|A|} \qquad y = \frac{\begin{vmatrix} 3 & -2 & 2 \\ 2 & 0 & 0 \\ 0 & -1 & 6 \end{vmatrix}}{|A|} \qquad z = \frac{\begin{vmatrix} 3 & 0 & -2 \\ 2 & -1 & 0 \\ 0 & 2 & -1 \end{vmatrix}}{|A|}$$

Expanding by cofactors we calculate $|A_1| = 10$, $|A_2| = 20$, and $|A_3| = -5$, obtaining

$$x = \frac{10}{10} = -1 \qquad y = \frac{20}{-10} = -2 \qquad z = \frac{-5}{-10} = \frac{1}{2}$$

■

 PROGRESS CHECK

Solve by Cramer's rule.

$$\begin{aligned} 3x - z &= 1 \\ -6x + 2y &= -5 \\ -4y + 3z &- 5 \end{aligned}$$

ANSWER

$$x = \frac{2}{3}, \ y = -\frac{1}{2}, \ z = 1$$

 WARNING

(a) Each equation of the linear system must be written in the form

$$Ax + By + Cz = k$$

before using Cramer's rule.

(b) If $|A| = 0$, Cramer's rule cannot be used.

Determinants have significant theoretical importance but are not of much use for computational purposes. The matrix methods discussed in this chapter provide the basis for techniques better suited for computer implementation.

EXERCISE SET 10.6

In Exercises 1–6 evaluate the given determinant.

1. $\begin{vmatrix} 2 & -3 \\ 4 & 5 \end{vmatrix}$

2. $\begin{vmatrix} 3 & 4 \\ -1 & 2 \end{vmatrix}$

3. $\begin{vmatrix} -4 & 1 \\ 0 & 2 \end{vmatrix}$

4. $\begin{vmatrix} 2 & 2 \\ 3 & 3 \end{vmatrix}$

5. $\begin{vmatrix} 0 & 0 \\ 1 & 3 \end{vmatrix}$

6. $\begin{vmatrix} -4 & -1 \\ -2 & 3 \end{vmatrix}$

In Exercises 7–10 let

$$A = \begin{bmatrix} 3 & -1 & 2 \\ 4 & 1 & -3 \\ 5 & -2 & 0 \end{bmatrix}$$

7. Compute the minor of each of the following elements.
   (a) a_{11} (b) a_{23} (c) a_{31} (d) a_{33}

8. Compute the minor of each of the following elements.
   (a) a_{12} (b) a_{22} (c) a_{23} (d) a_{32}

9. Compute the cofactor of each of the following elements.
   (a) a_{11} (b) a_{23} (c) a_{31} (d) a_{33}

10. Compute the cofactor of each of the following elements.
    (a) a_{12} (b) a_{22} (c) a_{23} (d) a_{32}

In Exercises 11–20 evaluate the given determinant.

11. $\begin{vmatrix} 4 & -2 & 5 \\ 5 & 2 & 0 \\ 2 & 0 & 4 \end{vmatrix}$

12. $\begin{vmatrix} 4 & 1 & 2 \\ 0 & 2 & 3 \\ 0 & 0 & -4 \end{vmatrix}$

13. $\begin{vmatrix} -1 & 2 & 0 \\ 3 & 4 & 1 \\ 6 & 5 & 2 \end{vmatrix}$

14. $\begin{vmatrix} -1 & 3 & 2 \\ 0 & 7 & 7 \\ 2 & 1 & 3 \end{vmatrix}$

15. $\begin{vmatrix} 2 & 2 & 4 \\ 3 & 8 & 1 \\ 1 & 1 & 2 \end{vmatrix}$

16. $\begin{vmatrix} 0 & 1 & 3 \\ 2 & 5 & -1 \\ 4 & 2 & -2 \end{vmatrix}$

17. $\begin{vmatrix} 3 & 2 & 1 & 0 \\ -1 & -3 & -1 & 0 \\ 0 & 0 & 2 & 2 \\ 4 & 1 & 3 & 3 \end{vmatrix}$

18. $\begin{vmatrix} -1 & 2 & 4 & 0 \\ 3 & -2 & -3 & 0 \\ 0 & 4 & 2 & 5 \\ 0 & -3 & 1 & 4 \end{vmatrix}$

19. $\begin{vmatrix} 2 & -3 & 2 & -4 \\ 0 & 4 & -1 & 9 \\ 0 & 1 & 2 & 0 \\ 0 & 1 & 3 & -1 \end{vmatrix}$

20. $\begin{vmatrix} 1 & 1 & 0 & 1 \\ 0 & -1 & 4 & -1 \\ -2 & 3 & 1 & -4 \\ 0 & 2 & 0 & 2 \end{vmatrix}$

In Exercises 21–28 solve the given linear system by using Cramer's rule.

21. $\begin{aligned} 2x + y + z &= -1 \\ 2x - y + 2z &= 2 \\ x + 2y + z &= -4 \end{aligned}$

22. $\begin{aligned} x - y + z &= -5 \\ 3x + y + 2z &= -5 \\ 2x - y - z &= -2 \end{aligned}$

23. $\begin{aligned} 2x + y - z &= 9 \\ x - 2y + 2z &= -3 \\ 3x + 3y + 4z &= 11 \end{aligned}$

24. $\begin{aligned} 2x + y - z &= -2 \\ -2x - 2y + 3z &= 2 \\ 3x + y - z &= -4 \end{aligned}$

25. $\begin{aligned} -x - y + 2z &= 7 \\ x + 2y - 2z &= -7 \\ 2x - y + z &= -4 \end{aligned}$

26. $\begin{aligned} 4x + y - z &= -1 \\ x - y + 2z &= 3 \\ -x + 2y - z &= 0 \end{aligned}$

27. $\begin{aligned} x + y - z + 2w &= 0 \\ 2x + y \quad - w &= -2 \\ 3x \quad + 2z \quad &= -3 \\ -x + 2y \quad + 3w &= 1 \end{aligned}$

28. $\begin{aligned} 2x + y \quad - 3w &= -7 \\ 3x \quad + 2z + w &= -1 \\ -x + 2y \quad + 3w &= 0 \\ -2x - 3y + 2z - w &= 8 \end{aligned}$

29. Show that
$$\begin{vmatrix} a_1 + b_1 & a_2 + b_2 \\ c & d \end{vmatrix} = \begin{vmatrix} a_1 & a_2 \\ c & d \end{vmatrix} + \begin{vmatrix} b_1 & b_2 \\ c & d \end{vmatrix}$$

30. Show that
$$\begin{vmatrix} ka_{11} & ka_{12} \\ a_{21} & a_{22} \end{vmatrix} = \begin{vmatrix} a_{11} & a_{12} \\ ka_{21} & ka_{22} \end{vmatrix} = k \begin{vmatrix} a_{11} & a_{12} \\ a_{21} & a_{22} \end{vmatrix}$$

31. Prove that if a row or column of a square matrix consists entirely of zeros, the determinant of the matrix is zero. (*Hint:* Expand by cofactors.)

32. Prove that if matrix B is obtained by multiplying each element of a row of a square matrix A by a constant k, then $|B| = k|A|$.

33. Prove that if A is an $n \times n$ matrix and $B = kA$, where k is a constant, then $|B| = k^n|A|$.

34. Prove that if matrix B is obtained from a square matrix A by interchanging the rows and columns of A, then $|B| = |A|$.

10.7 LINEAR PROGRAMMING

Let's pose the following problem.

A lot is zoned for an apartment building to consist of no more than 40 apartments, totaling no more than 45,000 square feet. A builder is planning to construct 1-bedroom apartments, each of which will require 1000 square feet and will rent for $200 per month, and 2-bedroom apartments, each of which will utilize 1500 square feet and will rent for $280 per month. If all available apartments can be rented, how many apartments of each type should be built to maximize the builder's monthly rental revenue?

If we let x denote the number of 1-bedroom units and y denote the number of 2-bedroom units, the accompanying table displays the information given in the problem.

	Number of units	Square feet	Rental
1-bedroom	x	1,000	$200
2-bedroom	y	1,500	280
Total	40	45,000	z

Using the methods of the previous section, we can translate the **constraints** or requirements upon the variables x and y into a system of inequalities. The total number of apartments is $x + y$, so we have

$$x + y \leq 40 \quad \text{number of units constraint}$$

Since each 1-bedroom apartment occupies 1000 square feet of space, x apartments will occupy $1000x$ square feet of space. Similarly, the 2-bedroom apartments will require $1500y$ square feet of space. The total amount of space needed is $1000x + 1500y$, so we must have

$$1000x + 1500y \leq 45,000 \quad \text{square footage constraint}$$

Moreover, since x and y denote the number of apartments to be built, we must have $x \geq 0$, $y \geq 0$. Thus, we have obtained the following system of inequalities:

$$
\begin{aligned}
x + \quad y &\leq \quad 40 \quad &\text{number of units constraint} \\
1000x + 1500y &\leq 45,000 \quad &\text{square footage constraint} \\
x &\geq \quad 0 \quad &\text{need for number of apartments} \\
y &\geq \quad 0 \quad &\text{to be nonnegative}
\end{aligned}
$$

We can graph the solution set of this system of linear inequalities as in Figure 14.

But the problem as stated asks that we *maximize* the monthly rental $z = 200x + 280y$, a requirement that we have never before seen in a mathematical problem of this sort! It is this requirement to **optimize,** that is, to seek a maximum or a minimum value of a linear expression, that characterizes a linear programming problem.

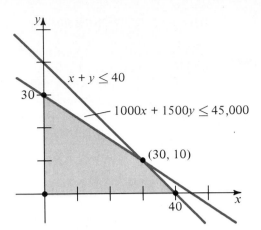

FIGURE 14 Graph of Linear Inequalities

Linear Programming Problem	A **linear programming problem** seeks the optimal (either the largest or the smallest) value of a linear expression called the **objective function** while satisfying constraints that can be formulated as a system of linear inequalities.

Returning to our apartment builder, we can state the linear programming problem in this way:

$$\text{maximize} \qquad z = 200x + 280y$$
$$\text{subject to} \qquad x + y \le 40$$
$$1000x + 1500y \le 45{,}000$$
$$x \ge 0$$
$$y \ge 0$$

Then the coordinates of each point of the solution set shown in Figure 14 are a **feasible solution,** that is, the coordinates give us ordered pairs (a, b) that satisfy the system of linear inequalities. But which points provide us with values of x and y that maximize the rental income z? For example, the points $(40, 0)$ and $(15, 20)$ are feasible solutions yielding these results for z:

x	y	$z = 200x + 280y$
40	0	8000
15	20	8600

Clearly, building 15 1-bedroom and 20 2-bedroom units yields a higher rental revenue than building 40 1-bedroom units, but is there a solution that will yield a still higher value for z?

Before providing the key to solving linear programming problems, we first must note that the solution set is bounded by straight lines, and we use the term **vertex** to denote an intersection point of any two boundary lines. We are then ready to state the following theorem.

Fundamental Theorem of Linear Programming	If a linear programming problem has an optimal solution, that solution occurs at a vertex of the set of feasible solutions.

With this result, the builder need only examine the vertices of the solution set of Figure 14, rather than considering each of the infinite number of feasible solutions—a bewildering task! We then evaluate the objective function $z - 200x + 280y$ for the coordinates of the vertices $(0, 0)$, $(0, 30)$, $(40, 0)$, and $(30, 10)$.

x	y	$z = 200x + 280y$
0	0	0
0	30	8400
40	0	8000
30	10	8800

Since the largest value of z is 8800 and this value corresponds to $x = 30$, $y = 10$, the builder finds that the optimal strategy is to build 30 1-bedroom and 10 2-bedroom units.

We can now illustrate the steps in solving a linear programming problem.

EXAMPLE 1

Solve the linear programming problem

$$\begin{aligned} \text{minimize} \qquad & z = x - 4y \\ \text{subject to} \qquad & x + 2y \leq 10 \\ & -x + 4y \leq 8 \\ & x \geq 0 \\ & y \geq 1 \end{aligned}$$

SOLUTION

Linear Programming	
Step 1. Sketch the solution set of the system of linear inequalities. (See figure 15)	*Step 1.*
Step 2. Determine all vertices of the solution set.	*Step 2.* The vertices $(0, 1)$ and $(0, 2)$ are the y-intercepts of the lines whose equations are $y = 1$ and $-x + 4y = 8$, respectively. The vertex B in Figure 15 is the

intersection of the lines $y = 1$ and $x + 2y = 10$ and is seen to be $(8, 1)$. The vertex A of Figure 15 is the intersection of the lines whose equations are

$$-x + 4y = 8$$

and

$$x + 2y = 10$$

Solving the system of equations (try elimination) yields the vertex $A(4, 3)$.

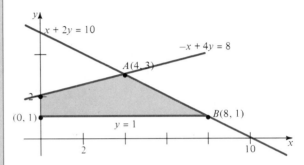

FIGURE 15   Graph of Linear Inequalities

Step 3. Evaluate the objective function for the coordinates of each vertex.

Step 3.

Vertex	x	y	$z = x - 4y$
$(0, 1)$	0	1	-4
$(0, 2)$	0	2	-8
$(8, 1)$	8	1	4
$(4, 3)$	4	3	-8

Step 4. The point or points providing the optimal value of the objective function are solutions of the linear programming problem.

Step 4. The minimal value of the objective function is -8, which occurs at the vertices $(0, 2)$ and $(4, 3)$. Thus, $x = 0$, $y = 2$ and $x = 4$, $y = 3$ are both solutions of the linear programming problem. ∎

Linear programming problems occur in real-life situations with great frequency. In certain industries these problems can involve thousands of variables and hundreds of constraints. Obviously, the method of graphical solution we presented for two variables cannot be used. A solution method known as the simplex algorithm was first devised by George Dantzig in 1947. Despite the sophistication of this approach, the number of calculations required becomes unmanageably large for hand computation for even relatively small numbers of constraints. Fortunately, the discovery of the simplex algorithm occurred at the time electronic computers made their initial appearance. Since then industries such as oil refining and steel production have used linear programming to determine the optimum use of their facilities.

EXERCISE SET 10.7

In Exercises 1–8 find the minimum value and the maximum value of the linear expression, subject to the given constraints. Indicate coordinates of the vertices at which the minimum and maximum values occur.

1. $x - \frac{1}{2}y$ subject to

$$3x - y \leq 1$$
$$x \geq 0$$
$$x \leq 5$$
$$y \geq 0$$

2. $2x + y$ subject to

$$x + y \leq 4$$
$$x \geq 1$$
$$y \geq 2$$

3. $\frac{1}{2}x - 2y$ subject to

$$x + 2y \leq 6$$
$$3y - 2x \leq 2$$
$$x \geq 0$$
$$y \geq 0$$

4. $0.2x + 0.8y$ subject to

$$3y + x \leq 8$$
$$3x - 5y \geq 2$$
$$x \geq 0$$
$$y > 0$$

5. $2x - y$ subject to

$$y - x \leq 0$$
$$4y + 3x \geq 6$$
$$x \leq 4$$

6. $x + 3y$ subject to

$$2x + y \geq 2$$
$$4x + 5y \leq 40$$
$$x \geq 0$$
$$y \geq 1$$
$$y \leq 6$$

7. $2x - y$ subject to

$$2y - x \leq 8$$
$$x + 2y \geq 12$$
$$5x + 2y \leq 44$$
$$x \geq 3$$

8. $y - x$ subject to

$$2y - 5x \leq 10$$
$$5x + 6y \leq 50$$
$$5x + y \leq 20$$
$$x \geq 0$$
$$y \geq 1$$

9. A firm has budgeted $1500 for display space at a toy show. Two types of display booths are available: "preferred space" costs $18 per square foot, with a minimum rental of 60 square feet, and "regular space" costs $12 per square foot, with a minimum rental of 30 square feet. It is estimated that there will be 120 visitors for each square foot of "preferred space" and 60 visitors for each square foot of "regular space." How should the firm allot its budget to maximize the number of potential clients that will visit the booths?

10. A company manufactures an 8-bit computer and a 16-bit computer. To meet existing orders, it must schedule at least 50 8-bit computers for the next production cycle

and can produce no more than 150 8-bit computers. The manufacturing facilities are adequate to produce no more than 300 16-bit computers, but the total number of computers that can be produced cannot exceed 400. The profit on each 8-bit computer is $310; on each 16-bit computer the profit is $275. Find the number of computers of each type that should be manufactured to maximize profit.

11. Swift Truckers is negotiating a contract with Better Spices, which uses two sizes of containers: large, 4-cubic-foot containers weighing 10 pounds and small, 2-cubic-foot containers weighing 8 pounds. Swift Truckers will use a vehicle that can handle a maximum load of 3280

pounds and a cargo size of up to 1000 cubic feet. The firms have agreed on a shipping rate of 50 cents for each large container and 30 cents for each small container. How many containers of each type should Swift place on a truck to maximize income?

12. A bakery makes both yellow cake and white cake. Each pound of yellow cake requires $\frac{1}{4}$ pound of flour and $\frac{1}{4}$ pound of sugar; each pound of white cake requires $\frac{1}{3}$ pound of flour and $\frac{1}{5}$ pound of sugar. The baker finds that 100 pounds of flour and 80 pounds of sugar are available. If yellow cake sells for $3 per pound and white cake sells for $2.50 per pound, how many pounds of each cake should the bakery produce to maximize income, assuming that all cakes baked can be sold?

13. A shop sells a mixture of Java and Colombian coffee beans for $4 per pound. The shopkeeper has allocated $1000 for buying fresh beans and finds that he must pay $1.50 per pound for Java beans and $2 per pound for Colombian beans. In a satisfactory mixture the weight of Colombian beans will be at least twice and no more than four times the weight of the Java beans. How many pounds of each type of coffee bean should be ordered to maximize the profit if all the mixture can be sold?

14. A pension fund plans to invest up to $50,000 in U.S. Treasury bonds yielding 12% interest per year and corporate bonds yielding 15% interest per year. The fund manager is told to invest a minimum of $25,000 in the Treasury bonds and a minimum of $10,000 in the corporate bonds, with no more than $\frac{1}{4}$ of the total investment to be in corporate bonds. How much should the manager invest in each type of bond to achieve a maximum amount of annual interest? What is the maximum interest?

15. A farmer intends to plant crops A and B on all or part of a 100-acre field. Seed for crop A costs $6 per acre, and

labor and equipment costs $20 per acre. For crop B, seed costs $9 per acre, and labor and equipment costs $15 per acre. The farmer cannot spend more than $810 for seed and $1800 for labor and equipment. If the income per acre is $150 for crop A and $175 for crop B, how many acres of each crop should be planted to maximize total income?

16. The farmer in Exercise 15 finds that a worldwide surplus in crop B reduces the income to $140 per acre while the income for crop A remains steady at $150 per acre. How many acres of each crop should be planted to maximize total income?

17. In preparing food for the college cafeteria, a dietitian will combine Volume Pack A and Volume Pack B. Each pound of Volume Pack A costs $2.50 and contains 4 units of carbohydrate, 3 units of protein, and 5 units of fat. Each pound of Volume Pack B costs $1.50 and contains 3 units of carbohydrate, 4 units of protein, and 1 unit of fat. If minimum monthly requirements are 60 units of carbohydrates, 52 units of protein, and 42 units of fat, how many pounds of each food pack will the dietitian use to minimize costs?

18. A lawn service uses a riding mower that cuts a 5000-square-foot area per hour and a smaller mower that cuts a 3000-square-foot area per hour. Surprisingly, each mower uses $\frac{1}{2}$ gallon of gasoline per hour. Near the end of a long summer day, the supervisor finds that both mowers are empty and that there remains 0.6 gallon of gasoline in the storage cans. To conclude the day at a sensible point, at least 4000 square feet of lawn must still be mowed. If the cost of operating the riding mower is $9 per hour and the cost of operating the smaller mower is $5 per hour, how much of the remaining gasoline should be allocated to each mower to do the job at the least possible cost?

TERMS AND SYMBOLS

KEY IDEAS FOR REVIEW

Topic	Page	Key Idea
Solving systems of equations	437	To solve a system of equations you must find a value for each variable that satisfies each equation of the system. This sequence of values is called a solution.
method of substitution	437	The method of substitution involves solving an equation for one variable and substituting the result into another equation.
method of elimination	440	The method of elimination involves multiplying an equation by a nonzero constant so that when it is added to a second equation, a variable drops out.
consistent and inconsistent systems	439	A consistent system of equations has one or more real solutions; an inconsistent system has no real solutions.
applications	442	It is often easier and more natural to set up word problems using two or more variables.
		The graph of a pair of linear equations in two variables is two straight lines, which may either (a) intersect in a point, (b) be parallel, or (c) be the same line. If the two straight lines intersect, the coordinates of the point of intersection are a solution of the system of linear equations. If the lines are distinct and do not intersect, the system is inconsistent.
Solving a system of linear inequalities	447	The solution of a system of linear inequalities can be found graphically as the region satisfying all the inequalities.
Gaussian elimination	452	Gaussian elimination is a systematic way of transforming a linear system to triangular form. A linear system in triangular form is easily solved by back-substitution.
Matrix	457	A matrix is a rectangular array of numbers.
sum and difference	466	The sum and difference of two matrices A and B can be formed only if A and B are of the same dimension.
product	469	The product AB can be formed only if the number of columns of A is the same as the number of rows of B.
Systems of linear equations and matrices	459	Systems of linear equations can be conveniently handled in matrix notation. By dropping the names of the variables, matrix notation focuses on the coefficients and the right-hand side of the system. The elementary row operations are then seen to be an abstraction of those operations that produce equivalent systems of equations.
Gaussian and Gauss–Jordan elimination	460	Gaussian elimination and Gauss–Jordan elimination both involve the use of elementary row operations on the augmented matrix corresponding to a linear system. In the case of a system of three equations with three unknowns and a

Topic	Page	Key Idea
		unique solution, the final matrices will be of this form:
		$$\begin{bmatrix} * & * & * & \vdots & * \\ 0 & * & * & \vdots & * \\ 0 & 0 & * & \vdots & * \end{bmatrix} \qquad \begin{bmatrix} 1 & 0 & 0 & \vdots & c_1 \\ 0 & 1 & 0 & \vdots & c_2 \\ 0 & 0 & 1 & \vdots & c_3 \end{bmatrix}$$ <div align="center">Gaussian elimination Gauss–Jordan elimination</div> If Gauss–Jordan elimination is used, the solution can be read from the final matrix; if Gaussian elimination is used, back-substitution is then performed with the final matrix.
matrix notation	471	A linear system can be written in the form $AX = B$ where A is the coefficient matrix, X is a column matrix of the unknowns, and B is the column matrix of the right-hand sides.
Determinants	473	Associated with every square matrix is a number called a determinant. The rule for evaluating a determinant of order 2 is $$\begin{vmatrix} a & b \\ c & d \end{vmatrix} = ad - bc$$
evaluation by cofactors	474	For determinants of order greater than 2, the method of expansion by cofactors may be used to reduce the problem to that of evaluating determinants of order 2. When expanding by cofactors, choose the row or column that contains the most zeros. This will ease the arithmetic burden.
Cramer's rule	478	Cramer's rule provides a means for solving a linear system by expressing the value of each unknown as a quotient of determinants.
Linear Programming Problem	485	To solve a linear programming problem, it is only necessary to consider the vertices of the region of feasible solutions.

REVIEW EXERCISES

Solutions to exercises whose numbers are in color are in the Solutions section in the back of the book.

10.1 In Exercises 1–5 solve the given system by the method of substitution.

1. $3x - 2y = 4$
   $2x + y = -2$

2. $2x + 3y = -7$
   $-x + 2y = 7$

3. $x^2 + y^2 = 25$
   $x + 3y = 5$

4. $y^2 = 2x - 1$
   $x - y = 2$

5. $2x - y = 0$
   $x - 3y = \dfrac{7}{4}$

In Exercises 6–10 solve the given system by the method of elimination.

6. $5x - 2y = 14$
   $-x - 3y = 4$

7. $2x + 3y = -1$
   $-3x + 4y = -\dfrac{11}{4}$

8. $x^2 + y^2 = 9$
   $y = x^2 + 3$

9. $y^2 = 4x$
   $y^2 + x - 2y = 12$

10. $\dfrac{1}{3}x + \dfrac{1}{2}y = -1$
    $\dfrac{1}{2}x - \dfrac{1}{4}y = \dfrac{5}{2}$

11. A manufacturer of faucets finds that the supply S and demand D are related to price p as follows.

$$S = 3p + 2$$
$$D = -2p + 17$$

Find the equilibrium price, that is, the price at which supply equals demand, and the number of faucets sold at that price.

10.2 In Exercises 12–17 graph the solution set of the linear inequality or system of linear inequalities.

12. $x - 2y \leq 5$ 13. $2x + y > 4$

14. $2x + 3y \leq 2$ 15. $x - 2y \geq 4$

$\ x - y \geq 1$ $\ 2x - y \leq 2$

16. $2x + 3y \leq 6$ 17. $2x + y \leq 4$

$\ x \geq 0$ $\ 2x - y \leq 3$

$\ y \geq 1$ $\ x \geq 0$

$\ y \geq 0$

10.3 In Exercises 18–20 use Gaussian elimination to solve the given linear systems.

18. $-3x - y + z = 12$ 19. $3x + 2y - z = -8$

$\ 2x + 5y - 2z = -9$ $\ 2x + 3z = 5$

$\ -x + 4y + 2z = 15$ $\ x - 4y = -4$

20. $5x - y + 2z = 10$

$\ -2x + 3y - z = -7$

$\ 3x + 2z = 7$

10.4 Exercises 21–24 refer to the matrix

$$A = \begin{bmatrix} -1 & 4 & 2 & 0 & 8 \\ 2 & 0 & -3 & -1 & 5 \\ 4 & -6 & 9 & 1 & -2 \end{bmatrix}$$

21. Determine the dimension of the matrix A.

22. Find a_{24}.   23. Find a_{31}.   24. Find a_{15}.

Exercises 25 and 26 refer to the linear system

$$3x - 7y = 14$$
$$x + 4y = 6$$

25. Write the coefficient matrix of the linear system.

26. Write the augmented matrix of the linear system.

In Exercises 27 and 28 write a linear system corresponding to the augmented matrix.

27. $\begin{bmatrix} 4 & -1 & \vdots & 3 \\ 2 & 5 & \vdots & 0 \end{bmatrix}$ 28. $\begin{bmatrix} -2 & 4 & 5 & \vdots & 0 \\ 6 & -9 & 4 & \vdots & 0 \\ 3 & 2 & -1 & \vdots & 0 \end{bmatrix}$

In Exercises 29 and 30 use back-substitution to solve the linear systems corresponding to the given augmented matrix.

29. $\begin{bmatrix} 1 & -2 & \vdots & 7 \\ 0 & 1 & \vdots & -4 \end{bmatrix}$ 30. $\begin{bmatrix} 1 & -2 & 2 & \vdots & -9 \\ 0 & 1 & 3 & \vdots & -8 \\ 0 & 0 & 1 & \vdots & -3 \end{bmatrix}$

In Exercises 31 and 32 use matrix methods to solve the given linear system.

31. $x + y = 2$ 32. $2x - y - 2z = 3$

$\ 2x - 4y = -5$ $\ -2x + 3y + z = 3$

$\ 2y - z = 6$

10.5 In Exercises 33 and 34 solve for x.

33. $\begin{bmatrix} 5 & -1 \\ 3 & 2x \end{bmatrix} = \begin{bmatrix} 5 & -1 \\ 3 & -6 \end{bmatrix}$

34. $\begin{bmatrix} 6 & x^2 \\ 4x & -2 \end{bmatrix} = \begin{bmatrix} 6 & 9 \\ -12 & -2 \end{bmatrix}$

Exercises 35–41 refer to the following matrices.

$$A = \begin{bmatrix} 2 & -1 \\ 3 & 2 \end{bmatrix} \qquad B = \begin{bmatrix} -1 & 5 \\ 4 & -3 \end{bmatrix}$$

$$C = \begin{bmatrix} -1 & 0 \\ 0 & 4 \\ 2 & -2 \end{bmatrix} \qquad D = \begin{bmatrix} 1 & 3 & 4 \\ -1 & 0 & -6 \end{bmatrix}$$

If possible, find the following.

35. $A + B$ 36. $B - A$ 37. $A + C$

38. $5D$ 39. CD 40. DC

41. $-AB$

10.6 In Exercises 42–45 evaluate the given determinant.

42. $\begin{vmatrix} 3 & 1 \\ -4 & 2 \end{vmatrix}$ 43. $\begin{vmatrix} -1 & 2 \\ 0 & 6 \end{vmatrix}$

44. $\begin{vmatrix} 1 & -1 & 2 \\ 0 & 5 & 4 \\ 2 & 3 & 8 \end{vmatrix}$ 45. $\begin{vmatrix} 1 & 2 & -1 \\ 0 & 3 & 4 \\ 0 & 0 & -1 \end{vmatrix}$

In Exercises 46–49 use Cramer's rule to solve the given linear system.

46. $2x - y = -3$ 47. $3x - y = 7$

$\ -2x + 3y = 11$ $\ 2x + 5y = -18$

48. $3x + z = 0$ 49. $2x + 3y + z = -5$

$\ x + y + z = 0$ $\ 2y + 2z = -3$

$\ -3y + 2z = -4$ $\ 4x + y - 2z = -2$

10.7 In Exercises 50 and 51 solve the given linear programming problem.

50. maximize $z = 5y - x$

subject to $8y - 3x \leq 36$

$6x + y \leq 30$

$y \geq 1$

$x \geq 0$

51. minimize $z = x + 4y$

subject to $4x - y \geq 8$

$4x + y \leq 24$

$5y + 4x \geq 32$

PROGRESS TEST 10A

In Problems 1–3 find all real solutions of the given system.

1. $y^2 = 5x$

$y^2 - x^2 = 6$

2. $\dfrac{1}{4}x + \dfrac{1}{2}y = 0$

$3x - 2y = 8$

3. $x^2 + y^2 = 25$

$4x^2 - y^2 = 20$ ·

4. An elegant men's shop is having a post-Christmas sale. All shirts are reduced to one low price, and all ties are reduced to an even lower price. A customer purchases 3 ties and 7 shirts, paying $135. Another customer selects 5 ties and 3 shirts, paying $95. What is the sale price of each tie and of each shirt?

In Problems 5 and 6 graph the solution set of the system of linear inequalities.

5. $x - 2y \leq 1$

$3x + 2y \geq 4$

6. $2x + y \leq 10$

$-x + 3y \leq 12$

$x \geq 0$

$y \geq 0$

7. Solve by Gaussian elimination.

$3x + 2y - z = -4$

$x - y + 3z = 12$

$2x - y - 2z = -20$

Problems 8 and 9 refer to the matrix

$$A = \begin{bmatrix} -1 & 2 \\ -2 & 4 \\ 0 & 7 \end{bmatrix}$$

8. Find the dimension of the matrix A.

9. Find a_{31}.

10. Write the augmented matrix of the linear system

$-7x \qquad + 6z = 3$

$2y - z = 10$

$x - y + z = 5$

11. Write a linear system corresponding to the augmented matrix

$$\begin{bmatrix} -5 & 2 & \vdots & 4 \\ 3 & -4 & \vdots & 4 \end{bmatrix}$$

12. Use back-substitution to solve the linear system corresponding to the augmented matrix

$$\begin{bmatrix} 1 & 1 & \vdots & 0 \\ 0 & 1 & \vdots & \frac{1}{2} \end{bmatrix}$$

13. Solve the linear system

$-x + 2y = -2$

$\dfrac{1}{2}x + 2y = -7$

by applying Gaussian elimination to the augmented matrix.

14. Solve the linear system

$2x - y + 3z = 2$

$x + 2y - z = 1$

$-x + y + 4z = 2$

by applying Gauss–Jordan elimination to the augmented matrix.

15. Solve for x.

$$\begin{bmatrix} 2x - 1 & 0 \\ 1 & -3 \end{bmatrix} = \begin{bmatrix} 5 & 0 \\ 1 & -3 \end{bmatrix}$$

Problems 16–19 refer to the matrices

$$A = \begin{bmatrix} -4 & 0 & 3 \\ 6 & 2 & -3 \end{bmatrix} \qquad B = \begin{bmatrix} -1 \\ -3 \end{bmatrix}$$

$$C = \begin{bmatrix} 4 & 2 \\ -2 & 0 \\ 3 & -1 \end{bmatrix} \qquad D = \begin{bmatrix} 1 & -6 \\ 0 & 2 \\ 4 & -1 \end{bmatrix}$$

If possible, find the following.

16. $C - 2D$ **17.** AC **18.** CB **19.** BA

In Problems 20 and 21 evaluate the given determinant.

20. $\begin{vmatrix} -6 & -2 \\ 2 & 1 \end{vmatrix}$ 21. $\begin{bmatrix} 0 & -1 & 2 \\ 2 & -2 & 3 \\ 1 & 4 & 5 \end{bmatrix}$

22. Use Cramer's rule to solve the linear system.

$$x + 2y = -2$$
$$-2x - 3y = 1$$

PROGRESS TEST 10B

In Problems 1–3 find all real solutions of the given system.

1. $x^2 - y^2 = 9$
   $x^2 + y^2 = 41$

2. $2x - 3y = 11$
   $3x + 5y = -12$

3. $x^2 + 3y^2 = 12$
   $x + 3y = 6$

4. A movie theater charges $3 admission for an adult and $1.50 for a child. If 600 tickets were sold and the total revenue received was $1350, how many tickets of each type were sold?

In Problems 5 and 6 graph the solution set of the system of linear inequalities.

5. $2x - 3y \geq 6$
   $3x + y \leq 3$

6. $2x + y \geq 4$
   $2x - 5y \leq 5$
   $y \geq 1$

7. Solve by Gaussian elimination.

$$x + 2z = 7$$
$$3y + 4z = -10$$
$$-2x + y - 2z = -14$$

Problems 8 and 9 refer to the matrix

$$B = \begin{bmatrix} -1 & 5 & 0 & 6 \\ 4 & -2 & 1 & -2 \end{bmatrix}$$

8. Determine the dimension of the matrix B.

9. Find b_{23}.

10. Write the coefficient matrix of the linear system.

$$2x - 6y = 5$$
$$x + 3y = -2$$

11. Write a linear system corresponding to the augmented matrix

$$\begin{bmatrix} 16 & 0 & 6 & | & 10 \\ -4 & -2 & 5 & | & 8 \\ 2 & 3 & -1 & | & 6 \end{bmatrix}$$

12. Use back-substitution to solve the linear system corresponding to the augmented matrix

$$\begin{bmatrix} 1 & 3 & -1 & | & 0 \\ 0 & 1 & -2 & | & 5 \\ 0 & 0 & 1 & | & -3 \end{bmatrix}$$

13. Solve the linear system

$$2x + 3y = -11$$
$$3x - 2y = 3$$

by applying Gaussian elimination to the augmented matrix.

14. Solve the linear system

$$2x + y - 2z = 7$$
$$-3x - 5y + 4z = -3$$
$$5x + 4y = 17$$

by applying Gauss–Jordan elimination to the augmented matrix.

15. Solve for y.

$$\begin{bmatrix} 2 & -5 & 1 \\ -3 & 1-y & 2 \end{bmatrix} = \begin{bmatrix} 2 & -5 & 1 \\ -3 & 6 & 2 \end{bmatrix}$$

Problems 16–19 refer to the matrices

$$A = \begin{bmatrix} 2 & -3 \\ -1 & 0 \\ 2 & 1 \end{bmatrix} \quad B = \begin{bmatrix} 1 & -2 & 3 \end{bmatrix} \quad C = \begin{bmatrix} 0 & -4 \\ 3 & 1 \\ -2 & 5 \end{bmatrix}$$

If possible, find the following.

16. BA

17. $2C + 3A$

18. CB

19. $BC - A$

In Problems 20 and 21 evaluate the given determinant.

20. $\begin{vmatrix} -2 & 4 \\ 3 & 5 \end{vmatrix}$ 21. $\begin{vmatrix} -2 & 0 & 1 \\ 1 & 2 & 0 \\ 2 & -1 & -1 \end{vmatrix}$

22. Use Cramer's rule to solve the linear system.

$$x + y = -1$$
$$2x - 4y = -5$$

CHAPTER 11

SEQUENCES AND SERIES

The topics in this chapter are related in that they all involve the set of natural numbers. As you might expect, despite our return to a simpler number system, the approach and results will be more advanced than in earlier chapters. For example, in discussing sequences we will be dealing with functions whose domain is the set of natural numbers. Yet, sequences lead to considerations of series, and the underlying concepts of infinite series can be used as an introduction to calculus.

Another of the topics, mathematical induction, provides a means of proving certain theorems involving the natural numbers that appear to resist other means of proof. As an example, we will use mathematical induction to prove that the sum of the first n consecutive positive integers is $n(n + 1)/2$.

Yet another topic is the binomial theorem, which gives us a way to expand the expression $(a + b)^n$ where n is a natural number. One of the earliest results obtained in a calculus course requires the binomial theorem in its derivation.

**11.1
SEQUENCES AND
SIGMA NOTATION**

INFINITE SEQUENCES

Can you see a pattern or relationship that describes this string of numbers?

$$1, 4, 9, 16, 25, \ldots$$

If we rewrite this string as

$$1^2, 2^2, 3^2, 4^2, 5^2, \ldots$$

it is clear that these are the squares of successive natural numbers. Each number in the string is called a **term.** We could write the nth term of the list as a function a defined by

$$a(n) = n^2$$

where n is a natural number. Such a string of numbers is called an infinite sequence, since the list is infinitely long.

> An **infinite sequence** (often called simply a **sequence**) is a function whose domain is the set of all natural numbers.

The range of the function a is

$$a(1), a(2), a(3), \ldots, a(n), \ldots$$

which we write as

$$a_1, a_2, a_3, \ldots, a_n, \ldots$$

That is, we indicate a sequence by using subscript notation rather than function notation. We say that a_1 is the **first term** of the sequence, a_2 is the **second term,** and so on, and we write the **nth term** as a_n where $a_n = a(n)$.

EXAMPLE 1 Finding Specific Terms of a Sequence
Write the first three terms and the tenth term of each of the sequences whose nth term is given.

(a) $a_n = n^2 + 1$ (b) $a_n = \dfrac{n}{n + 1}$ (c) $a_n = 2^n - 1$

SOLUTION
The first three terms are found by substituting $n = 1, 2,$ and 3 in the formula for a_n. The tenth term is found by substituting $n = 10$.

(a) $a_1 = 1^2 + 1 = 2$ $a_2 = 2^2 + 1 = 5$ $a_3 = 3^2 + 1 = 10$
  $a_{10} = 10^2 + 1 = 101$

(b) $a_1 = \dfrac{1}{1 + 1} = \dfrac{1}{2}$ $a_2 = \dfrac{2}{2 + 1} = \dfrac{2}{3}$ $a_3 = \dfrac{3}{3 + 1} = \dfrac{3}{4}$

  $a_{10} = \dfrac{10}{10 + 1} = \dfrac{10}{11}$

(c) $a_1 = 2^1 - 1 = 1$ $a_2 = 2^2 - 1 = 3$ $a_3 = 2^3 - 1 = 7$
  $a_{10} = 2^{10} - 1 = 1023$

PROGRESS CHECK

Write the first three terms and the twelfth term of each of the sequences whose nth term is given.

(a) $a_n = 3(1 - n)$ **(b)** $a_n = n^2 + n + 1$ **(c)** $a_n = 5$

ANSWERS

(c) $a_1 = a_2 = a_3 = a_{12} = 5$

(b) $a_1 = 3, a_2 = 7, a_3 = 13, a_{12} = 157$

(a) $a_1 = 0, a_2 = -3, a_3 = -6, a_{12} = -33$

An infinite sequence is often defined by a formula expressing the nth term by reference to preceding terms. Such a sequence is said to be defined by a **recursive** formula.

EXAMPLE 2   Finding Terms of a Recursive Sequence

Find the first four terms of the sequence defined by

$$a_n = a_{n-1} + 3 \qquad \text{with} \qquad a_1 = 2 \qquad \text{and} \qquad n \geq 2$$

SOLUTION

Any term of the sequence can be obtained if the preceding term is known. Of course, this recursive formulation requires a starting point, and we are indeed given a_1. Then

$$a_1 = 2$$
$$a_2 = a_1 + 3 = 2 + 3 = 5$$
$$a_3 = a_2 + 3 = 5 + 3 = 8$$
$$a_4 = a_3 + 3 = 8 + 3 = 11$$

∎

PROGRESS CHECK

Find the first four terms of the infinite sequence

$$a_n = 2a_{n-1} - 1 \qquad \text{with} \qquad a_1 = -1 \qquad \text{and} \qquad n \geq 2$$

ANSWER

$-1, -3, -7, -15$

SUMMATION NOTATION

In the following sections of this chapter, we will seek the sum of terms of a sequence such as

$$a_1 + a_2 + a_3 + \cdots + a_m$$

Since sums occur frequently in mathematics, a special notation has been developed that is defined in the following way.

Summation Notation	$\displaystyle\sum_{k=1}^{m} a_k = a_1 + a_2 + a_3 + \cdots + a_m$

This is often referred to as **sigma notation,** since the Greek letter Σ indicates a sum of terms of the form a_k. The letter k is called the **index of summation** and always assumes successive integer values, starting with the value written under the Σ sign and ending with the value written above the Σ sign.

EXAMPLE 3 Using Sigma Notation

Evaluate **(a)** $\displaystyle\sum_{k=1}^{3} 2^k(k+1)$ **(b)** $\displaystyle\sum_{i=2}^{4} (i^2 + 2)$.

SOLUTION

(a) The terms are of the form

$$a_k = 2^k(k+1)$$

and the sigma notation indicates that we want the sum of the terms a_1 through a_3. Forming the terms and adding.

$$\sum_{k=1}^{3} 2^k(k+1) = 2^1(1+1) + 2^2(2+1) + 2^3(3+1)$$

$$= 4 + 12 + 32 = 48$$

(b) Any letter may be used for the index of summation. Here, the letter i is used, and

$$\sum_{i=2}^{4} (i^2 + 2) = (2^2 + 2) + (3^2 + 2) + (4^2 + 2)$$

$$= 6 + 11 + 18 = 35$$

Note that the index of summation can begin with an integer value other than 1. ∎

EXAMPLE 4 Using Sigma Notation

Write each sum using summation notation.

(a) $\dfrac{1}{2} + \dfrac{1}{2 \cdot 2} + \dfrac{1}{2 \cdot 3} + \dfrac{1}{2 \cdot 4} + \dfrac{1}{2 \cdot 5}$ **(b)** $\dfrac{2}{3} + \dfrac{3}{4} + \dfrac{4}{5} + \dfrac{5}{6}$

SOLUTION

(a) The denominator of each term is of the form $2 \cdot k$, where k assumes integer values from 1 to 5. Then

$$\sum_{k=1}^{5} \frac{1}{2 \cdot k}$$

expresses the desired sum.

(b) If the value of the numerator of a term is k, then the denominator is $k + 1$. Letting k range from 2 to 5,

$$\sum_{k=2}^{5} \frac{k}{k+1}$$

expresses the desired sum.

∎

PROGRESS CHECK

Write each sum using summation notation.

(a) $x_1^2 + x_2^2 + x_3^2 + \cdots + x_{20}^2$ **(b)** $2^3 + 3^4 + 4^5 + 5^6$

ANSWERS

(a) $\sum_{k=1}^{20} x_k^2$ (b) $\sum_{k=2}^{5} k^{k+1}$

If a sequence is defined by $a_n = c$, where c is a real constant, then

$$\sum_{k=1}^{r} a_k = a_1 + a_2 + \cdots + a_r$$

$$= c + c + \cdots + c$$

$$= rc$$

This leads to the rule:

For any real constant c,

$$\sum_{k=1}^{n} c = nc$$

EXAMPLE 5 Using Sigma Notation with a Constant

Evaluate **(a)** $\sum_{j=1}^{20} 5$ **(b)** $\sum_{k=1}^{4} (k^2 - 2)$.

SOLUTION

(a) $\sum_{j=1}^{20} 5 = 20 \cdot 5 = 100$

(b) $\sum_{k=1}^{4} (k^2 - 2) = (1^2 - 2) + (2^2 - 2) + (3^2 - 2) + (4^2 - 2)$

$$= -1 + 2 + 7 + 14 = 22$$

The following are properties of sums expressed in sigma notation.

Properties of Sums

For the sequences $a_1, a_2, \ldots,$ and $b_1, b_2, \ldots,$

(i) $\sum_{k=1}^{n} (a_k + b_k) = \sum_{k=1}^{n} a_k + \sum_{k=1}^{n} b_k$

(ii) $\sum_{k=1}^{n} (a_k - b_k) = \sum_{k=1}^{n} a_k - \sum_{k=1}^{n} b_k$

(iii) $\sum_{k=1}^{n} ca_k = c \sum_{k=1}^{n} a_k, \quad c$ a constant

AREAS BY RECTANGLES

Many textbooks introduce the integral calculus by the use of rectangles to approximate area. In the accompanying figure, we are interested in calculating the area under the curve of the function $f(x) = x^2$ that is bounded by the x-axis and the lines $x = a$ and $x = b$. The interval $[a, b]$ is divided into n subintervals of equal width, and a rectangle is erected in each interval as shown. We then seek to use the sum of the areas of the rectangles as an approximation to the area under the curve.

To calculate the area of a rectangle, we need to know the height and the width. Since the interval $[a, b]$ has been divided into n parts of equal width, we see that

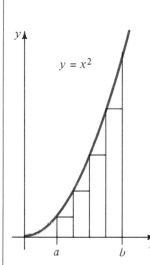

$$\text{width of rectangle} = \frac{b - a}{n}$$

Next, note that the height of the rectangle whose left endpoint is at x_k is determined by the value of the function at that point; that is,

$$\text{height of rectangle} = f(x_k) = x_k^2$$

The area of a "typical" rectangle is then

$$\left(\frac{b - a}{n}\right) f(x_k) = \left(\frac{b - a}{n}\right) x_k^2$$

and the sum of the areas of the rectangles is neatly expressed in summation notation by

$$\sum_{k=1}^{n} \left(\frac{b - a}{n}\right) x_k^2 = \left(\frac{b - a}{n}\right) \sum_{k=1}^{n} x_k^2$$

Intuitively, we see that the greater the number of rectangles, the better our approximation will be, and this concept is pursued in calculus. The student is urged to let $a = 1$ and $b = 3$ in the accompanying figure and to use the method of approximating rectangles with $n = 2$, $n = 4$, and $n = 8$. The *exact* answer is $26/3$ square units, and the approximations improve as n grows larger.

EXAMPLE 6   Using Sigma Notation with a Constant

Use the properties of sums to evaluate $\sum_{k=1}^{4} (k^2 - 2)$.

SOLUTION

Rather than write out the terms as was done in Example 5b, we may write

$$\sum_{k=1}^{4} (k^2 - 2) = \sum_{k=1}^{4} k^2 - \sum_{k=1}^{4} 2 = \sum_{k=1}^{4} k^2 - 8$$
$$= 1^2 + 2^2 + 3^2 + 4^2 - 8 = 30 - 8 = 22$$

∎

SERIES

If we begin with the sequence

$$a_1, a_2, a_3, \ldots, a_n, \ldots$$

we say that the sum of terms

$$a_1 + a_2 + a_3 + \cdots + a_n + \cdots$$

is the related **infinite series.** For this same sequence, we also define

$$S_1 = a_1$$
$$S_2 = a_1 + a_2$$
$$S_3 = a_1 + a_2 + a_3$$

and, in general,

$$S_n = \sum_{k=1}^{n} a_k = a_1 + a_2 + \cdots + a_n$$

The number S_n is called the **nth partial sum** of the sequence.

EXAMPLE 7   Finding the nth Partial Sum

Given the infinite sequence

$$a_n = n^2 - 1$$

find S_4.

SOLUTION

The first four terms of the sequence are

$$a_1 = 0 \qquad a_2 = 3 \qquad a_3 = 8 \qquad a_4 = 15$$

Then the sum S_4 is given by

$$S_4 = \sum_{k=1}^{4} a_k = 0 + 3 + 8 + 15 = 26$$

■

If a series alternates in sign, then a multiplicative factor of $(-1)^k$ or $(-1)^{k+1}$ can be used to obtain the proper sign. For example, the series

$$-1^2 + 2^2 - 3^2 + 4^2$$

can be written in sigma notation as

$$\sum_{k=1}^{4} (-1)^k k^2$$

while the series

$$1^2 - 2^2 + 3^2 - 4^2$$

can be written as

$$\sum_{k=1}^{4} (-1)^{k+1} k^2$$

EXAMPLE 8 Finding the General Term of an Alternating Sequence

The terms of a sequence are of the form $a_k = \sqrt{k}$, and the terms are negated when k is even. Write an expression for the general term a_n and for the sum S_n in summation notation.

SOLUTION

If we multiply each term by $(-1)^{k+1}$, then the odd terms will be positive and the even terms will be negative. The general term a_n is then

$$a_n = (-1)^{n+1}\sqrt{n}$$

and the sequence is

$$\sqrt{1}, -\sqrt{2}, \sqrt{3}, -\sqrt{4}, \ldots, (-1)^{n+1}\sqrt{n}, \ldots$$

Finally, the sum S_n is given by

$$S_n - \sum_{k=1}^{n} (-1)^{k+1}\sqrt{k}$$

■

EXAMPLE 9 Working with an Infinite Series

Given the sequence

$$a_n = \frac{3}{10^n}$$

find the first five partial sums. Use these results to predict the sum of the infinite series

$$\sum_{k=1}^{\infty} \frac{3}{10^k}$$

SOLUTION

The terms of the sequence can be written as

$$0.3, 0.03, 0.003, 0.0003, 0.00003, \ldots$$

and the partial sums are then seen to be

$$S_1 = 0.3$$
$$S_2 = 0.3 + 0.03 = 0.33$$
$$S_3 = 0.3 + 0.03 + 0.003 = 0.333$$
$$S_4 = 0.3 + 0.03 + 0.003 + 0.0003 = 0.3333$$
$$S_5 = 0.3 + 0.03 + 0.003 + 0.0003 + 0.00003 = 0.33333$$

It would appear reasonable to conclude that, as n grows large, S_n approaches $\frac{1}{3}$ and that

$$\sum_{k=1}^{\infty} \frac{3}{10^k} = \frac{1}{3}$$

although this does not constitute a *proof* in a mathematical sense.

■

EXERCISE SET 11.1

In Exercises 1–12 find the first four terms and the twentieth term of the sequence whose nth term is given.

1. $a_n = 2n$
2. $a_n = 2n + 1$
3. $a_n = 4n - 3$
4. $a_n = 3n - 1$

5. $a_n = 5$
6. $a_n = 1 - \dfrac{1}{n}$
7. $a_n = \dfrac{n}{n + 1}$
8. $a_n = \sqrt{n}$

9. $a_n = 2 + 0.1^n$
10. $a_n = \dfrac{n^2 - 1}{n^2 + 1}$
11. $a_n = \dfrac{n^2}{2n + 1}$
12. $a_n = \dfrac{2n + 1}{n^2}$

In Exercises 13–18 a sequence is defined recursively. Find the indicated term of the sequence.

13. $a_n = 2a_{n-1} - 1$, $a_1 = 2$; find a_4
14. $a_n = 3 - 3a_{n-1}$, $a_1 = -1$; find a_3

15. $a_n = \dfrac{1}{a_{n-1} + 1}$, $a_3 = 2$; find a_6
16. $a_n = \dfrac{n}{a_{n-1}}$, $a_2 = 1$; find a_5

17. $a_n = (a_{n-1})^2$, $a_1 = 2$; find a_4
18. $a_n = (a_{n-1})^{n-1}$, $a_1 = 2$; find a_4

In Exercises 19–26 find the indicated sum.

19. $\displaystyle\sum_{k=1}^{5} (3k - 1)$
20. $\displaystyle\sum_{k=1}^{5} (3 - 2k)$
21. $\displaystyle\sum_{k=1}^{6} (k^2 + 1)$
22. $\displaystyle\sum_{k=0}^{4} \dfrac{k}{k^2 + 1}$

23. $\displaystyle\sum_{k=3}^{5} \dfrac{k}{k - 1}$
24. $\displaystyle\sum_{k=2}^{4} 4(2^k)$
25. $\displaystyle\sum_{j=1}^{4} 20$
26. $\displaystyle\sum_{i=1}^{10} 50$

In Exercises 27–36 use summation notation to express the sum. (The answer is not unique.)

27. $1 + 3 + 5 + 7 + 9$
28. $2 + 5 + 8 + 11 + 14$

29. $1 + 4 + 9 + 16 + 25$
30. $1 - 4 + 9 - 16 + 25$

31. $-1 + \dfrac{1}{\sqrt{2}} - \dfrac{1}{\sqrt{3}} + \dfrac{1}{\sqrt{4}}$
32. $\dfrac{1}{2 \cdot 4} + \dfrac{1}{2 \cdot 5} + \dfrac{1}{2 \cdot 6} + \dfrac{1}{2 \cdot 7}$

33. $\dfrac{1}{1^2 + 1} - \dfrac{2}{2^2 + 1} + \dfrac{3}{3^2 + 1} - \dfrac{4}{4^2 + 1}$
34. $2 - 4 + 8 - 16$

35. $1 + \dfrac{1}{x} + \dfrac{1}{x^2} + \dfrac{1}{x^3} + \cdots + \dfrac{1}{x^n}$
36. $\dfrac{1}{1 \cdot 2} + \dfrac{1}{2 \cdot 3} + \dfrac{1}{3 \cdot 4} + \dfrac{1}{4 \cdot 5} + \cdots + \dfrac{1}{49 \cdot 50}$

11.2 ARITHMETIC SEQUENCES

The sequence

$$2, 5, 8, 11, 14, 17, \ldots$$

.s an example of a special type of sequence in which each successive term is obtained by adding a fixed number to the previous term.

Arithmetic Sequence	In an **arithmetic sequence** there is a real number d such that $$a_n = a_{n-1} + d$$ for all $n > 1$. The number d is called the **common difference.**

An arithmetic sequence is also called an **arithmetic progression.** Returning to the sequence

$$2, 5, 8, 11, 14, 17, \ldots$$

the nth term can be defined recursively by

$$a_n = a_{n-1} + 3, \qquad a_1 = 2$$

This is an arithmetic progression with the first term equal to 2 and a common difference of 3.

EXAMPLE 1   Finding the Terms of an Arithmetic Sequence

Write the first four terms of an arithmetic sequence whose first term is -4 and whose common difference is -3.

SOLUTION

Beginning with -4, we add the common difference -3 to obtain

$$-4 + (-3) = -7 \qquad -7 + (-3) = -10 \qquad -10 + (-3) = -13$$

Alternatively, we note that the sequence is defined by

$$a_n = a_{n-1} - 3, \qquad a_1 = -4$$

which leads to the terms

$$a_1 = -4, \qquad a_2 = -7, \qquad a_3 = -10, \qquad a_4 = -13$$

∎

PROGRESS CHECK

Write the first four terms of an arithmetic sequence whose first term is 4 and whose common difference is $-\frac{1}{3}$.

ANSWER

$$4, \frac{11}{3}, \frac{10}{3}, \ldots$$

EXAMPLE 2   Finding the Common Difference of an Arithmetic Sequence

Show that the sequence

$$a_n = 2n - 1$$

is an arithmetic sequence, and find the common difference.

SOLUTION

We must show that the sequence satisfies

$$a_n - a_{n-1} = d$$

for some real number d. We have

$$a_n = 2n - 1$$
$$a_{n-1} = 2(n - 1) - 1 = 2n - 3$$

so

$$a_n - a_{n-1} = 2n - 1 - (2n - 3) = 2$$

This demonstrates that we are dealing with an arithmetic sequence whose common difference is 2.

■

For a given arithmetic sequence, it's easy to find a formula for the nth term a_n in terms of n and the first term a_1. Since

$$a_2 = a_1 + d$$

and

$$a_3 = a_2 + d$$

we see that

$$a_3 = (a_1 + d) + d = a_1 + 2d$$

Similarly, we can show that

$$a_4 = a_3 + d = (a_1 + 2d) + d = a_1 + 3d$$
$$a_5 = a_4 + d = (a_1 + 3d) + d = a_1 + 4d$$

In general,

The nth term a_n of an arithmetic sequence is given by

$$a_n = a_1 + (n - 1)d$$

EXAMPLE 3 Finding a Specific Term of an Arithmetic Sequence

Find the seventh term of the arithmetic progression whose first term is 2 and whose common difference is 4.

SOLUTION
We substitute $n = 7$, $a_1 = 2$, $d = 4$ in the formula

$$a_n = a_1 + (n - 1)d$$

obtaining

$$a_7 = 2 + (7 - 1)4 = 2 + 24 = 26$$

■

PROGRESS CHECK
Find the 16th term of the arithmetic progression whose first term is -5 and whose common difference is $\frac{1}{2}$.

ANSWER
$\frac{5}{2}$

EXAMPLE 4 Finding a Specific Term of an Arithmetic Sequence

Find the 25th term of the arithmetic sequence whose first and 20th terms are -7 and 31, respectively.

SOLUTION

We can apply the given information to find d.

$$a_n = a_1 + (n - 1)d$$
$$a_{20} = a_1 + (20 - 1)d$$
$$31 = -7 + 19d$$
$$d = 2$$

Now we use the formula for a_n to find a_{25}.

$$a_n = a_1 + (n - 1)d$$
$$a_{25} = -7 + (25 - 1)2$$
$$a_{25} = 41$$

∎

 PROGRESS CHECK

Find the 60th term of the arithmetic sequence whose first and 10th terms are 3 and $-\frac{3}{2}$, respectively.

ANSWER

$-\frac{53}{2}$

ARITHMETIC SERIES

The series associated with an arithmetic sequence is called an **arithmetic series.** Since an arithmetic sequence has a common difference d, we can write the nth partial sum S_n as

$$S_n = a_1 + (a_1 + d) + (a_1 + 2d) + \cdots + (a_n - 2d) + (a_n - d) + a_n \qquad (1)$$

where we write $a_2, a_3, \ldots$ in terms of a_1 and we write $a_{n-1}, a_{n-2}, \ldots$ in terms of a_n. Rewriting the right-hand side of Equation (1) in reverse order, we have

$$S_n = a_n + (a_n - d) + (a_n - 2d) + \cdots + (a_1 + 2d) + (a_1 + d) + a_1 \qquad (2)$$

Summing the corresponding sides of Equations (1) and (2),

$$2S_n = (a_1 + a_n) + (a_1 + a_n) + (a_1 + a_n) + \cdots \qquad \text{Repeated } n \text{ times}$$
$$= n(a_1 + a_n)$$

Thus,

$$S_n = \frac{n}{2}(a_1 + a_n)$$

Since $a_n = a_1 + (n - 1)d$, we see that

$$S_n = \frac{n}{2}[a_1 + a_1 + (n - 1)d] \qquad \text{Substituting for } a_n$$

$$= \frac{n}{2}[2a_1 + (n - 1)d]$$

We now have two useful formulas.

Arithmetic Series

For an arithmetic series,

$$S_n = \frac{n}{2}(a_1 + a_n)$$

$$S_n = \frac{n}{2}[2a_1 + (n - 1)d]$$

The choice of which formula to use depends on the available information. The following examples illustrate the use of the formulas.

EXAMPLE 5   Working with an Arithmetic Series
Find the sum of the first 30 terms of an arithmetic sequence whose first term is -20 and whose common difference is 3.

SOLUTION
We know that $n = 30$, $a_1 = -20$, and $d = 3$. Substituting in

$$S_n = \frac{n}{2}[2a_1 + (n - 1)d]$$

we obtain

$$S_{30} = \frac{30}{2}[2(-20) + (30 - 1)3]$$

$$= 15(-40 + 87)$$

$$= 705$$

■

    PROGRESS CHECK
Find the sum of the first 10 terms of the arithmetic sequence whose first term is 2 and whose common difference is $-\frac{1}{2}$.

ANSWER
$-\dfrac{5}{2}$

EXAMPLE 6 Working with an Arithmetic Series

The first term of an arithmetic series is 2, the last term is 58, and the sum is 450. Find the number of terms and the common difference.

SOLUTION

We have $a_1 = 2$, $a_n = 58$, and $S_n = 450$. Substituting in

$$S_n = \frac{n}{2}(a_1 + a_n)$$

we have

$$450 = \frac{n}{2}(2 + 58)$$
$$900 = 60n$$
$$n = 15$$

Now we substitute in

$$a_n = a_1 + (n - 1)d$$
$$58 = 2 + (14)d$$
$$56 = 14d$$
$$d = 4$$

■

PROGRESS CHECK

The first term of an arithmetic series is 6, the last term is 1, and the sum is 77/2. Find the number of terms and the common difference.

ANSWER

$n = 11, \quad d = -\dfrac{1}{2}$

EXERCISE SET 11.2

In Exercises 1–8 write the next two terms of each of the following arithmetic sequences.

1. $3, 6, 9, 12, \ldots$

2. $2, -2, -6, -10, \ldots$

3. $0, \dfrac{1}{4}, \dfrac{1}{2}, \dfrac{3}{4}, \ldots$

4. $y - 4, y, y + 4, y + 8, \ldots$

5. $0, \log 10, \log 100, \log 1000, \ldots$

6. $4, \dfrac{11}{2}, 7, \dfrac{17}{2}, \ldots$

7. $\sqrt{5} - 2, \sqrt{5}, \sqrt{5} + 2, \sqrt{5} + 4, \ldots$

8. $12, 8, 4, 0, \ldots$

In Exercises 9–14 write the first four terms of the arithmetic sequence whose first term is a_1 and whose common difference is d.

9. $a_1 = 2, d = 4$

10. $a_1 = -2, d = -5$

11. $a_1 = 3, d = -\dfrac{1}{2}$

12. $a_1 = \dfrac{1}{2}, d = 2$

13. $a_1 = \dfrac{1}{3}, d = -\dfrac{1}{3}$

14. $a_1 = 6, d = \dfrac{5}{2}$

In Exercises 15–18 find the specified terms of the arithmetic sequence whose first term is a_1 and whose common difference is d.

15. $a_1 = 4, d = 3$; 8th term

16. $a_1 = -3, d = \dfrac{1}{4}$; 14th term

17. $a_1 = 14, d = -2$; 12th term

18. $a_1 = 6, d = -\dfrac{1}{3}$; 9th term

In Exercises 19–24, given two terms of an arithmetic sequence, find the specified term.

19. $a_1 = -2, a_{20} = -2$; 24th term

20. $a_1 = \dfrac{1}{2}, a_{12} = 6$; 30th term

21. $a_1 = 0, a_{61} = 20$; 20th term

22. $a_1 = 23, a_{15} = -19$; 6th term

23. $a_1 = -\dfrac{1}{4}, a_{41} = 10$; 22nd term

24. $a_1 = -3, a_{18} = 65$; 30th term

In Exercises 25–30 find the sum of the specified number of terms of the arithmetic sequence whose first term is a_1 and whose common difference is d.

25. $a_1 = 3, d = 2$; 20 terms

26. $a_1 = -4, d = \dfrac{1}{2}$; 24 terms

27. $a_1 = \dfrac{1}{2}, d = -2$; 12 terms

28. $a_1 = -3, d = -\dfrac{1}{3}$; 18 terms

29. $a_1 = 82, d = -2$; 40 terms

30. $a_1 = 6, d = 4$; 16 terms

31. How many terms of the arithmetic progression 2, 4, 6, 8,… add up to 930?

32. How many terms of the arithmetic progression 44, 41, 38, 35,… add up to 340?

33. The first term of an arithmetic series is 3, the last term is 90, and the sum is 1395. Find the number of terms and the common difference.

34. The first term of an arithmetic series is -3, the last term is $\frac{5}{2}$, and the sum is -3. Find the number of terms and the common difference.

35. The first term of an arithmetic series is $\frac{1}{2}$, the last term is $\frac{7}{4}$, and the sum is $\frac{27}{4}$. Find the number of terms and the common difference.

36. The first term of an arithmetic series is 20, the last term is -14, and the sum is 54. Find the number of terms and the common difference.

37. Find the sum of the first 16 terms of an arithmetic progression whose 4th and 10th terms are $-\frac{5}{4}$ and $\frac{1}{4}$, respectively.

38. Find the sum of the first 12 terms of an arithmetic progression whose 3rd and 6th terms are 9 and 18, respectively.

39. Show that the sum of the first n natural numbers is $n(n + 1)/2$.

40. Show that

$$1 + 3 + 5 + \cdots + (2n - 1) = n^2.$$

11.3 GEOMETRIC SEQUENCES

The sequence

$$3, 6, 12, 24, 48, \ldots$$

in which each term after the first is obtained by multiplying the preceding one by 2, is an example of a geometric sequence.

FIBONACCI COUNTS THE RABBITS

Here is a problem that was first published in the year 1202.

A pair of newborn rabbits begins breeding at age one month and thereafter produces one pair of offspring per month. If we start with a newly born pair of rabbits, how many rabbits will there be at the beginning of each month?

The problem was posed by Leonardo Fibonacci of Pisa, and the resulting sequence is known as a **Fibonacci sequence.**

The accompanying figure helps in analyzing the problem. At the beginning of month zero, we have the pair of newborn rabbits P_1. At the beginning of month 1, we still have the pair P_1, since the rabbits do not breed until age 1 month. At the beginning of month 2, the pair of P_1 has the pair of offspring P_2. At the beginning of month 3, P_1 again has offspring, P_3, but P_2 does not breed during its first month. At the beginning of month 4, P_1 has offspring P_4, P_2 has offspring P_5, and P_3 does not breed during its first month.

Month	Pairs of Rabbits		
0	P_1		
1	P_1		
2	$P_1 \rightarrow P_2$		
3	$P_1 \rightarrow P_3$	P_2	
4	$P_1 \rightarrow P_4$	$P_2 \rightarrow P_5$	
	P_3		
5	$P_1 \rightarrow P_6$	$P_2 \rightarrow P_7$	
	$P_3 \rightarrow P_8$	P_4	P_5

If we let u_n denote the number of pairs of rabbits at the beginning of month n, we see that

$$a_0 = 1, \quad a_1 = 1, \quad a_2 = 2, \quad a_3 = 3, \quad a_4 = 5, \quad a_5 = 8, \ldots$$

The sequence has the interesting property that each term is the sum of the two preceding terms; that is,

$$a_n = a_{n-1} + a_{n-2}$$

Strange as it seems, nature appears to be aware of the Fibonacci sequence. For example, arrangements of seeds on sunflowers and leaves on some trees are related to Fibonacci numbers. Stranger still, some researchers believe that cycle analysis, such as analysis of stock market prices, is also related in some way to Fibonacci numbers.

Geometric Sequence

In a **geometric sequence** there is a real number r such that

$$a_n = ra_{n-1}$$

for all $n > 1$. The number r is called the **common ratio.**

A geometric sequence is also called a **geometric progression.** The common ratio r can be found by dividing any term a_k by the preceding term, a_{k-1}.

In a geometric sequence, the common ratio r is given by

$$r = \frac{a_k}{a_{k-1}}$$

Let's look at successive terms of a geometric sequence whose first term is a_1 and whose common ratio is r. We have

$$a_2 = ra_1$$
$$a_3 = ra_2 = r(ra_1) = r^2a_1$$
$$a_4 = ra_3 = r(r^2a_1) = r^3a_1$$

The pattern suggests that the exponent of r is one less than the subscript of a in the left-hand side.

The nth term of a geometric sequence is given by

$$a_n = a_1r^{n-1}$$

Once again, mathematical induction is required to prove that the formula holds for all natural numbers.

EXAMPLE 1   Finding a Specific Term of a Geometric Sequence

Find the seventh term of the geometric sequence $-4, -2, -1, \ldots$.

SOLUTION

Since

$$r = \frac{a_k}{a_{k-1}}$$

we see that

$$r = \frac{a_3}{a_2} = \frac{-1}{-2} = \frac{1}{2}$$

Substituting $a_1 = -4$, $r = \frac{1}{2}$, and $n = 7$, we have

$$a_n = a_1r^{n-1}$$
$$a_7 = (-4)\left(\frac{1}{2}\right)^{7-1} = (-4)\left(\frac{1}{2}\right)^6$$
$$= (-4)\left(\frac{1}{64}\right) = -\frac{1}{64}$$

∎

PROGRESS CHECK

Find the sixth term of the geometric sequence $2, -6, 18, \ldots$.

ANSWER

-486

GEOMETRIC MEAN

In a geometric sequence, the terms between the first and last terms are called **geometric means.** We will illustrate the method of calculating such means.

EXAMPLE 2   Inserting a Geometric Mean
Insert three geometric means between 3 and 48.

SOLUTION
The geometric sequence must look like this.

$$3, a_2, a_3, a_4, 48,\dots$$

Thus, $a_1 = 3$, $a_5 = 48$, and $n - 5$. Substituting in

$$a_n = a_1 r^{n-1}$$
$$48 = 3r^4$$
$$r^4 = 16$$
$$r = \pm 2$$

Thus there are two geometric sequences with three geometric means between 3 and 48.

$$3, 6, 12, 24, 48,\dots \quad r = 2$$
$$3, -6, 12, -24, 48,\dots \quad r = -2$$

■

PROGRESS CHECK
Insert two geometric means between 5 and $\frac{8}{25}$.

ANSWER
2, $\frac{5}{4}$

GEOMETRIC SERIES

If $a_1, a_2,\dots$ is a geometric sequence, then the nth partial sum

$$S_n = a_1 + a_2 + \cdots + a_n \tag{1}$$

is called a **geometric series.** Since each term of the series can be rewritten as $a_k = a_1 r^{k-1}$, we can rewrite Equation (1) as

$$S_n = a_1 + a_1 r + a_1 r^2 + \cdots + a_1 r^{n-2} + a_1 r^{n-1} \tag{2}$$

Multiplying each term in Equation (2) by r, we have

$$rS_n = a_1 r + a_1 r^2 + a_1 r^3 + \cdots + a_1 r^{n-1} + a_1 r^n \tag{3}$$

Subtracting Equation (2) from Equation (3) produces

$$rS_n - S_n = a_1 r^n - a_1$$
$$(r - 1)S_n = a_1(r^n - 1) \qquad \text{Factoring}$$
$$S_n = \frac{a_1(r^n - 1)}{r - 1} \qquad \begin{array}{l}\text{Dividing by } r - 1 \\ \text{(if } r \neq 1\text{)}\end{array}$$

Changing the signs in both the numerator and denominator gives us the following formula for the nth partial sum.

| **Geometric Series** | In a geometric series with first term a_1 and common ratio $r \neq 1$. $$S_n = \frac{a_1(1 - r^n)}{1 - r}$$ |

EXAMPLE 3 Working with a Geometric Series

Find the sum of the first six terms of the geometric sequence whose first three terms are 12, 6, 3.

SOLUTION

The common ratio can be found by dividing any term by the preceding term.

$$r = \frac{a_k}{a_{k-1}} = \frac{a_2}{a_1} = \frac{6}{12} = \frac{1}{2}$$

Substituting $a_1 = 12$, $r = \frac{1}{2}$, $n = 6$ in the formula for S_n, we have

$$S_n = \frac{a_1(1 - r^n)}{1 - r} = \frac{12\left[1 - \left(\dfrac{1}{2}\right)^6\right]}{1 - \dfrac{1}{2}} = \frac{189}{8}$$

∎

PROGRESS CHECK

Find the sum of the first five terms of the geometric sequence whose first three terms are $2, -\frac{4}{3}, \frac{8}{9}$.

ANSWER

$\dfrac{18}{011}$

EXAMPLE 4 An Application of Geometric Series

A father promises to give each child 2 cents on the first day and 4 cents on the second day and to continue doubling the amount each day for a total of 8 days. How much will each child receive on the last day? How much will each child have received in total after 8 days?

SOLUTION

The daily payout to each child forms a geometric sequence $2, 4, 8, \ldots$ with $a_1 = 2$ and $r = 2$. The term a_8 is given by substituting in

$$a_n = a_1 r^{n-1}$$
$$a_8 = a_1 r^{8-1} = 2 \cdot 2^7 = 256$$

Thus, each child will receive $2.56 on the last day. The total received by each child is given by

$$S_n = \frac{a_1(1 - r^n)}{1 - r}$$

$$S_8 = \frac{a_1(1 - r^8)}{1 - r} = \frac{2(1 - 2^8)}{1 - 2}$$

$$= \frac{2(1 - 256)}{-1} = 510$$

Each child will have received a total of $5.10 after 8 days.

∎

PROGRESS CHECK

A ball is dropped from a height of 64 feet. On each bounce, it rebounds half the height it fell (Figure 1). How high is the ball at the top of the fifth bounce? What is the total distance the ball has traveled at the top of the fifth bounce?

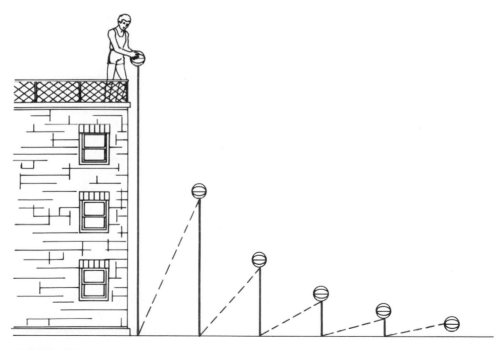

FIGURE 1 Diagram for Progress Check

ANSWER

2 feet; 186 feet

INFINITE GEOMETRIC SERIES

We now want to focus on a geometric series for which $|r| < 1$, say

$$\frac{1}{2} + \frac{1}{4} + \frac{1}{8} + \cdots + \frac{1}{2^n} + \cdots$$

To see how the sum increases as n increases, let's form a table of values of S_n.

n	1	2	3	4	5	6	7	8	9
S_n	0.500	0.750	0.875	0.938	0.969	0.984	0.992	0.996	0.998

We begin to suspect that S_n gets closer and closer to 1 as n increases. To see that this is really so, let's look at the formula

$$S_n = \frac{a_1(1 - r^n)}{1 - r}$$

when $|r| < 1$. When a number r that is less than 1 in absolute value is raised to higher and higher positive integer powers, the value of r^n gets smaller and smaller. Thus, the term r^n can be made as small as we like by choosing n sufficiently large. Since we are dealing with an infinite series, we say that "r^n approaches zero as n approaches infinity." We then replace r^n with 0 in the formula and denote the sum by S.

Sum of an Infinite Geometric Series

The sum S of the **infinite geometric series**

$$\sum_{k=0}^{\infty} a_1 r^k = a_1 + a_1 r + a_1 r^2 + \cdots + a_1 r^n + \cdots$$

is given by

$$S = \frac{a_1}{1 - r} \quad \text{when } |r| < 1$$

Applying this formula to the preceding series, we see that

$$S = \frac{\dfrac{1}{2}}{1 - \dfrac{1}{2}} = 1$$

which justifies the conjecture resulting from the examination of the above table. It is appropriate to remark that the ideas used in deriving the formula for an infinite geometric series have led us to the very border of the beginning concepts of calculus.

EXAMPLE 5 Finding the Sum of an Infinite Geometric Series

Find the sum of the infinite geometric series

$$\frac{3}{2} + 1 + \frac{2}{3} + \frac{4}{9} + \cdots$$

SOLUTION

The common ratio $r = \frac{2}{3}$. The sum of the infinite geometric series, with $|r| < 1$, is given by

$$S = \frac{a_1}{1 - r} = \frac{\dfrac{3}{2}}{1 - \dfrac{2}{3}} = \frac{9}{2}$$

■

PROGRESS CHECK

Find the sum of the infinite geometric series

$$4 - 1 + \frac{1}{4} - \frac{1}{16} + \cdots$$

ANSWER

$\dfrac{\text{ς}}{9\text{l}}$

The notation

$$0.6525\overline{252}$$

indicates a repeating decimal with a pattern in which 52 is repeated indefinitely. Every repeating decimal can be written as a rational number. We will apply the formula for the sum of an infinite geometric series to find the rational number equal to a repeating decimal.

EXAMPLE 6 Repeating Decimals as an Infinite Geometric Series

Find the rational number that is equal to $0.6525\overline{252}$.

SOLUTION

Note that

$$0.6525\overline{252} = 0.6 + 0.052 + 0.00052 + 0.0000052 + \cdots$$

We treat the sum

$$0.052 + 0.00052 + 0.0000052 + \cdots$$

as an infinite geometric series with $a = 0.052$ and $r = 0.01$. Then

$$S = \frac{a}{1 - r} = \frac{0.052}{1 - 0.01} = \frac{0.052}{0.99} = \frac{52}{990}$$

and the repeating decimal is equal to

$$0.6 + \frac{52}{990} = \frac{6}{10} + \frac{52}{990} = \frac{646}{990} = \frac{323}{495}$$

■

PROGRESS CHECK
Write the repeating decimal $2.54\overline{5454}$ as a rational number.

ANSWER

$\dfrac{66}{252}$ (upside down)

EXERCISE SET 11.3

In Exercises 1–6 find the next term of the given geometric sequence.

1. $3, 6, 12, 24,\ldots$

2. $-4, 12, -36, 108,\ldots$

3. $-4, 3, -\dfrac{9}{4}, \dfrac{27}{16},\ldots$

4. $2, -1, \dfrac{1}{2}, -\dfrac{1}{4},\ldots$

5. $1.2, 0.24, 0.048,\ldots$

6. $\dfrac{1}{8}, \dfrac{1}{2}, 2, 8,\ldots$

In Exercises 7–12 write the first four terms of the geometric sequence whose first term is a_1 and whose common ratio is r.

7. $a_1 = 3, r = 3$

8. $a_1 = -4, r = 2$

9. $a_1 = 4, r = \dfrac{1}{2}$

10. $a_1 = 16, r = -\dfrac{3}{2}$

11. $a_1 = -3, r = 2$

12. $a_1 = 3, r = -\dfrac{2}{3}$

In Exercises 13–24 use the information given about a geometric sequence to find the requested item.

13. $a_1 = 3, r = -2$; find a_8

14. $a_1 = 18, r = -\dfrac{1}{2}$; find a_6

15. $a_1 = 16, a_2 = 8$; find a_7

16. $a_1 = 15, a_2 = -10$; find a_6

17. $a_1 = 3, a_5 = \dfrac{1}{27}$; find a_7

18. $a_1 = 2, a_6 = \dfrac{1}{16}$; find a_3

19. $a_1 = \dfrac{16}{81}, a_6 = \dfrac{3}{2}$; find a_8

20. $a_4 = \dfrac{1}{4}, a_7 = 1$; find r

21. $a_2 = 4, a_8 = 256$; find r

22. $a_3 = 3, a_6 = -81$; find a_8

23. $a_1 = \dfrac{1}{2}, r = 2, a_n = 32$; find n

24. $a_1 = -2, r = 3, a_n = -162$; find n

25. Insert two geometric means between $\dfrac{1}{3}$ and 9.

26. Insert two geometric means between -3 and 192.

27. Insert two geometric means between 1 and $\dfrac{1}{64}$.

28. Insert three geometric means between $\dfrac{2}{3}$ and $\dfrac{32}{243}$.

In Exercises 29–32 find the requested partial sum for the geometric sequence whose first three terms are given.

29. $3, 1, \dfrac{1}{3}$; find S_7

30. $\dfrac{1}{3}, 1, 3$; find S_6

31. $-3, \dfrac{6}{5}, -\dfrac{12}{25}$; find S_5

32. $2, \dfrac{4}{3}, \dfrac{8}{9}$; find S_6

In Exercises 33–36 use the information given about a geometric sequence to find the requested partial sum.

33. $a_1 = 4, r = 2$; find S_8

34. $a_1 = -\dfrac{1}{2}, r = -3$; find S_{10}

35. $a_1 = 2, a_4 = -\dfrac{54}{8}$; find S_5

36. $a_1 = 64, a_7 = 1$; find S_6

37. A Christmas Club calls for savings of \$5 in January, and twice as much in each successive month as in the previous month. How much money will have been saved by the end of November?

 39. A city had 30,000 people in 1980. If the population increases 25% every 10 years, how many people will the city have in the year 2010?

 40. For good behavior a child is offered a reward consisting of 1 cent on the first day, 2 cents on the second day, 4 cents on the third day, and so on. If the child behaves properly for two weeks, what is the total amount that the child will receive?

38. A city had 20,000 people in 1980. If the population increases 5% per year, how many people will the city have in 1990?

In Exercises 41–49 evaluate the sum of each infinite geometric series.

41. $1 + \dfrac{1}{2} + \dfrac{1}{4} + \dfrac{1}{8} + \cdots$

42. $\dfrac{4}{5} + \dfrac{1}{5} + \dfrac{1}{20} + \dfrac{1}{80} + \cdots$

43. $1 - \dfrac{1}{3} + \dfrac{1}{9} - \dfrac{1}{27} + \cdots$

44. $\dfrac{1}{2} - \dfrac{1}{4} + \dfrac{1}{8} - \dfrac{1}{16} + \cdots$

45. $2 + \dfrac{1}{2} + \dfrac{1}{8} + \dfrac{1}{32} + \cdots$

46. $1 + 0.1 + 0.01 + 0.001 + \cdots$

47. $0.5 + (0.5)^2 + (0.5)^3 + (0.5)^4 + \cdots$

48. $\dfrac{2}{5} + \dfrac{4}{25} + \dfrac{8}{125} + \dfrac{16}{625} + \cdots$

49. $\dfrac{1}{3} - \dfrac{2}{9} + \dfrac{4}{27} - \dfrac{8}{81} + \cdots$

50. Find the rational number equal to $3.666\overline{6}$.

51. Find the rational number equal to $0.367\overline{67}$.

52. Find the rational number equal to $4.141\overline{41}$.

53. Find the rational number equal to $0.325\overline{325}$.

11.4 MATHEMATICAL INDUCTION

Mathematical induction is a method of proof that serves as one of the most powerful tools available to the mathematician. Viewed another way, mathematical induction is a property of the natural numbers that enables us to prove theorems that would otherwise appear unmanageable.

We begin by considering the sums of consecutive odd integers.

$$1 = 1$$
$$1 + 3 = 4$$
$$1 + 3 + 5 = 9$$
$$1 + 3 + 5 + 7 = 16$$
$$1 + 3 + 5 + 7 + 9 = 25$$

We instantly recognize that the sequence

$$1, 4, 9, 16, 25$$

consists of the squares of the integers 1, 2, 3, 4, and 5. Is this coincidental or do we have a general rule? Is the sum of the first n consecutive odd integers always equal to n^2? Curiosity leads us to try yet one more case.

$$1 + 3 + 5 + 7 + 9 + 11 = 36 = 6^2$$

Indeed, the sum of the first six odd integers is 6^2. This strengthens our *suspicion* that the result may hold in general, but we cannot possibly verify a theorem for *all* positive integers by testing one integer at a time. At this point we need to turn to the principle of mathematical induction.

Principle of Mathematical Induction	If a statement involving a natural number n **(I)** is true when $n = 1$ and **(II)** whenever it is true for $n = k$, is also true for $n = k + 1$, then the statement is true for all positive integer values of n.

What does it mean when we say that a statement is true if it satisfies the principle of mathematical induction? Part (I) says that we must verify the statement for $n = 1$. Then, by Part (II), the statement is also true for $n = 1 + 1 = 2$. But Part (II) then implies that the statement must also be true for $n = 2 + 1 = 3$, and so on. The effect is similar to an endless string of dominoes whereby each domino causes the next to fall. Thus, it is plausible that the principle has established the validity of the statement for *all* positive integer values of n.

We outline the steps involved in applying the principle of mathematical induction in the following example.

EXAMPLE 1   Applying the Principle of Mathematical Induction

Prove that the sum of the first n consecutive integers is given by $n(n + 1)/2$.

SOLUTION

Mathematical Induction	
Step 1. Verify that the statement is true for $n = 1$.	*Step 1.* The "sum" of the first integer is 1. Evaluating the formula for $n = 1$ yields $$\frac{1(1 + 1)}{2} = \frac{2}{2} = 1$$ which verifies the formula for $n = 1$.
Step 2. Assume the statement is true for $n = k$. Show it is true for $n = k + 1$.	*Step 2.* For $n = k$ we have $$1 + 2 + 3 + \cdots + k = \frac{k(k + 1)}{2}$$ Adding the next consecutive integer, $k + 1$, to both

sides, we obtain

$$1 + 2 + \cdots + k + (k + 1) = \frac{k(k + 1)}{2} + (k + 1)$$

$$= (k + 1)\left(\frac{k}{2} + 1\right)$$

$$= (k + 1)\left(\frac{k + 2}{2}\right)$$

$$= \frac{1}{2}(k + 1)(k + 2)$$

Thus, the formula holds for $n = k + 1$. By the principle of mathematical induction, it is then true for all positive integer values of n.

■

EXAMPLE 2 Applying the Principle of Mathematical Induction

Prove that the sum of the first n consecutive odd integers is given by n^2.

SOLUTION

To verify the formula for $n = 1$, we need only observe that $1 = 1^2$.

The following table shows the correspondence between the natural numbers and the odd integers. We see that when $n = k$, the value of the nth consecutive

n	1	2	3	4	$\cdots$	k
nth odd integer	1	3	5	7	$\cdots$	$2k - 1$

odd integer is $2k - 1$. Since the formula ia assumed to be true for $n = k$, we have

$$1 + 3 + 5 + \cdots + (2k - 1) = k^2$$

Adding the next consecutive odd integer, $2k + 1$, to both sides, we obtain

$$1 + 3 + \cdots + (2k - 1) + (2k + 1) = k^2 + (2k + 1)$$

or

$$1 + 3 + \cdots + (2k + 1) = (k + 1)^2$$

Thus, the sum of the first $k + 1$ consecutive odd integers is $(k + 1)^2$. By the principle of mathematical induction, the formula is true for all positive integer values of n.

■

The student should be aware that many of the theorems that were used in this book can be proved formally by using mathematical induction. Here is an example of a basic property of positive integer exponents that yields to this type of proof.

EXAMPLE 3   Applying the Principle of Mathematical Induction

Prove that $(xy)^n = x^n y^n$ for all positive integer values of n.

SOLUTION

For $n = 1$, we have

$$(xy)^1 = xy = x^1 y^1$$

which verifies the validity of the statement for $n = 1$. Assuming the statement holds for $n = k$, we have

$$(xy)^k = x^k y^k$$

To show that the statement holds for $n = k + 1$, we write

$$
\begin{aligned}
(xy)^{k+1} &= (xy)^k (xy) && \text{Definition of exponents} \\
&= (x^k y^k)(xy) && \text{Statement holds for } n = k \\
&= (x^k x)(y^k y) && \text{Associative and commutative laws} \\
&= x^{k+1} y^{k+1} && \text{Definition of exponents}
\end{aligned}
$$

Thus, the statement holds for $n = k + 1$, and by the principle of mathematical induction the statement holds for all integer values of n.

∎

EXERCISE SET 11.4

In Exercises 1–10 prove that the statement is true for all positive integer values of n by using the principle of mathematical induction.

1. $2 + 4 + 6 + \cdots + 2n = n(n + 1)$

2. $1^2 + 3^2 + 5^2 + \cdots + (2n - 1)^2 = \dfrac{n(2n + 1)(2n - 1)}{3}$

3. $2 + 5 + 8 + \cdots + (3n - 1) = \dfrac{n(3n + 1)}{2}$

4. $4 + 8 + 12 + \cdots + 4n = 2n(n + 1)$

5. $5 + 10 + 15 + \cdots + 5n = \dfrac{5n(n + 1)}{2}$

6. $1^2 + 2^2 + 3^2 + \cdots + n^2 = \dfrac{n(n + 1)(2n + 1)}{6}$

7. $1 \cdot 2 + 2 \cdot 3 + 3 \cdot 4 + \cdots + n(n + 1) = \dfrac{n(n + 1)(n + 2)}{3}$

8. $1^3 + 2^3 + 3^3 + \cdots + n^3 = \dfrac{n^2(n + 1)^2}{4}$

9. $1 + 5 + 9 + \cdots + (4n - 3) = n(2n - 1)$

10. $\left(\dfrac{x}{y}\right)^n = \dfrac{x^n}{y^n}$

11. Prove that the nth term a_n of an arithmetic progression whose first term is a_1 and whose common difference is d is given by $a_n = a_1 + (n - 1)d$.

12. Prove that the nth term a_n of a geometric progression whose first term is a_1 and whose common ratio is r is given by $a_n = a_1 r^{n-1}$.

13. Prove that $2 + 2^2 + 2^3 + \cdots + 2^n = 2^{n+1} - 2$.

14. Prove that $a + ar + ar^2 + \cdots + ar^{n-1} = \dfrac{a(1 - r^n)}{1 - r}$.

15. Prove that $x^n - 1$ is divisible by $x - 1$, $x \neq 1$. [*Hint:* Recall that divisibility requires the existence of a polynomial $Q(x)$ such that $x^n - 1 = (x - 1)Q(x)$.]

16. Prove that $x^n - y^n$ is divisible by $x - y$, $x \neq y$. [*Hint:* Note that $x^{n+1} - y^{n+1} = (x^{n+1} - xy^n) + (xy^n - y^{n+1})$.]

17. If α is a real number such that $\alpha > -1$, prove that

$$(1 + \alpha)^n \geq 1 + n\alpha \qquad \text{(Bernoulli's inequality)}$$

for every positive integer value of n.

11.5 THE BINOMIAL THEOREM

By sequential multiplication by $(a + b)$ you may verify that

$$(a + b)^1 = a + b$$
$$(a + b)^2 = a^2 + 2ab + b^2$$
$$(a + b)^3 = a^3 + 3a^2b + 3ab^2 + b^3$$
$$(a + b)^4 = a^4 + 4a^3b + 6a^2b^2 + 4ab^3 + b^4$$
$$(a + b)^5 = a^5 + 5a^4b + 10a^3b^2 + 10a^2b^3 + 5ab^4 + b^5$$

The expression on the right-hand side of the equation is called the **expansion** of the left-hand side. If we were to predict the form of the expansion of $(a + b)^n$, where n is a natural number, the preceding example would lead us to conclude that it has the following properties.

(a) The expansion has $n + 1$ terms.

(b) The first term is a^n and the last term is b^n.

(c) The sum of the exponents of a and b in each term is n.

(d) In each successive term after the first, the exponent of a decreases by 1, and the exponent of b increases by 1.

(e) The coefficients may be obtained from the following array, which is known as **Pascal's triangle.** Each number, with the exception of those at the ends of the rows, is the sum of the two nearest numbers in the row above. The numbers at the ends of the rows are always 1.

$$
\begin{array}{ccccccc}
 & & & 1 & & 1 & \\
 & & 1 & & 2 & & 1 \\
 & 1 & & 3 & & 3 & & 1 \\
1 & & 4 & & 6 & & 4 & & 1 \\
1 & & 5 & & 10 & & 10 & & 5 & & 1
\end{array}
$$

Pascal's triangle is not a convenient means for determining the coefficients of the expansion when n is large. Here is an alternative method.

(e′) The coefficient of any term (after the first) can be found by the following rule: In the preceding term, multiply the coefficient by the exponent of a and then divide by one more than the exponent of b.

EXAMPLE 1 Binomial Expansion
Write the expansion of $(a + b)^6$.

SOLUTION
From Property (b) we know that the first term is a^6. Thus,

$$(a + b)^6 = a^6 + \cdots$$

From Property (e') the next coefficient is

$$\frac{1 \cdot 6}{1} = 6$$

(since the exponent of b is 0). By Property (d) the exponents of a and b in this term are 5 and 1, respectively, so we have

$$(a + b)^6 = a^6 + 6a^5b + \cdots$$

Applying Property (e') again, we find that the next coefficient is

$$\frac{6 \cdot 5}{2} = 15$$

and by Property (d) the exponents of a and b in this term are 4 and 2, respectively. Thus,

$$(a + b)^6 = a^6 + 6a^5b + 15a^4b^2 + \cdots$$

Continuing in this manner, we see that

$$(a + b)^6 = a^6 + 6a^5b + 15a^4b^2 + 20a^3b^3 + 15a^2b^4 + 6ab^5 + b^6$$

∎

PROGRESS CHECK
Write the first five terms in the expansion of $(a + b)^{10}$.

ANSWER
$a^{10} + 10a^9b + 45a^8b^2 + 120a^7b^3 + 210a^6b^4$

The expansion of $(a + b)^n$ that we have described is called the **binomial theorem** or **binomial formula** and can be written

The Binomial Formula	$(a + b)^n = a^n + \dfrac{n}{1}a^{n-1}b + \dfrac{n(n - 1)}{1 \cdot 2}a^{n-2}b^2 + \dfrac{n(n - 1)(n - 2)}{1 \cdot 2 \cdot 3}a^{n-3}b^3$ $+ \cdots + \dfrac{n(n - 1)(n - 2) \cdots (n - r + 1)}{1 \cdot 2 \cdot 3 \cdots r}a^{n-r}b^r + \cdots + b^n$

The binomial formula can be proved by the method of mathematical induction discussed in the preceding section.

EXAMPLE 2 Applying the Binomial Formula
Find the expansion of $(2x - 1)^4$.

SOLUTION
Let $a = 2x$, $b = -1$, and apply the binomial formula.

$$(2x - 1)^4 = (2x)^4 + \frac{4}{1}(2x)^3(-1) + \frac{4 \cdot 3}{1 \cdot 2}(2x)^2(-1)^2$$

$$+ \frac{4 \cdot 3 \cdot 2}{1 \cdot 2 \cdot 3}(2x)(-1)^3 + (-1)^4$$

$$= 16x^4 - 32x^3 + 24x^2 - 8x + 1$$

■

☑ **PROGRESS CHECK**
Find the expansion of $(x^2 - 2)^4$.

ANSWER
$x^8 - 8x^6 + 24x^4 - 32x^2 + 16$

FACTORIAL NOTATION   Note that the denominator of the coefficient in the binomial formula is always the product of the first n natural numbers. We use the symbol $n!$, which is read as **n factorial,** to indicate this type of product. For example,

$$4! = 4 \cdot 3 \cdot 2 \cdot 1 = 24$$
$$6! = 6 \cdot 5 \cdot 4 \cdot 3 \cdot 2 \cdot 1 = 720$$

and, in general,

n Factorial	$n! = n(n - 1)(n - 2) \cdots 4 \cdot 3 \cdot 2 \cdot 1$

Since

$$(n - 1)! = (n - 1)(n - 2)(n - 3) \cdots 4 \cdot 3 \cdot 2 \cdot 1$$

we see that for $n > 1$

$$n! = n(n - 1)!$$

For convenience, we define $0!$ by

$$0! = 1$$

EXAMPLE 3   Working with Factorial Notation
Evaluate each of the following.

(a) $\dfrac{5!}{3!}$

Since $5! = 5 \cdot 4 \cdot 3!$ we may write

$$\frac{5!}{3!} = \frac{5 \cdot 4 \cdot 3!}{3!} = 5 \cdot 4 = 20$$

(b) $\dfrac{9!}{8!} = \dfrac{9 \cdot 8!}{8!} = 9$

(c) $\dfrac{10!4!}{12!} = \dfrac{10!4!}{12 \cdot 11 \cdot 10!} = \dfrac{4!}{12 \cdot 11} = \dfrac{4 \cdot 3 \cdot 2 \cdot 1}{12 \cdot 11} = \dfrac{2}{11}$

(d) $\dfrac{n!}{(n-2)!} = \dfrac{n(n-1)(n-2)!}{(n-2)!} = n(n-1) = n^2 - n$

(e) $\dfrac{(2-2)!}{3!} = \dfrac{0!}{3 \cdot 2} = \dfrac{1}{6}$

∎

PROGRESS CHECK

Evaluate each of the following.

(a) $\dfrac{12!}{10!}$

(b) $\dfrac{6!}{4!2!}$

(c) $\dfrac{10!8!}{9!7!}$

(d) $\dfrac{n!(n-1)!}{(n+1)!(n-2)!}$

(e) $\dfrac{8!}{6!(3-3)!}$

ANSWERS

(a) 132   (b) 15   (c) 80   (d) $\dfrac{n-1}{n+1}$ (e) 56

Here is what the binomial formula looks like in factorial notation.

$$(a+b)^n = a^n + \frac{n!}{1!(n-1)!} a^{n-1}b + \frac{n!}{2!(n-2)!} a^{n-2}b^2$$

$$+ \frac{n!}{3!(n-3)!} a^{n-3}b^3 + \cdots + \frac{n!}{r!(n-r)!} a^{n-r}b^r + \cdots + b^n$$

The symbol $\dbinom{n}{r}$, called the **binomial coefficient,** is defined in this way:

Binomial Coefficient	$\dbinom{n}{r} = \dfrac{n!}{r!(n-r)!}$

This symbol is useful in denoting the coefficients of the binomial expansion. Using this notation, the binomial formula can be written as

$$(a+b)^n = a^n + \binom{n}{1} a^{n-1}b + \binom{n}{2} a^{n-2}b^2 + \binom{n}{3} a^{n-3}b^3$$

$$+ \cdots + \binom{n}{r} a^{n-r}b^r + \cdots + b^n$$

Sometimes we merely want to find a certain term in the expansion of $(a+b)^n$. We shall use the following observation to answer this question. In the binomial formula for the expansion of $(a + b)^n$, b occurs in the second term, b^2 occurs in the third term, b^3 occurs in the fourth term, and, in general, b^k occurs in the $(k + 1)$th term. The exponents of a and b must add up to n in each term. Since the exponent of b in the $(k + 1)$th term is k, we conclude that the exponent of a must be $n - k$.

EXAMPLE 4 Finding a Specific Term of an Expansion
Find the fourth term in the expansion of $(x - 1)^5$.

SOLUTION
The exponent of b in the fourth term is 3, and the exponent of a is then $5 - 3 = 2$. From the binomial formula we see that the coefficient of the term a^2b^3 is

$$\binom{n}{3} = \binom{5}{3} = \frac{5!}{3!2!}$$

Since $a = x$ and $b = -1$, the fourth term is

$$\frac{5!}{3!2!} x^2(-1)^3 = -10x^2$$

■

PROGRESS CHECK
Find the third term in the expansion of

$$\left(\frac{x}{2} - 1\right)^8$$

ANSWER
$7x^6$

EXAMPLE 5 Finding a Specific Term of an Expansion
Find the term in the expansion of $(x^2 - y^2)^6$ that involves y^8.

SOLUTION
Since $y^8 = (-y^2)^4$, we seek that term which involves b^4 in the expansion of $(a + b)^6$. Thus, $b^4 = (-y^2)^4 = y^8$ occurs in the fifth term. In this term the exponent of a is $6 - 4 = 2$. By the binomial formula the corresponding coefficient is

$$\binom{6}{4} = \frac{6!}{4!2!} = 15$$

Since $a = x^2$ and $b = -y^2$, the desired term is

$$15(x^2)^2(-y^2)^4 = 15x^4y^8$$

■

PROGRESS CHECK

Find the term in the expansion of $(x^3 - \sqrt{2})^5$ that involves x^6.

ANSWER

$-20\sqrt{2}x^6$

EXERCISE SET 11.5

In Exercises 1–12 expand and simplify.

1. $(3x + 2y)^5$

2. $(2a - 3b)^6$

3. $(4x - y)^4$

4. $\left(3 + \dfrac{1}{2}x\right)^4$

5. $(2 - xy)^5$

6. $(3a^2 + b)^4$

7. $(a^2b + 3)^4$

8. $(x - y)^7$

9. $(a - 2b)^8$

10. $\left(\dfrac{x}{y} + y\right)^6$

11. $\left(\dfrac{1}{3}x + 2\right)^3$

12. $\left(\dfrac{x}{y} + \dfrac{y}{x}\right)^5$

In Exercises 13–20 find the first four terms in the given expansion and simplify.

13. $(2 + x)^{10}$

14. $(x - 3)^{12}$

15. $(3 - 2a)^9$

16. $(a^2 + b^2)^{11}$

17. $(2x - 3y)^{14}$

18. $\left(a - \dfrac{1}{a^2}\right)^8$

19. $(2x - yz)^{13}$

20. $\left(x - \dfrac{1}{y}\right)^{15}$

In Exercises 21–32 evaluate the expression.

21. $5!$

22. $7!$

23. $\dfrac{12!}{11!}$

24. $\dfrac{13!}{12!}$

25. $\dfrac{11!}{8!}$

26. $\dfrac{7!}{9!}$

27. $\dfrac{10!}{6!}$

28. $\dfrac{9!}{6!}$

29. $\dfrac{6!}{3!}$

30. $\dbinom{8}{5}$

31. $\dbinom{10}{6}$

32. $\dfrac{(n + 1)!}{(n - 1)!}$

In Exercises 33–46 find the specified term in the expression.

33. The fourth term in $(2x - 4)^7$.

34. The third term in $(4a + 3b)^{11}$.

35. The fifth term in $\left(\dfrac{1}{2}x - y\right)^{12}$.

36. The sixth term in $(3x - 2y)^{10}$.

37. The fifth term in $\left(\dfrac{1}{x} - 2\right)^9$.

38. The next to last term in $(a + 4b)^5$.

39. The middle term in $(x - 3y)^6$.

40. The middle term in $\left(2a + \dfrac{1}{2}b\right)^6$.

41. The term involving x^4 in $(3x + 4y)^7$.

42. The term involving x^6 in $(2x^2 - 1)^9$.

43. The term involving x^6 in $(2x^3 - 1)^9$.

44. The term involving x^8 in $\left(x^2 + \dfrac{1}{y}\right)^8$.

45. The term involving x^{12} in $\left(x^3 + \dfrac{1}{2}\right)^7$.

46. The term involving x^{-4} in $\left(y + \dfrac{1}{x^2}\right)^8$.

47. Evaluate $(1.3)^6$ to four decimal places by writing it as $(1 + 0.3)^6$ and using the binomial formula.

48. Using the method of Example 47, evaluate
    (a) $(3.4)^4$ **(b)** $(48)^5$ (*Hint:* $48 = 50 - 2$.)

TERMS AND SYMBOLS

KEY IDEAS FOR REVIEW

Topic	Page	Key Idea
Infinite sequence	495	An infinite sequence is a function whose domain is restricted to the set of natural numbers. We generally write a sequence by using subscript notation; that is, a_n replaces $a(n)$.
recursive definition	496	An infinite sequence is defined recursively if each term is defined by reference to preceding terms.
Sigma notation	496	Sigma (Σ) or summation notation is a handy means of indicating a sum of terms. The values written below and above the Σ indicate the starting and ending values, respectively, of the index of summation.
Arithmetic sequence	509	An arithmetic sequence has a common difference d between terms. We can define an arithmetic sequence recursively by writing $u_n = u_{n-1} + d$ and specifying a_1.
nth term a_n	504	$$a_n = a_1 + (n - 1)d$$
sum S_n of the first n terms	506	$$S_n = \frac{n}{2}(a_1 + a_n)$$ or $$S_n = \frac{n}{2}[2a_1 + (n - 1)d]$$
Geometric sequence	509	A geometric sequence has a common ratio r between terms. We can define a geometric sequence recursively by writing $a_n = ra_{n-1}$ and specifying a_1.
nth term a_n	510	$$a_n = a_1 r^{n-1}$$
sum S_n of the first n terms	512	$$S_n = \frac{a_1(1 - r^n)}{1 - r}$$
Infinite geometric series	514	If the common ratio r satisfies $-1 < r < 1$, then the infinite geometric series has the sum S given by $$S = \frac{a_1}{1 - r}$$ where a_1 is the first term of the series.

Topic	Page	Key Idea
Mathematical induction	518	Mathematical induction is useful in proving certain types of theorems involving natural numbers.
Factorial notation	523	The notation $n!$ indicates the product of the natural numbers 1 through n. $$n! = n(n-1)(n-2)\cdots 2\cdot 1 \qquad \text{for } n \geq 1$$ $$0! = 1$$
Binomial formula	522	The binomial formula provides the terms of the expansion of $(a+b)^n$. $$(a+b)^n = a^n + \frac{n!}{1!(n-1)!}a^{n-1}b + \frac{n!}{2!(n-2)!}a^{n-2}b^2$$ $$+ \frac{n!}{3!(n-3)!}a^{n-3}b^3 + \cdots + \frac{n!}{r!(n-r)!}a^{n-r}b^r + \cdots + b^n$$
binomial coefficient $\binom{n}{r}$	524	The notation $\binom{n}{r}$ is defined by $$\binom{n}{r} = \frac{n!}{r!(n-r)!}$$ and is useful in writing out the binomial formula.

REVIEW EXERCISES

Solutions to exercises whose numbers are in color are in the Solutions section in the back of the book.

11.1   In Exercises 1 and 2 write the first three terms and the tenth term of the sequence whose nth term is given.

1. $a_n = n^2 + n + 1$ 2. $a_n = \dfrac{n^3 - 1}{n + 1}$

In Exercises 3 and 4 find the fifth term of the recursive sequence.

3. $a_n = n - a_{n-1}, \quad a_1 = 0$

4. $a_n = na_{n-1}, \quad a_1 = 1$

In Exercises 5–7 find the indicated sum.

5. $\displaystyle\sum_{k=1}^{4} (1 - 2k)$ 6. $\displaystyle\sum_{k=3}^{5} k(k+1)$

7. $\displaystyle\sum_{i=1}^{5} 10$

In Exercises 8–10 express the sum in sigma notation.

8. $\dfrac{1}{3} + \dfrac{2}{4} + \dfrac{3}{5} + \dfrac{4}{6}$

9. $1 - x + x^2 - x^3 + x^4$

10. $\sin x + \sin 2x + \sin 3x + \cdots + \sin nx$

11.2   In Exercises 11 and 12 find the specified term of the arithmetic sequence whose first term is a_1 and whose common difference is d.

11. $a_1 = -2, d = 2$; 21st term

12. $a_1 = 6, d = -1$; 16th term

In Exercises 13 and 14, given two terms of an arithmetic sequence, find the specified term.

13. $a_1 = 4, a_{16} = 9$; 13th term

14. $a_1 = -4, a_{23} = -15$; 26th term

In Exercises 15 and 16 find the sum of the first 25 terms of the arithmetic sequence whose first term is a_1 and whose common difference is d.

15. $a_1 = -\dfrac{1}{3}, d = \dfrac{1}{3}$ 16. $a_1 = 6, d = -2$

11.3   In Exercises 17 and 18 determine the common ratio of the given geometric sequence.

17. $2, -6, 18, -54, \ldots$ 18. $-\dfrac{1}{2}, \dfrac{3}{4}, -\dfrac{9}{8}, \dfrac{27}{16}, \ldots$

In Exercises 19 and 20, write the first four terms of the geometric sequence whose first term is a_1 and whose common ratio is r.

19. $a_1 = 5, r = \dfrac{1}{5}$ **20.** $a_1 = -2, r = -1$

21. Find the sixth term of the geometric sequence $-4, 6, -9, \ldots$.

22. Find the eighth term of a geometric sequence for which $a_1 = -2$ and $a_5 = -32$.

23. Insert two geometric means between 3 and $\frac{1}{72}$.

24. Find the sum of the first six terms of the geometric progression whose first three terms are $\frac{1}{3}, \frac{1}{6}, \frac{1}{12}$.

25. Find the sum of the first six terms of the geometric progression for which $a_1 = -2$ and $r = 3$.

In Exercises 26 and 27 find the sum of the infinite geometric series.

26. $5 + \dfrac{5}{2} + \dfrac{5}{4} + \cdots$ **27.** $3 - 2 + \dfrac{4}{3} - \cdots$

11.4 **28.** Use the principle of mathematical induction to show that

$$3 + 6 + 9 + \cdots + 3n = \frac{3n(n+1)}{2}$$

is true for all positive integer values of n.

11.5 In Exercises 29–31 expand and simplify.

29. $(2x - y)^4$ **30.** $\left(\dfrac{x}{2} - 2\right)^4$

31. $(x^2 + 1)^2$

In Exercises 32–37 evaluate the expression.

32. $6!$ **33.** $\dfrac{13!}{11!2!}$

34. $\dfrac{(n-1)!(n+1)!}{n!n!}$ **35.** $\dbinom{6}{4}$

36. $\dbinom{3}{0}$ **37.** $\dbinom{10}{8}$

PROGRESS TEST 11A

1. Write the first four terms of a sequence whose nth term is $a_n = n/(n+1)^2$.

2. Evaluate $\displaystyle\sum_{j=2}^{4} \frac{j}{j-1}$.

3. Write the first four terms of the arithmetic sequence whose first term is -1 and whose common difference is $\frac{3}{2}$.

4. Find the 25th term of the arithmetic sequence whose first term is -4 and whose common difference is $\frac{1}{2}$.

5. Find the 15th term of an arithmetic sequence whose first and tenth terms are -1 and 26, respectively.

6. Find the sum of the first 10 terms of an arithmetic sequence whose first term is -4 and whose ninth term is 8.

7. Find the common ratio of the geometric sequence $12, 4, \dfrac{4}{3}, \dfrac{4}{9}, \ldots$.

8. Write the first four terms of the geometric sequence whose first term is $-\frac{2}{3}$ and whose common ratio is 2.

9. Find the tenth term of the sequence $2, -2, 2, \ldots$.

10. Insert two geometric means between -4 and 32.

11. Find the sum of the first seven terms of the geometric sequence whose first three terms are $-8, 4, -2$.

12. Find the sum of the infinite geometric series

$$-4 - \frac{4}{3} - \frac{4}{9} - \cdots$$

13. Use the principle of mathematical induction to show that $2 + 6 + 10 + \cdots + (4n - 2) = 2n^2$ is true for all positive integer values of n.

14. Find the first four terms in the expansion of $\left(a + \dfrac{1}{b}\right)^{10}$.

15. Evaluate $\dfrac{12!}{10!3!}$.

PROGRESS TEST 11B

1. Write the first four terms of a sequence whose nth term is

$$a_n = n^2 + \frac{2n}{n+2}$$

2. Write the sum $2! + 3! + 4! + \cdots + n!$ in summation notation.

3. Write the first four terms of the arithmetic sequence whose first term is 6 and whose common difference is $-\frac{2}{3}$.

4. Find the sixth term of the arithmetic sequence whose first term is -5 and whose common difference is 3.

5. Find the 30th term of an arithmetic sequence whose first and 20th terms are 3 and -35, respectively.

6. The first term of an arithmetic series is -5, the last term is -2, and the sum is $-\frac{91}{2}$. Find the number of terms and the common difference.

7. Find the common ratio of the geometric sequence 20, 4, 0.8, 0.16.

8. Write the first four terms of the geometric sequence whose first term is -1 and whose common ratio is $-\frac{1}{4}$.

9. Find the sixth term of a geometric sequence for which $a_1 = 3$ and $a_4 = -\frac{1}{9}$.

10. Insert two geometric means between -6 and $-\frac{16}{9}$.

11. Find the sum of the first five terms of a geometric sequence if $a_1 = -8$ and $a_4 = -1$.

12. Find the sum of the infinite geometric progression

$$5 - 2 + \frac{4}{5} - \cdots .$$

13. Use the principle of mathematical induction to show that

$$\frac{1}{1 \cdot 2} + \frac{1}{2 \cdot 3} + \frac{1}{3 \cdot 4} + \cdots + \frac{1}{n(n+1)} = \frac{n}{n+1}$$

is true for all positive integer values of n.

14. Find the third term in the expansion of $(2x - 1)^{10}$.

15. Evaluate $\dfrac{n \cdot n!}{(n+1)!}$.

UNIT 3 CUMULATIVE PROGRESS TEST
CHAPTERS 8–11

1. Write in the form $a + bi$.

   (a) $(3 - 2i)^3$ (b) $\dfrac{3 - 4i}{2 + i}$

2. Determine the absolute value of the complex number.
   (a) $3 + 2i$ (b) $6 - 3i$

3. Write the complex number in trigonometric form.
   (a) $-3i$ (b) $4 - 4i$

4. Find the quotient.

$$\frac{35(\cos 210° + i \sin 210°)}{7(\cos 180° + i \sin 180°)}$$

5. Find a polynomial that has the given roots and no others: -1 of multiplicity 2, $\frac{1}{2}$ of multiplicity 2.

6. Analyze the nature of the roots:

$$-2x^5 + x^3 - x - 3 = 0$$

7. The coordinates of the endpoint A of the line segment AD are $(2, -1)$. Let B be the midpoint of AD and C be the midpoint of AB. If the coordinates of C are $(3, 0)$, find the coordinates of D.

8. Find the equation of the circle with center at $(3, -1)$ and radius 5.

9. Describe the set of points determined by the equation.
   (a) $x^2 + y^2 + 4x - 8y + 7 = 0$
   (b) $2x^2 + 2y^2 - 12x - 20y + 75 = 0$

10. Find the equation of the parabola with vertex at $(4, 0)$ and directrix $y = \frac{1}{8}$.

11. Find the foci and vertices of the ellipse whose equation is

$$9x^2 - 36x + 4y^2 - 8y = -36$$

Sketch the graph.

12. Determine the foci, vertices, and asymptotes of the hyperbola

$$\frac{(y-2)^2}{2} - \frac{2x^2}{3} = 1$$

Sketch the graph.

13. Identify the conic section.
    (a) $x^2 - \sqrt{3}\,xy + 2x = 60$
    (b) $3x^2 - 3xy + 3y^2 - 5x + 2y = 100$
    (c) $r^2 \cos 2\theta = 1$

14. Show that the curve determined by the parametric equations

$$x = \sin t, \qquad y = 2\cos t, \qquad 0 \le t \le \pi$$

is an ellipse.

15. Solve by Gaussian elimination:

$$\begin{aligned}
4x + \ y - \ z &= \ \ 0 \\
-x + 3y + 2z &= \ \ 3 \\
3x - 2y - 2z &= -3
\end{aligned}$$

16. Given the system of equations

$$x - \ \ y = 4$$
$$3x + 2ky = 3$$

   (a) for what values of k is there no solution?

   (b) for what values of k are there infinitely many solutions?

17. Given the system of linear inequalities

$$x + y \leq 2$$
$$2x + y \geq 0$$
$$y \geq 0$$

   draw the graph of the solution set.

18. Given the matrices

$$A = \begin{bmatrix} 1 & -1 & 2 \\ 0 & 2 & 3 \\ -2 & 1 & -2 \end{bmatrix} \quad B = \begin{bmatrix} 2 & -1 \\ 0 & 2 \\ -1 & 4 \end{bmatrix} \quad C = \begin{bmatrix} -1 \\ -3 \\ 0 \end{bmatrix}$$

   (a) compute AB (b) compute $|A|$
   (c) solve $AX = C$

19. The sum of the digits of a two-digit number is 10. The square of the tens digit is one more than 16 times the ones digit. Find the number.

20. In a certain two-digit number, three times the tens digit is two more than five times the ones digit. If the digits are reversed, the new number is 18 less than the original number. Find the new number.

21. Evaluate:

   (a) $\sum\limits_{k=1}^{3} (2k - 1)^2$ (b) $\sum\limits_{k=0}^{3} k(k - 1)$

22. Find the first term a_1 of an arithmetic progression if $a_4 = 4$ and $a_7 = 3$.

23. If the first three terms of an arithmetic progression are $a_1 = 3 + 2k$, $a_2 = 2k - 1$, and $a_3 = -10k + 1$, find k.

24. If the first and third terms of a geometric progression are 5 and 2, respectively, find the second term.

25. Find the sum of the first four terms of a geometric progression if the first and second terms are 3 and 1, respectively.

26. If the sum of an infinite geometric progression is -6 and the first term is -5, find the common ratio r.

27. Find the third term in the binomial expansion of $\left(1 - \dfrac{1}{x}\right)^5$.

28. Use the method of induction to prove that $n < 2^n$ for all nonnegative integers.

APPLICATIONS: FROM WORDS TO ALGEBRA

Why the stress on learning algebra? Why are you required to learn the algebraic techniques presented in this book? One reason is that algebra provides the basic tools of mathematical manipulation that you will need in higher-level courses. In particular, a course in calculus cannot go smoothly for you if you are struggling with algebraic steps while the instructor is introducing new concepts.

But there is yet another reason for studying algebra. Many practical problems lead to equations or inequalities that must be solved. Once you have mastered the techniques of solving various algebraic forms, you have the potential to solve the applied problem.

Of course, you must first be able to translate a problem from words to algebra; that is, you must be able to go from words to an equation or inequality or a system of equations or inequalities. In reality, that's the tough part, since finding the solution is more or less a matter of rote application of the methods you have learned. But going from words to algebraic expressions is never a rote procedure. It requires careful interpretation of the words, a skill that comes only after practice, practice, and more practice.

We can't provide a set of rules that will enable you to translate *any* applied problem into algebraic expressions. We can, however, offer you a set of steps that will serve as a guide in "setting up" word problems.

Step 1. Read the problem carefully to understand what is required.
Step 2. Separate what is known from what is to be found.
Step 3. In many problems, an unknown quantity is the answer to a question such as "how much" or "how many." Let a symbol, say x, represent the unknown.
Step 4. If possible, represent other quantities in the problem in terms of x.
Step 5. Find the relationship in the problem that you can express as an equation (or an inequality).
Step 6. Solve and check.

The words and phrases in Table 1 should prove helpful in translating a word problem into an algebraic expression that can be solved.

Each problem in this appendix will lead to a linear or quadratic equation or inequality.

TABLE 1

Word or phrase	Algebraic symbol	Example	Algebraic expression
Sum	+	Sum of two numbers	$a + b$
Difference	−	Difference of two numbers	$a - b$
		Difference of a number and 3	$x - 3$
Product	× or ·	Product of two numbers	$a \cdot b$
Quotient	: or /	Quotient of two numbers	$\dfrac{a}{b}$ or a/b
Exceeds		a exceeds b by 3.	$a = b + 3$
More than		a is 3 more than b.	or
More of		There are 3 more of a than of b.	$a - 3 = b$
Twice		Twice a number	$2x$
		Twice the difference of x and 3	$2(x - 3)$
		3 more than twice a number	$2x + 3$
		3 less than twice a number	$2x - 3$
Is or equals	=	The sum of a number and 3 is 15.	$x + 3 = 15$

SIMPLE INTEREST

If a principal P is borrowed at a simple interest rate r, then the interest due at the end of each year is Pr, and the total interest I due at the end of t years is

$$I = Prt$$

Consequently, if S is the total amount owed at the end of t years, then

$$S = P + Prt$$

since both the principal and interest are to be repaid.

The basic formulas that we have derived for simple interest calculations are

$$I = Prt$$
$$S = P + Prt$$

EXAMPLE 1

A part of \$7000 was borrowed at a 6% simple annual interest and the remainder at 8%. If the total amount of interest due after 3 years is \$1380, how much was borrowed at each rate?

SOLUTION

Let

$$n = \text{the amount borrowed at } 6\%$$

Then

$$7000 - n = \text{the amount borrowed at } 8\%$$

since the total amount is \$7000. We can display the information in table form using the equation $I = Prt$.

	P	$\times$	r	$\times$	t	$=$	Interest
6% portion	n		0.06		3		$0.18n$
8% portion	$7000 - n$		0.08		3		$0.24(7000 - n)$

Note that we write the rate r in its decimal form, so $6\% = 0.06$ and $8\% = 0.08$.

Since the total interest of \$1380 is the sum of the interest from the two portions, we have

$$1380 = 0.18n + 0.24(7000 - n)$$

$$1380 = 0.18n + 1680 - 0.24n$$

$$0.06n = 300$$

$$n = 5000$$

We conclude that \$5000 was borrowed at 6% and \$2000 was borrowed at 8%.

∎

DISTANCE (UNIFORM MOTION) PROBLEMS

Here is the key to the solution of distance problems.

$$\text{Distance} = \text{rate} \times \text{time}$$

or

$$d = r \cdot t$$

The relationships that permit you to write an equation are sometimes obscured by the words. Here are some questions to ask as you set up a distance problem.

(a) Are there two distances that are equal? (Will two objects have traveled the same distance? Is the distance on a return trip the same as the distance going?)

(b) Is the sum (or difference) of two distances equal to a constant? (When two objects are traveling toward each other, they meet when the sum of the distances traveled by them equals the original distance between them.)

EXAMPLE 2

Two trains leave New York City for Chicago. The first train travels at an average speed of 60 miles per hour. The second train, which departs an hour later, travels at an average speed of 80 miles per hour. How long will it take the second train to overtake the first train?

SOLUTION

Since we are interested in the time the second train travels, we choose to let

$$t = \text{the number of hours the second train travels}$$

Then

$$t + 1 = \text{the number of hours the first train travels}$$

since the first train departs 1 hour earlier.

	Rate	$\times$	Time	$=$	Distance
First train	60		$t + 1$		$60(t + 1)$
Second train	80		t		$80t$

At the moment the second train overtakes the first, they must both have traveled the *same* distance. Thus

$$60(t + 1) = 80t$$
$$60t + 60 = 80t$$
$$60 = 20t$$
$$3 = t$$

It takes the second train 3 hours to catch up with the first train.

■

LITERAL EQUATIONS

It is sometimes convenient to be able to turn a formula around, that is, to be able to solve for a different variable. For example, if we want to express the radius of a circle in terms of the circumference, we have

$$C = 2\pi r$$

Dividing by 2π, we get

$$\frac{C}{2\pi} = \frac{2\pi r}{2\pi}$$
$$\frac{C}{2\pi} = r$$

Now, given a value of C, we can determine a value of r.

EXAMPLE 3

If an amount P is borrowed at the simple annual interest rate r, then the amount S due at the end of t years is

$$S = P + Prt$$

Solve for P.

SOLUTION

$$S = P + Prt$$
$$S = P(1 + rt) \quad \text{Common factor } P$$
$$\frac{S}{1 + rt} = P \qquad \text{Dividing both sides by } (1 + rt)$$

■

ECONOMIC ANALYSIS

A common application in business problems involves determining the best of two or more strategies. Deciding whether to purchase or to lease a piece of equipment is a good illustration.

EXAMPLE 4

A business can purchase a copier at a cost of $4000, or it can rent the same copier at a cost of $75 per month plus 2 cents per copy. In preparing a business plan, the comptroller estimates that the copier has a life of 5 years and would cost $20 per month to maintain. Above what volume of monthly use would it cost less to purchase the copier?

SOLUTION
Let x denote the number of copies made each month. Then the monthly cost of rental (in dollars) is

$$75 + 0.02x$$

and the total rental cost for 5 years (or 60 months) is

$$60(75 + 0.02x) \qquad \text{or} \qquad 4500 + 1.2x$$

The cost of purchase and maintenance is

$$4000 + 60(20) = 5200$$

so that purchase will cost less when

$$5200 < 4500 + 1.2x$$

Solving, we have

$$700 < 1.2x$$
$$583 < x$$

The firm should purchase the copier if it expects to make more than 583 copies per month.   ∎

WORK PROBLEMS

Work problems typically involve two or more people or machines working on the same task. The key to these problems is to express the *rate of work per unit of time*, whether it is an hour, a day, a week, or some other unit. For example, if a machine can do a job in 5 days, then

$$\text{rate of machine} = \frac{1}{5} \text{ job per day}$$

If this machine were used for 2 days, it would perform 2/5 of the job. In summary,

---

If a machine (or person) can complete a job in n days, then

$$\text{Rate of machine (or person)} = \frac{1}{n} \text{ job per day}$$

$$\text{Work done} = \text{Rate} \times \text{Time}$$

EXAMPLE 5
Working together, two cranes can unload a ship in 4 hours. The slower crane, working alone, requires 6 more hours than the faster crane to do the job. How long does it take each crane to do the job by itself?

SOLUTION
Let x = number of hours for the faster crane to do the job. Then $x + 6$ = number of hours for the slower crane to do the job. The rate of the faster crane is $1/x$, since this is the portion of the whole job completed in 1 hour; similarly, the rate of the slower crane is $1/(x + 6)$. We display this information in a table.

	Rate × Time = Work		
Faster crane	$\dfrac{1}{x}$	4	$\dfrac{4}{x}$
Slower crane	$\dfrac{1}{x+6}$	4	$\dfrac{4}{x+6}$

When the two cranes work together, we must have

$$\frac{\text{work done by}}{\text{fast crane}} + \frac{\text{work done by}}{\text{slow crane}} = 1 \text{ whole job}$$

or

$$\frac{4}{x} + \frac{4}{x+6} = 1$$

To solve, we multiply by $x(x + 6)$, obtaining

$$4(x + 6) + 4x = x^2 + 6x$$
$$0 = x^2 - 2x - 24$$
$$0 = (x + 4)(x - 6)$$
$$x = -4 \qquad \text{or} \qquad x = 6$$

The solution $x = -4$ is rejected, since it makes no sense to speak of negative hours of work. Summarizing, we see that

$x = 6$ is the number of hours in which the fast crane can do the job alone.

$x + 6 = 12$ is the number of hours in which the slow crane can do the job alone.

■

EXERCISE SET FOR APPENDIX A

1. A bicycle store is closing out its entire stock of a certain brand of 3-speed and 10-speed models. The profit on a 3-speed bicycle is 11% of the sale price, and the profit on a 10-speed model is 22% of the sale price. If the entire stock will be sold for $16,000 and the profit on the entire stock will be 19%, how much will be obtained from the sale of each type of bicycle?

2. A film shop carrying black-and-white film and color film has $4000 in inventory. The profit on black-and-white film is 12%. The profit on color film is 21%. If all the film is sold, and if the profit on color film is $150 less than the profit on black-and-white film, how much was invested in each type of film?

3. A firm borrowed $12,000 at a simple annual interest rate of 8% for a period of 3 years. At the end of the first year, the firm found that its needs were reduced. The firm returned a portion of the original loan and retained the remainder until the end of the 3-year period. If the total interest paid was $1760, how much was rented at the end of the first year?

4. A finance company lent a certain amount of money to Firm A at 7% annual interest. An amount $100 less than the amount lent to Firm A was lent to Firm B at 8%, and an amount $200 more than the amount lent to Firm A was lent to Firm C at 8.5% for one year. If the total annual income is $126.50, how much was lent to each firm?

5. Two trucks leave Philadelphia for Miami. The first truck to leave travels at an average speed of 50 kilometers per hour. The second truck, which leaves 2 hours later, travels at an average speed of 55 kilometers per hour. How long will it take the second truck to overtake the first truck?

6. Jackie either drives or bicycles from home to school. Her average speed when driving is 36 miles per hour, and her average speed when bicycling is 12 miles per hour. If it takes her 1/2 hour less to drive to school than to bicycle, how long does it take her to go to school, and how far is the school from her home?

7. Professors Roberts and Jones, who live 676 miles apart, are exchanging houses and jobs for the summer. They start out for their new locations at exactly the same time, and they meet after 6.5 hours of driving. If their average speeds differ by 4 miles per hour, what are their average speeds?

8. Steve leaves school by moped for spring vacation. Forty minutes later his roommate, Frank, notices that Steve forgot to take his camera, so Frank decides to try to catch up with Steve by car. If Steve's average speed is 25 miles per hour and Frank averages 45 miles per hour, how long does it take Frank to overtake Steve?

9. A certain number is three times another. If the difference of their reciprocals is 8, find the numbers.

10. If 1/3 is subtracted from 3 times the reciprocal of a certain number, the result is 25/6. Find the number.

11. A 12-meter-long steel beam is to be cut into two pieces so that one piece will be 4 meters longer than the other. How long will each piece be?

12. A rectangular field whose length is 10 meters longer than its width is to be enclosed with exactly 100 meters of fencing material. What are the dimensions of the field?

13. An express train and a local train start out from the same point at the same time and travel in opposite directions. The express train travels twice as fast as the local train. If after 4 hours they are 480 kilometers apart, what is the average speed of each train?

14. A boat travels 20 kilometers upstream in the time that it would take the same boat to travel 30 kilometers downstream. If the rate of the stream is 5 kilometers per hour, find the speed of the boat in still water.

15. An airplane flying against the wind travels 300 miles in the time that it would take the same plane to travel 400 miles with the wind. If the wind speed is 20 miles per hour, find the speed of the airplane in still air.

In Exercises 16–23 solve for the indicated variable in terms of the remaining variables.

16. $V = \frac{1}{3}\pi r^2 h$ for h

17. $F = \frac{9}{5}C + 32$ for C

18. $S = \frac{1}{2}gt^2 + vt$ for v

19. $A = \frac{1}{2}h(b + b')$ for b

20. $A = P(1 + rt)$ for r

21. $\frac{1}{f} = \frac{1}{f_1} + \frac{1}{f_2}$ for f_2

22. $a = \frac{v_1 - v_0}{t}$ for v_0

23. $S = \frac{a - rL}{L - r}$ for L

24. An auto rental agency offers two options: $35 per day with no charge for mileage or $28 per day and 12 cents per mile. If m is the number of miles traveled per day, for what values of m is the fixed $35 daily rate the better deal?

25. The Green Lawn Company offers a seasonal lawn feeding program at a cost of 3 cents per square foot. To attract owners of larger properties, Green Lawn also offers an alternative rate of $600 plus 2 cents per square foot. If s is the number of square feet of lawn, for what values of s is the 2-cent rate preferable?

26. A graphic designer and her assistant working together can complete an advertising layout in 6 days. The assistant working alone could complete the job in 16 more days than the designer working alone. How long would it take each person to do the job alone?

27. A roofer and his assistant working together can finish a roofing job in 4 hours. The roofer working alone could finish the job in 6 fewer hours than the assistant working alone. How long would it take each person to do the job alone?

28. Working together, computers A and B can complete a data-processing job in 2 hours. Computer A working alone can do the job in 3 fewer hours than computer B working alone. How long does it take each computer to do the job by itself?

29. A number of students rented a car for $160 for a one-week camping trip. If another student had joined the original group, each person's share of expenses would have been reduced by $8. How many students were there in the original group?

APPENDIX B

LIMITS: A PREVIEW OF CALCULUS

Calculus was developed in the seventeenth century by the British scientist Sir Issaac Newton and by the German mathematician Gottfried Wilhelm von Leibniz. One of the outstanding developments in the history of science and mathematics, calculus is the mathematics of change. Since everything around us is changing, calculus can be applied to the solution of a great many pressing problems in almost every field of human endeavor.

In this appendix we introduce the concept of a limit. This idea, which lies at the very heart of calculus, has important consequences.

In many applied problems, it is necessary to describe the behavior of a function f when x is near, but different from, a value a. For example, if x denotes the distance from an airplane to the ground, and $f(x)$ denotes the speed of the airplane at distance x, we are interested in knowing $f(a)$, the speed of the airplane at the time of landing. Moreover, we are interested in knowing how the speed $f(x)$ varies when x is near but different from a, since this information will enable the pilot to take the necessary steps for a smooth landing.

Let's begin by focusing on the function f defined by

$$f(x) = 2x + 1$$

How do the values $f(x)$ behave as x gets closer to (but remains different from) $x = 2$? As shown in Figure 1, we consider two different ways in which x can get closer and closer to, or approach, the value 2: from the right of 2 (that is, from values greater than 2) and from the left of 2 (values less than 2).

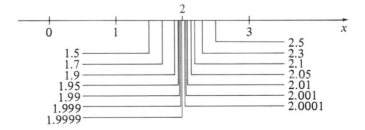

FIGURE 1

Table 1 shows the value of $f(x)$ as x approaches 2 from the right, and Table 2 shows these values as x approaches 2 from the left. We have plotted the data in Tables 1 and 2 on the graph shown in Figure 2.

TABLE 1   x Approaches 2 from the Right

x	3	2.5	2.3	2.1	2.05	2.01	2.001	2.0001
$f(x) = 2x + 1$	7	6	5.6	5.2	5.1	5.02	5.002	5.0002

TABLE 2   x Approaches 2 from the Left

x	1	1.5	1.7	1.9	1.95	1.99	1.999	1.9999
$f(x) = 2x + 1$	3	4	4.4	4.8	4.9	4.98	4.998	4.9998

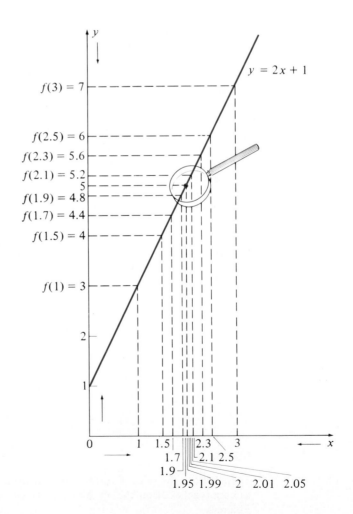

FIGURE 2

It is intuitively clear from Figure 2 that the values of $f(x) = 2x + 1$ get closer and closer to 5 as x approaches 2 from either side. The number 5 is called the **limit** of f as x approaches 2, and we write

$$\lim_{x \to 2} f(x) = 5 \qquad \text{or} \qquad \lim_{x \to 2} (2x + 1) = 5 \tag{1}$$

Here $x \to 2$ means "x approaches 2 (both from the right and from the left) but $x \neq 2$." Equation 1 is read, "the limit of $f(x)$ as x approaches, but remains different from, 2 is 5."

It is important to note that the value of the limit of f as $x \to 2$ in Equation (1) was obtained *intuitively*. It is possible to give a mathematically more precise definition of limit and then to carefully show that the above limit is indeed 5, but the intuitive approach is adequate for this introduction.

Let's examine the notion of a limit a little more closely.

> A function f is said to **approach the limit L** as x approaches a if the values of $f(x)$ get closer and closer to the unique real number L as x gets closer and closer to (but remains different from) a. We write this statement as
>
> $$\lim_{x \to a} f(x) = L$$

The definition of

$$\lim_{x \to a} f(x) = L$$

enables us to make the value of $f(x)$ as close to L as we want by taking x sufficiently close to (but different from) a.

 PROGRESS CHECK

(a) Find $\lim_{x \to 3} (3x - 2)$ by completing the following tables of values.

x	3.4	3.3	3.2	3.1	3.05	3.01	3.001
$3x - 2$							

x	2.6	2.7	2.8	2.9	2.95	2.99	2.999
$3x - 2$							

 (b) Find $\lim_{x \to -2} (x^2 + x)$ by forming appropriate tables.

ANSWERS

(a) 7   (b) 2

You may have notice that in Equation (1) we had $f(2) = 5$, so in *that* case, $\lim_{x \to 2} f(x)$ was simply $f(2)$. As we shall see in the following example, it is not always true that $\lim_{x \to a} f(x)$ is $f(a)$.

EXAMPLE 1

Given the function f defined by

$$f(x) = \frac{x^2 - 9}{x - 3}$$

find $\lim_{x \to 3} f(x)$.

SOLUTION

First, observe that this function is not defined for $x = 3$, because we cannot divide by zero. Thus, we cannot possibly find the limit as x approaches 3 merely by evaluating $f(3)$. The behavior of $f(x)$ as $x \to 3$ from the right and left is shown in Tables 3 and 4, respectively. Table 3 shows that the values of $f(x)$ approach 6 as x approaches 3 from the right. Table 4 shows that the values of $f(x)$ approach 6 as x approaches 3 from the left. Consequently,

$$\lim_{x \to 3} f(x) = 6 \qquad \text{or} \qquad \lim_{x \to 3} \frac{x^2 - 9}{x - 3} = 6$$

To see what is happening geometrically, we graph the function f. Observe that

$$f(x) = \frac{x^2 - 9}{x - 3} = \frac{(x + 3)(x - 3)}{(x - 3)} = x + 3, \qquad x \neq 3 \tag{2}$$

Thus, if $x \neq 3$, the graph of f is the graph of $y = x + 3$, a straight line. If $x = 3$, $f(x)$ is not defined. Hence, the graph of f, shown in Figure 3, is a straight line with a hole at $x = 3$. As usual, we have marked the point where $x = 3$ with an open circle to indicate that it is not on the line. As the graph shows, as x approaches 3, the values of $f(x)$ approach the number 6.

∎

TABLE 3 $x \to 3$ from the Right

x	4	3.5	3.1	3.01	3.001	3.0001
$f(x) = \dfrac{x^2 - 9}{x - 3}$	7	6.5	6.1	6.01	6.001	6.0001

TABLE 4 $x \to 3$ from the Left

x	2	2.5	2.9	2.99	2.999	2.9999
$f(x) = \dfrac{x^2 - 9}{x - 3}$	5	5.5	5.9	5.99	5.999	5.9999

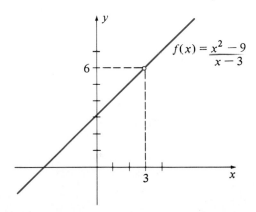

FIGURE 3

In the last two examples the value approached by $f(x)$ was the same finite number, whether x approached a from the right or from the left. This need not be the case, as the following examples show.

EXAMPLE 2

Let f be defined by

$$f(x) = \frac{|x|}{x}$$

Discuss $\lim_{x \to 0} f(x)$.

SOLUTION

First, observe that $f(0)$ is undefined. Moreover, we can write the given function as

$$f(x) = \begin{cases} \dfrac{x}{x} = 1 & \text{if } x > 0 \\[2mm] \dfrac{-x}{x} = -1 & \text{if } x < 0 \end{cases}$$

As $x \to 0$ from the right, then $f(x)$ is always 1, so $f(x)$ approaches 1. If $x \to 0$ from the left, then $f(x)$ is always -1, so $f(x)$ approaches -1. Hence, the values of $f(x)$ do not approach a single finite number as $x \to 0$. This behavior is evident in the graph of f, which is shown in Figure 4.

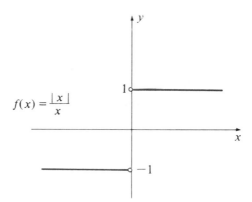

FIGURE 4

If the values of $f(x)$ do not approach a unique real number as x approaches the value a, we say that

$$\lim_{x \to a} f(x)$$

does not exist. Thus, for the function in Example 2,

$$\lim_{x \to 0} f(x) = \lim_{x \to 0} \frac{|x|}{x}$$

does not exist.

EXAMPLE 3

Consider the function f defined by

$$f(x) = \begin{cases} x + 2 & \text{if } x \leq 3 \\ \frac{1}{3}x + 5 & \text{if } x > 3 \end{cases} \tag{3}$$

Discuss $\lim_{x \to 3} f(x)$.

SOLUTION

The graph of this function is shown in Figure 5. Although $f(3)$ is defined, this limit does not exist. For, as $x \to 3$ from the right, then $f(x)$ approaches 6, and as $x \to 3$ from the left, then $f(x)$ approaches 5.

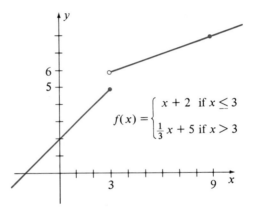

FIGURE 5

 PROGRESS CHECK

Consider the function f defined by

$$f(x) = \begin{cases} 2x + 1 & \text{if } x \leq 1 \\ x + 3 & \text{if } x > 1 \end{cases}$$

Verify that $\lim_{x \to 1} f(x)$ does not exist.

One might be inclined to conjecture that when the rule for a function is given by several equations, the limit of the function does not exist at the values in the domain where the rule changes. This was the case in Example 3, where the rule for the function in Equation (3) changed at $x = 3$ and the limit as $x \to 3$ did not exist. We shall see in Example 4, however, that a function defined by two or more equations may have a limit at a value in the domain where the rule changes.

EXAMPLE 4

Consider the function f defined by

$$f(x) = \begin{cases} -\frac{2}{3}x + 6 & \text{if } x \geq 3 \\ x + 1 & \text{if } x < 3 \end{cases}$$

Discuss $\lim_{x \to 3} f(x)$.

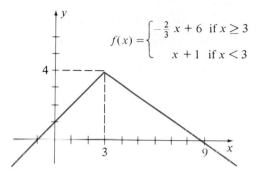

$$f(x) = \begin{cases} -\frac{2}{3}x + 6 & \text{if } x \geq 3 \\ x + 1 & \text{if } x < 3 \end{cases}$$

FIGURE 6

SOLUTION

The graph of this function is shown in Figure 6. By constructing appropriate tables, you can verify that

$$\lim_{x \to 3} f(x) = 4$$

It is possible also that, as $x \to a$, the values of $f(x)$ will not approach any real number. This phenomenon occurs in Example 5.

EXAMPLE 5

Determine whether $\lim_{x \to 0} f(x)$ exists for the function f defined by

$$f(x) = \frac{1}{x^2}$$

SOLUTION

The graph is shown in Figure 7.

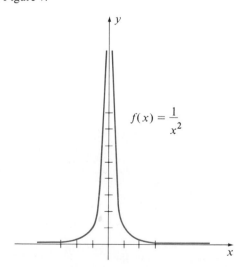

$$f(x) = \frac{1}{x^2}$$

FIGURE 7

To find $\lim_{x \to 0} f(x)$, we observe that as $x \to 0$ from the right, $f(x)$ gets larger and larger, and as $x \to 0$ from the left, $f(x)$ also gets larger and larger. This can also be seen from Tables 5 and 6. Since $f(x)$ approaches no real number as x approaches 0, we conclude that $\lim_{x \to 0} f(x)$ does not exist.

TABLE 5 $x \to 0$ from the Right

x	2	1	0.7	0.5	0.2	0.1	0.05	0.01	0.001
$f(x) = \dfrac{1}{x^2}$	0.25	1	2.04	4	25	100	400	1×10^4	1×10^6

TABLE 6 $x \to 0$ from the Left

x	-2	-1	-0.7	-0.5	-0.2	-0.1	-0.05	-0.01	-0.001
$f(x) = \dfrac{1}{x^2}$	0.25	1	2.04	4	25	100	400	1×10^4	1×10^6

EXERCISE SET FOR APPENDIX B

1. **(a)** Find $\lim_{x \to 4} (2x - 3)$ by completing the following tables of values.

x	4.4	4.2	4.1	4.01	4.001
$2x - 3$					

x	3.6	3.8	3.9	3.99	3.999
$2x - 3$					

   (b) Sketch the graph of $f(x) = 2x - 3$ and verify your conclusion.

2. **(a)** Find $\lim_{x \to 2} 1/(x - 2)$ by completing the following tables of values.

x	2.5	2.2	2.1	2.05	2.01	2.001	2.0001
$\dfrac{1}{x-2}$							

x	1.5	1.7	1.9	1.95	1.99	1.999	1.999
$\dfrac{1}{x-2}$							

   (b) Sketch the graph of $f(x) = 1/(x - 2)$ and verify your conclusion.

In Exercises 3–14 find the limit, if it exists.

3. $\lim_{x \to 3} 2x$

4. $\lim_{x \to -1} (2x + 3)$

5. $\lim_{x \to 2} (3x^2 + 2x - 5)$

6. $\lim_{x \to 0} \dfrac{2}{x}$

7. $\lim_{x \to -1} \dfrac{x^2 - 2x - 3}{x + 1}$

8. $\lim_{x \to 2} \dfrac{x^2 - 16}{x - 4}$

9. $\lim_{x \to -2} (16 - x^2)$

10. $\lim_{x \to 4} \dfrac{x + 1}{x^2 - 3x - 4}$

11. $\lim_{x \to 0} \dfrac{x^2 - 2x}{x}$

12. $\lim_{x \to 0} |x|$

13. $\lim_{x \to -6} \dfrac{x^2 - 36}{x + 6}$

14. $\lim_{h \to 0} \dfrac{(x + h)^2 - x^2}{h}$

15. Let f be defined by

$$f(x) = \begin{cases} 2x + 1 & \text{if } x \le 2 \\ -\frac{5}{2}x + 10 & \text{if } x > 2 \end{cases}$$

(a) Find $\lim_{x \to 2} f(x)$.

(b) Sketch the graph of $f(x)$.

16. Let f be defined by

$$f(x) = \begin{cases} x & \text{if } x \le 1 \\ 2x + 1 & \text{if } x > 1 \end{cases}$$

(a) Find $\lim_{x \to 1} f(x)$.

(b) Sketch the graph of $f(x)$.

In Exercises 17–23 find $\lim_{x \to a} f(x)$, if it exists, and sketch the graph of $f(x)$.

17. $f(x) = \dfrac{x^2 - 16}{x - 4}$, $a = 4$

18. $f(x) = \dfrac{x^2 - 25}{x + 5}$, $a = -5$

19. $f(x) = \dfrac{x^2 + x - 2}{x + 2}$, $a = -1$

20. $f(x) = \dfrac{x - 4}{x^2 - 3x - 4}$, $a = -1$

21. $f(x) = \dfrac{1}{2x - 1}$, $a = \dfrac{1}{2}$

22. $f(x) = |x - 1|$, $a = 2$

23. $f(x) = \dfrac{|x - 2|}{x - 2}$, $a = 2$

In Exercises 24–29 use the graph to find the limit, if it exists.

24. $\lim_{x \to 5} f(x)$

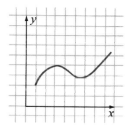

26. $\lim_{x \to 3} f(x)$

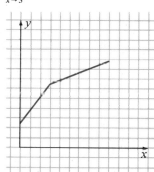

25. $\lim_{x \to 3} f(x)$

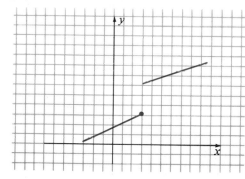

27. $\lim_{x \to 2} f(x)$

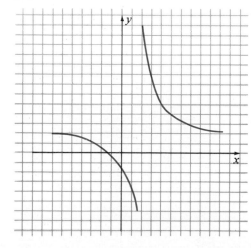

TABLES APPENDIX

TABLE I   Exponentials and Their Reciprocals

x	e^x	e^{-x}	x	e^x	e^{-x}
0.00	1.0000	1.0000	1.4	4.0552	0.2466
0.01	1.0101	0.9900	1.5	4.4817	0.2231
0.02	1.0202	0.9802	1.6	4.9530	0.2019
0.03	1.0305	0.9704	1.7	5.4739	0.1827
0.04	1.0408	0.9608	1.8	6.0496	0.1653
0.05	1.0513	0.9512	1.9	6.6859	0.1496
0.06	1.0618	0.9418	2.0	7.3891	0.1353
0.07	1.0725	0.9324	2.1	8.1662	0.1225
0.08	1.0833	0.9231	2.2	9.0250	0.1108
0.09	1.0942	0.9139	2.3	9.9742	0.1003
0.10	1.1052	0.9048	2.4	11.023	0.0907
0.11	1.1163	0.8958	2.5	12.182	0.0821
0.12	1.1275	0.8869	2.6	13.464	0.0743
0.13	1.1388	0.8781	2.7	14.880	0.0672
0.14	1.1503	0.8694	2.8	16.445	0.0608
0.15	1.1618	0.8607	2.9	18.174	0.0550
0.16	1.1735	0.8521	3.0	20.086	0.0498
0.17	1.1853	0.8437	3.1	22.198	0.0450
0.18	1.1972	0.8353	3.2	24.533	0.0408
0.19	1.2092	0.8270	3.3	27.113	0.0369
0.20	1.2214	0.8187	3.4	29.964	0.0334
0.21	1.2337	0.8106	3.5	33.115	0.0302
0.22	1.2461	0.8025	3.6	36.598	0.0273
0.23	1.2586	0.7945	3.7	40.447	0.0247
0.24	1.2712	0.7866	3.8	44.701	0.0224
0.25	1.2840	0.7788	3.9	49.402	0.0202
0.26	1.2969	0.7711	4.0	54.598	0.0183
0.27	1.3100	0.7634	4.1	60.340	0.0166
0.28	1.3231	0.7558	4.2	66.686	0.0150
0.29	1.3364	0.7483	4.3	73.700	0.0136
0.30	1.3499	0.7408	4.4	81.451	0.0123
0.35	1.4191	0.7047	4.5	90.017	0.0111
0.40	1.4918	0.6703	4.6	99.484	0.0101
0.45	1.5683	0.6376	4.7	109.95	0.0091
0.50	1.6487	0.6065	4.8	121.51	0.0082
0.55	1.7333	0.5769	4.9	134.29	0.0074
0.60	1.8221	0.5488	5	148.41	0.0067
0.65	1.9155	0.5220	6	403.43	0.0025
0.70	2.0138	0.4966	7	1,096.6	0.0009
0.75	2.1170	0.4724	8	2,981.0	0.0003
0.80	2.2255	0.4493	9	8,103.1	0.0001
0.85	2.3396	0.4274	10	22,026	0.00005
0.90	2.4596	0.4066	11	59,874	0.00002
0.95	2.5857	0.3867	12	162,754	0.000006
1.0	2.7183	0.3679	13	442,413	0.000002
1.1	3.0042	0.3329	14	1,202,604	0.0000008
1.2	3.3201	0.3012	15	3,269,017	0.0000003
1.3	3.6693	0.2725			

TABLE II Common Logarithms 549

N	0	1	2	3	4	5	6	7	8	9
1.0	.0000	.0043	.0086	.0128	.0170	.0212	.0253	.0294	.0334	.0374
1.1	.0414	.0453	.0492	.0531	.0569	.0607	.0645	.0682	.0719	.0755
1.2	.0792	.0828	.0864	.0899	.0934	.0969	.1004	.1038	.1072	.1106
1.3	.1139	.1173	.1206	.1239	.1271	.1303	.1335	.1367	.1399	.1430
1.4	.1461	.1492	.1523	.1553	.1584	.1614	.1644	.1673	.1703	.1732
1.5	.1761	.1790	.1818	.1847	.1875	.1903	.1931	.1959	.1987	.2014
1.6	.2041	.2068	.2095	.2122	.2148	.2175	.2201	.2227	.2253	.2279
1.7	.2304	.2330	.2355	.2380	.2405	.2430	.2455	.2480	.2504	.2529
1.8	.2553	.2577	.2601	.2625	.2648	.2672	.2695	.2718	.2742	.2765
1.9	.2788	.2810	.2833	.2856	.2878	.2900	.2923	.2945	.2967	.2989
2.0	.3010	.3032	.3054	.3075	.3096	.3118	.3139	.3160	.3181	.3201
2.1	.3222	.3243	.3263	.3284	.3304	.3324	.3345	.3365	.3385	.3404
2.2	.3424	.3444	.3464	.3483	.3502	.3522	.3541	.3560	.3579	.3598
2.3	.3617	.3636	.3655	.3674	.3692	.3711	.3729	.3747	.3766	.3784
2.4	.3802	.3820	.3838	.3856	.3874	.3892	.3909	.3927	.3945	.3692
2.5	.3979	.3997	.4014	.4031	.4048	.4065	.4082	.4099	.4116	.4133
2.6	.4150	.4166	.4183	.4200	.4216	.4232	.4249	.4265	.4281	.4298
2.7	.4314	.4330	.4346	.4362	.4378	.4393	.4409	.4425	.4440	.4456
2.8	.4472	.4487	.4502	.4518	.4533	.4548	.4564	.4579	.4594	.4609
2.9	.4624	.4639	.4654	.4669	.4683	.4698	.4713	.4728	.4742	.4757
3.0	.4771	.4786	.4800	.4814	.4829	.4843	.4857	.4871	.4886	.4900
3.1	.4914	.4928	.4942	.4955	.4969	.4983	.4997	.5011	.5024	.5038
3.2	.5051	.5065	.5079	.5092	.5105	.5119	.5132	.5145	.5159	.5172
3.3	.5185	.5198	.5211	.5224	.5237	.5250	.5263	.5276	.5289	.5302
3.4	.5315	.5328	.5340	.5353	.5366	.5378	.5391	.5403	.5416	.5428
3.5	.5441	.5453	.5465	.5478	.5490	.5502	.5514	.5527	.5539	.5551
3.6	.5563	.5575	.5587	.5599	.5611	.5623	.5635	.5647	.5658	.5670
3.7	.5682	.5694	.5705	.5717	.5729	.5740	.5752	.5763	.5775	.5786
3.8	.5798	.5809	.5821	.5832	.5843	.5855	.5866	.5877	.5888	.5899
3.9	.5911	.5922	.5933	.5944	.5955	.5966	.5977	.5988	.5999	.6010
4.0	.6021	.6031	.6042	.6053	.6064	.6075	.6085	.6096	.6107	.6117
4.1	.6128	.6138	.6149	.6160	.6170	.6180	.6191	.6201	.6212	.6222
4.2	.6232	.6243	.6253	.6263	.6274	.6284	.6294	.6304	.6314	.6325
4.3	.6335	.6345	.6355	.6365	.6375	.6385	.6395	.6405	.6415	.6425
4.4	.6435	.6444	.6454	.6464	.6474	.6484	.6493	.6503	.6513	.6522
4.5	.6532	.6542	.6551	.6561	.6571	.6580	.6590	.6599	.6609	.6618
4.6	.6628	.6637	.6646	.6656	.6665	.6675	.6684	.6693	.6702	.6712
4.7	.6721	.6730	.6739	.6749	.6758	.6767	.6776	.6785	.6794	.6803
4.8	.6812	.6821	.6830	.6839	.6848	.6857	.6866	.6875	.6884	.6893
4.9	.6902	.6911	.6920	.6928	.6937	.6946	.6955	.6964	.6972	.6981
5.0	.6990	.6998	.7007	.7016	.7024	.7033	.7042	.7050	.7059	.7067
5.1	.7076	.7084	.7093	.7101	.7110	.7118	.7126	.7135	.7143	.7152
5.2	.7160	.7168	.7177	.7185	.7193	.7202	.7210	.7218	.7226	.7235
5.3	.7243	.7251	.7259	.7267	.7275	.7284	.7292	.7300	.7308	.7316
5.4	.7324	.7332	.7340	.7348	.7356	.7364	.7372	.7380	.7388	.7396

TABLE II (*continued*)

N	0	1	2	3	4	5	6	7	8	9
5.5	.7404	.7412	.7419	.7427	.7435	.7443	.7451	.7459	.7466	.7474
5.6	.7482	.7490	.7497	.7505	.7513	.7520	.7528	.7536	.7543	.7551
5.7	.7559	.7566	.7574	.7582	.7589	.7597	.7604	.7612	.7619	.7627
5.8	.7634	.7642	.7649	.7657	.7664	.7672	.7679	.7686	.7694	.7701
5.9	.7709	.7716	.7723	.7731	.7738	.7745	.7752	.7760	.7767	.7774
6.0	.7782	.7789	.7796	.7803	.7810	.7818	.7825	.7832	.7839	.7846
6.1	.7853	.7860	.7868	.7875	.7882	.7889	.7896	.7903	.7910	.7917
6.2	.7924	.7931	.7938	.7945	.7952	.7959	.7966	.7973	.7980	.7987
6.3	.7993	.8000	.8007	.8014	.8021	.8028	.8035	.8041	.8048	.8055
6.4	.8062	.8069	.8075	.8082	.8089	.8096	.8102	.8109	.8116	.8122
6.5	.8129	.8136	.8142	.8149	.8156	.8162	.8169	.8176	.8182	.8189
6.6	.8195	.8202	.8209	.8215	.8222	.8228	.8235	.8241	.8248	.8254
6.7	.8261	.8267	.8274	.8280	.8287	.8293	.8299	.8306	.8312	.8319
6.8	.8325	.8331	.8338	.8344	.8351	.8357	.8363	.8370	.8376	.8382
6.9	.8388	.8395	.8401	.8407	.8414	.8420	.8426	.8432	.8439	.8445
7.0	.8451	.8457	.8463	.8470	.8476	.8482	.8488	.8494	.8500	.8506
7.1	.8513	.8519	.8525	.8531	.8537	.8543	.8549	.8555	.8561	.8567
7.2	.8573	.8579	.8585	.8591	.8597	.8603	.8609	.8615	.8621	.8627
7.3	.8633	.8639	.8645	.8651	.8657	.8663	.8669	.8675	.8681	.8686
7.4	.8692	.8698	.8704	.8710	.8716	.8722	.8727	.8733	.8739	.8745
7.5	.8751	.8756	.8762	.8768	.8774	.8779	.8785	.8791	.8797	.8802
7.6	.8808	.8814	.8820	.8825	.8831	.8837	.8842	.8848	.8854	.8859
7.7	.8865	.8871	.8876	.8882	.8887	.8893	.8899	.8904	.8910	.8915
7.8	.8921	.8927	.8932	.8938	.8943	.8949	.8954	.8960	.8965	.8971
7.9	.8976	.8982	.8987	.8993	.8998	.9004	.9009	.9015	.9020	.9025
8.0	.9031	.9036	.9042	.9047	.9053	.9058	.9063	.9069	.9074	.9079
8.1	.9085	.9090	.9096	.9101	.9106	.9112	.9117	.9122	.9128	.9133
8.2	.9138	.9143	.9149	.9154	.9159	.9165	.9170	.9175	.9180	.9186
8.3	.9191	.9196	.9201	.9206	.9212	.9217	.9222	.9227	.9232	.9238
8.4	.9243	.9248	.9253	.9258	.9263	.9269	.9274	.9279	.9284	.9289
8.5	.9294	.9299	.9304	.9309	.9315	.9320	.9325	.9330	.9335	.9340
8.6	.9345	.9350	.9355	.9360	.9365	.9370	.9375	.9380	.9385	.9390
8.7	.9395	.9400	.9405	.9410	.9415	.9420	.9425	.9430	.9435	.9440
8.8	.9445	.9450	.9455	.9460	.9465	.9469	.9474	.9479	.9484	.9489
8.9	.9494	.9499	.9504	.9509	.9513	.9518	.9523	.9528	.9533	.9538
9.0	.9542	.9547	.9552	.9557	.9562	.9566	.9571	.9576	.9581	.9586
9.1	.9590	.9595	.9600	.9605	.9609	.9614	.9619	.9624	.9628	.9633
9.2	.9638	.9643	.9647	.9652	.9657	.9661	.9666	.9671	.9675	.9680
9.3	.9685	.9689	.9694	.9699	.9703	.9708	.9713	.9717	.9722	.9727
9.4	.9731	.9736	.9741	.9745	.9750	.9754	.9759	.9763	.9768	.9773
9.5	.9777	.9782	.9786	.9791	.9795	.9800	.9805	.9809	.9814	.9818
9.6	.9823	.9827	.9832	.9836	.9841	.9845	.9850	.9854	.9859	.9863
9.7	.9868	.9872	.9877	.9881	.9886	.9890	.9894	.9899	.9903	.9908
9.8	.9912	.9917	.9921	.9926	.9930	.9934	.9939	.9943	.9948	.9952
9.9	.9956	.9961	.9965	.9969	.9974	.9978	.9983	.9987	.9991	.9996

TABLE III Natural Logarithms   551

N	$\ln N$	N	$\ln N$	N	$\ln N$
		4.5	1.5041	9.0	2.1972
0.1	-2.3026	4.6	1.5261	9.1	2.2083
0.2	-1.6094	4.7	1.5476	9.2	2.2192
0.3	-1.2040	4.8	1.5686	9.3	2.2300
0.4	-0.9163	4.9	1.5892	9.4	2.2407
0.5	-0.6931	5.0	1.6094	9.5	2.2513
0.6	-0.5108	5.1	1.6292	9.6	2.2618
0.7	-0.3567	5.2	1.6487	9.7	2.2721
0.8	-0.2231	5.3	1.6677	9.8	2.2824
0.9	-0.1054	5.4	1.6864	9.9	2.2925
1.0	0.0000	5.5	1.7047	10	2.3026
1.1	0.0953	5.6	1.7228	11	2.3979
1.2	0.1823	5.7	1.7405	12	2.4849
1.3	0.2624	5.8	1.7579	13	2.5649
1.4	0.3365	5.9	1.7750	14	2.6391
1.5	0.4055	6.0	1.7918	15	2.7081
1.6	0.4700	6.1	1.8083	16	2.7726
1.7	0.5306	6.2	1.8245	17	2.8332
1.8	0.5878	6.3	1.8405	18	2.8904
1.9	0.6419	6.4	1.8563	19	2.9444
2.0	0.6931	6.5	1.8718	20	2.9957
2.1	0.7419	6.6	1.8871	25	3.2189
2.2	0.7885	6.7	1.9021	30	3.4012
2.3	0.8329	6.8	1.9169	35	3.5553
2.4	0.8755	6.9	1.9315	40	3.6889
2.5	0.9163	7.0	1.9459	45	3.8067
2.6	0.9555	7.1	1.9601	50	3.9120
2.7	0.9933	7.2	1.9741	55	4.0073
2.8	1.0296	7.3	1.9879	60	4.0943
2.9	1.0647	7.4	2.0015	65	4.1744
3.0	1.0986	7.5	2.0149	70	4.2485
3.1	1.1314	7.6	2.0281	75	4.3175
3.2	1.1632	7.7	2.0412	80	4.3820
3.3	1.1939	7.8	2.0541	85	4.4427
3.4	1.2238	7.9	2.0669	90	4.4998
3.5	1.2528	8.0	2.0794	95	4.5539
3.6	1.2809	8.1	2.0919	100	4.6052
3.7	1.3083	8.2	2.1041		
3.8	1.3350	8.3	2.1163		
3.9	1.3610	8.4	2.1282		
4.0	1.3863	8.5	2.1401		
4.1	1.4110	8.6	2.1518		
4.2	1.4351	8.7	2.1633		
4.3	1.4586	8.8	2.1748		
4.4	1.4816	8.9	2.1861		

TABLE IV   Interest Rates

	$i = \frac{1}{2}\%$				$i = 1\%$				$i = 1\frac{1}{2}\%$		
n	$(1 + i)^n$	n	$(1 + i)^n$	n	$(1 + i)^n$	n	$(1 + i)^n$	n	$(1 + i)^n$	n	$(1 + i)^n$
1	1.0050 0000	51	1.2896 4194	1	1.0100 0000	51	1.6610 7814	1	1.0150 0000	51	2.1368 2106
2	1.0100 2500	52	1.2960 9015	2	1.0201 0000	52	1.6776 8892	2	1.0302 2500	52	2.1688 7337
3	1.0150 7513	53	1.3025 7060	3	1.0303 0100	53	1.6944 6581	3	1.0456 7838	53	2.2014 0647
4	1.0201 5050	54	1.3090 8346	4	1.0406 0401	54	1.7114 1047	4	1.0613 6355	54	2.2344 2757
5	1.0252 5125	55	1.3156 2887	5	1.0510 1005	55	1.7285 2457	5	1.0772 8400	55	2.2679 4398
6	1.0303 7751	56	1.3222 0702	6	1.0615 2015	56	1.7458 0982	6	1.0934 4326	56	2.3019 6314
7	1.0355 2940	57	1.3288 1805	7	1.0721 3535	57	1.7632 6792	7	1.1098 4491	57	2.3364 9259
8	1.0407 0704	58	1.3354 6214	8	1.0828 5671	58	1.7809 0060	8	1.1264 9259	58	2.3715 3998
9	1.0459 1058	59	1.3421 3946	9	1.0936 8527	59	1.7987 0960	9	1.1433 8998	59	2.4071 1308
10	1.0511 4013	60	1.3488 5015	10	1.1046 2213	60	1.8166 9670	10	1.1605 4083	60	2.4432 1978
11	1.0563 9583	61	1.3555 9440	11	1.1156 6835	61	1.8348 6367	11	1.1779 4894	61	2.4798 6807
12	1.0616 7781	62	1.3623 7238	12	1.1268 2503	62	1.8532 1230	12	1.1956 1817	62	2.5170 6609
13	1.0669 8620	63	1.3691 8424	13	1.1380 9328	63	1.8717 4443	13	1.2135 5244	63	2.5548 2208
14	1.0723 2113	64	1.3760 3016	14	1.1494 7421	64	1.8904 6187	14	1.2317 5573	64	2.5931 4442
15	1.0776 8274	65	1.3829 1031	15	1.1609 6896	65	1.9093 6649	15	1.2502 3207	65	2.6320 4158
16	1.0830 7115	66	1.3898 2486	16	1.1725 7864	66	1.9284 6015	16	1.2689 8555	66	2.6715 2221
17	1.0884 8651	67	1.3967 7399	17	1.1843 0443	67	1.9477 4475	17	1.2880 2033	67	2.7115 9504
18	1.0939 2894	68	1.4037 5785	18	1.1961 4748	68	1.9672 2220	18	1.3073 4064	68	2.7522 6896
19	1.0993 9858	69	1.4107 7664	19	1.2081 0895	69	1.9868 9442	19	1.3269 5075	69	2.7935 5300
20	1.1048 9558	70	1.4178 3053	20	1.2201 9004	70	2.0067 6337	20	1.3468 5501	70	2.8354 5629
21	1.1104 2006	71	1.4249 1968	21	1.2323 9194	71	2.0268 3100	21	1.3670 5783	71	2.8779 8814
22	1.1159 7216	72	1.4320 4428	22	1.2447 1586	72	2.0470 9931	22	1.3875 6370	72	2.9211 5796
23	1.1215 5202	73	1.4392 0450	23	1.2571 6302	73	2.0675 7031	23	1.4083 7715	73	2.9649 7533
24	1.1271 5978	74	1.4464 0052	24	1.2697 3465	74	2.0882 4601	24	1.4295 0281	74	3.0094 4996
25	1.1327 9558	75	1.4536 3252	25	1.2824 3200	75	2.1091 2847	25	1.4509 4535	75	3.0545 9171
26	1.1384 5955	76	1.4609 0069	26	1.2952 5631	76	2.1302 1975	26	1.4727 0953	76	3.1004 1059
27	1.1441 5185	77	1.4682 0519	27	1.3082 0888	77	2.1515 2195	27	1.4948 0018	77	3.1469 1674
28	1.1498 7261	78	1.4755 4622	28	1.3212 9097	78	2.1730 3717	28	1.5172 2218	78	3.1941 2050
29	1.1556 2197	79	1.4829 2395	29	1.3345 0388	79	2.1947 6754	29	1.5399 8051	79	3.2420 3230
30	1.1614 0008	80	1.4903 3857	30	1.3478 4892	80	2.2167 1522	30	1.5630 8022	80	3.2906 6279
31	1.1672 0708	81	1.4977 9026	31	1.3613 2740	81	2.2388 8237	31	1.5865 2642	81	3.3400 2273
32	1.1730 4312	82	1.5052 7921	32	1.3749 4068	82	2.2612 7119	32	1.6103 2432	82	3.3901 2307
33	1.1789 0833	83	1.5128 0561	33	1.3886 9009	83	2.2838 8390	33	1.6344 7918	83	3.4409 7492
34	1.1848 0288	84	1.5203 6964	34	1.4025 7699	84	2.3067 2274	34	1.6589 9637	84	3.4925 8954
35	1.1907 2689	85	1.5279 7148	35	1.4166 0276	85	2.3297 8997	35	1.6838 8132	85	3.5449 7838
36	1.1966 8052	86	1.5356 1134	36	1.4307 6878	86	2.3530 8787	36	1.7091 3954	86	3.5981 5306
37	1.2026 6393	87	1.5432 8940	37	1.4450 7647	87	2.3766 1875	37	1.7347 7663	87	3.6521 2535
38	1.2086 7725	88	1.5510 0585	38	1.4595 2724	88	2.4003 8494	38	1.7607 9828	88	3.7069 0723
39	1.2147 2063	89	1.5587 6087	39	1.4741 2251	89	2.4243 8879	39	1.7872 1025	89	3.7625 1084
40	1.2207 9424	90	1.5665 5468	40	1.4888 6373	90	2.4486 3267	40	1.8140 1841	90	3.8189 4851
41	1.2268 9821	91	1.5743 8745	41	1.5037 5237	91	2.4731 1900	41	1.8412 2868	91	3.8762 3273
42	1.2330 3270	92	1.5822 5939	42	1.5187 8989	92	2.4978 5019	42	1.8688 4712	92	3.9343 7622
43	1.2391 9786	93	1.5901 7069	43	1.5339 7779	93	2.5228 2869	43	1.8968 7982	93	3.9933 9187
44	1.2453 9385	94	1.5981 2154	44	1.5493 1757	94	2.5480 5698	44	1.9253 3302	94	4.0532 9275
45	1.2516 2082	95	1.6061 1215	45	1.5648 1075	95	2.5735 3755	45	1.9542 1301	95	4.1140 9214
46	1.2578 7892	96	1.6141 4271	46	1.5804 5885	96	2.5992 7293	46	1.9835 2621	96	4.1758 0352
47	1.2641 6832	97	1.6222 1342	47	1.5962 6344	97	2.6252 6565	47	2.0132 7910	97	4.2384 4057
48	1.2704 8916	98	1.6303 2449	48	1.6122 2608	98	2.6515 1831	48	2.0434 7829	98	4.3020 1718
49	1.2768 4161	99	1.6384 7611	49	1.6283 4834	99	2.6780 3349	49	2.0741 3046	99	4.3665 4744
50	1.2832 2581	100	1.6466 6849	50	1.6446 3182	100	2.7048 1383	50	2.1052 4242	100	4.4320 4565

TABLE IV (*continued*)   553

	$i = 2\%$				$i = 2\frac{1}{2}\%$				$i = 3\%$
n	$(1 + i)^n$	n	$(1 + i)^n$	n	$(1 + i)^n$	n	$(1 + i)^n$	n	$(1 + i)^n$
1	1.0200 0000	51	2.7454 1979	1	1.0250 0000	51	3.5230 3644	1	1.0300 0000
2	1.0404 0000	52	2.8003 2819	2	1.0506 2500	52	3.6111 1235	2	1.0609 0000
3	1.0612 0800	53	2.8563 3475	3	1.0768 9063	53	3.7013 9016	3	1.0927 2700
4	1.0824 3216	54	2.9134 6144	4	1.1038 1289	54	3.7939 2491	4	1.1255 0881
5	1.1040 8080	55	2.9717 3067	5	1.1314 0821	55	3.8887 7303	5	1.1592 7407
6	1.1261 6242	56	3.0311 6529	6	1.1596 9342	56	3.9859 9236	6	1.1940 5230
7	1.1486 8567	57	3.0917 8859	7	1.1886 8575	57	4.0856 4217	7	1.2298 7387
8	1.1716 5938	58	3.1536 2436	8	1.2184 0290	58	4.1877 8322	8	1.2667 7008
9	1.1950 9257	59	3.2166 9685	9	1.2488 6297	59	4.2924 7780	9	1.3047 7318
10	1.2189 9442	60	3.2810 3079	10	1.2800 8454	60	4.3997 8975	10	1.3439 1638
11	1.2433 7431	61	3.3466 5140	11	1.3120 8666	61	4.5097 8449	11	1.3842 3387
12	1.2682 4179	62	3.4135 8443	12	1.3448 8882	62	4.6225 2910	12	1.4257 6089
13	1.2936 0663	63	3.4818 5612	13	1.3785 1104	63	4.7380 9233	13	1.4685 3371
14	1.3194 7876	64	3.5514 9324	14	1.4129 7382	64	4.8565 4464	14	1.5125 8972
15	1.3458 6834	65	3.6225 2311	15	1.4482 9817	65	4.9779 5826	15	1.5579 6742
16	1.3727 8571	66	3.6949 7357	16	1.4845 0562	66	5.1024 0721	16	1.6047 0644
17	1.4002 4142	67	3.7688 7304	17	1.5216 1826	67	5.2299 6739	17	1.6528 4763
18	1.4282 4625	68	3.8442 5050	18	1.5596 5872	68	5.3607 1658	18	1.7024 3306
19	1.4568 1117	69	3.9211 3551	19	1.5986 5019	69	5.4947 3449	19	1.7535 0605
20	1.4859 4740	70	3.9995 5822	20	1.6386 1644	70	5.6321 0286	20	1.8061 1123
21	1.5156 6634	71	4.0795 4939	21	1.6795 8185	71	5.7729 0543	21	1.8602 9457
22	1.5459 7967	72	4.1611 4038	22	1.7215 7140	72	5.9172 2806	22	1.9161 0341
23	1.5768 9926	73	4.2443 6318	23	1.7646 1068	73	6.0651 5876	23	1.9735 8651
24	1.6084 3725	74	4.3292 5045	24	1.8087 2595	74	6.2167 8773	24	2.0327 9411
25	1.6406 0599	75	4.4158 3546	25	1.8539 4410	75	6.3722 0743	25	2.0937 7793
26	1.6734 1811	76	4.5041 5216	26	1.9002 9270	76	6.5315 1261	26	2.1565 9127
27	1.7068 8648	77	4.5942 3521	27	1.9478 0002	77	6.6948 0043	27	2.2212 8901
28	1.7410 2421	78	4.6861 1991	28	1.9964 9502	78	6.8621 7044	28	2.2879 2768
29	1.7758 4469	79	4.7798 4231	29	2.0464 0739	79	7.0337 2470	29	2.3565 6551
30	1.8113 6158	80	4.8754 3916	30	2.0975 6758	80	7.2095 6782	30	2.4272 6247
31	1.8475 8882	81	4.9729 4794	31	2.1500 0677	81	7.3898 0701	31	2.5000 8035
32	1.8845 4059	82	5.0724 0690	32	2.2037 5694	82	7.5745 5219	32	2.5750 8276
33	1.9222 3140	83	5.1738 5504	33	2.2588 5086	83	7.7639 1599	33	2.6523 3524
34	1.9606 7603	84	5.2773 3214	34	2.3153 2213	84	7.9580 1389	34	2.7319 0530
35	1.9998 8955	85	5.3828 7878	35	2.3732 0519	85	8.1569 6424	35	2.8138 6245
36	2.0398 8734	86	5.4905 3636	36	2.4325 3532	86	8.3608 8834	36	2.8982 7833
37	2.0806 8509	87	5.6003 4708	37	2.4933 4870	87	8.5699 1055	37	2.9852 2668
38	2.1222 9879	88	5.7123 5402	38	2.5556 8242	88	8.7841 5832	38	3.0747 8348
39	2.1647 4477	89	5.8266 0110	39	2.6195 7448	89	9.0037 6228	39	3.1670 2698
40	2.2080 3966	90	5.9431 3313	40	2.6850 6384	90	9.2288 5633	40	3.2620 3779
41	2.2522 0046	91	6.0619 9579	41	2.7521 9043	91	9.4595 7774	41	3.3598 9893
42	2.2972 4447	92	6.1832 3570	42	2.8209 9520	92	9.6960 6718	42	3.4606 9589
43	2.3431 8936	93	6.3069 0042	43	2.8915 2008	93	9.9384 6886	43	3.5645 1677
44	2.3900 5314	94	6.4330 3843	44	2.9638 0808	94	10.1869 3058	44	3.6714 5227
45	2.4378 5421	95	6.5616 9920	45	3.0379 0328	95	10.4416 0385	45	3.7815 9584
46	2.4866 1129	96	6.6929 3318	46	3.1138 5086	96	10.7026 4395	46	3.8950 4372
47	2.5363 4352	97	6.8267 9184	47	3.1916 9713	97	10.9702 1004	47	4.0118 9503
48	2.5870 7039	98	6.9633 2768	48	3.2714 8956	98	11.2444 6530	48	4.1322 5188
49	2.6388 1179	99	7.1025 9423	49	3.3532 7680	99	11.5255 7693	49	4.2562 1944
50	2.6915 8803	100	7.2446 4612	50	3.4371 0872	100	11.8137 1635	50	4.3839 0602

	$i = 4\%$		$i = 5\%$		$i = 6\%$		$i = 7\%$		$i = 8\%$
n	$(1+i)^n$	n	$(1+i)^n$	n	$(1+i)^n$	n	$(1+i)^n$	n	$(1+i)^n$
1	1.0400 0000	1	1.0500 0000	1	1.0600 0000	1	1.0700 0000	1	1.0800 0000
2	1.0816 0000	2	1.1025 0000	2	1.1236 0000	2	1.1449 0000	2	1.1664 0000
3	1.1248 6400	3	1.1576 2500	3	1.1910 1600	3	1.2250 4300	3	1.2597 1200
4	1.1698 5856	4	1.2155 0625	4	1.2624 7696	4	1.3107 9601	4	1.3604 8896
5	1.2166 5290	5	1.2762 8156	5	1.3382 2558	5	1.4025 5173	5	1.4693 2808
6	1.2653 1902	6	1.3400 9564	6	1.4185 1911	6	1.5007 3035	6	1.5868 7432
7	1.3159 3178	7	1.4071 0042	7	1.5036 3026	7	1.6057 8148	7	1.7138 2427
8	1.3685 6905	8	1.4774 5544	8	1.5938 4807	8	1.7181 8618	8	1.8509 3021
9	1.4233 1181	9	1.5513 2822	9	1.6894 7896	9	1.8384 5921	9	1.9990 0463
10	1.4802 4428	10	1.6288 9463	10	1.7908 4770	10	1.9671 5136	10	2.1589 2500
11	1.5394 5406	11	1.7103 3936	11	1.8982 9856	11	2.1048 5195	11	2.3316 3900
12	1.6010 3222	12	1.7958 5633	12	2.0121 9647	12	2.2521 9159	12	2.5181 7012
13	1.6650 7351	13	1.8856 4914	13	2.1329 2826	13	2.4098 4500	13	2.7196 2373
14	1.7316 7645	14	1.9799 3160	14	2.2609 0396	14	2.5785 3415	14	2.9371 9362
15	1.8009 4351	15	2.0789 2818	15	2.3965 5819	15	2.7590 3154	15	3.1721 6911
16	1.8729 8125	16	2.1828 7459	16	2.5403 5168	16	2.9521 6375	16	3.4259 4264
17	1.9479 0050	17	2.2920 1832	17	2.6927 7279	17	3.1588 1521	17	3.7000 1805
18	2.0258 1652	18	2.4066 1923	18	2.8543 3915	18	3.3799 3228	18	3.9960 1950
19	2.1068 4918	19	2.5269 5020	19	3.0255 9950	19	3.6165 2754	19	4.3157 0106
20	2.1911 2314	20	2.6532 9771	20	3.2071 3547	20	3.8696 8446	20	4.6609 5714
21	2.2787 6807	21	2.7859 6259	21	3.3995 6360	21	4.1405 6237	21	5.0338 3372
22	2.3699 1879	22	2.9252 6072	22	3.6035 3742	22	4.4304 0174	22	5.4365 4041
23	2.4647 1554	23	3.0715 2376	23	3.8197 4966	23	4.7405 2986	23	5.8714 6365
24	2.5633 0416	24	3.2250 9994	24	4.0489 3464	24	5.0723 6695	24	6.3411 8074
25	2.6658 3633	25	3.3863 5494	25	4.2918 7072	25	5.4274 3264	25	6.8484 7520
26	2.7724 6978	26	3.5556 7269	26	4.5493 8296	26	5.8073 5292	26	7.3963 5321
27	2.8833 6858	27	3.7334 5632	27	4.8223 4594	27	6.2138 6763	27	7.9880 6147
28	2.9987 0332	28	3.9201 2914	28	5.1116 8670	28	6.6488 3836	28	8.6271 0639
29	3.1186 5145	29	4.1161 3560	29	5.4183 8790	29	7.1142 5705	29	9.3172 7490
30	3.2433 9751	30	4.3219 4238	30	5.7434 9117	30	7.6122 5504	30	10.0626 5689
31	3.3731 3341	31	4.5380 3949	31	6.0881 0064	31	8.1451 1290	31	10.8676 6944
32	3.5080 5875	32	4.7649 4147	32	6.4533 8668	32	8.7152 7080	32	11.7370 8300
33	3.6483 8110	33	5.0031 8854	33	6.8405 8988	33	9.3253 3975	33	12.6760 4964
34	3.7943 1634	34	5.2533 4797	34	7.2510 2528	34	9.9781 1354	34	13.6901 3361
35	3.9460 8899	35	5.5160 1537	35	7.6860 8679	35	10.6765 8148	35	14.7853 4429
36	4.1039 3255	36	5.7918 1614	36	8.1472 5200	36	11.4239 4219	36	15.9681 7184
37	4.2680 8986	37	6.0814 0694	37	8.6360 8712	37	12.2236 1814	37	17.2456 2558
38	4.4388 1345	38	6.3854 7729	38	9.1542 5235	38	13.0792 7141	38	18.6252 7563
39	4.6163 6599	39	6.7047 5115	39	9.7035 0749	39	13.9948 2041	39	20.1152 9768
40	4.8010 2063	40	7.0399 8871	40	10.2857 1794	40	14.9744 5784	40	21.7245 2150
41	4.9930 6145	41	7.3919 8815	41	10.9028 6101	41	16.0226 6989	41	23.4624 8322
42	5.1927 8391	42	7.7615 8756	42	11.5570 3267	42	17.1442 5678	42	25.3394 8187
43	5.4004 9527	43	8.1496 6693	43	12.2504 5463	43	18.3443 5475	43	27.3666 4042
44	5.6165 1508	44	8.5571 5028	44	12.9854 8191	44	19.6284 5959	44	29.5559 7166
45	5.8411 7568	45	8.9850 0779	45	13.7646 1083	45	21.0024 5176	45	31.9204 4939
46	6.0748 2271	46	9.4342 5818	46	14.5904 8748	46	22.4726 2338	46	34.4740 8534
47	6.3178 1562	47	9.9059 7109	47	15.4659 1673	47	24.0457 0702	47	37.2320 1217
48	6.5705 2824	48	10.4012 6965	48	16.3938 7173	48	25.7289 0651	48	40.2105 7314
49	6.8333 4937	49	10.9213 3313	49	17.3775 0403	49	27.5299 2997	49	43.4274 1899
50	7.1066 8335	50	11.4673 9979	50	18.4201 5427	50	29.4570 2506	50	46.9016 1251

TABLE V   Trigonometric Functions of Radians and Real Numbers[a]   555

t	$\sin t$	$\cos t$	$\tan t$	$\cot t$	$\sec t$	$\csc t$
.00	.0000	1.0000	.0000	—	1.000	—
.01	.0100	1.0000	.0100	99.997	1.000	100.00
.02	.0200	.9998	.0200	49.993	1.000	50.00
.03	.0300	.9996	.0300	33.323	1.000	33.34
.04	.0400	.9992	.0400	24.987	1.001	25.01
.05	.0500	.9988	.0500	19.983	1.001	20.01
.06	.0600	.9982	.0601	16.647	1.002	16.68
.07	.0699	.9976	.0701	14.262	1.002	14.30
.08	.0799	.9968	.0802	12.473	1.003	12.51
.09	.0899	.9960	.0902	11.081	1.004	11.13
.10	.0998	.9950	.1003	9.967	1.005	10.02
.11	.1098	.9940	.1104	9.054	1.006	9.109
.12	.1197	.9928	.1206	8.293	1.007	8.353
.13	.1296	.9916	.1307	7.649	1.009	7.714
.14	.1395	.9902	.1409	7.096	1.010	7.166
.15	.1494	.9888	.1511	6.617	1.011	6.692
.16	.1593	.9872	.1614	6.197	1.013	6.277
.17	.1692	.9856	.1717	5.826	1.015	5.911
.18	.1790	.9838	.1820	5.495	1.016	5.586
.19	.1889	.9820	.1923	5.200	1.018	5.295
.20	.1987	.9801	.2027	4.933	1.020	5.033
.21	.2085	.9780	.2131	4.692	1.022	4.797
.22	.2182	.9759	.2236	4.472	1.025	4.582
.23	.2280	.9737	.2341	4.271	1.027	4.386
.24	.2377	.9713	.2447	4.086	1.030	4.207
.25	.2474	.9689	.2553	3.916	1.032	4.042
.26	.2571	.9664	.2660	3.759	1.035	3.890
.27	.2667	.9638	.2768	3.613	1.038	3.749
.28	.2764	.9611	.2876	3.478	1.041	3.619
.29	.2860	.9582	.2984	3.351	1.044	3.497
.30	.2955	.9553	.3093	3.233	1.047	3.384
.31	.3051	.9523	.3203	3.122	1.050	3.278
.32	.3146	.9492	.3314	3.018	1.053	3.179
.33	.3240	.9460	.3425	2.920	1.057	3.086
.34	.3335	.9428	.3537	2.827	1.061	2.999
.35	.3429	.9394	.3650	2.740	1.065	2.916
.36	.3523	.9359	.3764	2.657	1.068	2.839
.37	.3616	.9323	.3879	2.578	1.073	2.765
.38	.3709	.9287	.3994	2.504	1.077	2.696
.39	.3802	.9249	.4111	2.433	1.081	2.630

t	$\sin t$	$\cos t$	$\tan t$	$\cot t$	$\sec t$	$\csc t$
.40	.3894	.9211	.4228	2.365	1.086	2.568
.41	.3986	.9171	.4346	2.301	1.090	2.509
.42	.4078	.9131	.4466	2.239	1.095	2.452
.43	.4169	.9090	.4586	2.180	1.100	2.399
.44	.4259	.9048	.4708	2.124	1.105	2.348
.45	.4350	.9004	.4831	2.070	1.111	2.299
.46	.4439	.8961	.4954	2.018	1.116	2.253
.47	.4529	.8916	.5080	1.969	1.122	2.208
.48	.4618	.8870	.5206	1.921	1.127	2.166
.49	.4706	.8823	.5334	1.875	1.133	2.125
.50	.4794	.8776	.5463	1.830	1.139	2.086
.51	.4882	.8727	.5594	1.788	1.146	2.048
.52	.4969	.8678	.5726	1.747	1.152	2.013
.53	.5055	.8628	.5859	1.707	1.159	1.987
.54	.5141	.8577	.5994	1.668	1.166	1.945
.55	.5227	.8525	.6131	1.631	1.173	1.913
.56	.5312	.8473	.6269	1.595	1.180	1.883
.57	.5396	.8419	.6410	1.560	1.188	1.853
.58	.5480	.8365	.6552	1.526	1.196	1.825
.59	.5564	.8309	.6696	1.494	1.203	1.797
.60	.5646	.8253	.6841	1.462	1.212	1.771
.61	.5729	.8196	.6989	1.431	1.220	1.746
.62	.5810	.8139	.7139	1.401	1.229	1.721
.63	.5891	.8080	.7291	1.372	1.238	1.697
.64	.5972	.8021	.7445	1.343	1.247	1.674
.65	.6052	.7961	.7602	1.315	1.256	1.652
.66	.6131	.7900	.7761	1.288	1.266	1.631
.67	.6210	.7838	.7923	1.262	1.276	1.610
.68	.6288	.7776	.8087	1.237	1.286	1.590
.69	.6365	.7712	.8253	1.212	1.297	1.571
.70	.6442	.7648	.8423	1.187	1.307	1.552
.71	.6518	.7584	.8595	1.163	1.319	1.534
.72	.6594	.7518	.8771	1.140	1.330	1.517
.73	.6669	.7452	.8949	1.117	1.342	1.500
.74	.6743	.7385	.9131	1.095	1.354	1.483
.75	.6816	.7317	.9316	1.073	1.367	1.467
.76	.6889	.7248	.9505	1.052	1.380	1.452
.77	.6961	.7179	.9697	1.031	1.393	1.437
.78	.7033	.7109	.9893	1.011	1.407	1.422
.79	.7104	.7038	1.009	.9908	1.421	1.408

TABLE V (*continued*)

557

t	$\sin t$	$\cos t$	$\tan t$	$\cot t$	$\sec t$	$\csc t$
.80	.7174	.6967	1.030	.9712	1.435	1.394
.81	.7243	.6895	1.050	.9520	1.450	1.381
.82	.7311	.6822	1.072	.9331	1.466	1.368
.83	.7379	.6749	1.093	.9146	1.482	1.355
.84	.7446	.6675	1.116	.8964	1.498	1.343
.85	.7513	.6600	1.138	.8785	1.515	1.331
.86	.7578	.6524	1.162	.8609	1.533	1.320
.87	.7643	.6448	1.185	.8437	1.551	1.308
.88	.7707	.6372	1.210	.8267	1.569	1.297
.89	.7771	.6294	1.235	.8100	1.589	1.287
.90	.7833	.6216	1.260	.7936	1.609	1.277
.91	.7895	.6137	1.286	.7774	1.629	1.267
.92	.7956	.6058	1.313	.7615	1.651	1.257
.93	.8016	.5978	1.341	.7458	1.673	1.247
.94	.8076	.5898	1.369	.7303	1.696	1.238
.95	.8134	.5817	1.398	.7151	1.719	1.229
.96	.8192	.5735	1.428	.7001	1.744	1.221
.97	.8249	.5653	1.459	.6853	1.769	1.212
.98	.8305	.5570	1.491	.6707	1.795	1.204
.99	.8360	.5487	1.524	.6563	1.823	1.196
1.00	.8415	.5403	1.557	.6421	1.851	1.188
1.01	.8468	.5319	1.592	.6281	1.880	1.181
1.02	.8521	.5234	1.628	.6142	1.911	1.174
1.03	.8573	.5148	1.665	.6005	1.942	1.166
1.04	.8624	.5062	1.704	.5870	1.975	1.160
1.05	.8674	.4976	1.743	.5736	2.010	1.153
1.06	.8724	.4889	1.784	.5604	2.046	1.146
1.07	.8772	.4801	1.827	.5473	2.083	1.140
1.08	.8820	.4713	1.871	.5344	2.122	1.134
1.09	.8866	.4625	1.917	.5216	2.162	1.128
1.10	.8912	.4536	1.965	.5090	2.205	1.122
1.11	.8957	.4447	2.014	.4964	2.249	1.116
1.12	.9001	.4357	2.066	.4840	2.295	1.111
1.13	.9044	.4267	2.120	.4718	2.344	1.106
1.14	.9086	.4176	2.176	.4596	2.395	1.101
1.15	.9128	.4085	2.234	.4475	2.448	1.096
1.16	.9168	.3993	2.296	.4356	2.504	1.091
1.17	.9208	.3902	2.360	.4237	2.563	1.086
1.18	.9246	.3809	2.427	.4120	2.625	1.082
1.19	.9284	.3717	2.498	.4003	2.691	1.077

TABLE V (*continued*)

t	sin t	cos t	tan t	cot t	sec t	csc t
1.20	.9320	.3624	2.572	.3888	2.760	1.073
1.21	.9356	.3530	2.650	.3773	2.833	1.069
1.22	.9391	.3436	2.733	.3659	2.910	1.065
1.23	.9425	.3342	2.820	.3546	2.992	1.061
1.24	.9458	.3248	2.912	.3434	3.079	1.057
1.25	.9490	.3153	3.010	.3323	3.171	1.054
1.26	.9521	.3058	3.113	.3212	3.270	1.050
1.27	.9551	.2963	3.224	.3102	3.375	1.047
1.28	.9580	.2867	3.341	.2993	3.488	1.044
1.29	.9608	.2771	3.467	.2884	3.609	1.041
1.30	.9636	.2675	3.602	.2776	3.738	1.038
1.31	.9662	.2579	3.747	.2669	3.878	1.035
1.32	.9687	.2482	3.903	.2562	4.029	1.032
1.33	.9711	.2385	4.072	.2456	4.193	1.030
1.34	.9735	.2288	4.256	.2350	4.372	1.027
1.35	.9757	.2190	4.455	.2245	4.566	1.025
1.36	.9779	.2092	4.673	.2140	4.779	1.023
1.37	.9799	.1994	4.913	.2035	5.014	1.021
1.38	.9819	.1896	5.177	.1913	5.273	1.018
1.39	.9837	.1798	5.471	.1828	5.561	1.017
1.40	.9854	.1700	5.798	.1725	5.883	1.015
1.41	.9871	.1601	6.165	.1622	6.246	1.013
1.42	.9887	.1502	6.581	.1519	6.657	1.011
1.43	.9901	.1403	7.055	.1417	7.126	1.010
1.44	.9915	.1304	7.602	.1315	7.667	1.009
1.45	.9927	.1205	8.238	.1214	8.299	1.007
1.46	.9939	.1106	8.989	.1113	9.044	1.006
1.47	.9949	.1006	9.887	.1011	9.938	1.005
1.48	.9959	.0907	10.983	.0910	11.029	1.004
1.49	.9967	.0807	12.350	.0810	12.390	1.003
1.50	.9975	.0707	14.101	.0709	14.137	1.003
1.51	.9982	.0608	16.428	.0609	16.458	1.002
1.52	.9987	.0508	19.670	.0508	19.695	1.001
1.53	.9992	.0408	24.498	.0408	24.519	1.001
1.54	.9995	.0308	32.461	.0308	32.476	1.000
1.55	.9998	.0208	48.078	.0208	48.089	1.000
1.56	.9999	.0108	92.620	.0108	92.626	1.000
1.57	1.0000	.0008	1255.8	.0008	1255.8	1.000

TABLE VI Trigonometric Functions of Angles in Degrees[a]

t degrees	sin t	cos t	tan t	cot t	sec t	csc t	
0°00′	.0000	1.0000	.0000	—	1.000	—	**90°00′**
10	.0029	1.0000	.0029	343.8	1.000	343.8	50
20	.0058	1.0000	.0058	171.9	1.000	171.9	40
30	.0087	1.0000	.0087	114.6	1.000	114.6	30
40	.0116	.9999	.0116	85.94	1.000	85.95	20
50	.0145	.9999	.0145	68.75	1.000	68.76	10
1°00′	.0175	.9998	.0175	57.29	1.000	57.30	**89°00′**
10	.0204	.9998	.0204	49.10	1.000	49.11	50
20	.0233	.9997	.0233	42.96	1.000	42.98	40
30	.0262	.9997	.0262	38.19	1.000	38.20	30
40	.0291	.9996	.0291	34.37	1.000	34.38	20
50	.0320	.9995	.0320	31.24	1.001	31.26	10
2°00′	.0349	.9994	.0349	28.64	1.001	28.65	**88°00′**
10	.0378	.9993	.0378	26.43	1.001	26.45	50
20	.0407	.9992	.0407	24.54	1.001	24.56	40
30	.0436	.9990	.0437	22.90	1.001	22.93	30
40	.0465	.9989	.0466	21.47	1.001	21.49	20
50	.0494	.9988	.0495	20.21	1.001	20.23	10
3°00′	.0523	.9986	.0524	19.08	1.001	19.11	**87°00′**
10	.0552	.9985	.0553	18.07	1.002	18.10	50
20	.0581	.9983	.0582	17.17	1.002	17.20	40
30	.0610	.9981	.0612	16.35	1.002	16.38	30
40	.0640	.9980	.0641	15.60	1.002	15.64	20
50	.0669	.9978	.0670	14.92	1.002	14.96	10
4°00′	.0698	.9976	.0699	14.30	1.002	14.34	**86°00′**
10	.0727	.9974	.0729	13.73	1.003	13.76	50
20	.0756	.9971	.0758	13.20	1.003	13.23	40
30	.0785	.9969	.0787	12.71	1.003	12.75	30
40	.0814	.9967	.0816	12.25	1.003	12.29	20
50	.0843	.9964	.0846	11.83	1.004	11.87	10
5°00′	.0872	.9962	.0875	11.43	1.004	11.47	**85°00′**
10	.0901	.9959	.0904	11.06	1.004	11.10	50
20	.0929	.9957	.0934	10.71	1.004	10.76	40
30	.0958	.9954	.0963	10.39	1.005	10.43	30
40	.0987	.9951	.0992	10.08	1.005	10.13	20
50	.1016	.9948	.1022	9.788	1.005	9.839	10
6°00′	.1045	.9945	.1051	9.514	1.006	9.567	**84°00′**
10	.1074	.9942	.1080	9.255	1.006	9.309	50
20	.1103	.9939	.1110	9.010	1.006	9.065	40
30	.1132	.9936	.1139	8.777	1.006	8.834	30
40	.1161	.9932	.1169	8.556	1.007	8.614	20
50	.1190	.9929	.1198	8.345	1.007	8.405	10
7°00′	.1219	.9925	.1228	8.144	1.008	8.206	**83°00′**
	cos t	sin t	cot t	tan t	csc t	sec t	t degrees

[a] Reprinted with permission of the publisher from *Fundamentals of Algebra and Trigonometry*, Fourth Edition, by Earl W. Swokowski, copyright © 1978, Prindle, Weber & Schmidt.

t degrees	sin t	cos t	tan t	cot t	sec t	csc t	
7°00′	.1219	.9925	.1228	8.144	1.008	8.206	**83°00′**
10	.1248	.9922	.1257	7.953	1.008	8.016	50
20	.1276	.9918	.1287	7.770	1.008	7.834	40
30	.1305	.9914	.1317	7.596	1.009	7.661	30
40	.1334	.9911	.1346	7.429	1.009	7.496	20
50	.1363	.9907	.1376	7.269	1.009	7.337	10
8°00′	.1392	.9903	.1405	7.115	1.010	7.185	**82°00′**
10	.1421	.9899	.1435	6.968	1.010	7.040	50
20	.1449	.9894	.1465	6.827	1.011	6.900	40
30	.1478	.9890	.1495	6.691	1.011	6.765	30
40	.1507	.9886	.1524	6.561	1.012	6.636	20
50	.1536	.9881	.1554	6.435	1.012	6.512	10
9°00′	.1564	.9877	.1584	6.314	1.012	6.392	**81°00′**
10	.1593	.9872	.1614	6.197	1.013	6.277	50
20	.1622	.9868	.1644	6.084	1.013	6.166	40
30	.1650	.9863	.1673	5.976	1.014	6.059	30
40	.1679	.9858	.1703	5.871	1.014	5.955	20
50	.1708	.9853	.1733	5.769	1.015	5.855	10
10°00′	.1736	.9848	.1763	5.671	1.015	5.759	**80°00′**
10	.1765	.9843	.1793	5.576	1.016	5.665	50
20	.1794	.9838	.1823	5.485	1.016	5.575	40
30	.1822	.9833	.1853	5.396	1.017	5.487	30
40	.1851	.9827	.1883	5.309	1.018	5.403	20
50	.1880	.9822	.1914	5.226	1.018	5.320	10
11°00′	.1908	.9816	.1944	5.145	1.019	5.241	**79°00′**
10	.1937	.9811	.1974	5.066	1.019	5.164	50
20	.1965	.9805	.2004	4.989	1.020	5.089	40
30	.1994	.9799	.2035	4.915	1.020	5.016	30
40	.2022	.9793	.2065	4.843	1.021	4.945	20
50	.2051	.9787	.2095	4.773	1.022	4.876	10
12°00′	.2079	.9781	.2126	4.705	1.022	4.810	**78°00′**
10	.2108	.9775	.2156	4.638	1.023	4.745	50
20	.2136	.9769	.2186	4.574	1.024	4.682	40
30	.2164	.9763	.2217	4.511	1.024	4.620	30
40	.2193	.9757	.2247	4.449	1.025	4.560	20
50	.2221	.9750	.2278	4.390	1.026	4.502	10
13°00′	.2250	.9744	.2309	4.331	1.026	4.445	**77°00′**
10	.2278	.9737	.2339	4.275	1.027	4.390	50
20	.2306	.9730	.2370	4.219	1.028	4.336	40
30	.2334	.9724	.2401	4.165	1.028	4.284	30
40	.2363	.9717	.2432	4.113	1.029	4.232	20
50	.2391	.9710	.2462	4.061	1.030	4.182	10
14°00′	.2419	.9703	.2493	4.011	1.031	4.134	**76°00′**
	cos t	sin t	cot t	tan t	csc t	sec t	t degrees

TABLE VI (*continued*) 561

t degrees	$\sin t$	$\cos t$	$\tan t$	$\cot t$	$\sec t$	$\csc t$	
14°00′	.2419	.9703	.2493	4.011	1.031	4.134	**76°00′**
10	.2447	.9696	.2524	3.962	1.031	4.086	50
20	.2476	.9689	.2555	3.914	1.032	4.039	40
30	.2504	.9681	.2586	3.867	1.033	3.994	30
40	.2532	.9674	.2617	3.821	1.034	3.950	20
50	.2560	.9667	.2648	3.776	1.034	3.906	10
15°00′	.2588	.9659	.2679	3.732	1.035	3.864	**75°00′**
10	.2616	.9652	.2711	3.689	1.036	3.822	50
20	.2644	.9644	.2742	3.647	1.037	3.782	40
30	.2672	.9636	.2773	3.606	1.038	3.742	30
40	.2700	.9628	.2805	3.566	1.039	3.703	20
50	.2728	.9621	.2836	3.526	1.039	3.665	10
16°00′	.2756	.9613	.2867	3.487	1.040	3.628	**74°00′**
10	.2784	.9605	.2899	3.450	1.041	3.592	50
20	.2812	.9596	.2931	3.412	1.042	3.556	40
30	.2840	.9588	.2962	3.376	1.043	3.521	30
40	.2868	.9580	.2994	3.340	1.044	3.487	20
50	.2896	.9572	.3026	3.305	1.045	3.453	10
17°00′	.2924	.9563	.3057	3.271	1.046	3.420	**73°00′**
10	.2952	.9555	.3089	3.237	1.047	3.388	50
20	.2979	.9546	.3121	3.204	1.048	3.356	40
30	.3007	.9537	.3153	3.172	1.049	3.326	30
40	.3035	.9528	.3185	3.140	1.049	3.295	20
50	.3062	.9520	.3217	3.108	1.050	3.265	10
18°00′	.3090	.9511	.3249	3.078	1.051	3.236	**72°00′**
10	.3118	.9502	.3281	3.047	1.052	3.207	50
20	.3145	.9492	.3314	3.018	1.053	3.179	40
30	.3173	.9483	.3346	2.989	1.054	3.152	30
40	.3201	.9474	.3378	2.960	1.056	3.124	20
50	.3228	.9465	.3411	2.932	1.057	3.098	10
19°00′	.3256	.9455	.3443	2.904	1.058	3.072	**71°00′**
10	.3283	.9446	.3476	2.877	1.059	3.046	50
20	.3311	.9436	.3508	2.850	1.060	3.021	40
30	.3338	.9426	.3541	2.824	1.061	2.996	30
40	.3365	.9417	.3574	2.798	1.062	2.971	20
50	.3393	.9407	.3607	2.773	1.063	2.947	10
20°00′	.3420	.9397	.3640	2.747	1.064	2.924	**70°00′**
10	.3448	.9387	.3673	2.723	1.065	2.901	50
20	.3475	.9377	.3706	2.699	1.066	2.878	40
30	.3502	.9367	.3739	2.675	1.068	2.855	30
40	.3529	.9356	.3772	2.651	1.069	2.833	20
50	.3557	.9346	.3805	2.628	1.070	2.812	10
21°00′	.3584	.9336	.3839	2.605	1.071	2.790	**69°00′**
	$\cos t$	$\sin t$	$\cot t$	$\tan t$	$\csc t$	$\sec t$	t degrees

t degrees	sin *t*	cos *t*	tan *t*	cot *t*	sec *t*	csc *t*	
21°00′	.3584	.9336	.3839	2.605	1.071	2.790	**69°00′**
10	.3611	.9325	.3872	2.583	1.072	2.769	50
20	.3638	.9315	.3906	2.560	1.074	2.749	40
30	.3665	.9304	.3939	2.539	1.075	2.729	30
40	.3692	.9293	.3973	2.517	1.076	2.709	20
50	.3719	.9283	.4006	2.496	1.077	2.689	10
22°00′	.3746	.9272	.4040	2.475	1.079	2.669	**68°00′**
10	.3773	.9261	.4074	2.455	1.080	2.650	50
20	.3800	.9250	.4108	2.434	1.081	2.632	40
30	.3827	.9239	.4142	2.414	1.082	2.613	30
40	.3854	.9228	.4176	2.394	1.084	2.596	20
50	.3881	.9216	.4210	2.375	1.085	2.577	10
23°00′	.3907	.9205	.4245	2.356	1.086	2.559	**67°00′**
10	.3934	.9194	.4279	2.337	1.088	2.542	50
20	.3961	.9182	.4314	2.318	1.089	2.525	40
30	.3987	.9171	.4348	2.300	1.090	2.508	30
40	.4014	.9159	.4383	2.282	1.092	2.491	20
50	.4041	.9147	.4417	2.264	1.093	2.475	10
24°00′	.4067	.9135	.4452	2.246	1.095	2.459	**66°00′**
10	.4094	.9124	.4487	2.229	1.096	2.443	50
20	.4120	.9112	.4522	2.211	1.097	2.427	40
30	.4147	.9100	.4557	2.194	1.099	2.411	30
40	.4173	.9088	.4592	2.177	1.100	2.396	20
50	.4200	.9075	.4628	2.161	1.102	2.381	10
25°00′	.4226	.9063	.4663	2.145	1.103	2.366	**65°00′**
10	.4253	.9051	.4699	2.128	1.105	2.352	50
20	.4279	.9038	.4734	2.112	1.106	2.337	40
30	.4305	.9026	.4770	2.097	1.108	2.323	30
40	.4331	.9013	.4806	2.081	1.109	2.309	20
50	.4358	.9001	.4841	2.066	1.111	2.295	10
26°00′	.4384	.8988	.4877	2.050	1.113	2.281	**64°00′**
10	.4410	.8975	.4913	2.035	1.114	2.268	50
20	.4436	.8962	.4950	2.020	1.116	2.254	40
30	.4462	.8949	.4986	2.006	1.117	2.241	30
40	.4488	.8936	.5022	1.991	1.119	2.228	20
50	.4514	.8923	.5059	1.977	1.121	2.215	10
27°00′	.4540	.8910	.5095	1.963	1.122	2.203	**63°00′**
10	.4566	.8897	.5132	1.949	1.124	2.190	50
20	.4592	.8884	.5169	1.935	1.126	2.178	40
30	.4617	.8870	.5206	1.921	1.127	2.166	30
40	.4643	.8857	.5243	1.907	1.129	2.154	20
50	.4669	.8843	.5280	1.894	1.131	2.142	10
28°00′	.4695	.8829	.5317	1.881	1.133	2.130	**62°00′**
	cos *t*	sin *t*	cot *t*	tan *t*	csc *t*	sec *t*	*t* degrees

TABLE VI (*continued*)   563

t degrees	sin t	cos t	tan t	cot t	sec t	csc t	
28°00′	.4695	.8829	.5317	1.881	1.133	2.130	**62°00′**
10	.4720	.8816	.5354	1.868	1.134	2.118	50
20	.4746	.8802	.5392	1.855	1.136	2.107	40
30	.4772	.8788	.5430	1.842	1.138	2.096	30
40	.4797	.8774	.5467	1.829	1.140	2.085	20
50	.4823	.8760	.5505	1.816	1.142	2.074	10
29°00′	.4848	.8746	.5543	1.804	1.143	2.063	**61°00′**
10	.4874	.8732	.5581	1.792	1.145	2.052	50
20	.4899	.8718	.5619	1.780	1.147	2.041	40
30	.4924	.8704	.5658	1.767	1.149	2.031	30
40	.4950	.8689	.5696	1.756	1.151	2.020	20
50	.4975	.8675	.5735	1.744	1.153	2.010	10
30°00′	.5000	.8660	.5774	1.732	1.155	2.000	**60°00′**
10	.5025	.8646	.5812	1.720	1.157	1.990	50
20	.5050	.8631	.5851	1.709	1.159	1.980	40
30	.5075	.8616	.5890	1.698	1.161	1.970	30
40	.5100	.8601	.5930	1.686	1.163	1.961	20
50	.5125	.8587	.5969	1.675	1.165	1.951	10
31°00′	.5150	.8572	.6009	1.664	1.167	1.942	**59°00′**
10	.5175	.8557	.6048	1.653	1.169	1.932	50
20	.5200	.8542	.6088	1.643	1.171	1.923	40
30	.5225	.8526	.6128	1.632	1.173	1.914	30
40	.5250	.8511	.6168	1.621	1.175	1.905	20
50	.5275	.8496	.6208	1.611	1.177	1.896	10
32°00′	.5299	.8480	.6249	1.600	1.179	1.887	**58°00′**
10	.5324	.8465	.6289	1.590	1.181	1.878	50
20	.5348	.8450	.6330	1.580	1.184	1.870	40
30	.5373	.8434	.6371	1.570	1.186	1.861	30
40	.5398	.8418	.6412	1.560	1.188	1.853	20
50	.5422	.8403	.6453	1.550	1.190	1.844	10
33°00′	.5446	.8387	.6494	1.540	1.192	1.836	**57°00′**
10	.5471	.8371	.6536	1.530	1.195	1.828	50
20	.5495	.8355	.6577	1.520	1.197	1.820	40
30	.5519	.8339	.6619	1.511	1.199	1.812	30
40	.5544	.8323	.6661	1.501	1.202	1.804	20
50	.5568	.8307	.6703	1.492	1.204	1.796	10
34°00′	.5592	.8290	.6745	1.483	1.206	1.788	**56°00′**
10	.5616	.8274	.6787	1.473	1.209	1.781	50
20	.5640	.8258	.6830	1.464	1.211	1.773	40
30	.5664	.8241	.6873	1.455	1.213	1.766	30
40	.5688	.8225	.6916	1.446	1.216	1.758	20
50	.5712	.8208	.6959	1.437	1.218	1.751	10
35°00′	.5736	.8192	.7002	1.428	1.221	1.743	**55°00′**
	cos t	sin t	cot t	tan t	csc t	sec t	t degrees

TABLE VI (*continued*)

t degrees	sin t	cos t	tan t	cot t	sec t	csc t	
35°00′	.5736	.8192	.7002	1.428	1.221	1.743	**55°00′**
10	.5760	.8175	.7046	1.419	1.223	1.736	50
20	.5783	.8158	.7089	1.411	1.226	1.729	40
30	.5807	.8141	.7133	1.402	1.228	1.722	30
40	.5831	.8124	.7177	1.393	1.231	1.715	20
50	.5854	.8107	.7221	1.385	1.233	1.708	10
36°00′	.5878	.8090	.7265	1.376	1.236	1.701	**54°00′**
10	.5901	.8073	.7310	1.368	1.239	1.695	50
20	.5925	.8056	.7355	1.360	1.241	1.688	40
30	.5948	.8039	.7400	1.351	1.244	1.681	30
40	.5972	.8021	.7445	1.343	1.247	1.675	20
50	.5995	.8004	.7490	1.335	1.249	1.668	10
37°00′	.6018	.7986	.7536	1.327	1.252	1.662	**53°00′**
10	.6041	.7969	.7581	1.319	1.255	1.655	50
20	.6065	.7951	.7627	1.311	1.258	1.649	40
30	.6088	.7934	.7673	1.303	1.260	1.643	30
40	.6111	.7916	.7720	1.295	1.263	1.636	20
50	.6134	.7898	.7766	1.288	1.266	1.630	10
38°00′	.6157	.7880	.7813	1.280	1.269	1.624	**52°00′**
10	.6180	.7862	.7860	1.272	1.272	1.618	50
20	.6202	.7844	.7907	1.265	1.275	1.612	40
30	.6225	.7826	.7954	1.257	1.278	1.606	30
40	.6248	.7808	.8002	1.250	1.281	1.601	20
50	.6271	.7790	.8050	1.242	1.284	1.595	10
39°00′	.6293	.7771	.8098	1.235	1.287	1.589	**51°00′**
10	.6316	.7753	.8146	1.228	1.290	1.583	50
20	.6338	.7735	.8195	1.220	1.293	1.578	40
30	.6361	.7716	.8243	1.213	1.296	1.572	30
40	.6383	.7698	.8292	1.206	1.299	1.567	20
50	.6406	.7679	.8342	1.199	1.302	1.561	10
40°00′	.6428	.7660	.8391	1.192	1.305	1.556	**50°00′**
10	.6450	.7642	.8441	1.185	1.309	1.550	50
20	.6472	.7623	.8491	1.178	1.312	1.545	40
30	.6494	.7604	.8541	1.171	1.315	1.540	30
40	.6517	.7585	.8591	1.164	1.318	1.535	20
50	.6539	.7566	.8642	1.157	1.322	1.529	10
41°00′	.6561	.7547	.8693	1.150	1.325	1.524	**49°00′**
10	.6583	.7528	.8744	1.144	1.328	1.519	50
20	.6604	.7509	.8796	1.137	1.332	1.514	40
30	.6626	.7490	.8847	1.130	1.335	1.509	30
40	.6648	.7470	.8899	1.124	1.339	1.504	20
50	.6670	.7451	.8952	1.117	1.342	1.499	10
42°00′	.6691	.7431	.9004	1.111	1.346	1.494	**48°00′**
	cos t	sin t	cot t	tan t	csc t	sec t	t degrees

TABLE VI (*continued*)   565

t degrees	sin t	cos t	tan t	cot t	sec t	csc t	
42°00′	.6691	.7431	.9004	1.111	1.346	1.494	**48°00′**
10	.6713	.7412	.9057	1.104	1.349	1.490	50
20	.6734	.7392	.9110	1.098	1.353	1.485	40
30	.6756	.7373	.9163	1.091	1.356	1.480	30
40	.6777	.7353	.9217	1.085	1.360	1.476	20
50	.6799	.7333	.9271	1.079	1.364	1.471	10
43°00′	.6820	.7314	.9325	1.072	1.367	1.466	**47°00′**
10	.6841	.7294	.9380	1.066	1.371	1.462	50
20	.6862	.7274	.9435	1.060	1.375	1.457	40
30	.6884	.7254	.9490	1.054	1.379	1.453	30
40	.6905	.7234	.9545	1.048	1.382	1.448	20
50	.6926	.7214	.9601	1.042	1.386	1.444	10
44°00′	.6947	.7193	.9657	1.036	1.390	1.440	**46°00′**
10	.6967	.7173	.9713	1.030	1.394	1.435	50
20	.6988	.7153	.9770	1.024	1.398	1.431	40
30	.7009	.7133	.9827	1.018	1.402	1.427	30
40	.7030	.7112	.9884	1.012	1.406	1.423	20
50	.7050	.7092	.9942	1.006	1.410	1.418	10
45°00′	.7071	.7071	1.0000	1.0000	1.414	1.414	**45°00′**
	cos t	sin t	cot t	tan t	csc t	sec t	t degrees

ANSWERS TO ODD-NUMBERED EXERCISES AND TO REVIEW EXERCISES AND PROGRESS TESTS

CHAPTER 1
EXERCISE SET 1.1

1. F   3. F   5. T   7. T
9. T   11. F   13. F

15. 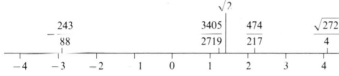   17.

19.

21. $a > 0$ 23. $b \geq 5$ 25. $b \leq -4$
27. multiplication by negative number   29. multiplication by negative number
31. multiplication by positive number
33. 1   35. 4   37. 2   39. 1/5
41. 3   43. 2   45. 8/5   47. $|x - 2| > 5$
49. $|x + 4| \geq 1$ 51. $|a - b| < 3$

EXERCISE SET 1.2

1. 13   3. (a) \$2160   (b) \$2080   (c) \$2106.67
5. 9.37   7. $3/2$ 9. b^7 11. $-20y^9$
13. $-3x^4$ 15. c, d 17. 2; 3   19. 3/5; 4
21. cost of all purchases   23. $5x + 3$
25. $xy^2z - 5x^2yz + xy + yz - x + 5$ 27. $-4y^5 - 2y^4 + 2y^3 - 5y^2 - 3y$
29. $4a^5 - 16a^4 + 14a^3 - 3a^2 - 14a + 15$ 31. $3b^4 + 3ab^3 + 2b^3 - 7ab^2 + 2b^2 - 4ab - 6a$

33. $-260x + 13y + 17z$ 35. $A = 2x^2 + 200x$ 37. (a) $\dfrac{5\pi r^3}{2}$ (b) $V = 5\pi r^{3/2}$

39. $5b(c + 5)$ 41. $-y^2(3 + 4y^3)$ 43. $3x^2(1 + 2y - 3z)$ 45. $(x + 1)(x + 3)$
47. $(x - 7)(x + 2)$ 49. $(2x + 1)(x - 2)$ 51. $(5x + 1)(2x - 3)$ 53. $(3x + 2)(2x + 3)$
55. $4(y + 2)(x - 1)$ 57. $-(x + 2)^2(5x + 31)$ 59. $(y - 1/3)(y + 1/3)$ 61. $(4 + 3xy)(4 - 3xy)$
63. $(x^2 + y^2)(x + y)(x - y)$ 65. $(x + 3y)(x^2 - 3xy + 9y^2)$ 67. $(3x - y)(9x^2 + 3xy + y^2)$
69. $(a + 2)(a^2 - 2a + 4)$ 71. $(\frac{1}{2}m - 2n)(\frac{1}{4}m^2 + mn + 4n^2)$
73. $(x + y - 2)(x^2 + 2xy + y^2 + 2x + 2y + 4)$ 75. $(2x^2 - 5y^2)(4x^4 + 10x^2y^2 + 25y^4)$
77. (a)

x	2.9	2.99	2.999	2.9999
$\dfrac{x^2 - 9}{x - 3}$	5.9	5.99	5.999	5.9999

x	3.1	3.01	3.001	3.0001
$\dfrac{x^2 - 9}{x - 3}$	6.1	6.01	6.001	6.0001

(b) 6   (c) 6

EXERCISE SET 1.3

1. y^{8n}

3. $-x^3/y^3$

5. 1

7. $(3/2)^n x^{2n} y^{3n}$

9. $4/3$

11. 3

13. 81

15. $a^9/3b^4$

17. $4a^{10}c^6/b^8$

19. $1/(a - 2b^2)$

21. $(b + a)/(b - a)$

25. 0.074

27. 8

29. $x^{5/36}$

31. $x^2 y^{12}$

33. $\sqrt[5]{1/16}$

35. $\sqrt[4]{a^3}$

37. $8^{3/4}$

39. $(-8)^{-2/5}$

41. not real

43. 5

45. $5/4$

47. $3, 4$

49. $4\sqrt{3}$

51. $3\sqrt[3]{2}$

53. $y^2 \sqrt[3]{y}$

55. $2x^2 \sqrt[4]{6\sqrt{x}}$

57. $\sqrt{5}/5$

59. $\sqrt{3y}/3y$

61. $3(3 - \sqrt{2})/7$

63. $-3(3\sqrt{a} - 1)/(9a - 1)$

65. $3 + 2\sqrt{2}$

67. $2 + \sqrt{6} + 3\sqrt{2} + 2\sqrt{3}$

69. $\dfrac{1}{\sqrt{x} - \sqrt{3}}, x \neq 3$

71. $\dfrac{1}{\sqrt{x} + 5}, x \neq 25$

73. $\dfrac{1}{2(\sqrt{x + 2} - \sqrt{2})}, x \neq 0$

75. $2(1 - x)\sqrt{2x - 1}$

77. $\dfrac{x^2 - 2}{\sqrt[3]{(x^2 - 1)^5}}$

79. $(x^2 - x + 1)\sqrt{x - 1}$

81. 1

83. 1

85. $-1/2 + 0i$

87. $0 + 5i$

89. $3 - 7i$

91. $0.3 - 7\sqrt{2}i$

93. $x = 2/3, y = -8$

95. $x = -1, y = -9/2$

97. $3 + i$

99. $5 + i$

101. $-1 - i/2$

103. $5 + 0i$

105. 0

107. 3

EXERCISE SET 1.4

1. $-10/3$

3. -2

5. 5

7. $-7/2$

9. 1

11. $8/(5 - k)$

13. $(6 + k)/5$

15. I

17. C

19. $x < 4$

21. $x < -6$

23. $x \geq 5$

25. $a > -1$

27. $y < -1/2$

29. $x \geq 0$

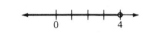

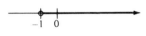

31. $r < 2$

33. $x \geq 1$

35. $x > 5/3$

37. $(-\infty, 2]$

39. $[2, \infty)$

41. $(-\infty, 3/2)$

43. $(-12, \infty)$

45. $(-\infty, 9/2)$ 47. $(-\infty, -6]$ 49. $(-\infty, 3/2)$ 51. $(-\infty, 7]$
53. $(-1/2, 5/4]$ 55. $[-3, -2]$ 57. $(-3, -1]$ 59. $[-1, 2)$
61. 98 63. 5 65. $x > 150$ 67. $L \le 20$

EXERCISE SET 1.5

1. $1, -5$ 3. $3, 1$ 5. $2, -4/3$
7. $3/2, -3$ 9. $4, -2$
11. $x < -4$ or $x > 2$ 13. $x < -1/2$ or $x > 1$ 15. $-1/3 < x < 1$

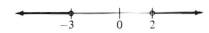 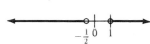 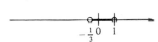

17. $(-\infty, -1], [7, \infty)$ 19. $(-7/2, 9/2)$ 21. no solution
23. $(-\infty, -7), (17, \infty)$ 27. $180,000 < u < 200,000$ 29. $|x - 100| \le 2; 98 \le x \le 102$

EXERCISE SET 1.6

1. $x < -3, x > -2$ 3. $-1/2 < x < 1$ 5. $x < 0, x > 2$
7. $-5 \le x < 3$ 9. $-1/2 \le x < 3$ 11. $s - 2/3, s > 1/2$
13. $-2 < x < 2/3, x > 1$
15. $x < -3, x > 2$ 17. $-1 < x < 5/2$

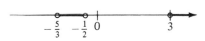 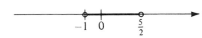

19. $-3/2 < x < 1/2$ 21. $x < -1, x \ge 1$

23. $x \le -1, x \ge -1/3$ 25. $x < -2, 2 \le x \le 3$

27. $-5/3 < x < -1/2, x > 3$

29. $x \le -1, x \ge 2$ 31. $x \le -2, x \ge 2$ 33. all x
35. $x \le -2, x \ge -3/2$ 37. 3 inches 39. $0 \le x < 100$

REVIEW EXERCISES, CHAPTER ONE

1. T 2. F 3. F 4. F
5. $\pi + (-\pi) = 0$ 6. $\pi \cdot 1/\pi = 1$ 7. $\sqrt{4} = 2$ 8. $a = 5, b = -5$
9. 10.

11.

12. -1

13. $3/2$

14. $51

15. c

16. $-0.5, 7$

17. $-7, 5$

18. $a^2b^2 - 3a^2b + 4b$

19. $2x^3 + 3x^2 - 2x$

20. $12x^3 + 12x^2 + 3x$

21. $2(x + 1)(x - 1)$

22. $(x + 5y)(x - 5y)$

23. $(2a + 3b)(a + 3)$

24. $(4x - 1)(x + 5)$

25. $(x + 1)(x - 1)(x^2 + 1)(x^4 + 1)$

26. $(3r^2 + 2s^2)(9r^4 - 6r^2s^2 + 4s^4)$

27. $b^9/8a^6$

28. 2

29. $1/x^4y^8$

30. x^3

31. $4\sqrt{5}$

32. $\sqrt{3}/3$

33. $(x - \sqrt{xy})/(x - y)$

34. $\dfrac{-1}{\sqrt{3} + \sqrt{x}}, x \neq 3$

35. $-2x\sqrt[3]{(2x + 1)^2}$

36. $x = -2, y = 4$

37. $-i$

38. $8 - i$

39. $3 + 4i$

40. $17 + 6i$

41. $8/3$

42. 0

43. $10/3$

44. $k/2(2k + 1)$

45. F

46. F

47. $x \geq 1$

48. $-9/2 \leq x < 5/2$

49. $(-\infty, 8)$

50. $(5/2, \infty)$

51. $[-9, \infty)$

52. $5/3, -3$

53. $x = -1, x = 3/2$

54. $x > 3$ or $x < -4$

55. $(1/5, 3/5)$

56. $(-\infty, -4/3], [8/3, \infty)$

57. $(-\infty, -5], [1, \infty)$

58. $(-\infty, -5), [-1/2, \infty)$

59. $(-2, -3/2), (3, \infty)$

PROGRESS TEST 1A

1. F

2. F

3. T

4. F

5.

6.

7. -1

8. 2

9. 25

10. $-7/3$

11. b

12. $-2.2, 5$

13. $14, 6$

14. $2xy + 3x + 4y + 1$

15. $3a^3 + 5a^2 + 3a + 10$

16. $4a^2b(2ab^4 - 3a^3b + 4)$

17. $(2 - 3x)(2 + 3x)$

18. $1/x^{17}$

19. y^{n+1}

20. -1

21. $4a^4/b^2$

22. $\dfrac{2(\sqrt{x} - 2)}{x - 4}, x \neq 4$

23. $\dfrac{2x}{\sqrt[3]{(2x - 1)^5}}$

24. $x > 2$

25. $-1 + 0i$

26. $16 - 11i$

27. $3/4$

28. $8/13$

29. T

30. $-2 \leq x < 1$

31. $(-\infty, \frac{15}{2}]$

32. $[-4, 4]$

33. $5/2, -2$

34. $-2 \leq x \leq 3$

35. $(-4/3, 2)$

36. $(-\infty, 1/2], [1, \infty)$

37. $[-2, 2/3], [1, \infty)$

PROGRESS TEST 1B

1. T
2. F
3. F
4. T
5.

6.

7. 1
8. 7
9. 14
10. 2
11. a, d
12. 4, 5
13. 1.5, 10
14. $2s^2t^3 - 3s^2t^2 + 2s^2t + 3st^2 + st - s + 2t - 3$
15. $-3b^3 - 7b^2 + 10b + 12$
16. $5r^3s^3(s - 8rt)$
17. $(2x - 1)(x + 4)$
18. $4/x$
19. b^{28}
20. x^6/y^9
21. -1
22. $\dfrac{1}{\sqrt{a} - \sqrt{b}}, a \neq b$
23. $\dfrac{x^2}{\sqrt{(x^2 - 1)^3}}$
24. $x \leq 2$
25. $2 - \frac{3}{2}i$
26. $4 - 7i$
27. -1
28. $k^2/(k + 3)$
29. F
30. $1 \leq x \leq 2$
31. $(-\infty, 4/5]$

32. $(3, \infty)$
33. $8/3, -2$
34. $x \leq 2, x \geq 6$
35. $(-\infty, -1), (1/5, \infty)$

36. $(-\infty, -5)$
37. $(-\infty, -4), (2/3, 1)$

CHAPTER 2
EXERCISE SET 2.1

1.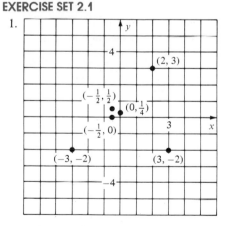
3. $3\sqrt{2}$

5. $4\sqrt{2}$
7. $\sqrt{1345}/6$
9. $\overline{BC} = \sqrt{37}$
11. $\overline{RS} = \sqrt{2}/2$
13. no
15. yes
21. $2\sqrt{10} + 7 + 5\sqrt{2} + \sqrt{37}$
25. any point satisfying $x^2 + y^2 - 10y - 6x + 29 = 0$

27. *x*-int.: -2
    y-int.: 4

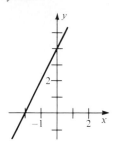

29. *x*-int.: 0
    y-int.: 0

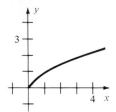

31. *x*-int.: -3
    y-int.: 3

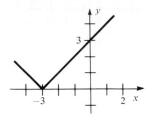

33. *x*-int.: $\pm\sqrt{3}$
    y-int.: 3

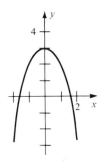

35. *x*-int.: -1
    y-int.: 1

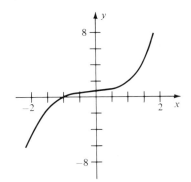

37. *x*-int.: -1
    y-int.: ± 1

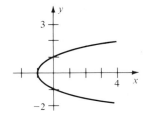

39. none   41. *x*-axis   43. *x*-axis   45. *x*-axis
47. none   49. *y*-axis   51. all   53. origin
55. $y = x + 2, x \neq 2$

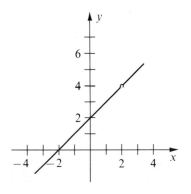

EXERCISE SET 2.2

1. domain: all reals
    range: all reals

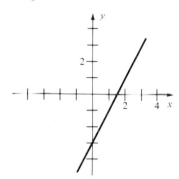

3. domain: all reals
    range: all reals

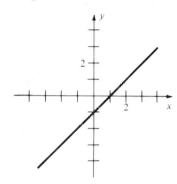

5. domain: $x \geq 1$
    range: $y \geq 0$

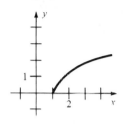

7. $[-3/2, \infty]$

9. $(2, \infty)$

11. $\{x \mid x \geq 1, x \neq 2\}$

13. $\{x \mid x \neq -1\}$

15. $7/2$

17. $3/2$

19. 5

21. $2a^2 + 5$

23. $6x^2 + 15$

25. -7

27. 3

29. $1/(x^2 + 2x)$

31. $a^2 + h^2 + 2ah + 2a + 2h$

33. -0.92

35. $(3x - 1)/(x^2 + 1)$

37. $2(4x^2 + 1)/(6x - 1)$

39. -0.21

41. $2(a - 1)/(4a^2 + 4a - 3)$

43. $(a - 1)/a(a + 4)$

45. $-2, h \neq 0$

47. $4x + 2h + 1, h \neq 0$

49. $\dfrac{-1}{2x(x + h)}, h \neq 0$

51. $3, x \neq 2$

53. $-\dfrac{1}{x - 1}, x \neq 2$

55. $I(x) = 0.28x$

57. $d(C) = C/\pi$

EXERCISE SET 2.3

1. increasing: $(-\infty, \infty)$

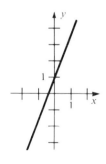

3. increasing: $x > 0$
    decreasing: $x \leq 0$

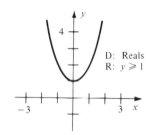

D: Reals
R: $y \geq 1$

5. increasing: $(-\infty, 2]$
    decreasing: $[2, \infty)$

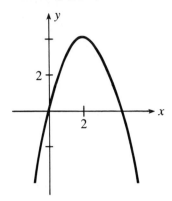

7. increasing: $x \geq -1/2$
    decreasing: $x \leq -1/2$

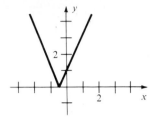

9. increasing: $x \geq 0$

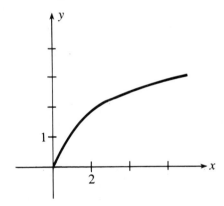

11. constant: $(-\infty, \infty)$

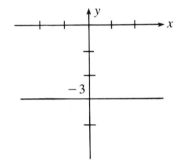

13. increasing: $x > -1$
     decreasing: $x \leq -1$

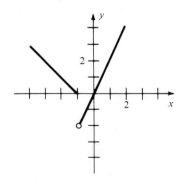

15. increasing: $x \leq 2$
     constant: $x \geq 2$

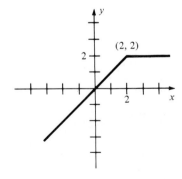

(2, 2)

17. increasing: $x \le 0$
decreasing: $0 \le x < 1,\, x > 2$
constant: $1 \le x \le 2$

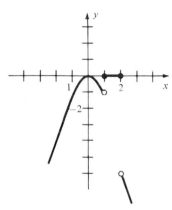

19. constant: $x < -2,\, -2 \le x \le -1,\, x > -1$

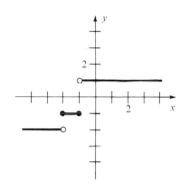

21. $C(u) = \begin{cases} 6.50, & 0 \le u \le 100 \\ 6.50 + 0.06(u - 100), & 100 < u \le 200 \\ 12.50 + 0.05(u - 200), & u > 200 \end{cases}$

23. $R(x) = \begin{cases} 30{,}000, & 0 \le x \le 100 \\ 400x - x^2, & 100 < x < 150 \end{cases}$

25. (a) $C(m) = 14 + 0.08m$ (b) $[0, \infty)$ (c) \$22

EXERCISE SET 2.4

1.

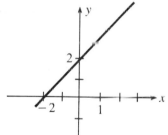

3

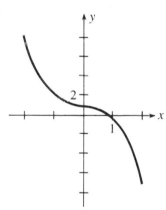

5.

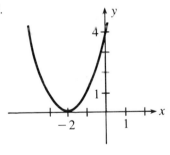

7.

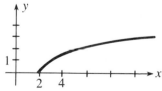

9.

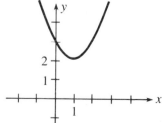

11.

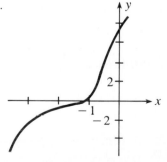

13.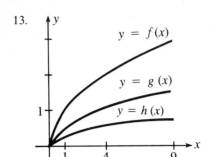

15.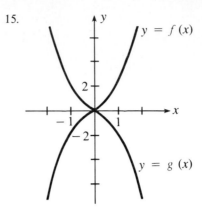

17.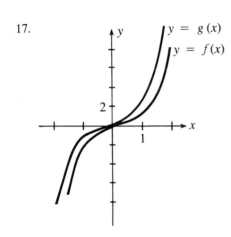

19.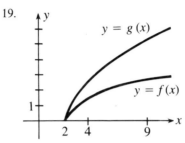

EXERCISE SET 2.5

1. 2; increasing
3. $-3/2$, decreasing
5. -1; decreasing
9. $2x - y + 5 = 0$
11. $3x - y = 0$
13. $2x - y = 0$
15. $2x - 3y = 0$
17. $2x - y = 0$
19. $f(x) = x - 7$
21. $f(x) = -x - 1/2$
23. $f(x) = 7x - 2$
25. $3x - y + 2 = 0$
27. $y - 2 = 0$
29. $x - 3y - 15 = 0$
31. $f(x) = -x + 2$
33. $f(x) = x/2 - 1/2$
35. $f(x) = 5$
37. $m = -3/4, b = 5/4$
39. $m = 0, b = 4$
41. $m = -3/4, b = -1/2$
43. (a) $y = 3$ (b) $x = -6$
45. (a) $y = 0$ (b) $x = -7$
47. (a) $y = -9$ (b) $x = 9$
49. (a) -3 (b) $1/3$
51. (a) $4/3$ (b) $-3/4$
53. (a) $3x + y - 6 = 0$ (b) $x - 3y + 8 = 0$

55. (a) $3x + 5y - 1 = 0$ (b) $5x - 3y + 21 = 0$
57. (a) $F = \dfrac{9}{5}C + 32$ (b) $68°F$

59. $\$1,000,000$
61. 5
63. $f(x) = 8x + 13$

EXERCISE SET 2.6

1. ± 3
3. $\pm 8i/3$
5. $-5/2 \pm \sqrt{2}$
7. $5/3 \pm 2\sqrt{2}/3$
9. 1, 2
11. $1, -2$
13. $-2, -4$
15. 0, 4
17. 1/2, 2
19. $4, -2$
21. $-1/2, 4$
23. $1/3, -3$
25. $-1/2 \pm i/2$
27. $1, -3/4$
29. $-1/3 \pm i\sqrt{2}/3$
43. $-1/2 \pm \sqrt{11}/2$
45. $-2, 2/3$
47. $\pm i\sqrt{2}$

49. $\pm\sqrt{c^2 - a^2}$

51. $\pm\sqrt{3V/\pi h}$

53. $(-v \pm \sqrt{v^2 + 2gs})/g$

55. two complex roots

57. double real root

59. two real roots

61. two real roots

63. 4

65. $0, -8$

67. 4

69. 3

71. $0, 4$

73. 5

75. $u = x^2;\ \pm i\sqrt{2},\ \pm\sqrt{3}/3$

77. $u = 1/x;\ 2,\ -3/2$

79. $u = x^{1/5};\ -32,\ -1/32$

81. $u = 1 + 1/x;\ 1/3,\ -2/7$

83. 10 feet

85. $L = 12$ feet, $W = 4$ feet

87. $L = 8$ cm, $W = 6$ cm

89. $3, 7$

91. 150 shares

93. 8 days

95. 4.93

99. $-1/2$

101. -6

103. ± 9

105. 6

EXERCISE SET 2.7

1. $(x - 3)^2 + 1$

3. $-2(x - 1)^2 - 3$

5. $2(x + 3/2)^2 + 1/2$

7. $-(x + 1/2)^2 + 1/4$

9. $-2(x - 0)^2 + 5$

11. vertex: $(1, -2)$
    $x = 0, 2$
    $y = 0$

13. vertex: $(1/2, 0)$
    $x = 1/2$
    $y = -1$

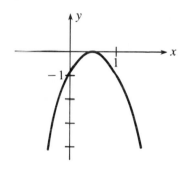

15. vertex: $(-2, 2)$
    $y = 4$

17. vertex: $(3, 1/2)$
    $x = 2, 4$
    $y = -4$

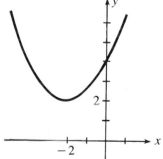

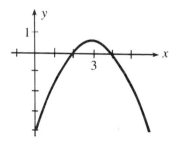

EXERCISE SET 2.8

	(a)	(b)	(c)
1.	min	1/3	11/3
3.	max	0	−5
5.	min	−5/2	−25/4
7.	min	1/8	−49/32

9. 10, 10

11. 25, 25

13. $(1/2, \sqrt{2}/2)$

15. 250 × 500

17. 10, 10

19. 2.5 sec., 100 ft.

21. $0.50

25. 100

27. $\dfrac{160}{4 + \pi}, \dfrac{40\pi}{4 + \pi}$

29. $V(x) = -4x^3 + 100x^2$

31. (a) $A(r) = \pi r^2 + 100/r$ (b) $A(r) = 2\pi r^2 + 100/r$

REVIEW EXERCISES, CHAPTER TWO

1. $\sqrt{61}$

2. $\sqrt{65}$

3.
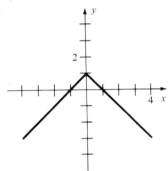

4.
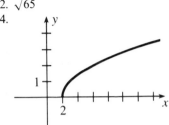

5. x-axis

6. all

7. yes

8. yes

9. $x \geq 5/3$

10. $x \neq -1$

11. 226

12. ± 3

13. 12

14. $y^2 - 3y + 2$

15. $3 + h$

16.
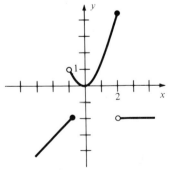

17. increasing: $x \leq -1, 0 \leq x \leq 2$
    decreasing: $-1 < x \leq 0$
    constant: $x > 2$

18. −5

19. −2

20.

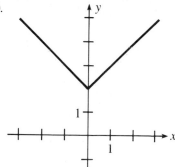

21.

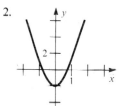

22. 3

25. $y = 3$

28. 5, −4

31. $(2 \pm i\sqrt{2})/2$

34. $\pm\sqrt{3\pi k}/k$

37. double root

40. $\pm\left(\dfrac{-5 \pm \sqrt{85}}{6}\right)^{1/2}$

23. $y - 3x - 6 = 0$

26. $y - 2x - 2 = 0$

29. 1/2, 4/3

32. −1, 1/3

35. −4, 3

38. two complex roots

41. −1/2, −1

24. $x = -4$

27. $y - 2x - 5 = 0$

30. $1 + i\sqrt{5}$

33. $\pm 3/7$

36. two real, rational roots

39. 4

42. 60

43. vertex: $(-2, 4)$; $x = 0, -4$; $y = 0$

45. (a) min   (b) 1/4   (c) 7/8

44. vertex: $(5/2, 3/4)$; no x-intercepts; $y = 7$

46. (a) max   (b) −3/2   (c) 23/4

PROGRESS TEST 2A

1. $3 + \sqrt{26} + \sqrt{41}$

2.

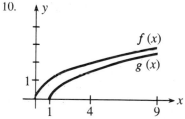

3. origin

7. increasing: $x \geq 0$
   decreasing: $-2 \leq x \leq 0$
   constant: $x < -2$

4. $x \geq 0, x \neq 1$

8. 0

5. 17

9. 2

6. $8t^2 + 3$

10.

11. $2y - 3x - 19 = 0$

15. $3y + x - 7 = 0$

19. two real roots

12. $x = -3$

16. −2, 7

20. two complex roots

13. $m = 1/2; b = 2$

17. $(1 \pm i\sqrt{79})/10$

21. −4

14. $y = -1$

18. −3/4, 1/3

22. $\pm\sqrt{2}i, \pm\sqrt{3}/3$

23.

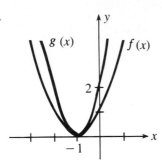

24. vertex: $(1/3, 2/3)$
    intercept: $(0, 1)$

PROGRESS TEST 2B

1. $6\sqrt{2}$

2.

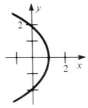

3. origin

4. $x \neq \pm 4$

5. 1

6. 1

7. increasing: $x > 3$
   decreasing: $x \leq -3$
   constant: $-3 < x \leq 3$

8. 10

9. 24

10.

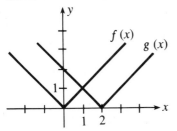

11. $-9/2$

12. $y = -5$

13. 1

14. -3

15. $x + 3y + 3 = 0$

16. $-5/2, 1/3$

17. $1/2, 2$

18. $(1 \pm i\sqrt{83})/6$

19. two complex roots

20. double real root

21. -3

22. $\pm i, \pm(-8)^{3/2}$

23.

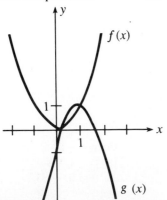

24. vertex: $(3/4, 1/8)$
    intercepts: $(0, -1), (1/2, 0), (1, 0)$

CHAPTER 3
EXERCISE SET 3.1

1. $Q(x) = x - 2$, $R(x) = 2$
3. $Q(x) = 2x - 4$, $R(x) = 8x - 4$
5. $Q(x) = 3x^3 - 9x^2 + 25x - 75$, $R(x) = 226$
7. $Q(x) = 2x - 3$, $R(x) = -4x + 6$
9. $Q(x) = x^2 - x + 1$, $R(x) = 0$
11. $Q(x) = x^2 - 3x$, $R = 5$
13. $Q(x) = x^3 + 3x^2 + 9x + 27$, $R = 0$
15. $Q(x) = 3x^2 - 4x + 4$, $R = 4$
17. $Q(x) = x^4 - 2x^3 + 4x^2 - 8x + 16$, $R = 0$
19. $Q(x) = 6x^3 + 18x^2 + 53x + 159$, $R = 481$
21. -7
23. -34
25. 0
27. -1
29. 0
31. -62
33. yes
35. no
37. yes
39. yes
41. yes
43. yes
45. no
47. $-1, 2$
49. 1, 1/2
51. 1/2, 1/2, $-1/2$
53. $r = 3, -1$
55. 5/2

EXERCISE SET 3.2

	Positive x	Negative x
9.	U	D
11.	D	U
13.	D	D
15.	U	U

17.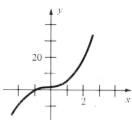

19.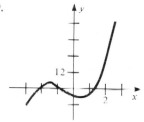

21.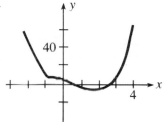

23. x-intercepts: -2, 1/2, 3
$P(x) > 0$: $(-2, 1/2), (3, \infty)$
$P(x) < 0$: $(-\infty, -2), (1/2, 3)$

25. x-intercepts: $-5/2$, 0, 1
$P(x) > 0$: $(-5/2, 0), (1, \infty)$
$P(x) < 0$: $(-\infty, -5/2), (0, 1)$

27. x-intercepts: -2, 0, 3
$P(x) > 0$: $(-\infty, -2), (3, \infty)$
$P(x) < 0$: $(-2, 0), (0, 3)$

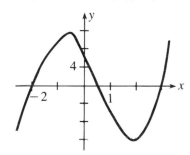

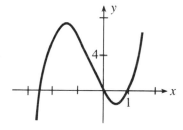

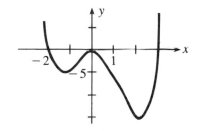

29. $x^3 - 2x^2 - 16x + 32$

31. $x^3 + 6x^2 + 11x + 6$

33. $x^3 - 6x^2 + 6x + 8$

EXERCISE SET 3.3

1. $((3x - 2)x + 5)x - 1$
3. $((((x + 2)x + 0)x + 0)x - 2)x - 3$
5. $(((2x + 0)x - 1)x + 1)x + 4$
7. -1.10
8. -0.56
9. 1.73
10. 2.05

11. 1.44

12. 1.14

13. -1.87

14. -0.55

15. 1.40

16. -2.31

17. 1.10

18. 0.76

EXERCISE SET 3.4

1. $1, -2, 3$

3. $2, -1, -1/2, 2/3$

5. $1, -1, -1, 1/5$

7. $1, -3/4$

9. $3, 3, 1/2$

11. $2, -1$

13. $(3 \pm i\sqrt{3})/2$

15. $-1, -2, 4$

17. $-1, 3/4, \pm i$

19. $3/5, \pm 2, \pm i\sqrt{2}$

21. $0, 1/2, 2/3, -1$

23. $1/2, -4, 2 \pm \sqrt{2}$

25. $k = 3, r = -2$

27. $k = 7, r = 1; k = -7, r = -1$

EXERCISE SET 3.5

	Domain	Intercepts
1.	$\{x \mid x \neq 1\}$	$(0,0)$
3.	$\{x \mid x \neq 0, 2\}$	none
5.	all real numbers	$(\pm\sqrt{3}, 0), (0, -1)$

7. $x = 4, y = 0$

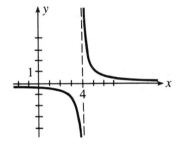

9. $x = -2, y = 0$

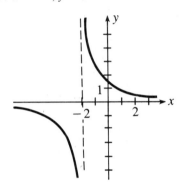

11. $x = -1, y = 0$

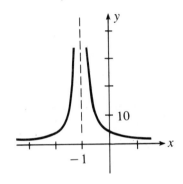

13. $x = 2, y = 1$

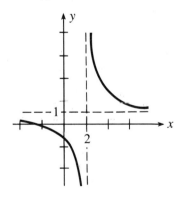

15. $x = 2, x = -2, y = 2$

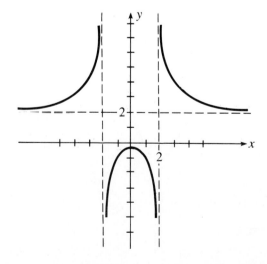

17. $x = 2, x = -3/2, y = 1/2$

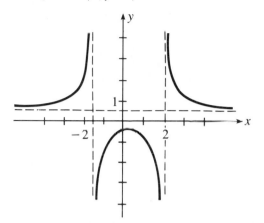

19. $x = 1$

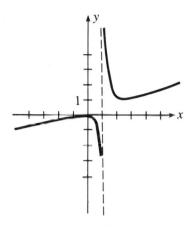

21. $x = 5, x = -5$

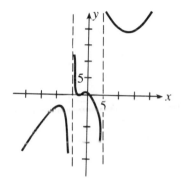

23. $x \neq -2$

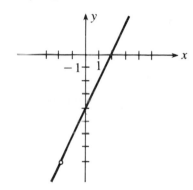

25. $x \neq 2$

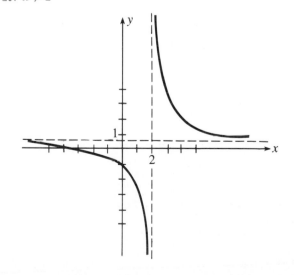

27. $x \neq 0, x \neq -1$

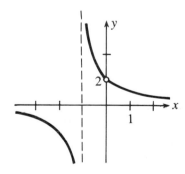

EXERCISE SET 3.6

1. $\dfrac{3}{x+2} - \dfrac{1}{x-3}$

3. $\dfrac{3}{3x-1} - \dfrac{1}{2x-1}$

5. $\dfrac{-2}{x} + \dfrac{2}{x-1} + \dfrac{1}{x+1}$

7. $\dfrac{2}{x} - \dfrac{1}{x^2} - \dfrac{2}{x+2}$

9. $\dfrac{\frac{1}{2}}{x-1} + \dfrac{\frac{1}{2}}{x+1} - \dfrac{2}{(x+1)^2}$

11. $\dfrac{\frac{1}{4}}{x} - \dfrac{\frac{1}{4}x+2}{x^2+4}$

13. $\dfrac{2x-1}{x^2+3} + \dfrac{3-5x}{(x^2+3)^2}$

15. $\dfrac{\frac{1}{3}}{x+1} + \dfrac{\frac{1}{3} - \frac{1}{3}x}{x^2-3x-1}$

17. $\dfrac{-1}{x+1} + \dfrac{2x-2}{x^2+2} + \dfrac{x-1}{(x^2+2)^2}$

19. $x + \dfrac{2x}{x^2+1} - \dfrac{2}{x+1}$

REVIEW EXERCISES, CHAPTER THREE

1. $Q(x) = 2x^2 + 2x + 8$, $R = 4$

2. $Q(x) = x^3 - 5x^2 + 10x - 18$, $R = 31$

3. $46, -8$

4. $4, 1$

7. large positive x: down
large negative x: up

8. large positive x: up
large negative x: down

9. $3, -2/3, -3/2$

10. $1, -2, 2/3, 3/2$

11. none

12. $-1/2, 3$

13. $-1 \pm \sqrt{2}$

14. $4, 2 + i, 2 - i$

15.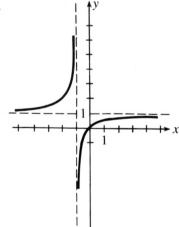

16.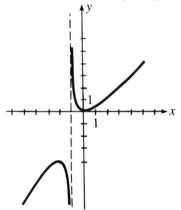

17. $\dfrac{3}{2x-1} - \dfrac{2}{x+2}$

18. $\dfrac{3x}{x^2+1} + \dfrac{2x-1}{(x^2+1)^2}$

19. $2x + 1 + \dfrac{2}{x-1} - \dfrac{1-2x}{(x-1)^2}$

PROGRESS TEST 3A

1. $Q(x) = 2x^2 - 5$, $R(x) = 11$

2. $Q(x) = 3x^3 - 7x^2 + 14x - 28$, $R(x) = 54$

3. -25

4. -165

6. extends indefinitely downward

7. extends indefinitely upward

8. 8

9. none

10. $1, 1, -1, -1, 1/2$

11. $1/2, 1/2$

12. $1, -1 \pm i$

13.

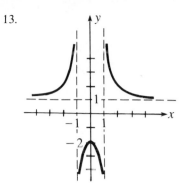

14. $\dfrac{3}{x+3} - \dfrac{2}{x-2}$

PROGRESS TEST 3B

1. $Q(x) = 3x^3/2 + 3x^2/4 + 17x/8 + 15/16$, $R(x) = 49x - 17$
2. $Q(x) = -2x^2 + x + 1$, $R(x) = 0$
3. -1
4. 24
6. extends indefinitely downward
7. extends indefinitely downward
8. 3
9. $-1, 2/3, -2$
10. $-1/2, 3/2, \pm i$
11. $(3 \pm \sqrt{17})/2$
12. $-2 \pm 2\sqrt{2}$
13.

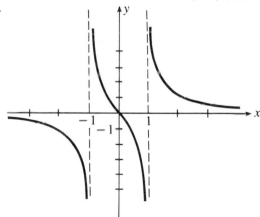

14. $\dfrac{1}{x} - \dfrac{2}{x^2} + \dfrac{4}{x+1}$

CHAPTER 4
EXERCISE SET 4.1

1. $x^2 + x - 1$
3. $x^2 - x + 3$
5. $x^3 - 2x^2 + x - 2$
7. $(x^2 + 1)/(x - 2)$
9. domain of f and of g: all reals
11. $4x^2 + 2x + 1$
13. 21
15. $4x^2 + 10x + 7$
17. $8x^2 - 6x + 1$
19. $x + 6$
21. 29
23. all reals
25. $(f \circ g)(x) = x + 1; (g \circ f)(x) = x + 1$
27. $(f \circ g)(x) = (x - 1)/x; (g \circ f)(x) = -(x + 1)/x$
29. $f(x) = x + 3; g(x) = x^2$
31. $f(x) = x^8; g(x) = 3x + 2$
33. $f(x) = x^{1/3}; g(x) = x^3 - 2x^2$
35. $f(x) = |x|; g(x) = x^2 - 4$
37. $f(x) = \sqrt{x}; g(x) = 4 - x$
45. (a) 4   (b) -5 (c) $y + 1$

47. $f^{-1}(x) = (x - 3)/2$

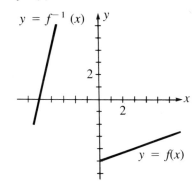

49. $f^{-1}(x) = (3 - x)/2$

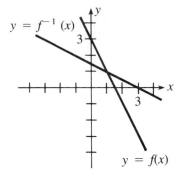

51. $f^{-1}(x) = 3x + 15$

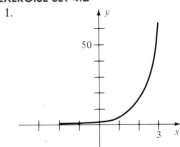

53. $f^{-1}(x) = (x - 1)^{1/3}$

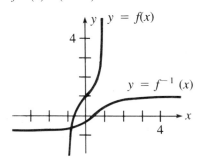

55. yes   57. no   59. yes   61. yes
63. yes   65. no   67. 0   71. $(x - b)/a$
73. (a) yes   (b) 5   (c) 0   (d)

x	17	20	-4	3	10
$G^{-1}(x)$	-10	-5	0	5	10

EXERCISE SET 4.2

1.

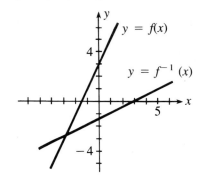

3.

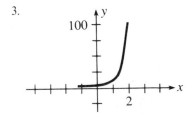

5.

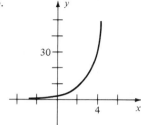

7.

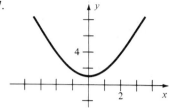

9.

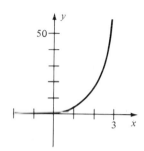

11.

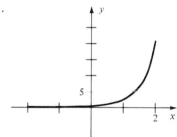

13. 3

15. 4

17. 2

19. 4

21. 2

23. 1

25. (a) 200   (b) 29,682   (c) 256.8, 543.7, 1478, 2436

27. 17.9 billion

29. 670.3 grams

31. \$41,611

33. \$45,417.50

35. \$173.33

37. \$2489

39. π^2

41. 2.725×10^3

43. 8.4×10^{-3}

45. 7.16×10^5

47. 2.962×10^2

EXERCISE SET 4.3

1. $2^2 = 4$

3. $9^{-2} = 1/81$

5. $e^3 = 20.09$

7. $10^3 = 1000$

9. $e^0 = 1$

11. $3^{-3} = 1/27$

13. $\log_5 25 = 2$

15. $\log_{10} 10,000 = 4$

17. $\log_2 1/8 = -3$

19. $\log_2 1 = 0$

21. $\log_{36} 6 = 1/2$

23. $\log_{16} 64 = 3/2$

25. $\log_{27} 1/3 = -1/3$

27. 25

29. 1/5

31. $e^2 \approx 7.39$

33. $e^{-1/2} \approx 0.61$

35. -2

37. 512

39. 124

41. 2

43. 3

45. 6

47. 2

49. 3

51. 1/2

53. 2

55. 1

57. 0

59. -2

61. 4

63. 2

65.

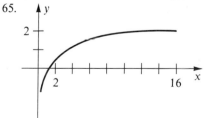

67.

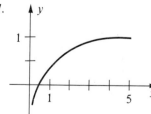

69.
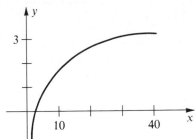

71.
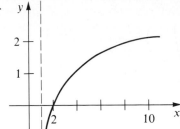

73. $(-\infty, 1)$ 75. $(-\infty, 0), (1, \infty)$ 77. all x 79. no x

EXERCISE SET 4.4

1. $\log_{10} 120 + \log_{10} 36$ 3. 4 5. $\log_a 2 + \log_a x + \log_a y$

7. $\log_a x - \log_a y - \log_a z$ 9. $5 \ln x$ 11. $2 \log_a x + 3 \log_a y$

13. $\frac{1}{2}(\log_a x + \log_a y)$ 15. $2 \ln x + 3 \ln y + 4 \ln z$ 17. $\frac{1}{2} \ln x + \frac{1}{3} \ln y$

19. $2 \log_a x + 3 \log_a y - 4 \log_a z$ 21. 0.77 23. 0.94

25. 1.07 27. 0.87 29. 0.435

31. $\log x^2 \sqrt{y}$ 33. $\ln \sqrt[3]{xy}$ 35. $\log_a \dfrac{x^{1/3} y^2}{z^{3/2}}$

37. $\log_a \sqrt{xy}$ 39. $\ln \dfrac{\sqrt[3]{x^2 y^4}}{z^3}$ 41. $\log_a \dfrac{\sqrt{x-1}}{(x+1)^2}$

43. $\log_a \dfrac{x^3 (x+1)^{1/6}}{(x-1)^2}$ 45. 1.2304 47. 4.5046

49. 2.3892

EXERCISE SET 4.5

1. $\log_5 18$ 3. $1 + \log_2 7$ 5. $(\log_3 46)/2$ 7. $(5 + \log_5 564)/2$

9. $(\log 2 + \log 3)/(\log 3 - 2 \log 2)$ 11. $-\log_2 15$

13. $(1 - \log_4 12)/2$ 15. $\ln 18$ 17. $(-3 + \ln 20)/2$ 19. 500

21. $1/2$ 23. 5 25. 3 27. 8

29. $-1 + \sqrt{17}$ 31. $\ln(y + \sqrt{y^2 + 1})$ 33. 36.62 years   35. 12.6 hours

37. 8.8 years   39. 27.47 days   41. 1.386 days

REVIEW EXERCISES, CHAPTER FOUR

1. $x^2 + x$ 2. 0 3. $(x+1)/(x^2 - 1)$ 4. $x \neq \pm 1$

5. $x^2 + 2x$ 6. 4 7. $|x| - 2$

8. $x + 4 - 4\sqrt{x}$ 9. 0 10. not defined

12.

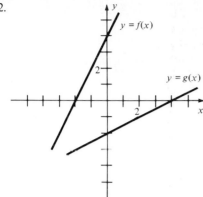

13.

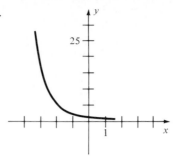

14. 3

15. 2

16. $12,750.78

17. $\log_9 27 = 3/2$

18. $8 = 64^{1/2}$

19. $1/8 = 2^{-3}$

20. $\log_6 1 = 0$

21. 2

22. -2

23. e^{-4}

24. 26

25. 5

26. $-1/3$

27. -1

28. 3

29.

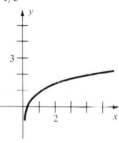

30. $\frac{1}{2}\log_a(x-1) - \log_a 2 - \log_a x$

31. $\log_a x + 2\log_a(2-x) - \frac{1}{2}\log_a(y+1)$

32. $4\ln(x+1) + 2\ln(y-1)$

33. $\frac{2}{5}\log y + \frac{1}{5}\log z - \frac{1}{5}\log(z+3)$

34. 1.15

35. 0.55

36. 0.4

37. -0.15

38. $\log_a \dfrac{\sqrt[3]{x}}{\sqrt{y}}$

39. $\log(x^2 - x)^{4/3}$

40. $\ln \dfrac{3xy^2}{z}$

41. $\log_a \dfrac{(x+2)^2}{(x+1)^{3/2}}$

42. 5/3

43. 15/7

44. 11.5 hours

45. $\dfrac{1}{3} + \dfrac{\log 14}{3\log 2}$

46. $\sqrt{5000}$

47. $\dfrac{199}{98}$

PROGRESS TEST 4A

1. -3

2. $x^2(x-1)$

3. 1/4

5.

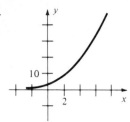

6. $-2/3$

7. $1/9 = 3^{-2}$

8. $\log_{16} 64 = 3/2$

9. 3

10. -1

11. $5/2$

12. $1/2$

13. $3\log_a x - 2\log_a y - \log_a z$

14. $2\log x + \frac{1}{2}\log(2y - 1) - 3\log y$

15. 0.7

16. 0.45

17. $\log\dfrac{x^2}{(y+1)^3}$

18. $\log\left(\dfrac{x+3}{x-3}\right)^{2/3}$

19. 34.6 hours

20. 200

21. 4

PROGRESS TEST 4B

1. 1

2. $\sqrt{2}/2$

3. $\sqrt{x}/x$

5.

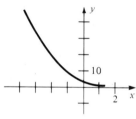

6. 8

7. $\log\dfrac{1}{1000} = -3$

8. $1 = 3^0$

9. $3/2$

10. $3/2$

11. 10

12. 4

13. $\log_a(x - 1) + \frac{5}{4}\log_a(y + 3)$

14. $\frac{1}{2}\ln x + \frac{1}{2}\ln y + \frac{1}{4}\ln 2z$

15. 0.85

16. 1.5

17. $\dfrac{1}{5}\ln\dfrac{(x-1)^3 y^2}{z}$

18. $\log\dfrac{x^2}{y^2}$

19. \$530.76

20. 3

21. 10

ANSWERS TO UNIT 1 PROGRESS TEST

1. $x^{11/6} - x^{5/2}$

2. $3\sqrt{3}$

3. $-\dfrac{2(x+1)}{x}, x \ne 1$

4. $-2 - 2i$

5. $x(3x - 2)(x + 1)$

6. $\dfrac{x+1}{(x-1)^{5/3}}$

7. $\dfrac{2(\sqrt{x} + \sqrt{2})}{x - 2}$

8. $x \le 5/2$

9. $h < -1/2$ or $h > 3/2$

10. $t \le -3/2$ or $t > 4$

11. $12 + 5i$

12. $\dfrac{1}{\sqrt{2x - 1}}$, domain: $x > 1/2$

13. $\dfrac{x}{x^2 - 1}$, domain: $|x| \ne 1$

14. $\sqrt{61}$

15. (a) $4a(1 - a)$ (b) $4, -4$ (c) $-(2t + h), h \ne 0$

16. (a) 0   (b) $1/2$ (c) -9

17. $3/2, -1$

18. $(1 \pm i\sqrt{31})/4$

19. 16

20. $\sqrt{2}$

21. vertex: $(-1/6, 25/12)$
    intercepts: $(-1, 0), (2/3, 0), (0, 2)$

22. (a) -1 (b) $y = -x + 1$ (c) $x = -1$ (d) $y = x + 3$
23. (a) 7   (b) $2x^2 + 3x - 3, 6$ 24. $1/5, 2, -2$ 25. $(1/2, 0), (\sqrt{3}, 0), (-\sqrt{3}, 0)$
26. $\dfrac{1}{3}\left[\dfrac{7}{x-2} - \dfrac{4}{x+1}\right]$
27. (a) $2x/(x-3), x \neq -3$
    (b) vert. asymptote: $x = 3$
        horiz. asymptote: $y = 2$

(c)   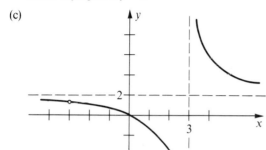   28.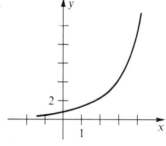

29. (a) $2t + s$ (b) $u - 2s$ (c) $t - s$ (d) $(u + t - s)/2$
30. \$960.79   31. $2\ln x + \frac{1}{2}\ln(x+1) - \ln(x-1)$ 32. $9/5$
33. 3   34. 9   35. 7.92 years

CHAPTER 5
EXERCISE SET 5.1

1. 0   3. $\pi/7$ 5. $3\pi/2$ 7. $5\pi/6$
9. π 11. $7\pi/5$
13.   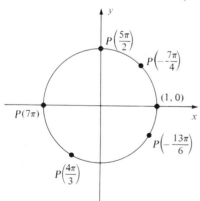   15.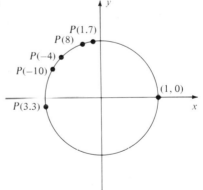

17. $(-1, 0)$ 19. $(\sqrt{2}/2, -\sqrt{2}/2)$ 21. $(-\sqrt{2}/2, -\sqrt{2}/2)$
23. $(-1/2, -\sqrt{3}/2)$ 25. $(-1/2, -\sqrt{3}/2)$ 27. $(-\sqrt{3}/2, -1/2)$
29. $(-\sqrt{3}/2, -1/2)$ 31. $(1/2, \sqrt{3}/2)$ 33. $\pi, -\pi$
35. $3\pi/4, -5\pi/4$ 37. $5\pi/6, -7\pi/6$ 39. $5\pi/3, -\pi/3$
41. $2\pi/3, -4\pi/3$ 43. (a) $(4/5, 3/5)$ (b) $(3/5, -4/5)$ (c) $(-4/5, 3/5)$ (d) $(4/5,\ 3/5)$

EXERCISE SET 5.2

1. $\sin t = -\sqrt{3}/2,\ \cos t = 1/2,\ \tan t = -\sqrt{3}$

3. $\sin t = 0,\ \cos t = -1,\ \tan t = 0$

5. $\sin t = -\sqrt{2}/2,\ \cos t = \sqrt{2}/2,\ \tan t = -1$

7. $\sin t = \sqrt{3}/2,\ \cos t = -1/2,\ \tan t = -\sqrt{3}$

9. $\sin t = -\sqrt{3}/2,\ \cos t = 1/2,\ \tan t = -\sqrt{3}$

11. $\sin t = -\sqrt{2}/2,\ \cos t = -\sqrt{2}/2,\ \tan t = 1$

13. II

15. III

17. III

19. $\pi/2,\ 3\pi/2$

21. $\pi/4,\ 5\pi/4$

23. $\pi/4,\ 3\pi/4$

25. $5\pi/6,\ 7\pi/6$

27. $\pi/3,\ 4\pi/3$

29. $\pi/3,\ 2\pi/3$

31. $7\pi/6,\ 11\pi/6$

33. none

35. $\pi/6,\ 5\pi/6$

37. $-3/4$

39. $-12/13$

41. $-3/5$

43. $4/5$

EXERCISE SET 5.3

1.

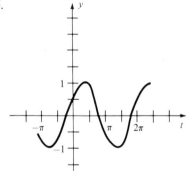

3.

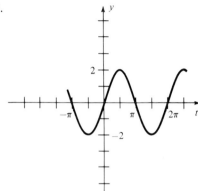

5.

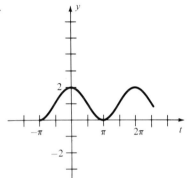

7.

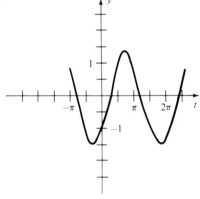

9.

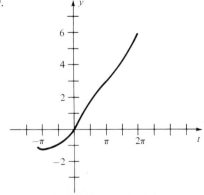

EXERCISE SET 5.4

1. 0.4357	3. −5.471	5. 0.3093	7. 0.8415
9. −0.9737	11. −0.1307	13. 0.7174	15. −0.1987
17. 0.1003	19. 1.4592	21. 0.9888	23. −0.3897
25. 0.7818	27. −0.4660	29. 0.4983	

EXERCISE SET 5.5

1. amplitude: 3
   period: 2π

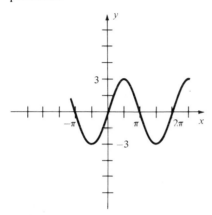

3. amplitude: 1
   period: $\pi/2$

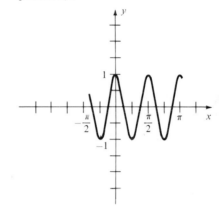

5. amplitude: 2
   period: $\pi/2$

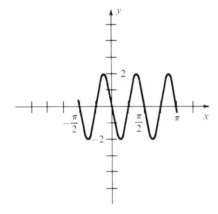

7. amplitude: 2
   period: 6π

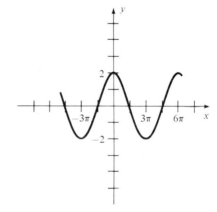

9. amplitude: 1/4
    period: 8π

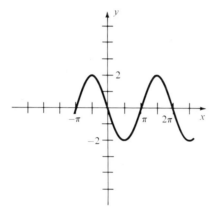

11. amplitude: 3
    period: $2\pi/3$

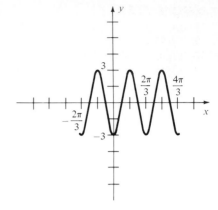

13. amplitude: 2
    period: 2π
    phase shift: π

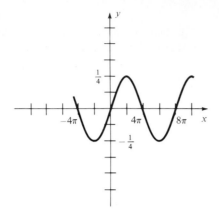

15. amplitude: 3
    period: π
    phase shift: $\pi/2$

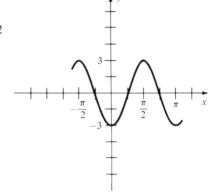

17. amplitude: 1/3
    period: $2\pi/3$
    phase shift: $-\pi/4$

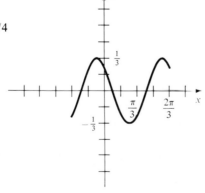

19. amplitude: 2
    period: 8π
    phase shift: 4π

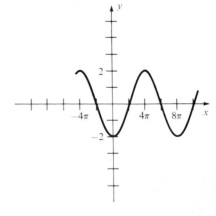

21. $y = 2\sin(2x - \pi)$

23. $y = 3\cos(x/3 - 2\pi/3)$

EXERCISE SET 5.6

1. $\sec t = 2$, $\csc t = 2\sqrt{3}/3$, $\cot t = \sqrt{3}/3$
5. $\sec t = -2\sqrt{3}/3$, $\csc t = 2$, $\cot t = -\sqrt{3}$
9. $\sec t = -\sqrt{2}$, $\csc t = \sqrt{2}$, $\cot t = -1$

3. $\sec t = \sqrt{2}$, $\csc t = \sqrt{2}$, $\cot t = 1$
7. $\sec t$ not defined, $\csc t = -1$, $\cot t = 0$
11. $\sec t = -\sqrt{2}$, $\csc t = -\sqrt{2}$, $\cot t = 1$

13. $0, 2\pi$
15. $7\pi/6, 11\pi/6$
17. $\pi/4, 5\pi/4$
19. $3\pi/4, 7\pi/4$
21. $\pi/4, 7\pi/4$
23. $5\pi/6, 11\pi/6$
25. III
27. II
29. III
31. III
33. $5\pi/6$
35. $2\pi/3$
37. $5\pi/4$
39. $3\pi/4$
41. 4.271
43. 1.000
45. 3.270

EXERCISE SET 5.7

1. $-\pi/6$
3. $\pi/3$
5. $-\pi/4$
7. $5\pi/6$
9. $-\pi/2$
11. $\pi/2$
13. 0
15. $-\pi/4$
17. $2\pi/3$
19. 0.38
21. 2.44
23. 1.30
25. $\sqrt{2}/2$
27. 0
29. $\pi/4$
31. $2\pi/3$
33. $\arcsin(\pm\sqrt{7}/7)$
35. $\cos^{-1}(1/3), \cos^{-1}(-1/4)$
37. $\sin^{-1}(2/3)$
39. 0
41. π
43. 0.6749
45. 0.8861

REVIEW EXERCISES, CHAPTER FIVE

1. $\pi/2$
2. $\pi/2$
3. 0
4. $5\pi/3$
5. $6\pi/5$
6. $(-\sqrt{3}/2, -1/2)$
7. $(-1/2, -\sqrt{3}/2)$
8. $(-\sqrt{3}/2, 1/2)$
9. $(\sqrt{2}/2, \sqrt{2}/2)$
10. $(\sqrt{3}/2, -1/2)$
11. $(-4/5, 3/5)$
12. $(3/5, 4/5)$
13. $(4/5, 3/5)$
14. $(-3/5, -4/5)$
15. $(-4/5, -3/5)$
16. $\sqrt{3}/2$
17. $-\sqrt{2}$
18. $-\sqrt{3}/3$
19. -2
20. $5\pi/4$
21. $11\pi/6$
22. $\pi/3$
23. $2\pi/3$
24. $3/4$
25. $-5/3$
26. $-12/5$
27. $13/12$
30. -1.8334
31. 0.4228
32.

33.

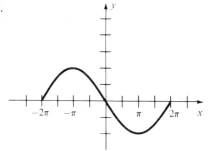

34. amplitude: 1; period: π; phase shift: $\pi/2$
35. amplitude: 4; period: 2π; phase shift: $\pi/2$
36. amplitude: 2; period: 6π; phase shift: $-\pi$
37. $-\pi/6$
38. 0
39. 5
40. $\cos^{-1}(\pm 2\sqrt{5}/5)$

PROGRESS TEST 5A

1. $\pi/3$
2. 0
3. $\pi/4$
4. $(-\sqrt{3}/2, 1/2)$
5. $(1/2, -\sqrt{3}/2)$
6. $(5/13, -12/13)$
7. $(12/13, 5/13)$
8. $(-5/13, -12/13)$
9. 1/2
10. $-2\sqrt{3}/3$
11. $5\pi/4$
12. $7\pi/4$

13. $-5/13$ 14. $-5/4$ 16. -0.5994 17. -0.2509

18.

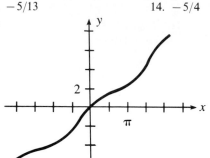

19. amplitude: 2; period: 2π; phase shift: π

21. $-\pi/3$ 22. $1/2$

20. amplitude: 2; period: 4π; phase shift: π

23. $\arctan(2/3)$, $\arctan(3/2)$

PROGRESS TEST 5B

1. 0 2. $\pi/5$ 3. $5\pi/6$ 4. $(\sqrt{3}/2, -1/2)$
5. $(-\sqrt{2}/2, -\sqrt{2}/2)$ 6. $(-4/5, 3/5)$ 7. $(3/5, -4/5)$ 8. $(4/5, -3/5)$
9. -1 10. 1 11. $\pi/3$ 12. $2\pi/3$
13. $-5/12$ 14. $-3/4$ 16. 0.6378 17. -3.1106
18.

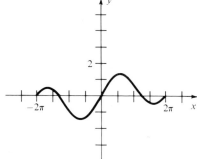

19. amplitude: 4; period: $2\pi/3$; phase shift: $\pi/3$

21. $\pi/6$ 22. 1

20. amplitude: $1/2$; period: π; phase shift: $-\pi/4$

23. $\arcsin\left(-\dfrac{3}{5}\right), \dfrac{\pi}{2}$

CHAPTER 6
EXERCISE SET 6.1

1. IV   3. I   5. II   7. I
9. III   11. II   13. II   15. III
17. I   19. $\pi/6$ 21. $-5\pi/6$ 23. $5\pi/12$
25. $-5\pi/2$ 27. $3\pi/4$ 29. $2\pi/3$ 31. 0.251π
33. $45°$ 35. $270°$ 37. $-90°$ 39. $240°$
41. $450°$ 43. $-300°$ 45. $98°33'$ 47. T
49. F   51. F   53. $50°$ 55. $20°$
57. $85°$ 59. $2\pi/5$ 61. $72°$ 63. $\pi/4$

65. 4/7; 32°44'   67. 6/π   69. 6.8 ft; ≈776.5 rotations   71. 10
73. πrd/180   75. 15/π

EXERCISE SET 6.2

	sin	cos	tan	csc	sec	cot
1. 135°	$\sqrt{2}/2$	$-\sqrt{2}/2$	-1	$\sqrt{2}$	$-\sqrt{2}$	-1
3. $-30°$	$-1/2$	$\sqrt{3}/2$	$-\sqrt{3}/3$	-2	$2\sqrt{3}/3$	$-\sqrt{3}$
5. 315°	$-\sqrt{2}/2$	$\sqrt{2}/2$	-1	$-\sqrt{2}$	$\sqrt{2}$	-1
7. 270°	-1	0		-1		0

9. 70°   11. 30°   13. π/5   15. 25°
17. 47°   19. π/7   21. 0.7353   23. 0.2504
25. 2.128   27. -24.54 29. -0.9147 31. 0.2700
33. -0.2419

EXERCISE SET 6.3

	$\sin\theta$	$\cos\theta$	$\tan\theta$	$\csc\theta$	$\sec\theta$	$\cot\theta$
1.	3/5	4/5	3/4	5/3	5/4	4/3
3.	4/5	3/5	4/3	5/4	5/3	3/4
5.	$2\sqrt{5}/5$	$\sqrt{5}/5$	2	$\sqrt{5}/2$	$\sqrt{5}$	1/2
7.	$\dfrac{\sqrt{x^2+1}}{x^2+1}$	$\dfrac{x\sqrt{x^2+1}}{x^2+1}$	$\dfrac{1}{x}$	$\sqrt{x^2+1}$	$\dfrac{\sqrt{x^2+1}}{x}$	x
9.	12/13	$-5/13$	$-12/5$	13/12	$-13/5$	$-5/12$
11.	$-\sqrt{2}/2$	$-\sqrt{2}/2$	1	$-\sqrt{2}$	$-\sqrt{2}$	1
13.	3/5	$-4/5$	$-3/4$	5/3	$-5/4$	$-4/3$
15.	$-5/13$	12/13	$-5/12$	13/5	13/12	$-12/5$
17.	$-5/13$	$-12/13$	5/12	$-13/5$	$-13/12$	12/5
19.	$\sqrt{5}/5$	$-2\sqrt{5}/5$	$-1/2$	$\sqrt{5}$	$-\sqrt{5}/2$	-2

21. $5\sin\theta$ 23. $6.5\cot\theta$ 25. $3.7\csc\theta$ 27. 36°50'
29. 62.2   31. 30.3   33. 33.9   35. $-5/12$
37. 3/5   39. $-4/3$ 41. 53°10'   43. 61°
45. 7767 feet   47. 970 meters   49. 39°10', 50°50'   51. 18.7 cm
53. 53 feet   55. 24.19 miles

EXERCISE SET 6.4

1. 41°10'   3. 14.4   5. 17.8   7. 62°30'
9. 90°   11. 82°10'   13. 52.5 miles, S29°W   15. 32.8 miles
17. 68.5

EXERCISE SET 6.5

1. 29.1   3. 7.2   5. 15.7   7. none
9. 10.9   11. 8.3, 1.6   13. 98.9 meters   15. 682 meters
17. 40.5 miles   19. 18.8 meters   21. 115 cm

REVIEW EXERCISES, CHAPTER SIX

1. $-\pi/3$ 2. 270°   3. $-75°$ 4. π/4
5. yes   6. no   7. yes   8. 50°
9. 5°   10. 45°   11. 1.4   12. 15/4 cm

13. $5/13$
14. $4/3$
15. $\sqrt{65}/7$
16. 2
17. -1
18. $-1/2$
19. $\sqrt{2}/2$
20. $39°50'$
21. 14.6
22. 25.4
23. 16.6
24. 5.4 meters
25. $68°10'$
26. $36°$
27. $51°50'$
28. $37°50'$

PROGRESS TEST 6A
1. $300°$
2. $-10\pi/9$
3. $5\pi/12$
4. $335°$
5. $45°$
6. $20°$
7. $\pi/4$
8. $4/5$
9. $7/5$
10. 3
11. -1
12. -1
13. 2
14. $56°30'$
15. 23.6
16. $53°10'$
17. 138 meters

PROGRESS TEST 6B
1. $-3\pi/4$
2. $135°$
3. $-150°$
4. $70°$
5. $240°$
6. $7\pi/16$
7. $15°$
8. $21/5$
9. $5/4$
10. $6/7$
11. undefined
12. $2\sqrt{3}/3$
13. 1
14. 10.3
15. $66°30'$
16. 12.2
17. 67.5 feet

CHAPTER 7
EXERCISE SET 7.1
47. $3\pi/2$
49. $\pi/4$
51. π

EXERCISE SET 7.2
1. $s = t = 0$
3. $s = \pi, t = \pi/2$
5. $s = t = \pi/4$
7. $(\sqrt{6} - \sqrt{2})/4$
9. $(\sqrt{2} + \sqrt{6})/4$
11. $-\sqrt{3}/2$
13. $\sqrt{3}$
15. $(\sqrt{6} - \sqrt{2})/4$
17. $(\sqrt{2} - \sqrt{6})/4$
19. $-1/2$
21. $2 - \sqrt{3}$
23. $\cos 43°$
25. $\cot \pi/3$
27. $\sin \pi/6$
29. $-4/5$
31. -7
33. -2.29
35. $-16/65$
37. $2/29$

EXERCISE SET 7.3
1. $7/25$
3. $-\sqrt{3}/2$
5. $-240/161$
7. $-24/25$
9. $-161/289$
11. 0.1022
13. $\sqrt{2 - \sqrt{3}}/2$
15. $\sqrt{(2 - \sqrt{2})}/\sqrt{(2 + \sqrt{2})}$
17. $2/\sqrt{2 - \sqrt{3}}$
19. $\sqrt{20}/5$
21. $\sqrt{6}/3$
23. -2
25. $-\sqrt{6}/3$

EXERCISE SET 7.4
1. $\sin 6\alpha + \sin 4\alpha$
3. $(\cos 5x - \cos x)/2$
5. $-(\cos 7\theta + \cos 3\theta)$
7. $(\cos 2\alpha + \cos 2\beta)/2$
9. $-(2 + \sqrt{2})/4$
11. $\sqrt{3}/4$
13. $2 \sin 3x \cos 2x$
15. $2 \cos 4\theta \cos 2\theta$
17. $2 \sin \alpha \cos \beta$
19. $2 \cos 5x \cos 2x$
21. $\sqrt{6}/2$
23. $-\sqrt{2}$
35. $\dfrac{\sin(a + b)x + \sin(a - b)x}{2}$

EXERCISE SET 7.5

1. $\pi/6, 5\pi/6$; $30°, 150°$
3. π; $180°$
5. $\pi/6, 5\pi/6, 7\pi/6, 11\pi/6$; $30°, 150°, 210°, 330°$
7. $\pi/6, 5\pi/6, 7\pi/6, 11\pi/6$; $30°, 150°, 210°, 330°$
9. $0°, 30°, 150°, 180°$; $0, \pi/6, 5\pi/6, \pi$
11. $0, \pi/3, 5\pi/3$; $0°, 60°, 300°$
13. $\pi/10, \pi/2, 9\pi/10, 13\pi/10, 17\pi/10$; $18°, 90°, 162°, 234°, 306°$
15. $\pi/3, 5\pi/3$; $60°, 300°$
17. $\pi/6, 5\pi/6, 3\pi/2$; $30°, 150°, 270°$
19. 0; $0°$
21. $\pi/6 + \pi n$; $5\pi/6 + \pi n$
23. $\pi/3 + \pi n, 2\pi/3 + \pi n$
25. $\pi/6 + \pi n$; $5\pi/6 + \pi n$
27. $\pi n/4$
29. $\pi/12 + \pi n/2, 5\pi/12 + \pi n/2$
31. $\pi/2 + \pi n$
33. $\pi/2 + 2\pi n/3$
35. $\pi n, \pi/4 + \pi n$
37. $\pi/6 + 2\pi n, 5\pi/6 + 2\pi n$
39. $0.83, 2.31, 3.71, 5.71$ radians
41. 6.05 radians, 5.32 radians

REVIEW EXERCISES, CHAPTER SEVEN

4. $(\sqrt{2} + \sqrt{6})/4$
5. $-\sqrt{2}/2$
6. $-2 - \sqrt{3}$
7. $(\sqrt{2} + \sqrt{6})/4$
8. $\sec 75°$
9. $\sin 67°$
10. $\cos 3\pi/8$
11. $\cot 3\pi/14$
12. $5/13$
13. $10(3 + 4\sqrt{3})/39$
14. $3/4$
15. 70
16. $-16/65$
17. $7/25$
18. $-24/25$
19. $24/25$
20. $-\sqrt{3}/2$
21. $120/169$
22. $-\sqrt{10}/10$
23. $-1/3$
24. $-\sqrt{30}/6$
25. $\sqrt{2 + \sqrt{3}}/2$
26. $\sqrt{2 - \sqrt{2}}/2$
27. $-\sqrt{2 + \sqrt{2}}/\sqrt{2 - \sqrt{2}}$
31. $(\cos\alpha - \cos 2\alpha)/2$
32. $-2\sin 2x \sin x$
33. $1/4$
34. $2(\cos\pi/2)(\cos\pi/4)$
35. $\pi/4, 3\pi/4, 5\pi/4, 7\pi/4$
36. $0, \pi/2, \pi, 3\pi/2$
37. $0, \pi/3, \pi, 5\pi/3$
38. $90° + 360°n, 270° + 360°n$
39. $45° + 60°n$
40. $30° + 90°n, 60° + 90°n$

PROGRESS TEST 7A

2. $1/2$
3. $\sqrt{3} - 2$
4. $\cos 43°$
5. $3/5$
6. $-81/76$
7. $-119/169$
8. $7/25$
9. $-\sqrt{2 - \sqrt{3}}/2$
10. $\sqrt{2 - \sqrt{3}}/\sqrt{2 + \sqrt{3}}$
12. $2(\sin 5x/2)(\cos x/2)$
13. $2(\cos 90°)(\cos 60°) = 0$
14. $\pi/3, 2\pi/3, 4\pi/3, 5\pi/3$
15. $45° + 90°n$

PROGRESS TEST 7B

2. 2
3. $(\sqrt{6} + \sqrt{2})/4$
4. $\cot 19°$
5. $-13\sqrt{2}/7$
6. 0
7. $24/25$
8. $336/625$
9. $-3\sqrt{13}/13$
10. $\sqrt{2 - \sqrt{2}}/2$
12. $(\sin 7\pi/12 - \sin \pi/12)/2$
13. $(\cos 90° + \cos 60°)/2$
14. $0, 3\pi/4, \pi, 7\pi/4$
15. $90° + 120°n$

ANSWERS TO UNIT 2 PROGRESS TEST

1. (a) $\sin \pi/5$ (b) $-\cos \pi/3$ (c) $-\tan 3\pi/7$
2. (a) 2 (b) $-\sqrt{3}$ (c) $-1/2$

3.

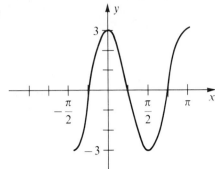

4. (a) $3\pi/4, 5\pi/4$ (b) $\pi/6, 5\pi/6$ (c) $2\pi/3, 5\pi/3$ 5. (a) $4/3$ (b) $13/12$ (c) π

6. (a) $4\pi/3$ (b) $5\pi/3$ (c) none

8.

	(a)	(b)
amplitude	1	3
period	4π	π
phase shift	$-\pi$	$\pi/4$

9. (a) $\sin^{-1}(\pm\sqrt{6}/3)$ (b) $\cos^{-1}(-1)$ 10. $73°41', 30°22'$

11. (a) $85°$ (b) $\pi/3$ (c) $60°$ 12. (a) $\pi/36$ (b) $-30°$ (c) $(360/\pi)°$

13. (a) $\sqrt{34}/5$ (b) $5\sqrt{34}/34$ (c) $5/3$

14. $120/\pi$ 15. $59°$ 16. $56°$

17. (a) $28°57'$ (b) $63°11', 16°49'$ 19. (a) $\dfrac{1}{2}(\sqrt{3}\cos\theta + \sin\theta)$ (b) $-\dfrac{1 - \tan\theta}{1 + \tan\theta}$

20. (a) $\dfrac{1}{2}\sqrt{2 + \sqrt{2}}$ (b) $\dfrac{3 + \sqrt{3}}{3 - \sqrt{3}}$ 21. (a) $\tan 28°$ (b) $\cos 5\pi/12$ 22. (a) $\sqrt{10}/10$ (b) $-24/7$

23. $2\cos 30° \cos 10°$ 24. $(\sin 5t - \sin t)/2$ 25. $\pi/6, 5\pi/6, 3\pi/2$

CHAPTER 8
EXERCISE SET 8.1

1. 5 3. 25 5. 20 7. $-13/10 + 11i/10$

9. $-7/25 - 24i/25$ 11. $8/5 - i/5$ 13. $5/3 - 2i/3$ 15. $4/5 + 8i/5$

17. $4/25 - 3i/25$ 19. $9/10 + 3i/10$ 21. $0 + i/5$ 25. $\sqrt{13}$

27. $\sqrt{2}$ 29. $2\sqrt{10}$

EXERCISE SET 8.2

1. $3\sqrt{2}(\cos 7\pi/4 + i\sin 7\pi/4)$ 3. $2(\cos 11\pi/6 + i\sin 11\pi/6)$ 5. $\sqrt{2}(\cos 3\pi/4 + i\sin 3\pi/4)$

7. $4(\cos\pi + i\sin\pi)$ 9. -4 11. $-1 + i$

13. $-5i$ 15. 6 17. $\sqrt{3} - i$

19. $\sqrt{2}(\cos 7\pi/4 + i\sin 7\pi/4), 2(\cos\pi/2 + i\sin\pi/2); 2 + 2i$

21. $4(\cos 2\pi/3 + i\sin 2\pi/3), 3\sqrt{2}(\cos\pi/4 + i\sin\pi/4), (-6 - 6\sqrt{3}) + (6\sqrt{3} - 6)i$

23. $5(\cos 0 + i\sin 0), 2\sqrt{2}(\cos 5\pi/4 + i\sin 5\pi/4), -10 - 10i$

25. $0 + 512i$ 27. $16 - 16i$ 29. $-8 + 8i$ 31. $\pm\sqrt{2} \pm \sqrt{2}i$

33. $\sqrt{2}(\cos 150° + i\sin 150°), \sqrt{2}(\cos 330° + i\sin 330°)$

35. $-2, 1 \pm \sqrt{3}i$ 37. $\pm 2, \pm 2i$

EXERCISE SET 8.3

1. $x^3 - 2x^2 - 16x + 32$
3. $x^3 + 6x^2 + 11x + 6$
5. $x^3 - 6x^2 + 6x + 8$
7. $x^3/3 + x^2/3 - 7x/12 + 1/6$
9. $x^3 - 4x^2 - 2x + 8$
11. $3, -1, 2$
13. $-2, 4, -4$
15. $-2, -1, 0, -1/2$
17. $5, 5, 5, -5, -5$
19. $x^3 + 6x^2 + 12x + 8$
21. $4x^4 + 4x^3 - 3x^2 - 2x + 1$
23. $x^2 + (1 - 3i)x - (2 + 6i)$
25. $x^2 - 3x + (3 + i)$
27. $x^3 + (1 + 2i)x^2 + (-8 + 8i)x + (-12 + 8i)$
29. $(x^2 - 6x + 10)(x - 1)$
31. $(x^2 + 2x + 5)(x^2 + 2x + 4)$
33. $(x - 2)(x + 2)(x - 3)(x^2 + 6x + 10)$
35. $x - (a + bi)$

EXERCISE SET 8.4

	Positive roots	Negative roots	Complex roots
1.	3	1	0
	1	1	2
3.	0	0	6
5.	3	2	0
	1	2	2
	3	0	2
	1	0	4
7.	1	2	0
	1	0	2
9.	2	0	2
	0	0	4
11.	1	1	6

13. $3, 3, \pm i$
15. $2, -2, \pm\sqrt{2}$
17. $-1/2, -1/2, \pm 1$

REVIEW EXERCISES, CHAPTER EIGHT

1. $6/25 - 17i/25$
2. $-1/5 + 2i/5$
3. $-5/2 + 5i/2$
4. $1/10 - 3i/10$
5. $i/4$
6. $2/29 + 5i/29$
7. $\sqrt{5}$
8. $\sqrt{13}$
9. $\sqrt{41}$
10. $3\sqrt{2}(\cos 3\pi/4 + i\sin 3\pi/4)$
11. $2i$
12. $1 - i$
13. $2(\cos \pi + i\sin \pi)$
14. $24(\cos 37° + i\sin 37°)$
15. $5(\cos 21° + i\sin 21°)/3$
16. $2(\cos 90° + i\sin 90°)$
17. $-972 + 972i$
18. $0 - 8i$
19. $3(\cos 90° + i\sin 90°), 3(\cos 270° + i\sin 270°)$
20. $1, -1/2 \pm \sqrt{3}i/2$
21. $x^3 + 6x^2 + 11x + 6$
22. $x^3 - 3x^2 + 3x - 9$
23. $x^4 + x^3 - 5x^2 - 3x + 6$
24. $4x^4 + 4x^3 - 3x^2 - 2x + 1$
25. $x^4 + 2x^2 + 1$
26. $x^4 - 6x^2 - 8x - 3$
27. 1 positive, 1 negative
28. 5 positive, 0 negative
29. 1 positive, 0 negative
30. 2 positive, 2 negative
31. $-1, (-9 \pm \sqrt{321})/12$
32. $2, 3/2, -1 \pm \sqrt{2}$

PROGRESS TEST 8A

1. $1/13 - 5i/13$
2. $-2/5 - i/5$
3. 5
4. $5(\cos 86° + i\sin 86°)$
5. $(\cos 77° + i\sin 77°)/2$
6. $-\dfrac{1}{2\cdot 5^4} + \dfrac{\sqrt{3}}{2\cdot 5^4}i$
7. $3(\cos 60° + i\sin 60°); 3(\cos 180° + i\sin 180°); 3(\cos 300° + i\sin 300°)$
8. $x^3 - 2x^2 - 5x + 6$
9. $x^4 - 6x^3 + 6x^2 + 6x - 7$
10. $2, \pm i$
11. $-1, -1, (3 \pm \sqrt{17})/2$
12. $x^5 + 3x^4 - 6x^3 - 10x^2 + 21x - 9$
13. $16x^5 - 8x^4 + 9x^3 - 9x^2 - 7x - 1$
14. $x^2 - (1 + 2i)x + (-1 + i)$
15. $(x^2 - 4x + 5)(x - 2)$
16. 2
17. 1
18. $2/3, -3, \pm i$

PROGRESS TEST 8B

1. $-1/4 + i/4$
2. $3/25 + 4i/25$
3. $2\sqrt{2}$
4. $5(\cos 250° + i \sin 250°)$
5. $2(\cos 55° + i \sin 55°)$
6. $-64 + 0i$
7. $-1, 1/2 \pm \sqrt{3}i/2$
8. $2x^4 - x^3 - 3x^2 + x + 1$
9. $x^3 - 4x^2 + 2x + 4$
10. $1, 2, 2, 2$
11. $(-3 \pm \sqrt{13})/2, -3, -3$
12. $8x^4 + 4x^3 - 18x^2 + 11x - 2$
13. $x^4 + 4x^3 - x^2 - 6x + 18$
14. $x^4 - 4x^3 - x^2 + 14x + 10$
15. $(x^2 - 2x + 2)(2x - 1)(x + 2)$
16. 1
17. 2
18. $0, 1/2, \pm\sqrt{2}$

CHAPTER 9

EXERCISE SET 9.1

1. $(5/2, 5)$
3. $(1, 5/2)$
5. $(-7/2, -1)$
7. $(0, -1/2)$
9. $(-1, 9/2)$
11. $(0, 0)$

EXERCISE SET 9.2

1. $(x - 2)^2 + (y - 3)^2 = 4$
3. $(x + 2)^2 + (y + 3)^2 = 5$
5. $x^2 + y^2 = 9$
7. $(x + 1)^2 + (y - 4)^2 = 8$
9. $(h, k) = (2, 3); r = 4$
11. $(h, k) = (2, -2); r = 2$
13. $(h, k) = (-4, -3/2); r = 3\sqrt{2}$
15. no graph
17. $(x + 2)^2 + (y - 4)^2 = 16; (h, k) = (-2, 4); r = 4$
19. $(x - 3/2)^2 + (y - 5/2)^2 = 11/2; (h, k) = (3/2, 5/2); r = \sqrt{22}/2$
21. $(x - 1)^2 + y^2 = 7/2; (h, k) = (1, 0); r = \sqrt{14}/2$
23. $(x - 2)^2 + (y + 3)^2 = 8; (h, k) = (2, -3); r = 2\sqrt{2}$
25. $(x - 3)^2 + (y + 4)^2 = 18; (h, k) = (3, -4); r = 3\sqrt{2}$
27. $(x + 3/2)^2 + (y - 5/2)^2 = 3/2; (h, k) = (-3/2, 5/2); r = \sqrt{6}/2$
29. $(x - 3)^2 + y^2 = 11; (h, k) = (3, 0); r = \sqrt{11}$
31. $(x - 3/2)^2 + (y - 1)^2 = 17/4; (h, k) = (3/2, 1); r = \sqrt{17}/2$
33. $(x + 2)^2 + (y - 2/3)^2 = 100/9; (h, k) = (-2, 2/3); r = 10/3$
35. neither

EXERCISE SET 9.3

1. focus: $(0, 1)$; directrix: $y = -1$

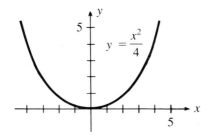

3. focus: $(1/2, 0)$; directrix: $x = -1/2$

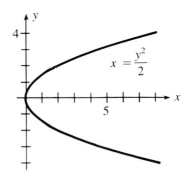

5. focus: $(0, -5/4)$; directrix: $y = 5/4$

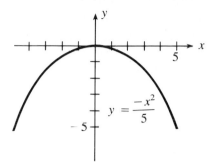

$$y = \frac{-x^2}{5}$$

7. focus: $(3, 0)$; directrix: $x = -3$

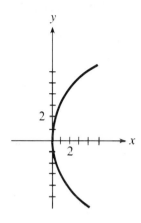

9. $y^2 = 4x$ 11. $y^2 - 6x$ 13. $y^2 = \frac{1}{2}x$ 15. $y^2 = -5x$
17. $y^2 = -4x$ 19. $y^2 = x$ 21. downward   23. to the left

EXERCISE SET 9.4

1. $\dfrac{x^2}{16} + \dfrac{y^2}{25} = 1$; foci: $(0, \pm 3)$; vertices:
   $(0, \pm 5), (\pm 4, 0)$

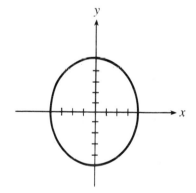

3. $\dfrac{x^2}{4} + \dfrac{y^2}{36} = 1$; foci: $(0, \pm 4\sqrt{2})$; vertices:
   $(0, \pm 6), (\pm 2, 0)$

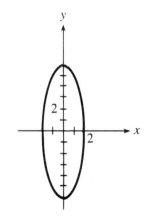

5. $\dfrac{x^2}{64} + \dfrac{y^2}{25} = 1$; foci: $(\pm\sqrt{39}, 0)$; vertices:
$(\pm 8, 0), (0, \pm 5)$

7. $\dfrac{x^3}{9/4} + \dfrac{y^2}{9} = 1$; foci: $(0, \pm 3\sqrt{3}/2)$; vertices:
$(0, \pm 3), (\pm 3/2, 0)$

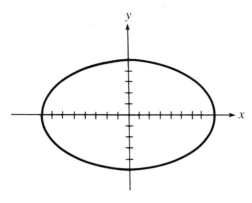

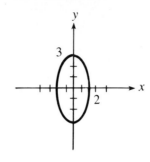

9. $\dfrac{x^2}{20} + \dfrac{y^2}{4} = 1$; foci: $(\pm 4, 0)$; vertices: $(\pm 2\sqrt{5}, 0), (0, \pm 2)$

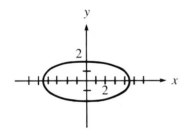

11. $\dfrac{x^2}{40} + \dfrac{y^2}{49} = 1$

13. $\dfrac{x^2}{4} + \dfrac{y^2}{9} = 1$

15. $\dfrac{x^2}{16} + \dfrac{y^2}{1} = 1$

17. $\dfrac{x^2}{1} + \dfrac{y^2}{9} = 1$

19. $\dfrac{x^2}{25} + \dfrac{y^2}{9} = 1$

21. $\dfrac{x^2}{81} + \dfrac{y^2}{1} = 1$

23. $\dfrac{x^2}{49} + \dfrac{y^2}{9/4} = 1$

25. y-axis

27. x-axis

EXERCISE SET 9.5

1. foci: $(\pm\sqrt{10}, 0)$; vertices: $(\pm 3, 0)$

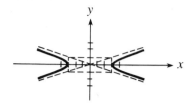

3. foci: $(0, \pm\sqrt{13})$; vertices: $(0, \pm 2)$

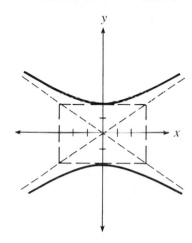

5. foci: $(\pm 3\sqrt{2}, 0)$; vertices: $(\pm 3, 0)$

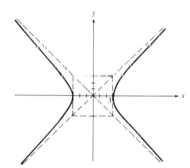

7. foci: $(0, \pm 5\sqrt{5})$; vertices: $(0, \pm 5)$

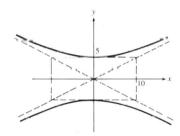

9. x-axis

11. y-axis

13. $\dfrac{y^2}{16} - \dfrac{x^2}{9} = 1$

15. $\dfrac{x^2}{4} - \dfrac{y^5}{5} = 1$

17. $\dfrac{y^2}{1} - \dfrac{x^2}{80} - 1$

19. $\dfrac{y^2}{9} - \dfrac{x^2}{9} = 1$

21. $\dfrac{y^2}{4} - \dfrac{x^2}{16} = 1$

23. $\dfrac{y^2}{81/10} - \dfrac{x^2}{9/10} = 1$

25. $\dfrac{y^2}{16} - \dfrac{x^2}{400/9} = 1$

27. $\dfrac{y^2}{9/2} - \dfrac{x^2}{9/2} = 1$

EXERCISE SET 9.6

1. $(1, -4)$

3. $(5, -1)$

5. $(-3, 4)$

7. $(1, 7)$

9.

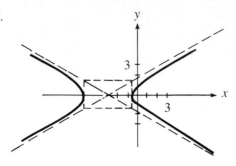

11.

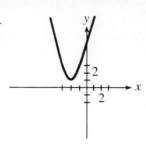

13.

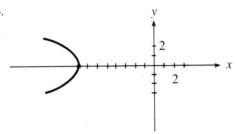

15.

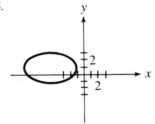

17.

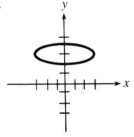

19. parabola   21. ellipse   23. hyperbola   25. hyperbola

EXERCISE SET 9.7

1. $2x'^2 + 8y'^2 = 36$

3. $y' = -2$

5. $9x'^2 - y'^2 - 18x' - 6y' - 9 = 0$

7. $y' = 6x'$

9. $-3x'^2 - 7y'^2 - 3x' - 3\sqrt{3}\,y' = 0$

11. $\dfrac{x'^2}{1/2} + y'^2 = 1$

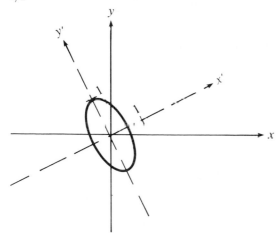

13. $(x' - 1)^2 = -4(y' + 2)$

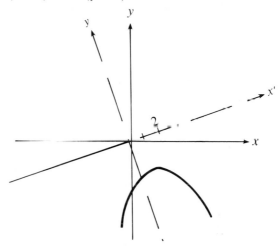

15 $\dfrac{(x' - 3)^2}{4} + y'^2 - 1$

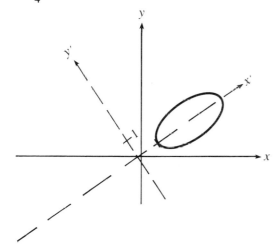

17. $\dfrac{(y' - 3)^2}{4} - \dfrac{(x' - 2)^2}{1} = 1$

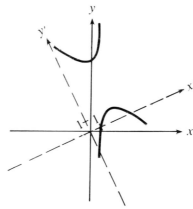

19. $\dfrac{(x' + 3)^2 + (y' - 1)^2}{4} = 1$

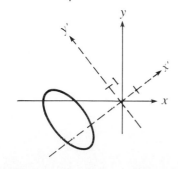

EXERCISE SET 9.8

1. (a)

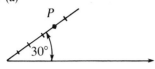

(b)

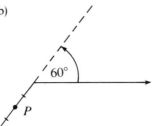

(c)

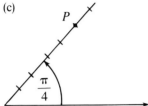

(d)

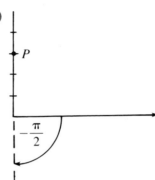

3. (a) $(6, 495°), (-6, 315°)$ (b) $(-2, 480°), (2, 300°)$ (c) $(4, 17\pi/6), (-4, 11\pi/6)$ (d) $(-4, \pi/4), (4, 5\pi/4)$

5. (a) $(2, 570°)$ (b) $(4, 480°)$ (c) $(3, 11\pi/3)$ (d) $(1, 11\pi/6)$

7. (a) yes   (b) no   (c) yes   (d) no

9. (a) $(5\sqrt{3}/2, -5/2)$ (b) $(0, -2)$ (c) $(2\sqrt{3}, 2)$ (d) $(3/2, 3\sqrt{3}/2)$

11. (a) $(2\sqrt{2}, 3\pi/4)$ (b) $(2, -\pi/3)$ (c) $(2, \pi/6)$ (d) $(4\sqrt{2}, -3\pi/4)$

13.

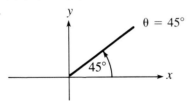

15.

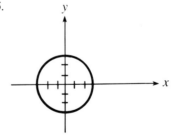

17.

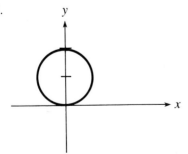

19.

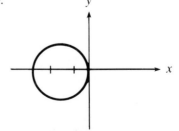

21.

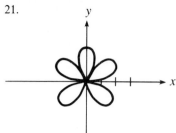

23.

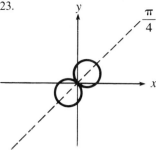

25.

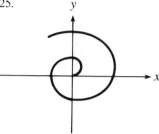

27. $x^2 + y^2 = 16$

33. $r = 5$

29. $x^2 + y^2 + 2y = 0$

35. $r = -5 \sec \theta$

31. $x = 2$

37. $\tan \theta = 3$

EXERCISE SET 9.9

1. $y = \frac{1}{2}x + \frac{1}{2}$

3. $y = x^{3/2}$

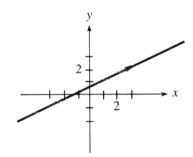

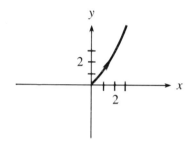

5. $y = 8x^3$

7. $x^3 + y^3 = 3xy$

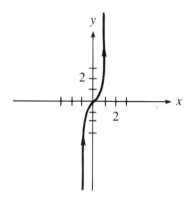

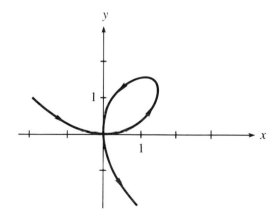

9. $\dfrac{x^2}{16} - \dfrac{y^2}{9} = 1$

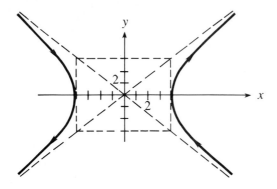

11. $\dfrac{x^2}{4} + \dfrac{y^2}{9} = 1$

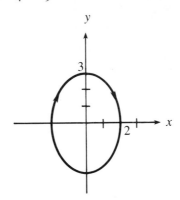

REVIEW EXERCISES, CHAPTER NINE

1. $(-1, -1)$
2. $(-5/2, 5/2)$
3. $(10, 7)$
6. $y = -\frac{5}{6}x - \frac{5}{4}$
7. $(x + 5)^2 + (y - 2)^2 = 16$
8. $(x + 3)^2 + (y - 3)^2 = 4$
9. center: $(2, -3)$; radius: 3
10. center: $(-2, 3)$; radius: $\sqrt{3}$
11. focus: $(0, -1/6)$; directrix: $y = 1/6$
12. focus: $(-1/6, 0)$; directrix: $x = 1/6$

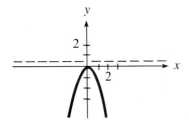

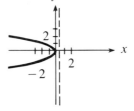

13. $x^2 = -7y$
14. $y = \frac{5}{2}x^2$
15. vertices: $(\pm 1/3, 0), (0, \pm 3/2)$; foci: $(0, \pm\sqrt{77}/6)$
16. vertices: $(-1, 4), (-1, -2), (-5/2, 1),$ and $(1/2, 1)$; foci: $(-1, 1 \pm 3\sqrt{3}/2)$

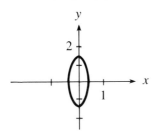

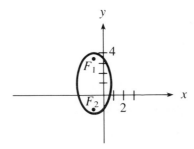

17. $\dfrac{x^2}{81} + \dfrac{y^2}{72} = 1$

18. $\dfrac{x^2}{1} + \dfrac{y^2}{5} = 1$

19. vertices: $(\pm\sqrt{3}/2, -1)$; foci: $(\pm\sqrt{15}/2, -1)$;
    asymptotes: $y + 1 = \pm 2x$

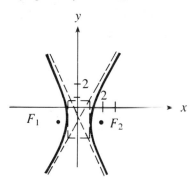

20. vertices: $(0, \pm 2/\sqrt{7})$; foci: $(0, \pm 4\sqrt{2}/7$;
    asymptotes: $y = \pm\sqrt{7}\,x/7$

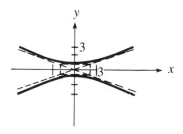

21. $\dfrac{y^2}{4} - \dfrac{x^2}{45} = 1$

22. $\dfrac{x^2}{4} - \dfrac{y^2}{4/5} = 1$

23. ellipse

24. parabola

25. no graph

26. hyperbola

27. two intersecting lines

28. hyperbola

29. hyperbola

30. ellipse

31. hyperbola

32. parabola

33. ellipse

34. ellipse

35.

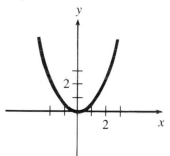

36.

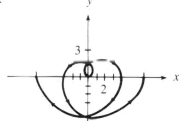

37. $y = 3$

38. $r^2 = \dfrac{1}{1 - (8\cos^2\theta)/9}$

39.

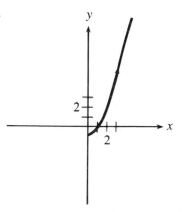

PROGRESS TEST 9A

1. $(0, 4)$

2. $(-1, 2)$

4. $(x - 2)^2 + (y + 3)^2 = 36$

5. center: $(1, -2)$: radius: $\sqrt{6}$

6. vertex: $(-3, 1)$; axis: $x = -3$

7. $\dfrac{x^2}{4} + \dfrac{y^2}{1} = 1$; intercepts: $(\pm 2, 0), (0, \pm 1)$

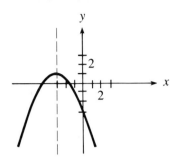

8. $\dfrac{y^2}{9} - \dfrac{x^2}{4} = 1$; intercept: $(0, \pm 3)$

9. $\dfrac{x^2}{1/4} - \dfrac{y^2}{1/4} = 1$; intercept: $(\pm 1/2, 0)$

10.

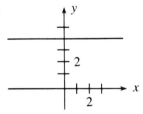

11. ellipse

12. parabola

13. parabola

14. two intersecting lines

15. ellipse

16.

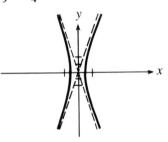

17. $4x + y^2 - 4 = 0$

18. $r^2 = 2/(2 + \sin 2\theta)$

19.

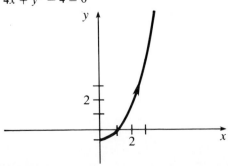

20. $\dfrac{(x - 1)^2}{16} + \dfrac{(y + 1)^2}{9} = 1$; ellipse

PROGRESS TEST 9B

1. $(-1/2, -1)$

2. $(-7, 1)$

4. $(x + 2)^2 + (y + 5)^2 = 25$

5. center: $(1/2, 1)$; radius: $\sqrt{10}$

6. vertex: $(1/3, -1/3)$; axis: $x = 1/3$

7. $\dfrac{x^2}{5} + \dfrac{y^2}{25/9} = 1$; intercepts: $(\pm\sqrt{5}, 0), (0, \pm 5/3)$

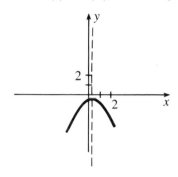

8. $\dfrac{x^2}{3} + \dfrac{y^2}{7/2}$; intercepts: $(\pm\sqrt{3}, 0), (0, \pm\sqrt{7/2})$

9. $\dfrac{y^2}{9} + \dfrac{x^2}{3} = 1$; intercept: $(0, \pm 3)$

10.

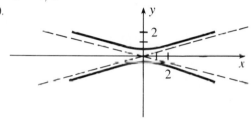

11. hyperbola

12. point$(-2, 3)$

13. hyperbola

14. two intersecting lines

15. hyperbola

16.

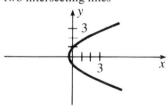

17. $x^2 + y^2 - 4x + y = 0$

18. $r^2 = 2/(2\cos 2\theta - \sin 2\theta)$

19.

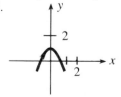

20. $x + 1 = (y - 2)^2$; parabola

CHAPTER 10
EXERCISE SET 10.1

1. $x = 2, y = -1$

3. $x = 3, y = 2; x = 1/5, y = -18/5$

5. $x = 1, y = 1; x = 9/16, y = -3/4$

7. $x = 1, y = 2; x = 13/5, y = -6/5$

9. $x = (-1 + \sqrt{5})/2, y = (1 + \sqrt{5})/2; x = (-1 - \sqrt{5})/2, y = (1 - \sqrt{5})/2$

11. $x = 3, y = -1$

13. no solution

15. $x = 3, y = 2; x = 3, y = -2$

17. none

19. $x = 3, y = 2; x = -3, y = 2; x = 3, y = -2; x = -3, y = -2$

21. I

23. I

25. I

27. C; all points on the line $3x - y = 18$

29. C; $x = 1, y = -1; x = 5/2, y = 1/2$

31. I

33. 22 nickels, 12 quarters

35. 6/5 pounds nuts, 4/5 pounds raisins

37. 125 type-A, 75 type-B

39. 6 and 8

41. 4 and 5

EXERCISE SET 10.2

1.

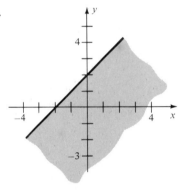

3.

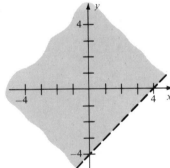

5.

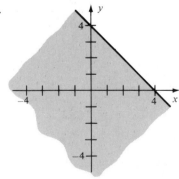

7.

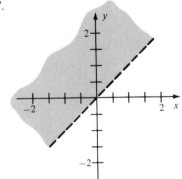

9.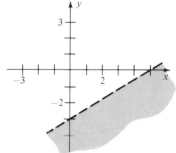

11.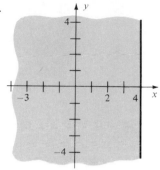

13.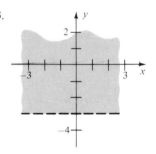

15.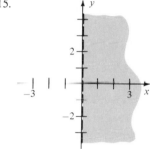

17.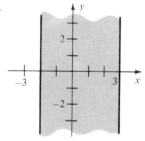

19. $2x + 5y \le 15; \ x \ge 0; \ y \ge 0$

21.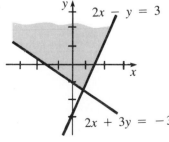

23.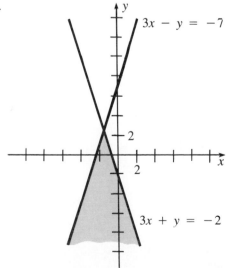

25.

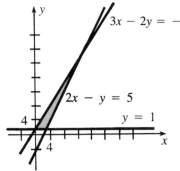

27.

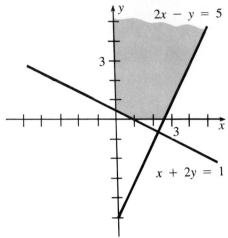

29. no solution

31.

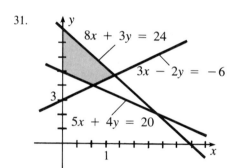

33.

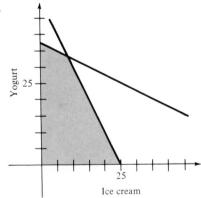

35.

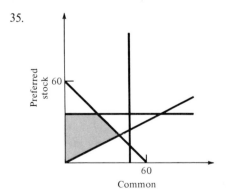

EXERCISE SET 10.3

1. $x = 2, y = -1, z = -2$

3. $x = 1, y = 2/3, z = -2/3$

5. no solution

7. $x = 1, y = 2, z = 2$

9. $x = 1, y = 1, z = 0$

11. $x = 1, y = 27/2, z = -5/2$

13. no solution

15. no solution

17. $x = 5, y = -5, z = -20$

19. A: 2; B: 3; C: 3

21. three 12″-scts, eight 16″-sets, five 19″-sets

EXERCISE SET 10.4

1. 2×2

3. 4×3

5. 3×3

7. (a) -4 (b) 7 (c) 6 (d) -3

9. $\begin{bmatrix} 3 & -2 \\ 5 & 1 \end{bmatrix}, \begin{bmatrix} 3 & -2 & \vdots & 12 \\ 5 & 1 & \vdots & -8 \end{bmatrix}$

11. $\begin{bmatrix} 1/2 & 1 & 1 \\ 2 & -1 & -4 \\ 4 & 2 & -3 \end{bmatrix}, \begin{bmatrix} 1/2 & 1 & 1 & \vdots & 4 \\ 2 & -1 & -4 & \vdots & 6 \\ 4 & 2 & -3 & \vdots & 8 \end{bmatrix}$

13. $\frac{3}{2}x + 6y = -1$
$4x + 5y - \quad 3$

15. $\quad x + \quad y + 3z = -4$
$\quad -3x + 4y \qquad = \quad 8$
$\quad 2x \qquad + 7z = \quad 6$

17. $x = -13, y = 8, z = 2$

19. $x = 35, y = 14, z = -4$

21. $x = 2, y - 3$

23. $x = 2, y = -1, z = 3$

25. $x = 3, y = 2, z = -1$

27. $x = -5, y = 2, z = 3$

29. $x = -5/7, y = -2/7, z = -3/7, w = 2/7$

EXERCISE SET 10.5

1. $a = 3, b = -4, c = 6, d = -2$

3. $\begin{bmatrix} 2 & -1 & 5 \\ 7 & 1 & 6 \\ 4 & 3 & 7 \end{bmatrix}$

5. $\begin{bmatrix} -2 & 18 & 8 \\ -3 & 8 & 11 \end{bmatrix}$

7. not possible

9. $\begin{bmatrix} 17 & 5 \\ 10 & 12 \end{bmatrix}$

11. not possible

13. $\begin{bmatrix} 10 & 4 \\ 12 & 28 \end{bmatrix}$

15. not possible

17. $\begin{bmatrix} 18 & 23 & 29 \\ 17 & -12 & 13 \end{bmatrix}$

19. $AB = \begin{bmatrix} 8 & -6 \\ -8 & 6 \end{bmatrix} \quad AC = \begin{bmatrix} 8 & 6 \\ -8 & 6 \end{bmatrix}$

25. The amount of pesticide 2 eaten by herbivore 3.

27. $A = \begin{bmatrix} 3 & 4 \\ 3 & -1 \end{bmatrix}, \quad X = \begin{bmatrix} x \\ y \end{bmatrix}, \quad B = \begin{bmatrix} -3 \\ 5 \end{bmatrix}$

29. $A = \begin{bmatrix} 3 & -1 & 4 \\ 2 & 2 & 3/4 \\ 1 & -1/4 & 1 \end{bmatrix}, \quad X = \begin{bmatrix} x \\ y \\ z \end{bmatrix}, \quad B = \begin{bmatrix} 5 \\ -1 \\ 1/2 \end{bmatrix}$

31. $x_1 - 5x_2 = 0$
$4x_1 + 3x_2 = 2$

33. $4x_1 + 5x_2 - 2x_3 = \quad 2$
$\quad 3x_2 - \quad x_3 = -5$
$\quad 2x_3 = \quad 4$

EXERCISE SET 10.6

1. 22

3. -8

5. 0

7. (a) -6 (b) -1 (c) 1 (d) 7

9. (a) -6 (b) 1 (c) 1 (d) 7

11. 52

13. -3

15. 0

17. -12

19. 0

21. $x = 1, y = -2, z = -1$

23. $x = 3, y = 2, z = -1$

25. $x = -3, y = 0, z = 2$

27. $x = -5/7, y = -2/7, z = -3/7, w = 2/7$

EXERCISE SET 10.7

	Minimum	Maximum
1.	$-2; (5, 4)$	$5; (5, 0)$
3.	$-3; (2, 2)$	$3; (6, 0)$
5.	$\dfrac{6}{7}; \left(\dfrac{6}{7}, \dfrac{6}{7}\right)$	$5; (4, 3)$
7.	$\dfrac{1}{2}; \left(3, \dfrac{11}{2}\right)$	$14; (8, 2)$

9. preferred: 190/3 square feet
   regular: 30 square feet

11. large: 120
    small: 260

13. Java: 2000/11 pounds
    Colombian: 4000/11 pounds

15. crop A: 30 acres
    crop B: 70 acres

17. pack A: 6 pounds
    pack B: 12 pounds

REVIEW EXERCISES, CHAPTER TEN

1. $x = 0, y = -2$

2. $x = -5, y = 1$

3. $x = 5, y = 0; x = -4, y = 3$

4. $x = 1, y = -1; x = 5, y = 3$

5. $x = -7/20, y = -7/10$

6. $x = 2, y = -2$

7. $x = 1/4, y = -1/2$

8. $x = 0, y = 3$

9. $x = 4, y = 4; x = 36/25, y = -12/5$

10. $x = 3, y = -4$

11. 3, 11

12.

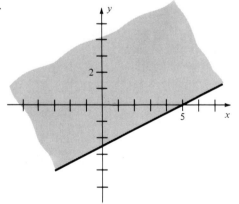

13.

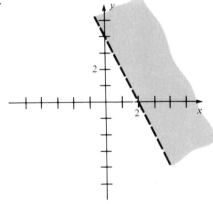

14.

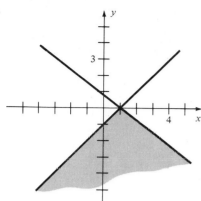

15.

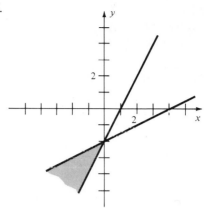

16.

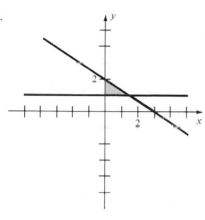

17.

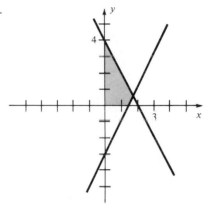

18. $x = -3, y = 1, z = 4$

19. $x = -2, y = 1/2, z = 3$

20. $x = 1, y = -1, z = 2$

21. 3×5

22. -1

23. 4

24. 8

25. $\begin{bmatrix} 3 & -7 \\ 1 & 4 \end{bmatrix}$

26. $\begin{bmatrix} 3 & -7 & | & 14 \\ 1 & 4 & | & 6 \end{bmatrix}$

27. $4x - y = 3$
    $2x + 5y = 0$

28. $-2x + 4y + 5z = 0$
    $6x - 9y + 4z = 0$
    $3x + 2y - z = 0$

29. $x = -1, y = -4$

30. $x = -1, y = 1, z = -3$

31. $x = 1/2, y = 3/2$

32. $x = 3 + 5t/4, y = 3 + t/2, z = t$

33. -3

34. -3

35. $\begin{bmatrix} 1 & 4 \\ 7 & -1 \end{bmatrix}$

36. $\begin{bmatrix} -3 & 6 \\ 1 & -5 \end{bmatrix}$

37. not possible

38. $\begin{bmatrix} 5 & 15 & 20 \\ -5 & 0 & -30 \end{bmatrix}$

39. $\begin{bmatrix} -1 & -3 & -4 \\ -4 & 0 & -24 \\ 4 & 6 & 20 \end{bmatrix}$

40. $\begin{bmatrix} 7 & 4 \\ -11 & 12 \end{bmatrix}$

41. $\begin{bmatrix} 6 & -13 \\ -5 & -9 \end{bmatrix}$

42. 10

43. -6

44. 0

45. -3

46. $x = 1/2, y = 4$

47. $x = 1, y = -4$

48. $x = 1/3, y = 2/3, z = -1$

49. $x = 1/4, y = -2, z = 1/2$

50. $x = 4, y = 6, z = 26$

51. $x = 11/12, y = 2, z = 27/2$

PROGRESS TEST 10A

1. $x = 2, y = \pm\sqrt{10}; x = 3, y = \pm\sqrt{15}$
3. $x = 3, y = \pm 4; x = -3, y = \pm 4$
5.

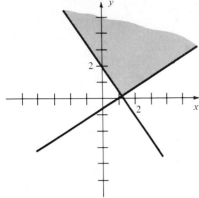

2. $x = 2, y = -1$
4. shirts: $15; ties: $10
6.

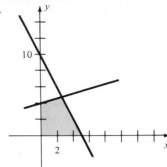

7. $x = -2, y = 4, z = 6$ 8. 3×2

9. 0

10. $\begin{bmatrix} -7 & 0 & 6 & \vdots & 3 \\ 0 & 2 & -1 & \vdots & 10 \\ 1 & -1 & 1 & \vdots & 5 \end{bmatrix}$

11. $-5x + 2y = 4$
    $3x - 4y = 4$

12. $x = -1/2, y = 1/2$

13. $x = -6, y = -2$

14. $x = 1/2, y = 1/2, z = 1/2$

15. 3

16. $\begin{bmatrix} 2 & 14 \\ -2 & -4 \\ -5 & 1 \end{bmatrix}$

17. $\begin{bmatrix} -7 & -11 \\ 11 & 15 \end{bmatrix}$

18. $\begin{bmatrix} -10 \\ 2 \\ 0 \end{bmatrix}$

19. not possible

20. -2

21. 27

22. $x = 4, y = -3$

PROGRESS TEST 10B

1. $x = 5, y = \pm 4; x = -5, y = \pm 4$
3. $x = 0, y = 2; x = 3, y = 1$
5.

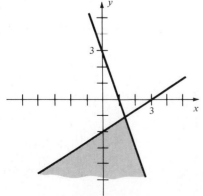

2. $x = 1, y = -3$
4. 300 of each
6.

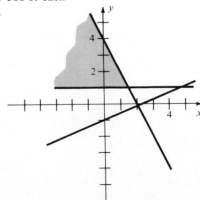

7. $x = -3, y = -10, z = 5$

8. 2×4

9. 1

10. $\begin{bmatrix} 2 & -6 & | & 5 \\ 1 & 3 & | & -2 \end{bmatrix}$

11. $16x \quad\quad + 6z = 10$
$\quad -4x - 2y + 5z = 8$
$\quad\quad 2x + 3y - z = -6$

12. $x = 0, y = -1, z = -3$

13. $x = -1, y = -3$

14. $x = 5, y = -2, z = 1/2$

15. -5

16. $[10, 0]$

17. $\begin{bmatrix} 6 & -17 \\ 3 & 2 \\ 2 & 13 \end{bmatrix}$

18. not possible

19. not possible

20. -22

21. -1

22. $x = -3/2, y = 1/2$

CHAPTER 11
EXERCISE SET 11.1

1. $2, 4, 6, 8; 40$

3. $1, 5, 9, 13; 77$

5. $5, 5, 5, 5; 5$

7. $1/2, 2/3, 3/4, 4/5; 20/21$

9. $2.1, 2.01, 2.001, 2.0001; 2 + 0.1^{20}$

11. $1/3, 4/5, 9/7, 16/9; 400/41$

13. 9

15. $4/7$

17. 256

19. 40

21. 97

23. $49/12$

25. 80

27. $\sum_{k=1}^{5} (2k - 1)$

29. $\sum_{k=1}^{5} k^2$

31. $\sum_{k=1}^{4} \frac{(-1)^k}{\sqrt{k}}$

33. $\sum_{k=1}^{4} \frac{(-1)^{k+1}k}{k^2 + 1}$

35. $\sum_{k=0}^{n} \frac{1}{x^k}$

EXERCISE SET 11.2

1. $15, 18$

3. $1, 5/4$

5. $4, 5$

7. $\sqrt{5} + 6, \sqrt{5} + 8$

9. $2, 6, 10, 14$

11. $3, 5/2, 2, 3/2$

13. $1/3, 0, -1/3, -2/3$

15. 25

17. -8

19. -2

21. $19/3$

23. $821/160$

25. 440

27. -126

29. 1720

31. 30

33. $n = 30, d = 3$

35. $n = 6, d = 1/4$

37. -2

EXERCISE SET 11.3

1. 48

3. $-81/64$

5. 0.0096

7. $3, 9, 27, 81$

9. $4, 2, 1, 1/2$

11. $-3, -6, -12, -24$

13. -384

15. $1/4$

17. $1/243$

19. $27/8$

21. 2

23. 7

25. $1, 3$

27. $1/4, 1/16$

29. $1093/243$

31. $-1353/625$

33. 1020

35. $55/8$

37. \$10,235

39. 58,594

41. 2

43. $3/4$

45. $8/3$

47. 1

49. $1/5$

51. $364/990$

53. $325/999$

EXERCISE SET 11.5

1. $243x^5 + 810x^4y + 1080x^3y^2 + 720x^2y^3 + 240xy^4 + 32y^5$

3. $256x^4 - 256x^3y + 96x^2y^2 - 16xy^3 + y^4$

5. $32 - 80xy + 80x^2y^2 - 40x^3y^3 + 10x^4y^4 - x^5y^5$

7. $a^8b^4 + 12a^6b^3 + 54a^4b^2 + 108a^2b + 81$

9. $a^8 - 16a^7b + 112a^6b^2 - 448a^5b^3 + 1120a^4b^4 - 1792a^3b^5 + 1792a^2b^6 - 1024ab^7 + 256b^8$

11. $\frac{1}{27}x^3 + \frac{2}{3}x^2 + 4x + 8$

13. $1024 + 5120x + 11{,}520x^2 + 15{,}360x^3$

15. $19{,}683 - 118{,}098a + 314{,}928a^2 - 489{,}888a^3$

17. $16,384x^{14} - 344,064x^{13}y + 3,354,624x^{12}y^2 - 20,127,744x^{11}y^3$
19. $8192x^{13} - 53,248x^{12}yz + 159,744x^{11}y^2z^2 - 292,864x^{10}y^3z^3$

21. 120   23. 12   25. 990   27. 5040
29. 120   31. 210   33. $-35,840x^4$ 35. $\frac{495}{256}x^8y^4$
37. $2016x^{-5}$ 39. $-540x^3y^3$ 41. $181,440x^4y^3$ 43. $-144x^6$
45. $\frac{35}{8}x^{12}$ 47. 4.8268

REVIEW EXERCISES, CHAPTER ELEVEN

1. 3, 7, 13; 111   2. 0, 7/3, 13/2; 999/11   3. 2   4. 120

5. -16 6. 62   7. 50   8. $\sum_{k=1}^{4} \frac{k}{k+2}$

9. $\sum_{k=0}^{4} (-1)^k x^k$ 10. $\sum_{k=1}^{n} \sin kx$ 11. 38   12. -9
13. 8   14. $-33/2$ 15. 275/3   16. -450
17. -3 18. $-3/2$ 19. 5, 1, 1/5, 1/25   20. $-2, 2, -2, 2$
21. 243/8   22. ± 256 23. 1/2, 1/12   24. 21/32
25. -728 26. 10   27. 9/5
29. $16x^4 - 32x^3y + 24x^2y^2 - 8xy^3 + y^4$ 30. $x^4/16 - x^3 + 6x^2 - 16x + 16$
31. $x^6 + 3x^4 + 3x^2 + 1$ 32. 720   33. 78
34. $(n+1)/n$ 35. 15   36. 1   37. 90

PROGRESS TEST 11A

1. 1/4, 2/9, 3/16, 4/25   2. 29/6   3. $-1, 1/2, 2, 7/2$
4. 8   5. 41   6. 55/2
7. 1/3   8. $-2/3, -4/3, -8/3, -16/3$ 9. -2
10. 8, -16 11. $-43/8$ 12. -6
14. $a^{10} + 10a^9/b + 45a^8/b^2 + 120a^7/b^3$ 15. 22

PROGRESS TEST 11B

1. 5/3, 5, 51/5, 52/3   2. $\sum_{k=2}^{n} k!$ 3. 6, 16/3, 14/3, 4

4. 10   5. -55 6. 13, 1/4
7. 0.2   8. $-1, 1/4, -1/16, 1/64$ 9. $-1/81$
10. $-4, -8/3$ 11. $-31/2$ 12. 25/7
14. $11520x^8$ 15. $n(n+1)$

ANSWERS TO UNIT 3 PROGRESS TEST

1. (a) $-9 - 46i$ (b) $\frac{2}{5} - \frac{11}{5}i$ 2. (a) $\sqrt{13}$ (b) $\sqrt{45}$ 3. $3(\cos 270° + i\sin 270°)$
4. $5(\sqrt{3} + i)/2$ 5. $4x^4 + 4x^3 - 3x^2 - 2x + 1 = 0$
6.

Positive	Negative	Complex
0	1	4
0	3	2
2	1	2
2	3	0

7. (6, 3)   8. $x^2 + y^2 - 6x + 2y - 15 = 0$
9. (a) circle with center at $(-2, 4)$ and radius $\sqrt{13}$ (b) None: no graph.
10. $(x - 4)^2 = -y/2$

12. Center: $(2, 1)$
 Foci: $(2, 1 \pm \sqrt{5}/3)$
 Vertices: $(2, 0), (2, 2)$

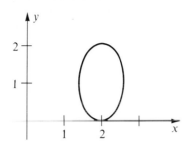

13. Foci: $(0, 9/2), (0, -1/2)$
 Vertices: $(0, 2 \pm \sqrt{2})$
 Asymptotes: $y = 2 \pm \dfrac{2\sqrt{3}}{3} x$

14. (a) parabola   (b) circle   (c) hyperbola
16. $x = -1, y = 2, z = -2$

17. (a) $k = -3/2$ (b) none

18.

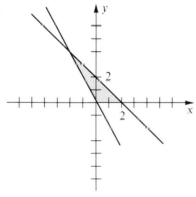

19. (a) $\begin{bmatrix} 0 & 5 \\ -3 & 16 \\ -2 & -4 \end{bmatrix}$ (b) 7 (c) $\begin{bmatrix} 1 & 0 & -1 \end{bmatrix}$

20. 73

21. 24

22. (a) 35   (b) 8

23. 2

24. 1/2

25. $\sqrt{10}$

26. 40/9

27. 1/6

28. $\dfrac{10}{x^2}$

APPENDIX A

1. $11,636.36 on 10 speeds, $4363.64 on 3-speeds

3. $7000

5. 20 hours

7. 50 and 54 mph

9. 1/12, 1/4; -1/4, -1/12

11. 4 meters and 8 meters

13. 40 kph, 80 kph

15. 140 mph

17. $(5F - 160)/9$

19. $(2A/h) - b'$

21. $f_1 f/(f_1 - f)$

23. $(a + rS)/(r + S)$

25. $s > 60{,}000$ sq. ft.

27. roofer: 6 hours; assistant: 12 hours

29. 4 students

APPENDIX B

1. (a)

x	4.4	4.2	4.1	4.01	4.001
$2x - 3$	5.8	5.4	5.2	5.02	5.002

$$\lim_{x \to 4} (2x - 3) = 5$$

x	3.6	3.8	3.9	3.99	3.999
$2x - 3$	4.2	4.6	4.8	4.98	4.998

(b)

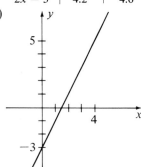

3. 6

5. 11

7. −4

9. 12

11. −2

13. −12

15. (a) 5

(b)

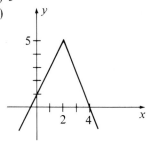

17.

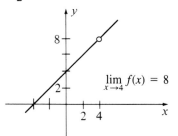

$$\lim_{x \to 4} f(x) = 8$$

19.

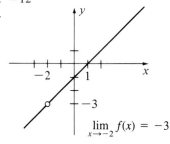

$$\lim_{x \to -2} f(x) = -3$$

21.

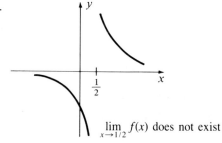

$$\lim_{x \to 1/2} f(x) \text{ does not exist}$$

23.

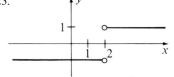

$$\lim_{x \to 2} f(x) \text{ does not exist}$$

25. does not exist

27. does not exist

SOLUTIONS TO SELECTED REVIEW EXERCISES

CHAPTER 1

1. T. (Irrational numbers are a subset of the real numbers.)

3. F. (The negative integers and zero are a subset of the integers.)

12. $|-3| - |1 - 5|$
$= |-3| - |-4|$ Performing operations within bars first
$= 3 - 4$ Definition of absolute value
$= -1$

13. $\overline{PQ} = |9/2 - 6| = |-3/2| = 3/2$

15. c. (Every exponent of a polynomial must be a nonnegative integer.)

19. $x(2x - 1)(x + 2)$
$= (2x^2 - x)(x + 2)$
$= 2x^3 + 3x^2 - 2x$

23. $2a^2 + 3ab + 6a + 9b$
$= (2a^2 + 6a) + (3ab + 9b)$ Grouping
$= 2a(a + 3) + 3b(a + 3)$ Common factors $2a$, $3b$
$= (a + 3)(2a + 3b)$ Common factor $a + 3$

25. $x^8 - 1 = (x^4)^2 - (1)^2$ Difference of squares: $a = x^4$, $b = 1$
$= (x^4 + 1)(x^4 - 1)$ Factoring $a^2 - b^2 = (a + b)(a - b)$
$= (x^4 + 1)(x^2 + 1)(x^2 - 1)$ Another difference of squares: $x^4 - 1$
$= (x^4 + 1)(x^2 + 1)(x + 1)(x - 1)$ Another difference of squares: $x^2 - 1$

27. $(2a^2b^{-3})^{-3} = (2)^{-3}(a^2)^{-3}(b^{-3})^{-3} = \frac{1}{8}a^{-6}b^9 = \dfrac{b^9}{8a^6}$

31. $\sqrt{80} = \sqrt{16 \cdot 5} = \sqrt{16} \cdot \sqrt{5} = 4\sqrt{5}$

33. $\dfrac{\sqrt{x}}{\sqrt{x} + \sqrt{y}} = \dfrac{\sqrt{x}}{\sqrt{x} + \sqrt{y}} \cdot \dfrac{\sqrt{x} - \sqrt{y}}{\sqrt{x} - \sqrt{y}} = \dfrac{x - \sqrt{xy}}{x - y}$

35. $(2x + 1)^{2/3} - (2x + 1)^{5/3}$
$= (2x + 1)^{2/3}[1 - (2x + 1)]$ Factor out expression with the smallest exponent.
$= 2x(2x + 1)^{2/3}$ Simplify.

36. Equate the real and the imaginary parts.

$$x - 2 = -4 \qquad 2y - 1 = 7$$
$$x = -2 \qquad y = 4$$

37. $i^{47} = i^{44} \cdot i^3 = i^3 = -i$

39. $(2 + i)(2 + i) = 4 + 2i + 2i + i^2$
$= 4 + 4i - 1$
$= 3 + 4i$

44. $k - 2x = 4kx$
$k = 4kx + 2x$
$k = x(4k + 2)$

$$x = \dfrac{k}{4k + 2} = \dfrac{k}{2(2k + 1)}$$

45. F. (The equation does not hold for $x = 0$, and therefore it does not hold for all real values of x.)

48. $-4 < -2x + 1 \le 10$
$-5 < -2x \le 9$

$$\dfrac{-5}{-2} > \dfrac{-2x}{-2} \ge \dfrac{9}{-2}$$

$$\dfrac{5}{2} > x \ge -\dfrac{9}{2}$$

or

$$-\dfrac{9}{2} \le x < \dfrac{5}{2}$$

50. Since the numerator is negative, the denominator must be positive if the quotient is to be negative. Note also that the denominator cannot equal 0.

$$2x - 5 > 0$$
$$x > 5/2 \quad \text{or} \quad (5/2, \infty)$$

52. $|3x + 2| = 7$

$3x + 2 = 7$	$-(3x + 2) = 7$
$3x = 5$	$-3x = 9$
$x = 5/3$	$x = -3$

55. $|2 - 5x| < 1$

$-1 < 2 - 5x < 1$

$-3 < -5x < -1$

$\dfrac{-3}{-5} > \dfrac{-5x}{-5} > \dfrac{-1}{-5}$

$\dfrac{3}{5} > x > \dfrac{1}{5}$

$\dfrac{1}{5} < x < \dfrac{3}{5}$ or $\left(\dfrac{1}{5}, \dfrac{3}{5}\right)$

58. $2x + 1$ $- - - - - - 0 + +$

$x + 5$ $- - 0 + + + + +$

$\begin{array}{ccc} & -5 & -\dfrac{1}{2} \end{array}$

$\dfrac{2x + 1}{x + 5}$ $+ + \quad - - - 0 + +$

$(-\infty, -5), [-1/2, \infty)$

(Exclude $x = -5$ since the denominator cannot equal 0.)

CHAPTER 2

1. $d = \sqrt{(x_2 - x_1)^2 + (y_2 - y_1)^2}$

$= \sqrt{(2 + 4)^2 + (-1 + 6)^2}$

$= \sqrt{36 + 25} = \sqrt{61}$

5.   *y*-axis test   *x*-axis test

Replace x with $-x$:   Replace y with $-y$:

$y^2 = 1 - (-x)^3$ $(-y)^2 = 1 - x^3$

$y^2 = 1 + x^3$ $y^2 = 1 - x^3$

   no   yes

   origin test

   Replace both:

$(-y)^2 = 1 - (-x)^3$

$y^2 = 1 + x^3$

   no

7. Yes. No vertical line meets the graph in more than one point.

9. The quantity under the radical cannot be negative.

$3x - 5 \geq 0$

$x \geq \dfrac{5}{3}$

11. Solve the equation:

$$f(x) = 15 = \sqrt{x - 1}$$

$$225 = x - 1$$

$$x = 226$$

Checking: $x = 226$

$$15 \overset{?}{=} \sqrt{x - 1}$$

$$\overset{?}{=} \sqrt{226 - 1}$$

$$\overset{\checkmark}{=} 15$$

The solution is $x = 226$.

14. Replace x with $y - 1$.

$$f(x) = x^2 - x = (y - 1)^2 - (y - 1)$$

$$= y^2 - 2y + 1 - y + 1$$

$$= y^2 - 3y + 2$$

18. $f(x) = x - 1$ when $x \leq -1$

$f(-4) = -4 - 1 = -5$

19. $f(x) = -2$ when $x > 2$

$f(4) = -2$

22. $m = \dfrac{y_2 - y_1}{x_2 - x_1} = \dfrac{3 - (-6)}{-1 - (-4)} = \dfrac{9}{3} = 3$

23. $y - y_1 = m(x - x_1)$

$y - (-6) = 3[x - (-4)]$

$y + 6 = 3x + 12$

$y = 3x + 6$

27. $2y + x - 5 = 0$

$y = -\dfrac{1}{2}x + \dfrac{5}{2}$

The slope of the given line is $m_1 = -1/2$. The slope m of any line perpendicular to the given line is

$$m = -\dfrac{1}{m_1} = 2$$

$$y - y_1 = m(x - x_1)$$

$$y - 3 = 2(x + 1)$$

$$y = 2x + 5$$

29. $6x^2 - 11x + 4 = (2x - 1)(3x - 4)$

$2x - 1 = 0$ $3x - 4 = 0$

$x = \dfrac{1}{2}$ $x = \dfrac{4}{3}$

30. $\quad x^2 - 2x = -6$
$x^2 - 2x + 1 = -6 + 1$
$\quad (x - 1)^2 = -5$
$\quad\quad x - 1 = \pm\sqrt{-5}$
$\quad\quad\quad x = 1 \pm i\sqrt{5}$

34. $kx^2 - 3\pi = 0$
$\quad kx^2 = 3\pi$

$$x^2 = \frac{3\pi}{k}$$

$$x = \pm\sqrt{\frac{3\pi}{k}} = \pm\frac{\sqrt{3\pi k}}{k}$$

36. $3r^2 - 2r - 5 = 0$
$a = 3, b = -2, c = -5$
$b^2 - 4ac = 64$
Since $b^2 - 4ac$ is positive and a square, the roots are real and rational.

39. $\sqrt{x} + 2 = x$
$\quad \sqrt{x} = x - 2$
$\quad\quad x = x^2 - 4x + 4$
$\quad\quad 0 = x^2 - 5x + 4$
$\quad\quad 0 = (x - 1)(x - 4)$

$\quad\quad x = 1 \quad\quad x = 4$

Checking:

$x = 1 \quad\quad\quad\quad x = 4$

$\sqrt{1} + 2 \overset{?}{=} 1 \quad\quad \sqrt{4} + 2 \overset{?}{=} 4$

$1 + 2 \overset{?}{=} 1 \quad\quad 2 + 2 \overset{?}{=} 4$

$3 \neq 1 \quad\quad\quad\quad 4 \overset{\checkmark}{=} 4$

The solution is 4.

41. Let $u = 1 - \dfrac{2}{x}$.

$u^2 - 8u + 15 = 0$
$(u - 3)(u - 5) = 0$
$\quad u = 3 \quad\quad u = 5$

Substituting:

$3 = 1 - \dfrac{2}{x} \quad\quad 5 = 1 - \dfrac{2}{x}$

$2 = -\dfrac{2}{x} \quad\quad 4 = -\dfrac{2}{x}$

$x = -1 \quad\quad\quad x = -\dfrac{1}{2}$

43. $f(x) = -x^2 - 4x$
$\quad\quad = -(x^2 + 4x \quad\quad)$
$\quad\quad = -(x^2 + 4x + 4) + 4$
$\quad\quad = -(x + 2)^2 + 4$

vertex: $(-2, 4)$

x-intercepts	y-intercept
Let $y = f(x) = 0$.	Let $x = 0$.
$0 = -x^2 - 4x$	Then $y = f(0) = 0$.
$0 = -x(x + 4)$	
$x = 0 \quad x = -4$	

45. $a = 2, b = -1$
  (a) Since $a > 0$, it is a minimum.
  (b) The minimum occurs at

$$x = -\frac{b}{2a} = -\frac{-1}{4} = \frac{1}{4}$$

  (c) The minimum value is

$$f\left(\frac{1}{4}\right) = 2\left(\frac{1}{4}\right)^2 - \frac{1}{4} + 1$$

$$= \frac{7}{8}$$

CHAPTER 3

1. $1 \underline{\big|} \quad 2 \quad\quad 0 \quad\quad 6 \quad\quad 4$
$\quad\quad\quad\quad\quad\quad 2 \quad\quad 2 \quad\quad 8$
$\quad\quad\quad\underbrace{2 \quad\quad 2 \quad\quad 8}_{Q(x)} \quad \underset{R}{4}$

$Q(x) = 2x^2 + 2x + 8; R = 4$

3. $-1\underline{\big|} \quad 7 \quad -3 \quad\quad 0 \quad\quad 2 \quad\quad\quad 2\underline{\big|} \quad 7 \quad -3 \quad\quad 0 \quad\quad 2$
$\quad\quad\quad\quad\quad\quad -7 \quad 10 \quad -10 \quad\quad\quad\quad\quad\quad\quad 14 \quad 22 \quad 44$
$\quad\quad\quad\quad 7 \quad -10 \quad 10 \quad -8 \quad\quad\quad\quad\quad 7 \quad 11 \quad 22 \quad 46$

$P(-1) = -8 \quad\quad\quad\quad\quad\quad\quad\quad P(2) = 46$

5. $-2\underline{\big|} \quad 2 \quad\quad 4 \quad\quad 3 \quad\quad 5 \quad -2$
$\quad\quad\quad\quad\quad\quad -4 \quad\quad 0 \quad -6 \quad\quad 2$
$\quad\quad\quad\quad 2 \quad\quad 0 \quad\quad 3 \quad -1 \quad\quad 0$

Since $P(-2) = 0$, $x + 2$ is a factor.

11. The only possible rational roots are ± 1. Using condensed synthetic division, we find

$\quad\quad\quad\quad\quad\quad\quad\quad 1 \quad 3 \quad 2 \quad 1 \quad -1$
$\quad\quad\quad\quad 1 \,\big|\, 1 \quad 4 \quad 6 \quad 7 \quad \boxed{6}$
$\quad\quad\quad -1 \,\big|\, 1 \quad 2 \quad 0 \quad 1 \quad \boxed{-2}$

Since neither remainder is zero, there are no rational roots.

12. Divide by $x + 2$ to find the depressed equation.

$$\underline{-2|}\ \ \begin{array}{rrrr} 2 & -1 & -13 & -6 \\ & -4 & 10 & 6 \\ \hline 2 & -5 & -3 & 0 \end{array}$$

depressed
equation

Solving $2x^2 - 5x - 3 = 0$, we have

$$(2x + 1)(x - 3) = 0$$

$$x = -\frac{1}{2} \qquad x = 3$$

17. $\dfrac{8 - x}{2x^2 + 3x - 2} = \dfrac{8 - x}{(2x - 1)(x + 2)} = \dfrac{A}{2x - 1} + \dfrac{B}{x + 2}$

$8 - x = A(x + 2) + B(2x - 1)$

$x = -2$: $10 = -5B$ or $B = -2$

$x = \dfrac{1}{2}$: $\dfrac{15}{2} = \dfrac{5}{2}A$ or $A = 3$

$\dfrac{8 - x}{2x^2 + 3x - 2} = \dfrac{3}{2x - 1} - \dfrac{2}{x + 2}$

18. $\dfrac{3x^3 + 5x - 1}{(x^2 + 1)^2} = \dfrac{Ax + B}{x^2 + 1} + \dfrac{Cx + D}{(x^2 + 1)^2}$

$3x^3 + 5x - 1 = (Ax + B)(x^2 + 1) + Cx + D$
$3x^3 + 5x - 1 = Ax^3 + Bx^2 + (A + C)x + (B + D)$
coeff. of x^3: $A = 3$
coeff. of x^2: $B = 0$
coeff. of x: $A + C = 5$
$\qquad\qquad 3 + C = 5$
$\qquad\qquad\quad C = 2$
coeff. of x^0: $B + D = -1$
$\qquad\qquad\quad\ D = -1$

$\dfrac{3x^3 + 5x - 1}{(x^2 + 1)^2} = \dfrac{3x}{x^2 + 1} + \dfrac{2x - 1}{(x^2 + 1)^2}$

CHAPTER 4

2. $(f \cdot g)(x) = (x + 1)(x^2 - 1)$
$\qquad\qquad = x^3 + x^2 - x - 1$
$(f \cdot g)(-1) = (-1)^3 + (-1)^2 - (-1) - 1 = 0$
5. $(g \circ f)(x) = g(x + 1) = (x + 1)^2 - 1 = x^2 + 2x$
6. $g(x) = x^2 - 1$
$g(2) = 2^2 - 1 = 3$
$(f \circ g)(2) = f(3) = 3 + 1 = 4$
7. $(f \circ g)(x) = f(x^2) = \sqrt{x^2 - 2} = |x| - 2$
9. $(f \circ g)(-2) = |-2| - 2 = 0$

11. $(f \circ g)(x) = f\left(\dfrac{x}{2} - 2\right) = 2\left(\dfrac{x}{2} - 2\right) + 4 = x$

$(g \circ f)(x) = g(2x + 4) = \dfrac{2x + 4}{2} - 2 = x$

14. $2^{2x} = 8^{x-1} = (2^3)^{x-1}$ Write in terms of same base.
$\quad 2^{2x} = 2^{3x-3}$ $\qquad\qquad\quad (a^m)^n = a^{mn}$
$\quad 2x = 3x - 3$ $\qquad\qquad\ \ $ If $a^u = a^v$, then $u = v$.
$\quad\ x = 3$ $\qquad\qquad\qquad\ $ Solve for x.

16. $S = P(1 + i)^n$ $\qquad\qquad$ Compound interest formula.

$i = \dfrac{r}{k} = \dfrac{0.12}{2} = 0.6$ Interest rate i per conversion period.

$n = 4 \times 2 = 8$ $\qquad\quad$ Number of conversion periods.
$S = 8000(1 + 0.6)^8$ Substitute for P, i, and n.
$\ = 8000(1.5938)$ $\qquad$ Table IV in Tables Appendix or a calculator.
$\ = \$12,750.40$

22. $\log_5 \dfrac{1}{125} = x - 1$

$5^{x-1} = \dfrac{1}{125}$ Equivalent exponential form.

$5^{x-1} = 5^{-3}$ Write in terms of same base.
$x - 1 = -3$ If $a^u = a^v$, then $u = v$.
$\quad\ x = -2$ Solve for x.

24. $\log_3(x + 1) = \log_3 27$
$\qquad\quad x + 1 = 27$ $\qquad$ If $\log_a u = \log_a v$, then $u = v$.
$\qquad\qquad\ x = 26$ $\qquad$ Solve for x.

25. $\qquad \log_3 3^5 = 5$ Since $\log_a a^x = x$.

or

$\qquad \log_3 3^5 = x$ Introduce unknown x.
$\qquad\qquad 3^x = 3^5$ Equivalent exponential form.
$\qquad\qquad\ x = 5$ If $a^u = a^v$, then $u = v$.

28. $\qquad e^{\ln 3} = 3$ Since $a^{\log_a x} = x$.

or

$\qquad e^{\ln 3} = x$ Introduce unknown x.
$\qquad \ln x = \ln 3$ Equivalent logarithmic form.
$\qquad\quad x = 3$ If $\log_a u = \log_a v$, then $u = v$.

30. $\log_a \dfrac{\sqrt{x - 1}}{2x} = \log_a \dfrac{(x - 1)^{1/2}}{2x}$ Exponent form of radical.

$= \log_a(x - 1)^{1/2} - \log_a 2x$ $\qquad$ Property 2.

$= \log_a(x - 1)^{1/2} - [\log_a 2 + \log_a x]$ Property 1.

$= \dfrac{1}{2}\log_a(x - 1) - \log_a 2 - \log_a x$ $\quad$ Property 3.

34. $\log 14 = \log(2 \cdot 7)$
$\qquad\quad = \log 2 + \log 7$ $\qquad$ Property 1.
$\qquad\quad = 0.30 + 0.85 = 1.15$ Substitute given data.

37. $\log 0.7 = \log \dfrac{7}{10}$

$\qquad = \log 7 - \log 10 \quad$ Property 2.
$\qquad = 0.85 - 1 \qquad \log_a a = 1.$
$\qquad = -0.15$

38. $\dfrac{1}{3}\log_a x - \dfrac{1}{2}\log_a y = \log_a x^{1/3} - \log_a y^{1/2} \quad$ Property 3.

$\qquad\qquad\qquad = \log_a \dfrac{x^{1/3}}{y^{1/2}} \qquad$ Property 2.

$\qquad\qquad\qquad = \log_a \dfrac{\sqrt[3]{x}}{\sqrt{y}} \qquad$ Radical form.

39. $\dfrac{4}{3}[\log x + \log(x - 1)]$

$\qquad - \dfrac{4}{3}\log(x)(x - 1) \quad$ Property 1.

$\qquad = \log(x^2 - x)^{4/3} \quad$ Property 3.

42. $\log_b x = \dfrac{\log_a x}{\log_a b} \qquad$ Change of base formula.

$\quad \log_8 32 = \dfrac{\log 32}{\log 8} \qquad x = 32, b = 8, a = 10.$

$\quad \log_8 32 = \dfrac{1.5}{0.9} = \dfrac{5}{3} \quad$ Substitute given data.

Checking: $\qquad 8^{5/3} = 32 \quad$ Write in equivalent exponent form.
$\qquad\qquad\qquad 32 = 32$

44. $Q(t) = q_0 e^{-kt} \quad$ Exponential decay model.

$\quad \dfrac{q_0}{2} = q_0 e^{-0.06t} \quad$ Substitute $k = 0.06$ and $Q(t) = \dfrac{1}{2}q_0.$

$\quad \dfrac{1}{2} = e^{-0.06t}$

$\quad \ln \dfrac{1}{2} = \ln e^{-0.06t} \quad$ Take logs of both sides.

$\quad \ln 0.5 = -0.06t(\ln e) = -0.06t \quad$ Property 3, $\ln e = 1.$

$\quad t = \dfrac{\ln 0.5}{-0.06} = \dfrac{-0.6931}{-0.06} = 11.5 \text{ hours} \quad$ Calculator.

45. $\qquad 2^{3x-1} = 14$
$\quad (3x - 1)\log 2 = \log 14 \qquad$ Take logs of both sides.

$\qquad x = \dfrac{1}{3} + \dfrac{\log 14}{3\log 2} \quad$ Solve for $x.$

46. $2 \log x - \log 5 = 3$
$\quad \log x^2 - \log 5 = 3 \qquad$ Property 3.

$\log \dfrac{x^2}{5} = 3 \qquad\qquad$ Property 2.

$\dfrac{x^2}{5} = 10^3 = 1000 \quad$ Equivalent exponent form.

$x = \sqrt{5000} \qquad$ Solve for $x.$

CHAPTER 5

1. $P\left(\dfrac{9\pi}{2}\right) = P\left(\dfrac{9\pi}{2} - 4\pi\right) = P\left(\dfrac{\pi}{2}\right)$

$\qquad t' = \dfrac{\pi}{2}$

2. $P\left(-\dfrac{15\pi}{2}\right) = P\left(-\dfrac{15\pi}{2} + 8\pi\right) = P\left(\dfrac{\pi}{2}\right)$

$\qquad t' = \dfrac{\pi}{2}$

6. $P(7\pi/6)$ is obtained by reflecting $P(\pi/6) = \sqrt{3}/2, 1/2$ about the origin. From the accompanying figure, we see that

$$P\left(\dfrac{7\pi}{6}\right) = \left(-\dfrac{\sqrt{3}}{2}, -\dfrac{1}{2}\right)$$

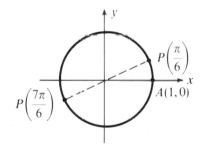

7. $\qquad P\left(-\dfrac{8\pi}{3}\right) = P\left(-\dfrac{8\pi}{3} + 4\pi\right) = P\left(\dfrac{4\pi}{3}\right)$

$P(4\pi/3)$ is obtained by reflecting $P(\pi/3) = (1/2, \sqrt{3}/2)$ about the origin. From the accompanying figure, we see

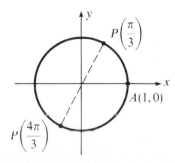

that

$$P\left(-\frac{8\pi}{3}\right) = P\left(\frac{4\pi}{3}\right) = \left(-\frac{1}{2}, -\frac{\sqrt{3}}{2}\right)$$

12. If $P(t) = (a, b)$, then $P(t + \pi/2) = (-b, a)$ or $P(t + \pi/2) = (b, -a)$. In the accompanying figure, $P(t + \pi/2)$ is in the first quadrant, so $P(t + \pi/2) = (3/5, 4/5)$.

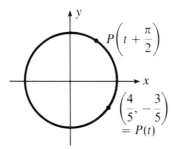

15. $P(-t)$ is the reflection of $P(t)$ about the x-axis. From the accompanying figure, we see that $P(-t) = (4/5, 3/5)$. Then $P(-t - \pi)$ is the reflection of $P(-t)$ about the origin, so $P(-t - \pi) = (-4/5, -3/5)$.

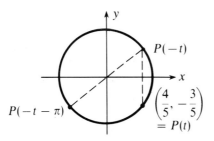

16. $P(2\pi/3)$ is the reflection about the y-axis of $P(\pi/3) = (1/2, \sqrt{3}/2)$. Therefore,

$$P\left(\frac{2\pi}{3}\right) = \left(-\frac{1}{2}, \frac{\sqrt{3}}{2}\right) = \left(\cos\frac{2\pi}{3}, \sin\frac{2\pi}{3}\right)$$

so $\sin 2\pi/3 = \sqrt{3}/2$.

17. $P(-5\pi/4)$ is the reflection of $P(\pi/4) = (\sqrt{2}/2, \sqrt{2}/2)$ about the y-axis. Then

$$P\left(-\frac{5\pi}{4}\right) = \left(-\frac{\sqrt{2}}{2}, \frac{\sqrt{2}}{2}\right)$$

$$= \left(\cos -\frac{5\pi}{4}, \sin -\frac{5\pi}{4}\right)$$

or

$$\cos\left(-\frac{5\pi}{4}\right) = -\frac{\sqrt{2}}{2}$$

$$\sec\left(-\frac{5\pi}{4}\right) = \frac{1}{\cos\left(-\frac{5\pi}{4}\right)} = -\frac{2}{\sqrt{2}} = -\sqrt{2}$$

20. $P(\pi/4) = \sqrt{2}/2, \sqrt{2}/2) = (\cos\pi/4, \sin\pi/4)$. Reflecting $P(\pi/4)$ about the origin brings us to the point $P(\pi/4 + \pi) = P(5\pi/4)$ in quadrant III. Then $P(5\pi/4) = (-\sqrt{2}/2, -\sqrt{2}/2) = (\cos 5\pi/4, \sin 5\pi/4)$, so $\sin 5\pi/4 = -\sqrt{2}/2$ and $t = 5\pi/4$.

21. $P(\pi/6) = (\sqrt{3}/2, 1/2) = (\cos\pi/6, \sin\pi/6)$. Reflecting $P(\pi/6)$ about the x-axis brings us to the point $P(-\pi/6) = P(2\pi - \pi/6) = P(11\pi/6)$ in quadrant IV. Then $P(11\pi/6) = (\sqrt{3}/2, -1/2) = (\cos 11\pi/6, \sin 11\pi/6)$, so $\cos 11\pi/6 = \sqrt{3}/2$ and $t = 11\pi/6$.

24. $\sin^2 t + \cos^2 t = 1$

$$\sin^2 t = 1 - \left(\frac{3}{5}\right)^2 = \frac{16}{25}$$

$$\sin t = -\frac{4}{5} \quad \text{(Since } P(t) \text{ is in quadrant IV, sine is negative.)}$$

$$\tan t = \frac{\sin t}{\cos t} = \frac{-4/5}{3/5} = -\frac{4}{3}$$

$$\cot t = \frac{1}{\tan t} = -\frac{3}{4}$$

25. $\sin^2 t + \cos^2 t = 1$

$$\cos^2 t = 1 - \left(-\frac{4}{5}\right)^2 = \frac{9}{25}$$

$$\cos t = -\frac{3}{5} \quad \text{(Since } \sin t < 0, \tan t > 0, P(t) \text{ must be in quadrant III.)}$$

$$\sec t = \frac{1}{\cos t} = -\frac{5}{3}$$

29. $\dfrac{\sin t}{\cos^2 t} = \dfrac{\sin t}{\cos t} \cdot \dfrac{1}{\cos t} = (\tan t)(\sec t)$

30. Since $\pi < 3.71 < 3\pi/2$, $P(3.71)$ is in quadrant III. The reference number is $3.71 - \pi = 3.71 - 3.14 = 0.57$, and

$$\left.\begin{array}{l} \cos 3.71 = -\cos 0.57 = -0.8419 \\ \sin 1.44 = 0.9915 \\ \cos 3.71 - \sin 1.44 = -1.8334 \end{array}\right\} \begin{array}{l}\text{Table V in} \\ \text{Tables Appendix.}\end{array}$$

34. $\qquad f(x) = -\cos(2x - \pi) = A\cos(Bx + C)$

Then $A = -1$, $B = 2$, $C = -\pi$.

$$\text{amplitude} = |A| = 1$$

$$\text{period} = \frac{2\pi}{B} = \frac{2\pi}{2} = \pi$$

$$\text{phase shift} = -\frac{C}{B} = \frac{\pi}{2}$$

Or set $2x - \pi = 0$ to find that $x = \pi/2$ is the phase shift.

35.
$$f(x) = 4\sin\left(-x + \frac{\pi}{2}\right) = A\sin(Bx + C)$$

Since we require $B > 0$, we use the identity $\sin(-t) = -\sin t$ to obtain

$$f(x) = -4\sin\left(x - \frac{\pi}{2}\right) = A\sin(Bx + C)$$

Then $A = -4$, $B = 1$, $C = -\pi/2$.

$$\text{amplitude} = |A| = 4$$

$$\text{period} = \frac{2\pi}{B} = 2\pi$$

$$\text{phase shift} = -\frac{C}{B} = \pi/2$$

Or set $x - \pi/2 = 0$ to find that $x = \pi/2$ is the phase shift.

38. Let $x = \cos^{-1} 1$. Then $\cos x = 1$, $0 \le x \le \pi$, so $x = 0$. Evaluating, we find

$$\tan(\cos^{-1} 1) = \tan 0 = 0$$

39. Let $x = \tan^{-1} 5$. Then $\tan x = 5$, $-\pi/2 < x < \pi/2$, and $\tan(\tan^{-1} 5) = \tan x = 5$. Alternatively, since $f[f^{-1}(x)] = x$, $\tan(\tan^{-1} 5) = 5$.

40. $5\cos^2 x = 4$

$$\cos^2 x = \frac{4}{5}$$

$$\cos x = \pm\sqrt{\frac{4}{5}} = \pm\frac{2\sqrt{5}}{5}$$

$$x = \arccos\left(\pm\frac{2\sqrt{5}}{5}\right)$$

CHAPTER 6

1. $\dfrac{-60°}{180°} = \dfrac{\theta}{\pi}$

$$\theta = -\frac{\pi}{3}\text{ radians}$$

3. $\dfrac{-\dfrac{5\pi}{12}}{\pi} = \dfrac{\theta}{180°}$

$$\theta = \left(-\frac{5}{12}\right)(180°) = -75°$$

5. Convert $\theta = 100°$ to radians:

$$\frac{100°}{180°} = \frac{\theta}{\pi}$$

$$\theta = \frac{5\pi}{9}\text{ radians}$$

The answer is yes.

7. Convert $5\pi/4$ to degrees:

$$\frac{\dfrac{5\pi}{4}}{\pi} = \frac{\theta}{180°}$$

$$\theta = 225°$$

But $-135°$ is coterminal with

$$-135° + 360° = 225°$$

The answer is yes.

9. $-185°$ is coterminal with

$$-185° + 360° = 175°$$

The reference angle for this second-quadrant angle is

$$\theta' = 180° - 175° = 5°$$

11. $\theta = \dfrac{s}{r} = \dfrac{14}{10} = 1.4$ radians

14. $b^2 = c^2 - a^2 = 25 - 9 = 16$
$b = 4$

$$\tan\beta = \frac{\text{opposite}}{\text{adjacent}} = \frac{4}{3}$$

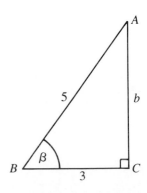

16. hypotenuse $\overline{OP} = \sqrt{3+1} = 2$

$$\csc \theta' = \frac{\text{hypotenuse}}{\text{opposite}} = \frac{2}{1} = 2$$

Since θ is in quadrant II and cosecant is positive in quadrant II,

$$\csc \theta = \csc \theta' = 2$$

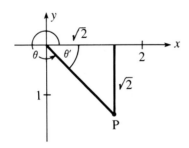

17. $\cot \theta' = \dfrac{\text{adjacent}}{\text{opposite}} = \dfrac{\sqrt{2}}{\sqrt{2}} = 1$

Since θ is in quadrant IV and cotangent is negative in quadrant IV,

$$\cot \theta = -\cot \theta' = -1$$

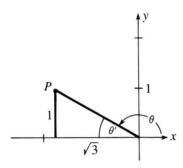

20. $\tan \alpha = \dfrac{50}{60} = 0.8333$

$$\alpha = 39°50'$$

22. $\sin 52° = \dfrac{20}{c}$

$$c = \frac{20}{\sin 52°} = 25.4$$

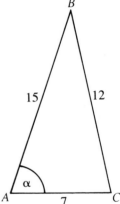

25. $\tan \theta = \dfrac{\text{tree height}}{\text{shadow length}} = \dfrac{25}{10} = 2.5$

$$\theta = 68°10'$$

27. $a^2 = b^2 + c^2 - 2bc \cos \alpha$
$144 = 49 + 225 - 2(7)(15) \cos \alpha$

$$\frac{-130}{-210} = 0.6190 = \cos \alpha$$

$$\alpha = 51°50'$$

CHAPTER 7

1. $\sin \sigma \sec \sigma + \tan \sigma$

$$= \sin \sigma \frac{1}{\cos \sigma} + \tan \sigma$$
$$= \tan \sigma + \tan \sigma$$
$$= 2 \tan \sigma$$

3. $\sin \alpha + \sin \alpha \cot^2 \alpha$
$$= \sin \alpha (1 + \cot^2 \alpha)$$
$$= \sin \alpha \csc^2 \alpha$$

$$= \frac{\csc^2 \alpha}{\csc \alpha} = \csc \alpha$$

5. $\cos(45° + 90°)$
$$= \cos 45° \cos 90° - \sin 45° \sin 90°$$

$$= \left(\frac{\sqrt{2}}{2}\right)(0) - \left(\frac{\sqrt{2}}{2}\right)(1) = -\frac{\sqrt{2}}{2}$$

7. $\sin\dfrac{7\pi}{12} = \sin\left(\dfrac{3\pi}{12} + \dfrac{4\pi}{12}\right) = \sin\left(\dfrac{\pi}{4} + \dfrac{\pi}{3}\right)$

$\qquad = \sin\left(\dfrac{\pi}{4}\right)\cos\left(\dfrac{\pi}{3}\right) + \cos\left(\dfrac{\pi}{4}\right)\sin\left(\dfrac{\pi}{3}\right)$

$\qquad = \left(\dfrac{\sqrt{2}}{2}\right)\left(\dfrac{1}{2}\right) + \left(\dfrac{\sqrt{2}}{2}\right)\left(\dfrac{\sqrt{3}}{2}\right)$

$\qquad = \dfrac{\sqrt{2} + \sqrt{6}}{4}$

8. $\cos 15° = \cos(90° - 15°) = \sin 75°$

10. $\sin\dfrac{\pi}{8} = \cos\left(\dfrac{\pi}{2} - \dfrac{\pi}{8}\right) = \cos\dfrac{3\pi}{8}$

13. $\csc\left(\sigma + \dfrac{\pi}{3}\right) = \dfrac{1}{\sin(\sigma + \pi/3)}$

$\qquad \sin\left(\sigma + \dfrac{\pi}{3}\right) = \sin\sigma\cos\dfrac{\pi}{3} + \cos\sigma\sin\dfrac{\pi}{3}$

$\qquad \cos\sigma = \dfrac{1}{\sec\sigma} = \dfrac{8}{10}$

$\qquad \sin^2\sigma = 1 - \cos^2\sigma = \dfrac{36}{100}$

$\qquad \sin\sigma = -\dfrac{6}{10}$ Since σ is in quadrant IV.

Substituting, we have

$\sin\left(\sigma + \dfrac{\pi}{3}\right) = \left(-\dfrac{6}{10}\right)\left(\dfrac{1}{2}\right) + \left(\dfrac{8}{10}\right)\left(\dfrac{\sqrt{3}}{2}\right)$

$\qquad\qquad = \dfrac{4\sqrt{3} - 3}{10}$

$\csc\left(\sigma + \dfrac{\pi}{3}\right) = \dfrac{10}{4\sqrt{3} - 3} = \dfrac{10(4\sqrt{3} + 3)}{39}$

15. $\tan(\alpha + \beta) = \dfrac{\tan\alpha + \tan\beta}{1 - \tan\alpha\tan\beta}, \qquad \tan\beta = -\dfrac{5}{2}$

$\qquad \tan^2\alpha = \sec^2\alpha - 1 = \dfrac{1}{\cos^2\alpha} - 1$

$\qquad\qquad = \left(-\dfrac{13}{12}\right)^2 - 1 = \dfrac{25}{144}$

$\qquad \tan\alpha = -\dfrac{5}{12}$ Since α is in quadrant II.

Substituting, we have

$\qquad\qquad \tan(\alpha + \beta) = 70$

17. $\cos 2u = \cos^2 u - \sin^2 u$

$\qquad \sin u = \dfrac{1}{\csc u} = -\dfrac{4}{5}$

$\qquad \cos^2 u = 1 - \sin^2 u = 1 - \left(-\dfrac{4}{5}\right)^2 = \dfrac{9}{25}$

Substituting, we have

$\qquad \cos 2u = \dfrac{9}{25} - \left(-\dfrac{4}{5}\right)^2 = -\dfrac{7}{25}$

19. $\sin 4t = \sin 2u$, where $u = 2t$

$\qquad \sin 2u = 2\sin u\cos u = 2\sin 2t\cos 2t$

$\qquad \sin 2t = \dfrac{3}{5}$

$\qquad \cos^2 2t = 1 - \sin^2 2t = \dfrac{16}{25}$

$\qquad \cos 2t = \dfrac{4}{5}$ Since $P(2t)$ is in quadrant I.

Substituting, we have

$\qquad \sin 4t = 2\left(\dfrac{3}{5}\right)\left(\dfrac{4}{5}\right) = \dfrac{24}{25}$

21. $\cos\dfrac{\sigma}{2} = \dfrac{12}{13} = \pm\sqrt{\dfrac{1 + \cos\sigma}{2}}$

$\qquad \cos\sigma = 2\left(\dfrac{12}{13}\right)^2 - 1 = \dfrac{119}{169}$

$\qquad \sin^2\sigma = 1 - \cos^2\sigma = \dfrac{14400}{28561}$

$\qquad \sin\sigma = \dfrac{120}{169}$ Since σ is acute (quadrant I).

23. $\tan\dfrac{t}{2} = \pm\sqrt{\dfrac{1 - \cos t}{1 + \cos t}}$

$\qquad \csc^2 t = \cot^2 t + 1 = \left(-\dfrac{4}{3}\right)^2 + 1 = \dfrac{25}{9}$

$\qquad \sin^2 t = \dfrac{1}{\csc^2 t} = \dfrac{9}{25}$

$\qquad \cos^2 t = 1 - \sin^2 t = \dfrac{16}{25}$

$\qquad \cos t = \dfrac{4}{5}$ Since $P(t)$ is in quadrant IV.

$$\tan\frac{t}{2} = \pm\sqrt{\frac{1 - \frac{4}{5}}{1 + \frac{4}{5}}} = \pm\sqrt{\frac{1}{9}} = \pm\frac{1}{3}$$

Since $270° < t < 360°$, $135° < t/2 < 180°$, so $t/2$ is in quadrant II. Then $\tan t/2 = -1/3$.

25. $\cos 15° = \cos\frac{30°}{2} = \sqrt{\frac{1 + \cos 30°}{2}}$

$\cos 15° = \frac{1}{2}\sqrt{2 + \sqrt{3}}$

27. $\tan 112.5° = \tan\frac{225°}{2} = -\sqrt{\frac{1 - \cos 225°}{1 + \cos 225°}}$

Since $\cos 225° = -\cos 45° = -\frac{\sqrt{2}}{2}$,

$$\tan 112.5° = -\sqrt{\frac{2 + \sqrt{2}}{2 - \sqrt{2}}}$$

28. Let $u = 15x$.
Then $2u = 30x$.

$$\cos 2u = \cos^2 u - \sin^2 u$$
$$= 1 - 2\sin^2 u$$

Since $u = 15x$,

$$\cos 30x = 1 - 2\sin^2 15x$$

30. $\tan\frac{\alpha}{2} = \pm\sqrt{\frac{1 - \cos\alpha}{1 + \cos\alpha}} \cdot \sqrt{\frac{1 - \cos\alpha}{1 - \cos\alpha}}$

$$= \pm\frac{1 - \cos\alpha}{\sqrt{1 - \cos^2\alpha}} = \pm\frac{1 - \cos\alpha}{\sqrt{\sin^2\alpha}}$$

$$= \pm\frac{1 - \cos\alpha}{\sin\alpha}$$

Since $1 - \cos\alpha \geq 0$ for all α, the sign of $\tan\alpha/2$ is determined by the sign of $\sin\alpha$. Therefore,

$$\tan\frac{\alpha}{2} = \frac{1 - \cos\alpha}{\sin\alpha}$$

31. $\sin\frac{3\alpha}{2}\sin\frac{\alpha}{2} = \frac{1}{2}\left[\cos\left(\frac{3\alpha}{2} - \frac{\alpha}{2}\right) - \cos\left(\frac{3\alpha}{2} + \frac{\alpha}{2}\right)\right]$

$$= \frac{1}{2}[\cos\alpha - \cos 2\alpha]$$

32. $\cos 3x - \cos x = -2\sin\frac{3x + x}{2}\sin\frac{3x - x}{2}$

$$= -2\sin 2x \sin x$$

36. $$2\sin\sigma\cos\sigma = 0$$
$$\sin\sigma = 0 \quad\text{or}\quad \cos\sigma = 0$$
$$\sigma = 0, \pi \qquad\qquad \sigma = \frac{\pi}{2}, \frac{3\pi}{2}$$

37. $$\sin 2t - \sin t = 0$$
$$2\sin t\cos t - \sin t = 0$$
$$\sin t(2\cos t - 1) = 0$$
$$\sin t = 0 \quad\text{or}\quad 2\cos t - 1 = 0$$
$$t = 0, \pi \qquad\qquad \cos t = \frac{1}{2}$$
$$t = \frac{\pi}{3}, \frac{5\pi}{3}$$

CHAPTER 8

2. $\dfrac{2 + i}{0 - 5i}\left(\dfrac{0 + 5i}{0 + 5i}\right) = \dfrac{10i + 5i^2}{-25i^2} = \dfrac{-5 + 10i}{25} = -\dfrac{1}{5} + \dfrac{2}{5}i$

7. $|2 - i| = \sqrt{2^2 + (-1)^2} = \sqrt{5}$

10. $r = \sqrt{a^2 + b^2} = \sqrt{(-3)^2 + (3)^2} = \sqrt{18} = 3\sqrt{2}$

$\tan\theta = \dfrac{3}{-3} = -1$

$\theta = 135°$
$-3 + 3i = r(\cos\theta + i\sin\theta)$
$\qquad = 3\sqrt{2}(\cos 135° + i\sin 135°)$

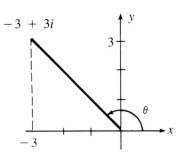

11. $2(\cos 90° + i\sin 90°) = 2(0 + i) = 0 + 2i$

15. $\dfrac{5(\cos 71° + i\sin 71°)}{3(\cos 50° + i\sin 50°)}$

$$= \frac{5}{3}[\cos(71° - 50°) + i\sin(71° - 50°)]$$

$$= \frac{5}{3}(\cos 21° + i\sin 21°)$$

17. $r = \sqrt{a^2 + b^2} = \sqrt{3^2 + (-3)^2} = \sqrt{18} = 3\sqrt{2}$

$$\tan \theta = \frac{-3}{3} = -1$$

$$\theta = 315°$$

$$(3 - 3i)^5 = [3\sqrt{2}(\cos 315° + i\sin 315°)]^5$$
$$= (3\sqrt{2})^5(\cos 1575° + i\sin 1575°)$$
$$= 972\sqrt{2}(\cos 135° + i\sin 135°)$$
$$= 972\sqrt{2}\left(-\frac{\sqrt{2}}{2} + i\frac{\sqrt{2}}{2}\right)$$
$$= -972 + 972i$$

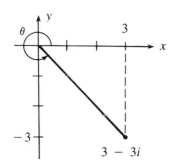

19. In trigonometric form.

$$-9 = 9(\cos 180° + i\sin 180°)$$

so $r = 9$, $\theta = 180°$, and $n = 2$.
The square roots are

$$\sqrt{9}\left[\cos\left(\frac{180° + 360°k}{2}\right) + i\sin\left(\frac{180° + 360°k}{2}\right)\right]$$

for $k = 0, 1$. Substituting for k, we have

$$3(\cos 90° + i\sin 90°) = 3i$$
$$3(\cos 270° + i\sin 270°) = -3i$$

22. With $\sqrt{-3} = \sqrt{3}\,i$, form the product:

$$(x - 3)(x - \sqrt{3}\,i)(x + \sqrt{3}\,i) = (x - 3)(x^2 + 3)$$
$$= x^3 - 3x^2 + 3x - 9$$

24. The number $1/2$ is a root of the linear factor $(2x - 1)$, and -1 is a root of the linear factor $(x + 1)$. Form the product:

$$(2x - 1)^2(x + 1)^2 = 4x^4 + 4x^3 - 3x^2 - 2x + 1$$

28. The polynomial

$$P(x) = x^5 - x^4 + 3x^3 - 4x^2 + x - 5$$

has 5 variations in sign and therefore has a maximum

of 5 positive real roots. The polynomial

$$P(-x) = -x^5 - x^4 - 3x^3 - 4x^2 - x - 5$$

has no variations in sign, and therefore there are no negative real roots.

30. The polynomial

$$P(x) = 3x^4 - 2x^2 + 1$$

has 2 variations in sign, so there can be at most 2 positive real roots. $P(-x) = P(x)$, so there can be at most 2 negative real roots.

31. Since the coefficients are all integers, the Rational Root Theorem restricts the possible rational roots to

$$\pm 1, \quad \pm\frac{1}{2}, \quad \pm\frac{1}{3}, \quad \pm\frac{1}{6}, \quad \pm 2, \quad \pm\frac{2}{3}, \quad \pm 5, \quad \pm\frac{5}{2},$$

$$\pm\frac{5}{3}, \quad \pm\frac{5}{6}, \quad \pm 10, \quad \pm\frac{10}{3}$$

Testing by synthetic division,

$$\begin{array}{r|rrrr} -1 & 6 & 15 & -1 & -10 \\ & & -6 & -9 & 10 \\ \hline & 6 & 9 & -10 & 0 \end{array}$$

we show that -1 is a root. The remaining roots are those of the depressed equation

$$6x^2 + 9x - 10 = 0$$

and are found by the quadratic formula:

$$x = \frac{-9 \pm \sqrt{81 + 240}}{12}$$
$$= \frac{-9 \pm \sqrt{321}}{12}$$

CHAPTER 9

1. $x = \dfrac{x_1 + x_2}{2} = \dfrac{-5 + 3}{2} = -1$

   $y = \dfrac{y_1 + y_2}{2} = \dfrac{4 + (-6)}{2} = -1$

   The midpoint is $(-1, -1)$.

3. By the midpoint formula,

$$2 = \frac{x - 6}{2} \quad \text{and} \quad 2 = \frac{y - 3}{2}$$

so $x = 10$ and $y = 7$.

6. We first find the coordinates of the midpoint of AB:

$$x = \frac{x_1 + x_2}{2} = \frac{-4 + 1}{2} = -\frac{3}{2}$$

$$y = \frac{y_1 + y_2}{2} = \frac{-3 + 3}{2} = 0$$

The slope of AB is

$$m_{AB} = \frac{y_2 - y_1}{x_2 - x_1} = \frac{3 - (-3)}{1 - (-4)} = \frac{6}{5}$$

Then the slope of the perpendicular bisector of AB is $-5/6$. Using the point-slope form for the equation of a line

$$y - y_1 = m(x - x_1)$$

we now obtain an equation of the perpendicular bisector:

$$y - 0 = -\frac{5}{6}\left[x - \left(-\frac{3}{2}\right)\right]$$

$$y = -\frac{5}{6}\left(x + \frac{3}{2}\right)$$

$$y = -\frac{5}{6}x - \frac{5}{4}$$

10. Completing the square in both x and y, we have

$$x^2 + 4x + y^2 - 6y = -10$$
$$(x + 2)^2 + (y - 3)^2 = -10 + 4 + 9 = 3$$

The center is at $(-2, 3)$, and the radius is $\sqrt{3}$.

11. We have $4p = -2/3$, so $p = -1/6$. Thus, the focus is at $(0, -1/6)$, and the directrix is $y = 1/6$.

13. Since the directrix is $y = 7/4$, $p = -7/4$ and $4p = -7$. The equation of the parabola is then

$$x^2 = 4py = -7y$$

16. Completing the squares in both x and y, we have

$$4(x^2 + 2x) + y^2 - 2y = 4$$
$$4(x + 1)^2 + (y - 1)^2 = 4 + 4 + 1 = 9$$
$$\frac{(x + 1)^2}{\frac{9}{4}} + \frac{(y - 1)^2}{9} = 1$$

The center of the ellipse is at $(-1, 1)$. Since $a > b$, $a^2 = 9$ and $b^2 = 9/4$, and the major axis is on the line $x = -1$. The vertices are at $(-1, 1 \pm a)$ and $(-1 \pm b, 1)$, or

$(-1, 4), (-1, -2), (1/2, 1)$, and $(-5/2, 1)$. Also,

$$c^2 = a^2 - b^2 = 9 - \frac{9}{4} = \frac{27}{4}$$

The foci are on the major axis at $(-1, 1 \pm c)$, or $(-1, 1 \pm 3\sqrt{3}/2)$.

17. Since the foci are at $(\pm 3, 0)$, the center must be at the origin. We are given $a^2 = 81$ and $c^2 = 9$. Then $b^2 = a^2 - c^2 = 81 - 9 = 72$. An equation of the ellipse is then

$$\frac{x^2}{81} + \frac{y^2}{72} = 1$$

19. Completing the squares in both x and y, we have

$$4x^2 - (y^2 + 2y) = 4$$
$$4x^2 - (y + 1)^2 = 4 - 1 = 3$$
$$\frac{x^2}{\frac{3}{4}} - \frac{(y + 1)^2}{3} = 1$$

The center of the hyperbola is at $(0, -1)$; $a^2 = 3/4$ and $b^2 = 3$. The vertices are at $(\pm\sqrt{3}/2, -1)$. Since $c^2 = a^2 + b^2$, we have $c^2 = 3/4 + 3 = 15/4$. Thus, the foci are at $(\pm\sqrt{15}/2, -1)$. The asymptotes are $y + 1 = \pm 2x$.

21. We are given $a^2 = 4$ and $c^2 = 49$, so $b^2 = c^2 - a^2 = 49 - 4 = 45$. An equation of the hyperbola is then

$$\frac{y^2}{4} - \frac{x^2}{45} = 1$$

29. We eliminate the xy term by performing a rotation of axes. Since $\cot 2\theta = 0$, $2\theta = \frac{\pi}{2}$ and $\theta = \frac{\pi}{4}$. Then

$$x = x'\cos\theta - y'\sin\theta = \frac{\sqrt{2}}{2}x' - \frac{\sqrt{2}}{2}y'$$

$$y = x'\sin\theta + y'\cos\theta = \frac{\sqrt{2}}{2}x' + \frac{\sqrt{2}}{2}y'$$

Then

$$\left(\frac{\sqrt{2}}{2}x' - \frac{\sqrt{2}}{2}y'\right)\left(\frac{\sqrt{2}}{2}x' + \frac{\sqrt{2}}{2}y'\right)$$
$$+ 2\left(\frac{\sqrt{2}}{2}x' - \frac{\sqrt{2}}{2}y'\right) - 2\left(\frac{\sqrt{2}}{2}x' + \frac{\sqrt{2}}{2}y'\right) = 6$$
$$x'^2 - y'^2 - 4\sqrt{2}\,y' = 12$$
$$x'^2 - (y' + 2\sqrt{2})^2 = 4$$
$$\frac{x'^2}{4} - \frac{(y' + 2\sqrt{2})^2}{4} = 1$$

The conic section is a hyperbola.

31. We eliminate the xy term by performing a rotation of axes. Since

$$\cot 2\theta = \frac{A - C}{B} = \frac{3}{4}$$

from the accompanying figure, $\cos 2\theta = 3/5$. Then

$$\cos \theta = \sqrt{\frac{1 + \cos 2\theta}{2}} = \sqrt{\frac{1 + 3/5}{2}} = \frac{2}{\sqrt{5}}$$

$$\sin \theta = \sqrt{\frac{1 - \cos 2\theta}{2}} = \sqrt{\frac{1 - 3/5}{2}} = \frac{1}{\sqrt{5}}$$

and

$$x = x' \cos \theta - y' \sin \theta = \frac{2}{\sqrt{5}} x' - \frac{1}{\sqrt{5}} y'$$

$$y = x' \sin \theta + y' \cos \theta = \frac{1}{\sqrt{5}} x' + \frac{2}{\sqrt{5}} y'$$

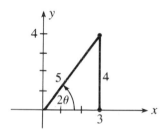

Substituting these equations in the given equation of the conic, we have

$$2\left(\frac{2}{\sqrt{5}} x' - \frac{1}{\sqrt{5}} y'\right)^2 - \left(\frac{1}{\sqrt{5}} x' + \frac{2}{\sqrt{5}} y'\right)^2$$

$$+ 4\left(\frac{2}{\sqrt{5}} x' - \frac{1}{\sqrt{5}} y'\right)\left(\frac{1}{\sqrt{5}} x' + \frac{2}{\sqrt{5}} y'\right)$$

$$- 2\left(\frac{2}{\sqrt{5}} x' - \frac{1}{\sqrt{5}} y'\right) + 3\left(\frac{1}{\sqrt{5}} x' + \frac{2}{\sqrt{5}} y'\right) = 6$$

$$3x'^2 - 2y'^2 - \frac{1}{\sqrt{5}} x' + \frac{8}{\sqrt{5}} y' = 6$$

$$3\left(x'^2 - \frac{1}{3\sqrt{5}} x'\right) - 2\left(y'^2 + \frac{4y'}{\sqrt{5}}\right) = 6$$

Completing the squares in both x' and y', we obtain

$$3\left(x' - \frac{1}{6\sqrt{5}}\right)^2 - 2\left(y' + \frac{2}{\sqrt{5}}\right)^2 = 6 + \frac{1}{60} - \frac{8}{5} = \frac{265}{60}$$

The conic section is a hyperbola.

37. Since $r = 3 \csc \theta$, $r = \dfrac{3}{\sin \theta}$, or $r \sin \theta = 3$, so $y = 3$.

38. We have

$$r^2 \cos^2 \theta + 9r^2 \sin^2 \theta = 9$$

$$r^2 = \frac{9}{\cos^2 \theta + 9 \sin^2 \theta}$$

40. We have

$$\frac{x}{y} = \sqrt{t - 1} \quad \text{or} \quad t = 1 + \frac{x^2}{y^2}$$

Then

$$y^2(t + 2) = 1$$

Substituting the expression for t in this last equation, we have

$$y^2(t + 2) = 1$$

$$y^2\left(3 + \frac{x^2}{y^2}\right) = 1$$

$$x^2 + 3y^2 = 1$$

$$x^2 + \frac{y^2}{\frac{1}{3}} = 1$$

which is the standard form of the equation of an ellipse with its center at the origin.

CHAPTER 10

3. Substituting $x = 5 - 3y$, we have

$$(5 - 3y)^2 + y^2 = 25$$

$$25 - 30y + 9y^2 + y^2 = 25$$

$$10y^2 - 30y = 0$$

$$10y(y - 3) = 0$$

$$y = 0 \qquad \text{or} \qquad y = 3$$

$$x = 5 - 3y = 5 \qquad x = 5 - 3y = -4$$

7. To eliminate x, multiply the first equation by 3 and the second equation by 2, and add:

$$6x + 9y = -3$$

$$-6x + 8y = -\frac{11}{2}$$

$$\overline{ \quad}$$

$$17y = -\frac{17}{2}$$

$$y = -\frac{1}{2}$$

$$2x + 3\left(-\frac{1}{2}\right) = -1$$

$$x = \frac{1}{4}$$

8. Rewriting the equations and adding,

$$x^2 + y^2 - 9 = 0$$
$$\underline{-x^2 + y\ - 3 = 0}$$
$$y^2 + y - 12 = 0$$
$$(y - 3)(y + 4) = 0$$

$$y = 3 \qquad \text{or} \qquad y = -4$$
$$x^2 = y - 3 = 0 \qquad x^2 = y - 3 = -7$$
$$x = 0 \qquad\qquad \text{no real solutions}$$

The circle and parabola are tangent at $(0, 3)$.

18. Interchange equations 1 and 3:

$$-x + 4y + 2z = 15$$
$$2x + 5y - 2z = -9$$
$$-3x - y + z = 12$$

Multiply equation 1 by 2 and add it to equation 2; multiply equation 1 by -3 and add it to equation 3:

$$-x + 4y + 2z = 15$$
$$13y + 2z = 21$$
$$-13y - 5z = -33$$

Add equation 2 to equation 3:

$$-x + 4y + 2z = 15$$
$$13y + 2z = 21$$
$$-3z = -12$$

Use back-substitution:

$$-3z = -12 \quad \text{or} \quad z = 4$$
$$13y + 2(4) = 21 \quad \text{or} \quad y = 1$$
$$-x + 4(1) + 2(4) = 15 \quad \text{or} \quad x = -3$$
$$x = -3, \quad y = 1, \quad z = 4$$

30. From the third row, $x_3 = -3$. Then, from row 2,

$$x_2 + 3x_3 = -8$$
$$x_2 + 3(-3) = -8$$
$$x_2 = 1$$

From row 1,

$$x_1 - 2x_2 + 2x_3 = -9$$
$$x_1 - 2(1) + 2(-3) = -9$$
$$x_1 = -1, \quad x_2 = 1, \quad x_3 = -3$$

31. In matrix form,

$$\begin{bmatrix} 1 & 1 & \vdots & 2 \\ 2 & -4 & \vdots & -5 \end{bmatrix}$$

Add -2 times row 1 to row 2:

$$\begin{bmatrix} 1 & 1 & \vdots & 2 \\ 0 & -6 & \vdots & -9 \end{bmatrix}$$

Multiply row 2 by $-1/6$:

$$\begin{bmatrix} 1 & 1 & \vdots & 2 \\ 0 & 1 & \vdots & 3/2 \end{bmatrix}$$

Add -1 times row 2 to row 1:

$$\begin{bmatrix} 1 & 0 & \vdots & 1/2 \\ 0 & 1 & \vdots & 3/2 \end{bmatrix}$$

The solution is $x = 1/2$, $y = 3/2$.

34. Two matrices of the same dimension are equal if corresponding entries are equal. This requires that

$$x^2 = 9 \quad \text{and} \quad 4x = -12$$

The only value satisfying both equations is $x = -3$.

36. $B - A = \begin{bmatrix} -1 & 5 \\ 4 & -3 \end{bmatrix} - \begin{bmatrix} 2 & -1 \\ 3 & 2 \end{bmatrix}$

$$= \begin{bmatrix} -1-2 & 5-(-1) \\ 4-3 & -3-2 \end{bmatrix} = \begin{bmatrix} -3 & 6 \\ 1 & -5 \end{bmatrix}$$

37. Addition of matrices is defined only when the matrices are of the same dimension.

44. Expanding by the cofactors of the first column, we have

$$1\begin{vmatrix} 5 & 4 \\ 3 & 8 \end{vmatrix} + 2\begin{vmatrix} -1 & 2 \\ 5 & 4 \end{vmatrix} = (40 - 12) + 2(-4 - 10) = 0$$

46.

$$D = \begin{vmatrix} 2 & -1 \\ -2 & 3 \end{vmatrix} = 6 - 2 = 4$$

$$x = \frac{\begin{vmatrix} -3 & -1 \\ 11 & 3 \end{vmatrix}}{4} = \frac{2}{4} = \frac{1}{2}$$

$$y = \frac{\begin{vmatrix} 2 & -3 \\ -2 & 11 \end{vmatrix}}{4} = \frac{16}{4} = 4$$

50. The figure shows the set of feasible solutions and the co-ordinates of the vertices. Evaluating the objective function at these points gives us the following information:

x	y	$z = 5y - x$
0	1	5
0	$\dfrac{9}{2}$	$\dfrac{45}{2}$
4	6	26
$\dfrac{29}{6}$	1	$\dfrac{1}{6}$

The maximum value, $z = 26$, occurs at $x = 4$, $y = 6$.

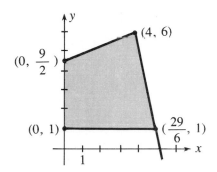

CHAPTER 11

3. $a_n = n - a_{n-1}$
   $a_1 = 0$
   $a_2 = 2 - 0 = 2$
   $a_3 = 3 - 2 = 1$
   $a_4 = 4 - 1 = 3$
   $a_5 = 5 - 3 = 2$

5. $\displaystyle\sum_{k=1}^{4}(1 - 2k) = (1 - 2) + (1 - 4) + (1 - 6) + (1 - 8)$
   $$= -16$$

11. $a_n = a_1 + (n - 1)d$
    $a_{21} = -2 + (21 - 1)(2) = 38$

15. $S_n = \dfrac{n}{2}[2a_1 + (n - 1)d]$
    $$= \dfrac{25}{2}\left[-\dfrac{2}{3} + 24\left(\dfrac{1}{3}\right)\right]$$
    $$= \dfrac{275}{3}$$

17. $r = \dfrac{a_2}{a_1} = \dfrac{-6}{2} = -3$

19. $a_2 = a_1 r = 5\left(\dfrac{1}{5}\right) = 1$

    $a_3 = a_2 r = 1\left(\dfrac{1}{5}\right) = \dfrac{1}{5}$

    $a_4 = a_3 r = \left(\dfrac{1}{5}\right)\left(\dfrac{1}{5}\right) = \dfrac{1}{25}$

21. $r = \dfrac{a_2}{a_1} = \dfrac{6}{-4} = -\dfrac{3}{2}$

    $a_n = a_1 r^{n-1}$

    $a_6 = (-4)\left(-\dfrac{3}{2}\right)^5$

    $$= \dfrac{243}{8}$$

23. The sequence is
    $$3, a_2, a_3, 1/72$$
    With $a_1 = 3$, $a_4 = 1/72$, and $n = 4$,
    $$a_n - a_1 r^{n-1}$$
    $$a_4 = a_1 r^3$$
    $$r^3 = \dfrac{1}{216}$$
    $$r = \dfrac{1}{6}$$
    Then
    $$a_2 = a_1 r = 3\left(\dfrac{1}{6}\right) = \dfrac{1}{2}$$
    $$a_3 = a_2 r = \left(\dfrac{1}{2}\right)\left(\dfrac{1}{6}\right) = \dfrac{1}{12}$$

24. $r = \dfrac{a_2}{a_1} = \dfrac{1}{2}$
    $$S_n - \dfrac{a_1(1 - r^n)}{1 - r}$$
    $$S_6 = \dfrac{\dfrac{1}{3}\left[1 - \left(\dfrac{1}{2}\right)^6\right]}{1 - \dfrac{1}{2}}$$
    $$= \dfrac{21}{32}$$

27. $r = \dfrac{a_2}{a_1} = -\dfrac{2}{3}$

$S = \dfrac{a_1}{1 - r} = \dfrac{3}{1 - \left(-\dfrac{2}{3}\right)} = \dfrac{9}{5}$

30. **By the binomial formula,**

$$\left(\frac{x}{2} - 2\right)^4 = \left(\frac{x}{2}\right)^4 + \frac{4}{1}\left(\frac{x}{2}\right)^3(-2) + \frac{4 \cdot 3}{1 \cdot 2}\left(\frac{x}{2}\right)^2(-2)^2$$
$$+ \frac{4 \cdot 3 \cdot 2}{1 \cdot 2 \cdot 3}\left(\frac{x}{2}\right)(-2)^3 + (-2)^4$$
$$= \frac{x^4}{16} - x^3 + 6x^2 - 16x + 16$$

33. $\dfrac{13!}{11!2!} = \dfrac{13 \cdot 12 \cdot 11!}{11!2!} = \dfrac{13 \cdot 12}{2} = 78$

34. $\dfrac{(n-1)!(n+1)!}{n!n!} = \dfrac{(n-1)!(n+1)n!}{n!n(n-1)!} = \dfrac{n+1}{n}$

35. $\dbinom{6}{4} = \dfrac{6!}{4!2!} = \dfrac{6 \cdot 5}{2} = 15$

INDEX

11^0 1 2^0

11^1 1   1 2^1

11^2 1   2   1 2^2

11^3 1   3   3   1 2^3

11^4 1   4   6   4   1 2^4

11^5 1 5   10   10 5   1 2^5